Risk Analysis of Complex and Uncertain Systems

INT. SERIES IN OPERATIONS RESEARCH & MANAGEMENT SCIENCE

Series Editor: Frederick S. Hillier, Stanford University
Special Editorial Consultant: Camille C. Price, Stephen F. Austin State University
Titles with an asterisk (*) were recommended by Dr. Price

Axsäter/ *INVENTORY CONTROL, 2nd Ed.*
Hall/ *PATIENT FLOW: Reducing Delay in Healthcare Delivery*
Józefowska & Węglarz/ *PERSPECTIVES IN MODERN PROJECT SCHEDULING*
Tian & Zhang/ *VACATION QUEUEING MODELS: Theory and Applications*
Yan, Yin & Zhang/ *STOCHASTIC PROCESSES, OPTIMIZATION, AND CONTROL THEORY APPLICATIONS IN FINANCIAL ENGINEERING, QUEUEING NETWORKS, AND MANUFACTURING SYSTEMS*
Saaty & Vargas/ *DECISION MAKING WITH THE ANALYTIC NETWORK PROCESS: Economic, Political, Social & Technological Applications with Benefits, Opportunities, Costs & Risks*
Yu/*TECHNOLOGY PORTFOLIO PLANNING AND MANAGEMENT: Practical Concepts and Tools*
Kandiller/ *PRINCIPLES OF MATHEMATICS IN OPERATIONS RESEARCH*
Lee & Lee/ *BUILDING SUPPLY CHAIN EXCELLENCE IN EMERGING ECONOMIES*
Weintraub/ *MANAGEMENT OF NATURAL RESOURCES: A Handbook of Operations Research Models, Algorithms, and Implementations*
Hooker/ *INTEGRATED METHODS FOR OPTIMIZATION*
Dawande et al./ *THROUGHPUT OPTIMIZATION IN ROBOTIC CELLS*
Friesz/ *NETWORK SCIENCE, NONLINEAR SCIENCE, and INFRASTRUCTURE SYSTEMS*
Cai, Sha & Wong/ *TIME-VARYING NETWORK OPTIMIZATION*
Mamon & Elliott/ *HIDDEN MARKOV MODELS IN FINANCE*
del Castillo/ *PROCESS OPTIMIZATION: A Statistical Approach*
Józefowska/*JUST-IN-TIME SCHEDULING: Models & Algorithms for Computer & Manufacturing Systems*
Yu, Wang & Lai/ *FOREIGN-EXCHANGE-RATE FORECASTING WITH ARTIFICIAL NEURAL NETWORKS*
Beyer et al./ *MARKOVIAN DEMAND INVENTORY MODELS*
Shi & Olafsson/ *NESTED PARTITIONS OPTIMIZATION: Methodology and Applications*
Samaniego/ *SYSTEM SIGNATURES AND THEIR APPLICATIONS IN ENGINEERING RELIABILITY*
Kleijnen/ *DESIGN AND ANALYSIS OF SIMULATION EXPERIMENTS*
Førsund/ *HYDROPOWER ECONOMICS*
Kogan & Tapiero/ *SUPPLY CHAIN GAMES: Operations Management and Risk Valuation*
Vanderbei/ *LINEAR PROGRAMMING: Foundations & Extensions, 3rd Edition*
Chhajed & Lowe/*BUILDING INTUITION: Insights from Basic Operations Mgmt. Models and Principles*
Luenberger & Ye/*LINEAR AND NONLINEAR PROGRAMMING, 3rd Edition*
Drew et al./ *COMPUTATIONAL PROBABILITY: Algorithms and Applications in the Mathematical Sciences**
Chinneck/ *FEASIBILITY AND INFEASIBILITY IN OPTIMIZATION: Algorithms and Computation Methods*
Tang, Teo & Wei/ *SUPPLY CHAIN ANALYSIS: A Handbook on the Interaction of Information, System, and Optimization*
Ozcan/ *HEALTH CARE BENCHMARKING AND PERFORMANCE EVALUATION: An Assessment Using Data Envelopment Analysis (DEA)*
Wierenga/*HANDBOOK OF MARKETING DECISION MODELS*
Agrawal & Smith/ *RETAIL SUPPLY CHAIN MANAGEMENT: Quantitative Models and Empirical Studies*

~A list of the early publications in the series is found at the end of the book~

Louis Anthony Cox, Jr.

Risk Analysis of Complex and Uncertain Systems

Springer

Louis Anthony Cox, Jr.
Cox Associates
503 Franklin Street
Denver CO 80218
USA
TCoxDenver@aol.com

ISBN 978-0-387-89013-5 e-ISBN 978-0-387-89014-2
DOI 10.1007/978-0-387-89014-2

Library of Congress Control Number: 2008940639

Printed on acid-free paper

springer.com

To Christine and Emeline

Preface

Why This Book?

This book is motivated by the following convictions:

1) Quantitative risk assessment (QRA) can be a powerful discipline for improving risk management decisions and policies.
2) Poorly conducted QRAs can produce results and recommendations that are worse than useless.
3) Sound risk assessment methods provide the benefits of QRA modeling – being able to predict and compare the probable consequences of alternative actions, interventions, or policies and being able to identify those that make preferred consequences more probable – while avoiding the pitfalls.

This book develops and illustrates QRA methods for complex and uncertain biological, engineering, and social systems. These systems have behaviors that are too complex or uncertain to be modeled accurately in detail with high confidence. Practical applications include assessing and managing risks from chemical carcinogens, antibiotic resistance, mad cow disease, terrorist attacks, and accidental or deliberate failures in telecommunications network infrastructure.

For Whom Is It Meant?

This book is intended primarily for practitioners who want to use rational quantitative risk analysis to support and improve risk management decisions in important health, safety, environmental, reliability, and security applications, but who have been frustrated in trying to apply traditional quantitative modeling methods by the high uncertainty and/or complexity of the systems involved. We emphasize methods and strategies for modeling causal relations in complex and uncertain systems well enough to make effective risk management decisions. The book is written for practitioners from multiple disciplines – decision and risk analysts, operations researchers and management scientists, quantitative policy analysts, economists, health and safety risk assessors, engineers, and modelers – who need practical ways to predict and manage risks in complex and uncertain systems.

What's in It?

Three introductory chapters describe QRA and compare it to less formal alternatives, such as taking prompt action to address current concerns, even if the consequences caused by the recommended action are unknown (Chapter 1). These chapters survey QRA methods for engineering risks (Chapter 2) and health risks (Chapter 3). Brief examples of applications such as flood control, software failures, chemical releases, and food safety illustrate the scope and capabilities of QRA for complex and uncertain systems.

Chapter 1 discusses a concept of *concern-driven risk management*, in which qualitative expert judgments about whether concerns warrant specified risk management interventions are used in preference to QRA to guide risk management decisions. Where QRA emphasizes the formal quantitative assessment and comparison of the probable consequences caused by recommended actions to the probable consequences of alternatives, including the status quo, concern-driven risk management instead emphasizes the perceived urgency or severity of the situation motivating recommended interventions. In many instances, especially those involving applications of a "Precautionary Principle" (popular in much European legislation), no formal quantification or comparison of probable consequences for alternative decisions is seen as being necessary (or, perhaps, possible or desirable) before implementing risk management measures that are intended to prevent serious or irreversible harm, even if the causal relations between the recommended measures and their probable consequences are unclear. Such concern-driven risk management has been recommended by critics of QRA in several areas of applied risk management.

Based on case studies and psychological literature on the empirical performance of judgment-based decision making under risk and uncertainty, we conclude that, although concern-driven risk management has several important potential political and psychological advantages over QRA, it often performs less well than QRA in identifying risk management interventions that successfully protect human health or achieve other desired consequences. Therefore, those who advocate replacing QRA with concern-driven alternatives, such as expert judgment and consensus decision processes, should assess whether their recommended alternatives truly outperform QRA, by the criterion of producing preferred consequences, before rejecting the QRA paradigm for practical applications.

Chapter 2 introduces methods of probabilistic risk assessment (PRA) for predicting and managing risks in complex engineered systems. It surveys methods for PRA and decision making in engineered systems, emphasizing progress in methods for dealing with uncertainties, communicating results effectively, and using the results to guide improved decision making by multiple parties. For systems operating under threats from intelligent adversaries, novel methods and game-theoretic ideas can help to identify effective risk reduction strategies and resource allocations. In hard decision problems, where the best course of action is unclear and data are sparse, ambiguous, or conflicting, state-of-the-art methodology can be critical for good risk management. This chapter discusses some of the most useful PRA methods and possible extensions and improvements.

Chapter 3 introduces methods of quantitative risk assessment (QRA) for public health risks. These arise from the operation of complex engineering, economic, medical, and social systems, ranging from food supply networks to industrial plants to administration of school vaccination programs and hospital infection control programs. The decisions and behaviors of multiple economic agents (e.g., the producers, distributors, retailers, and consumers of a product) or other decision makers (e.g., parents, physicians, and schools involved in vaccination programs) affect risks that, in turn, typically affect many other people. Health risks are commonly different for different subpopulations (e.g., infants, the elderly, and the immunocompromised, for a microbial hazard; or customers, employees, and neighbors of a production process). Thus, public health risk analysis often falls in the intersection of politics, business, law, economics, ethics, science, and technology, with different participants and stakeholders favoring different risk management alternatives. In this politicized context, QRA seeks to clarify the probable consequences of different risk management decisions.

Chapters 4 and 5 (as well as Chapter 15, which deals specifically with terrorism risk assessment) emphasize that *sound risk assessment requires developing sound risk models* in enough detail to represent correctly the (often probabilistic) *causal relations* between a system's controllable inputs and the outputs or consequences that decision makers care about. "Sound" does not imply completely accurate, certain, or detailed. Imperfect and high-level risk models, or sets of alternative risk models that are contingent on explicitly stated assumptions, can still be sound and useful for improving decision making. But a sound model must describe causal relations correctly, even if not in great detail, and even if contingent on stated assumptions. Incorrect causal models, or models with hidden false assumptions about cause and effect, can lead to poor risk management recommendations and decisions.

Chapters 4 and 5 warn against popular shortcut methods of risk analysis that try to avoid the work required to develop and validate sound risk models. These include replacing empirically estimated and validated causal risk models (e.g., simulation models) with much simpler ratings of risky prospects using terms such as *high*, *medium*, and *low* for attributes such as the frequency and severity of adverse consequences. Other shortcut methods use highly aggregate risk models or scoring formulas (such as "*risk = potency × exposure,*" or "*risk = threat × vulnerability × consequence*") in place of more detailed causal models. Many professional consultants, risk assessors, and regulatory agencies use such methods today. However, *these attempted shortcuts do not work well in general.* As discussed in Chapters 4 and 5, they can produce results, recommendations, and priorities that are worse than useless: they are even less effective, on average, than making decisions randomly! Poor risk management decisions, based on false predictions and assumptions, result from these shortcut methods.

Fortunately, it is possible to do much better. Building and validating sound causal risk models leads to QRA models and analyses that can greatly improve risk management decisions. Chapters 6 through 16 explain how. They introduce and illustrate techniques for testing causal hypotheses and for identifying potential causal relations from data (Chapters 6 and 7), for developing (and empirically testing and validating) risk models to predict the responses of complex, uncertain, and nonlinear systems to changes in controllable inputs (Chapters 8-13), and for making more effective risk management decisions, despite uncertainties and complexities (Chapters 14-16). These chapters pose a variety of important risk analysis challenges for complex and uncertain systems, and propose and illustrate methods for solving them in important real-world applications.

Key challenges, methods and applications in Chapters 6 through 16 include the following:

- *Information-theory and data-mining algorithms.* Chapter 6 shows how to detect initially unknown, possibly nonlinear (including *u*-shaped) causal relations in epidemiological data sets, using food poisoning data as an example. A combination of information theory and nonparametric modeling methods (especially, classification tree algorithms) provide constructive ways to identify potential causal relations (including nonlinear and multivariate ones with high-order interactions) in multivariate epidemiological data sets.
- *Testing causal hypotheses and discovering causal relations.* Chapter 7, building on the methods in Chapter 6, discusses how to test causal hypotheses using data, how to discover new causal relations directly from data without any a priori hypotheses, and how to use data mining and other statistical methods to avoid imposing one's own prior beliefs on the interpretation of data – a perennial challenge in risk assessment and other quantitative modeling disciplines. An application to antibiotic-resistant bacterial infections illustrates these techniques.
- *Use of new molecular-biological and "-omics" information in risk assessment.* Chapter 8 shows how to use detailed biological data (arising from advances in genomics, proteomics, metabolomics, and other low-level biological data) to predict the fraction of illnesses, diseases, or other unwanted effects in a population that could be prevented by removing specific hazards or sources of exposure. This challenge is addressed by using conditional probability formulas and conservative upper bounds on the observed occurrence and co-occurrence rates of events in a causal network to obtain useful upper bounds on unknown causal fractions. Bounding calculations are illustrated by quantifying the preventable fraction of smoking-associated lung cancers in smokers caused by – and preventable by blocking – a particular causal pathway (involving polycyclic aromatic hydrocarbons forming adducts with DNA in a critical tumor suppressor gene) that has attracted great recent interest.
- *Upper-bounding methods.* Chapters 8 through 12 consider how to use available knowledge and information about causal pathways in complex systems, even if very imperfect and incomplete (e.g., biomarker data for complex diseases), to estimate upper bounds on the *preventable fractions* of disease that could be

eliminated by removing specific hazardous exposures. (Analogous strategies for using partial information to bound the preventable risks of adverse outcomes can be used for other complex systems.) The applications in these chapters focus on antibiotic-resistant bacterial infections and on smoking-related lung cancers as examples of partly understood complex systems with large and important knowledge and data gaps, but with enough available knowledge about causal pathways to be useful.

- *Identification of a discrete set of possible risks.* Using dose-response relations for lung cancer risk as an extended example, Chapters 10 and 11 show how to quantify several different input-output relations for a complex system that are consistent with available knowledge and data about uncertain causal mechanisms. Chapter 10 addresses how to identify promising leads for R&D on designing a less hazardous cigarette. It uses a portfolio of causal mechanisms to identify removing cadmium as a promising (but uncertain) way to reduce total risk, despite the complexity of the mixture of chemicals to which smokers are exposed, the complex and uncertain biological pathways by which these chemicals affect lung cancer risk, and the many scientific uncertainties that remain. Chapter 11 shows that sometimes the response of a complex system to a change in inputs can be identified as one of a small number of equally probable alternatives, all of which are consistent with past data.
- *Systems dynamics analysis and simulation.* Chapters 10 through 13 illustrate how to predict input-output relations of dynamic systems using simulation modeling and mathematical analysis (solution of systems of ordinary differential equations and algebraic equations), derived from empirical data and knowledge of the causal processes being simulated. Systems dynamics models can benefit from other techniques demonstrated in these chapters, including modeling only the steady-state levels of subprocesses that adjust relatively quickly and that affect slower processes primarily through time-averaged values (so that hard-to-model but brief, bounded transients can safely be ignored) and using Markov's inequality to relate deterministic simulations of mean values to bounds on probable values of underlying stochastic processes.
- *Comparative statics analysis and reduction of complex models.* Chapter 13 discusses how to reduce large dynamic models, represented by networks of interacting dynamic processes, to much smaller ones that predict the same equilibrium behaviors in response to changes in inputs.
- *Decision tree, sequential decision optimization, and value of information (VOI) analysis.* Chapter 14 estimates the economic value of information from tracking country-of-origin information for cattle imported into the United States from Canada (or other countries with "mad cow" disease). Deliberately using worse-than-realistic probability distributions for scenarios yields a lower bound on the economic value of information (VOI) from tracking. [The author has long believed that the USDA's policy of allowing Canadian cattle – especially, older cattle – into the United States is inconsistent with the policy goal of keeping mad cow disease (bovine spongiform encephalitis, BSE) out of the United States; he has served as an expert in litigation intended to force the USDA to reconsider and

revise this policy.] Assuming that the USDA continues to allow these imports, Chapter 14 considers how to manage the resulting economic risks to the United States created by the increased probability that another case of BSE in an animal imported from Canada will be discovered. The analytic methods demonstrated in Chapter 14 are also useful for many other public risk management and policy optimization applications in which future events and decisions affect the eventual outcomes of present decisions.

- *Game-theory and hierarchical optimization models.* Modeling the behaviors of intelligent attackers and intelligent defenders of a facility (or other target) and optimizing the allocation of defensive resources, taking into account how attackers may respond, are crucial topics in terrorism risk analysis. Methods currently in widespread use for these challenges have serious limitations, and improved methods are urgently needed. Chapter 15 considers both the limitations and ways to improve upon current methods of terrorism risk analysis.
- *Mathematical optimization and phase-transition modeling.* Chapter 16 surveys methods for predicting the resilience of complex systems (e.g., telecommunications networks) to deliberate attacks, and for designing systems to make them resilient to attack. One of the key ideas in this chapter is that the dynamic behaviors of large networks can be extremely simple. For example, simple statistical ("scale-free") models of telecommunications networks predict almost complete resilience to attacks that are limited to knocking out at most a small number of nodes (or links) simultaneously, provided that each node has "enough" (at least a certain critical percentage) of surplus routing capacity to handle the displaced traffic. (Here, "resilient" means that at most only a small fraction of traffic between other nodes, approaching zero percent in large networks, will be made unroutable by such an attack.) At the same time, these simple models predict that networks may be highly vulnerable to such attacks (meaning that most of the traffic in the network will become unroutable after the initial attacks cause node overloads and failures to cascade through the network) if each node has less than the critical amount of surplus capacity. Such a "phase transition" (with a transition threshold determined by the critical amount of surplus capacity) from resilient to vulnerable is characteristic of many highly idealized models of scale-free networks. Assuming that real networks have similar phase-transition behavior – which is currently an important unknown – individual network owners and operators may still lack incentives to invest in increasing resilience, even if doing so would benefit them collectively.

Some Specific Risk Models and Applications for Interested Specialists

In addition to general risk modeling methods, several chapters present specific risk models and results that may be of independent interest to scientists and researchers in cancer risk analysis, bioinformatics and toxicology, microbial and antimicrobial risk assessment, food safety, and terrorism risk analysis. For example, Chapters 11 and 12 develop and apply a new model of lung carcinogenesis. Exposure-related carcinogenesis is often modeled by assuming that cells progress between successive stages – possibly undergoing proliferation at some of them – at rates that depend (usually linearly) on biologically effective doses. Biologically effective doses, in turn, may depend nonlinearly on administered doses, due to pharmacokinetic non-linearities. Chapter 11 provides a mathematical analysis of the expected number of cells in the last ("malignant") stage of a "multistage clonal expansion" (MSCE) model as a function of dose rate and age. The solution displays symmetries such that several distinct sets of parameter values fit past epidemiological data equally well. These different possible sets of parameter values make identical predictions about how changing exposure levels or timing would affect risk. Yet they make significantly different predictions about how changing the composition of exposure would affect risk. Biological data, revealing which rate parameters describe which specific stages, are required to yield unambiguous predictions. From epidemiological data alone, only a set of equally likely alternative predictions can be made for the effects on risk of such interventions.

Chapter 12 asks the following question: If a specific biological mechanism could be discovered by which a carcinogen increases lung cancer risk, how might this knowledge be used to improve risk assessment? For example, suppose that arsenic in cigarette smoke increases lung cancer risk by hypermethylating the promoter region of a specific gene (p16INK4a), leading to more rapid entry of altered (initiated) cells into a clonal expansion phase. How could the potential impact on lung cancer of removing arsenic be quantified in light of such knowledge (assuming, for purposes of illustration, that this proposed mechanism is correct)? Chapter 12 provides an answer, using a three-stage version of the MSCE model from Chapter 11. [This refines a more usual two-stage clonal expansion (TSCE) model of carcinogenesis by resolving its intermediate or "initiated" cell compartment into two subcompartments, representing experimentally observed "patch" and "field" cells. This refinement allows p16 methylation effects to be represented as speeding transitions of cells from the patch state to the clonally expanding field state.] Given these assumptions, removing arsenic might greatly reduce the number of non-small cell lung cancer cells produced in smokers, by up to two thirds, depending on the fraction (between 0 and 1) of the smoking-induced increase in the patch-to-field transition rate prevented if arsenic were removed. At present, this fraction is unknown (and could be as low as zero), but the possibility that it could be high (close to 1) cannot be ruled out without further data.

Chapter 13 presents a dynamic disease model for chronic obstructive pulmonary disease (COPD), a family of smoking-associated diseases having complex causes and consequences. It shows how improved understanding of interactions among biological processes, and of how exposures (in this case, to cigarette smoke) affect these processes and their interactions, can be used to better predict health risks caused by exposures. COPD, although the fourth-leading cause of death worldwide, has a puzzling etiology. It is a smoking-associated disease, but only a minority of smokers develop it. Moreover, in people (but not in animals), unresolved inflammation of the lung and destruction of lung tissue, once started, continue even after smoking ceases. Chapter 13 proposes a biologically based risk assessment model of COPD that offers a possible explanation of these and other features of the disease. COPD causation is modeled as resulting from a dynamic imbalance between protein-digesting enzymes (proteases) and the antiproteases that inhibit them in the lung. This leads to ongoing proteolysis (digestion) of lung tissue by excess proteases. The model is formulated as a system of seven ordinary differential equations (ODEs) with 18 parameters to describe the network of interacting homeostatic processes regulating the levels of key proteases and antiproteases. Mathematical analysis shows that this system can be simplified to a single quadratic equation to predict the equilibrium behavior of the entire network. There are two possible equilibrium behaviors: a unique stable "normal" (healthy) equilibrium, or a "COPD" equilibrium with elevated levels of lung macrophages and neutrophils (and their elastases) and reduced levels of antiproteases. The COPD equilibrium is induced only if cigarette smoking increases the average production of macrophage elastase (MMP-12) per alveolar macrophage above a certain threshold. Following smoking cessation, the COPD equilibrium levels of MMP-12 and other disease markers decline but do not return to their original (presmoking) levels. These and other predictions of the model are consistent with limited available human data.

Chapters 14, 15, and 16 present risk models for systems in which the future decisions of multiple participants affect the final consequences of current decisions. These chapters present several example models and results for "mad cow" disease (BSE) risk management, terrorist risk analysis, and risk analysis of telecommunications network infrastructure.

Why Do These Models and Methods Matter?

The main purpose of the specific models and applications in the later chapters, as well as of the general QRA methods in earlier chapters, is to show how *QRA can be carried out successfully for uncertain, complex, and nonlinear systems of great practical importance.* Some skeptics have argued that QRA modeling is impractical and/or too laden with uncertain assumptions to give useful and trustworthy results in practice (see Chapter 1). This book seeks to show, both through general modeling principles and by means of constructive examples, how QRA can successfully be carried out and used today to improve risk management in a variety of important real-world applications.

Acknowledgments

It is a great pleasure to acknowledge and thank several colleagues, friends, and coauthors whose ideas, suggestions, and collaborations have contributed to this book.

Dr. Douglas Popken, of Systems View and Cox Associates, has been an invaluable collaborator on Chapters 5, 7, 9, and 14. He and Dr. Jerry Mathers of Alpharma coauthored the article on which Chapter 9 is based. Doug is also coauthor of articles used for the aggregate exposure metric material in Chapter 5 and technical analyses in Chapters 7, 9, and 14. Doug's passion for excellence in obtaining and analyzing real-world data to illuminate complex risk management policy issues is a continuing inspiration and has made our decade-long collaboration on applied risk assessment fun and productive.

Professor Vicki Bier of the University of Wisconsin-Madison coauthored much of Chapter 2, including material on dependence, risk communication, and game theory. Chapter 2 is an extension and update of a chapter that we wrote together a few years ago (Bier and Cox, 2007). Material from that chapter is reprinted in Chapter 2 with the kind permission of Cambridge University Press. Chapter 16 is based on a chapter that I wrote for Vicki's recent book on game theory and security risk analysis (Bier and Azaiez, 2009). Material from that chapter is reprinted in Chapter 16 with the kind permission of Springer. In addition, Vicki generously read and commented on new material in Chapters 2, 3, 5, and 15. I am grateful for her many insights and suggested improvements.

Professor William Huber of Haverford College and Quantitative Decisions coauthored the article on which Chapter 11 is based (material reprinted with permission from Wiley-Blackwell, publishers of *Risk Analysis: An International Journal*). Bill tremendously improved upon my initial approach and provided the elegant analysis and proofs in the appendix to Chapter 11. Bill and I have also collaborated on mathematical and algorithmic research related to risk matrices. Although Chapter 4 of this book shows that risk matrices have many limitations, its concluding suggestion, that designing risk matrices to minimize the maximum possible size of classification errors may be useful, reflects joint research with Bill on how to limit the sizes and frequencies of errors in special situations, such as classifying prospects as having risks greater or less than a specified threshold.

Dr. Edward (Ted) Sanders of Philip Morris International (PMI) coauthored the paper on which Chapter 8 is based (Cox and Sanders, 2006). Ted has also been a constant source of fascinating applied research problems and stimulating and informative discussions and insights on points of biology and epidemiological methodology. Chapters 8, 10, 11, 12, and 13 grew out of applied research supported primarily by PMI (and also by the EPA, for Chapter 11). The challenging problems suggested by Ted have advanced my understanding of what quantitative risk assessment models can accomplish, and it has been a pleasure discussing problems and solutions with Ted and his team at PMI.

The research leading to Chapter 7 was supported primarily by Phibro Animal Health, a manufacturer of the animal antibiotic virginiamycin. The research in Chapter 9 was supported primarily by Alpharma, also a manufacturer of animal antibiotics. I thank Drs. Ken Bafundo and Richard Coulter of Phibro Animal Health and Dr. Jerry Mathers of Alpharma (who coauthored Chapter 9) for their dedication to making better use of science and data to improve quantitative antimicrobial risk assessment. My friends and colleagues Drs. Michael Vaughn and Tom Shryock and Professors Randy Singer, Ian Phillips, Paolo Ricci, and Scott Hurd have discussed many aspects of microbial risk assessment and risk analysis methodology with me over the years. I am grateful to them for stimulating discussions that contributed to the approaches and examples in Chapters 3, 7, and 9.

Chapter 14 is based on research carried out for R-CALF (the Ranchers-Cattlemen Action Legal Fund, United Stockgrowers of America), a national cattle producer organization that has studied marketing and trade issues and advocated various policies in the live cattle industry. I have advised both R-CALF and the USDA on matters related to BSE ("mad cow" disease) risk and have supported R-CALF's efforts to use risk analysis principles to assess risks to the United States from importing Canadian cattle. My own view, that importing cattle from Canada is statistically almost certain to introduce BSE into the United States, perhaps greatly undermining the value of the domestic herd, is reflected in examples in Chapters 1 and 2.

My interest in RAMCAP and infrastructure risk analysis, discussed in Chapter 15, grew out of background reading for a National Research Council of the National Academy of Sciences (NAS) project on methods for improving bioterrorism risk assessment. I have greatly enjoyed discussions and collaboration with Professors Gerald (Jerry) Brown of the Naval Postgraduate School and Steve Pollock of the University of Michigan on some limitations of probabilistic risk assessment techniques and possible ways to do better. I also thank Jerry Brown and Vicki Bier for many stimulating conversations on game theory, optimization, and alternatives for protecting the United States against terrorist attacks. Jerry's thoughtful comments on Chapter 15 and parts of Chapter 5 improved the substance and exposition, and inspired several of the examples used to illustrate key points.

Most of this book is based on recent journal articles. Material from the following articles has been used with the kind permission of Wiley-Blackwell, the publishers of *Risk Analysis: An International Journal.*

Cox LA Jr. Some limitations of "*Risk* = *Threat* × *Vulnerability* × *Consequence*" for risk analysis of terrorist attacks. *Risk Analysis* 2009 (in press). Material from this article is used in Chapter 15.

Cox LA Jr. A mathematical model of protease-anti-protease homeostasis failure in chronic obstructive pulmonary disease (COPD). *Risk Analysis* 2009 (in press). Material from this article is used in Chapter 13.

Cox LA Jr. Could removing arsenic from tobacco smoke significantly reduce smoker risks of lung cancer? *Risk Analysis* 2008 (in press). Material from this article is used in Chapter 12.

Cox LA Jr. Some limitations of frequency as a component of risk: An expository note. *Risk Analysis* 2009. Material from this article is used in Chapter 5.

Cox LA Jr., Popken DA. Overcoming confirmation bias in causal attribution: A case study of antibiotic resistance risks. *Risk Analysis* 2008 (in press). Material from this article is used in Chapter 7.

Cox LA Jr.Why risk is not variance: An expository note. *Risk Analysis* 2008 Aug 28(4):925–928. http://www.ncbi.nlm.nih.gov/pubmed/18554271 Material from this article is used in Chapter 2.

Cox LA Jr. What's wrong with risk matrices? *Risk Analysis* 2008 Apr; 28(2): 497–512. Material from this article is used in Chapter 4.

Cox LA Jr, Huber WA. Symmetry, identifiability, and prediction uncertainties in multistage clonal expansion (MSCE) models of carcinogenesis. *Risk Analysis* 2007 Dec; 27(6):1441–53. Material from this article is used in Chapter 11.

Cox LA Jr., Popken DA. Some limitations of aggregate exposure metrics. *Risk Analysis* 2007 Apr; 27(2):439–45. Material from this article is used in Chapter 5.

Cox LA Jr. Does concern-driven risk management provide a viable alternative to QRA? *Risk Analysis.* 2007 Feb; 27(1):27–43. Material from this article is used in Chapter 1.

Cox LA Jr. Quantifying potential health impacts of cadmium in cigarettes on smoker risk of lung cancer: A portfolio-of-mechanisms approach. *Risk Analysis* 2006 Dec; 26(6):1581–99. Material from this article is used in Chapter 10.

Cox LA Jr., Sanders E. Estimating preventable fractions of disease caused by a specified biological mechanism: PAHs in smoking lung cancers as an example. *Risk Analysis* 2006 Aug; 6(4):881–92. Material from this article is used in Chapter 8.

Cox LA Jr., Popken DA, VanSickle JJ, Sahu R. Optimal tracking and testing of U.S. and Canadian herds for BSE: A value-of-information (VOI) approach. *Risk Analysis*, 2005; 25(4): 827–40. Material from this article is used in Chapter 14.

Chapters 6 and 2 use material from the following two papers, respectively, reprinted with permission from the University of Massachusetts at Amherst, which publishes *Dose-Response*:

Cox LA. Detecting causal nonlinear exposure-response relations in epidemiological data. *Dose Response.* 2006 Aug 19;4(2):119–32.

Cox LA. Universality of J-shaped and U-shaped dose-response relations as emergent properties of stochastic transition systems. *Dose-Response* 2006 May 1; 3(3): 353–68.

Chapters 2 and 3 use material from the following two chapters, respectively, reprinted with permission from Cambridge University Press:

Bier V, Cox LA Jr. Probabilistic risk analysis for engineered systems. Chapter 15 in *Advances in Decision Analysis*. W. Edwards, R. Miles, D. von Winterfeldt, Eds. Cambridge University Press. 2007. www.cambridge.org/us/catalogue/catalogue.asp?isbn=0521682304.

Cox LA Jr. Health risk analysis for risk management decision-making. Chapter 17 in *Advances in Decision Analysis*. W. Edwards, R. Miles, D. von Winterfeldt, Eds. Cambridge University Press. 2007. www.cambridge.org/us/catalogue/catalogue.asp?isbn=0521682304.

Chapter 16 uses material from the following chapter, reprinted with permission from Springer:

Cox, LA Jr. Making telecommunications networks resilient against terrorist attacks. Chapter 8 in *Game Theoretic Risk Analysis of Security Threats.* VM Bier, MN Azaiez, Eds. Springer, New York. 2009.

Contents

Part I
Introduction to Risk Analysis

Chapter 1
Quantitative Risk Assessment Goals and Challenges

The Quantitative Risk Assessment (QRA) Paradigm

How should societies, organizations, and individuals manage risks from activities with unknown or uncertain consequences? Many regulators and scientists advocate quantitative risk assessment (QRA) as providing both a logical framework and a systematic procedure for organizing and applying scientific and engineering knowledge to improve "rational" (consequence-driven) decision making when the consequences of alternative decisions are uncertain. It seeks to do so by using predictive models to identify and recommend choices (typically, among alternative risk management interventions, policies, or plans) that are predicted to make preferred consequences more likely. This typically involves clarifying the following:

- The *probable consequences* of alternative decisions. QRA models typically present results by showing the conditional probabilities of different consequences occurring if each decision alternative is adopted, given specified current information and a probabilistic risk model incorporating uncertainty and variability in outcomes.
- *Preferences* for consequences. These include value trade-offs among different consequences. For financial risk analyses, engineering reliability risk analyses, and health risk analyses, preferences are often simple: Larger profits, higher reliability, and fewer illnesses are preferred to smaller profits, lower reliability, and more illnesses, respectively. Choices that require trading off such desirable outcomes against each other are more difficult, but QRA can help to identify the necessary trade-offs and structure deliberation, by identifying questions of preference and clearly distinguishing them from questions of fact about causes and probable effects.
- The set of *undominated choices*. These choices have the desirable property that no other choices are clearly superior (e.g., always yielding preferred outcomes, no matter how current uncertainties are resolved). The best choice, no matter how value trade-offs are made, should be one of the subset of undominated choices.

L.A. Cox, Jr., *Risk Analysis of Complex and Uncertain Systems*,
International Series in Operations Research & Management Science 129,
DOI 10.1007/978-0-387-89014-2_1,

- How *current uncertainties about probable consequences might change* as more information is gathered. The potential for further information to change the currently best decision (based in what is known now) is usually represented via value of information (VOI) calculations (Chapter 14).

QRAs usually use explicit (documented and often published) predictive risk models to predict the probable consequences of alternative actions and to help identify undominated actions. These risk models allow different users and stakeholders to trace how changing input assumptions and data affects the outputs predicted by the models.

Example: A Simple QRA Risk Assessment Model

The following formula for sporadic exposure-related illnesses in a population is an example of a simple risk assessment model:

change in expected excess illnesses per year caused by exposure to hazard X if action A is taken = (change in units of exposure to hazard X received per year if action A is taken) × (expected excess illnesses per incremental unit of exposure to hazard X).

This may be abbreviated as

$$\Delta risk = \Delta exposure \times (dose\text{-}response\ slope\ factor),$$

where *Δexposure* is the change in exposure that would be caused by the intervention being evaluated, and *dose-response slope factor* is the expected number of additional illnesses per year caused by an additional unit of exposure. (When interindividual heterogeneities and/or dose-response nonlinearities are important, the data-mining techniques in Chapters 6 and 7 can be used to extend this simple formula. Even if the dose-response slope factor varies significantly with the current level of exposure to which the increment *Δexposure* is added and with covariates such as sex and age, this formula can be summed over relatively homogeneous subpopulations of individuals, having similar values of *Δexposure* and *dose-response slope factor*, to estimate the change in total expected illnesses per year for each alternative risk management action being evaluated.)

The *process* of quantitative risk assessment for such health risk assessment models, discussed in Chapter 3, is usually described as consisting of the following four steps:

1. *Hazard identification.* This identifies potential *causal relations* between exposures to hazards (i.e., sources of risk) and resulting increases in risk (the frequency or severity of adverse consequences). Using a combination of statistical tests for potential causation, such as conditional independence tests (see Chapters

6 and 7), and insights from biological mechanisms and molecular epidemiological data (see Chapters 7–12), QRA focuses on identifying and quantifying causal relations, rather than only statistical associations, among actions, exposures, and their health consequences. This emphasis on causality reflects a pragmatic concern with identifying actions that will increase the probabilities of desired outcomes and reduce the probabilities of undesired ones.

2. *Exposure assessment,* such as quantification of $\Delta exposure$ in the above example. This predicts the change in exposures that would be caused by each risk management act being evaluated.
3. *Dose-response modeling* estimates the dose-response slope factors for these changes. These factors quantify the changes in the frequency and severity of illnesses (or other adverse consequences) caused by changes in exposure. Chapters 9–13 discuss and illustrate methods for estimating useful bounds on slope factors and for quantifying dose-response relations.
4. *Risk characterization* describes the change in aggregate population risk caused by changes in exposures, as well as the interindividual variability or frequency distributions of changes in individual risks in the population [e.g., the frequency distribution of the product $\Delta exposure \times (dose\text{-}response\ slope\ factor)$ in the population]. *Uncertainty and sensitivity analyses* are used to show where additional information could reduce uncertainty about population and individual risks and where more information might change the current best decision. These concepts are discussed further in Chapter 3.

Advocates of QRA claim that using explicitly documented assumptions, knowledge, facts, and data (encapsulated in risk models) to assess the predicted changes in risks caused by alternative risk management interventions has many potential benefits in improving societal risk management decisions. Among these are correcting misperceptions about the sizes of different risks (Emmons et al., 2004) and about the relative contributions of different preventable causes (e.g., environmental vs. diet and exercise) to adverse health effects, such as cancers (Wold et al., 2005); focusing resources and priorities where they are likely to be most productive in improving outcomes (Allio et al., 2005; Gerrard, 2000); anticipating and managing the otherwise unforeseen consequences of current and proposed policies; and bringing a valuable "rational" perspective to concerns and anxieties over risks and to deliberations and politicized debates over risk management policies.

As a political process, in this view, QRA invites and empowers a participatory "democracy of science" by enabling stakeholders to calculate for themselves (and others) the probable consequences of alternative risk management decisions, using the best science and data sources that they can find, together with explicitly stated models, calculations, and input assumptions that are open to public inspection. This may give stakeholders who wish to change current risk management policies both the incentive to produce improved scientific information and the means to use it effectively to change policy (if the new information shows that a different policy dominates the current one). Decision-analytic calculations of the potential "value of information" (VOI) for new data in improving current and future decisions can provide insights on when it is worth collecting more information before taking action, when the cost of waiting further is expected to outweigh the benefits, and what risk management actions or interventions should be taken when it is time to act (see Chapter 14).

Example: Explicit QRA Reasoning Can Be Checked and Debated

In 2005, the U.S. Food and Drug Administration (FDA) withdrew approval for an antibiotic used in chickens (enrofloxacin, a member of the fluoroquinolone family of antibiotics). Following the logic used by the FDA Center for Veterinary Medicine (FDA-CVM, 2001) to justify its decision, a QRA calculation estimated that continued enrofloxacin use in poultry could compromise response to antibiotics in over 24,000 persons per year (made sick by fluoroquinolone-resistant *Campylobacter* bacteria) in the United States (Collignon, 2005). Because all assumptions and calculations supporting this number were explicitly stated, those who disagree with the assessment and its resulting decision recommendation can identify exactly where they believe different data values should have been used and where updated data and corrections are needed. For example, one set of proposed corrections (Cox, 2006c) indicates that

- Not attributing resistance from foreign travel and human ciprofloxacin use to the domestic use of enrofloxacin in poultry reduces the estimated risk by about 1/3 (from 19% assumed in the calculation based on all cases to about 6.4% for domestically acquired cases) (Cox, 2006b).
- Updating the estimated fraction of human foodborne *Campylobacter* infections caused by poultry to reflect declines in microbial loads on chicken carcasses since 1992 reduces the estimated risk by a factor of about 1/10 (Stern and Robach, 2003).
- Replacing an assumption that 10% of infected persons would benefit from antimicrobial drug therapy with a value of 0.6% based on the fraction of hospitalized cases (Buzby et al., 1996) that are most likely to be ill enough to warrant antibiotic treatment reduces the estimated risk by a factor of $0.6/10 = 0.06$.
- Replacing an assumption that *all* affected patients receiving antibiotic treatment are prescribed fluoroquinolones (rather than, say, erythromycin) by a more realistic value of perhaps 50% of patients being prescribed fluoroquinolones (FDA-CVM, 2001) reduces the estimated risk by a factor of 1/2.
- Replacing an assumption that *all* such cases lead to compromised responses with a more data-driven estimate that perhaps about 17% of patients have compromised responses (Sanders et al., 2002) reduces the estimated risk by a factor of 1/6.

Together, such changes could easily reduce the estimated risk by a factor of as much as $(1/3)^*(1/10)^*(0.6/10)^*(1/2)^*(1/6) = 0.00017$, or by more than 99.9%, to about four cases per year. (Multiplication is justified, in principle, if the value of each factor is conditioned on all of its predecessors. The numerical values in this example are intended to be realistic but are not expected to be completely accurate.) Whether or not these suggested corrections are accepted – after all, someone else might produce further improved data and estimates for some of these factors – explicitly documenting all assumptions and calculations makes it possible to identify specific areas of disagreement and to either resolve them or note how different input assumptions affect the results of the risk assessment.

QRA encourages a particular view of the roles of agencies, experts, and the public. The public determines preferences for *consequences*, such as reducing the number of fatalities [or illness-days, quality-adjusted life-years (QALYs lost, etc.)] per year caused by a preventable exposure to a hazard. The agency tries to take *actions* to achieve these preferred consequences. To this end, it draws on experts with specialized knowledge and techniques to help identify actions that are likely to bring about preferred consequences. The experts use *risk assessment models* to predict the probabilities of different consequences if alternative actions are taken. They present this information to agencies and other stakeholders, who consider it via a structured and documented *analytic-deliberative* process and make a final choice of action. The agency *implements* the selected actions (e.g., by publishing and enforcing new regulatory requirements), monitors the results, and feeds new information back into the decision process to improve risk assessments and risk management decisions. The agency is ultimately accountable to the public for taking actions that achieve preferred consequences, and the technical experts are accountable to the agency for identifying effective risk management actions and policies.

Against QRA: Toward Concern-Driven Risk Management

Dissatisfactions with QRA

Many scholars, activists, members of the public, and authoritative public health and regulatory agencies have expressed skepticism, disillusionment, distrust, and dissatisfaction with the QRA paradigm (e.g., Healy, 2001; Ball, 2002; Frewer, 2004), arousing concern among professional risk analysts who perceive a great potential practical value in QRA (Thompson et al., 2005). Common criticisms of QRA are that it omits key social, cultural, and political realities (Martuzzi, 2005); that it neglects emotional responses that importantly affect perceptions, judgments, and behaviors in response to real or perceived risks (Slovic et al., 2004); that it cannot or does not deal adequately with realistic uncertainties, complexities, and value judgments (Klinke and Renn, 2002; WHO, 2003); that it is too easily made a political tool for furthering hidden agendas (Ball, 2002); and that simpler and more popular techniques such as the Precautionary Principle should be used instead of or in addition to QRA to arrive at realistic decisions (e.g., Klinke and Renn, 2002; Hayes, 2005).

Observers untrained (or mistrained) in QRA methods also sometimes object that it requires unrealistically perfect knowledge of inputs and/or produces spuriously precise, hence meaningless, numerical outputs. In reality, QRA can work with highly uncertain inputs (e.g., via conditioning, bounding, Bayesian model averaging, and sensitivity analyses, as explained in detail in Chapters 7–9) and usually produces interval estimates or probability distributions as outputs (Chapters 2–3); hence, such objections are not addressed further here.

QRA proponents believe that many of the preceding criticisms and concerns about QRA reflect deep misunderstandings of the nature of QRA (Thompson et al., 2005). They point out that QRA methods have been developed specifically to help achieve preferred consequences in situations with high uncertainty and complexity (typically represented via statistical models or stochastic simulation models that model interactions among multiple factors). QRA has also proved useful in many real-world cases that require a combination of value-focused thinking (Arvai et al., 2001) with well-documented, open, explicit processes and rationales, all supported by the effective (informed and informative) participation of community members and other stakeholders (Jardine et al., 2003; Arvai, 2003).

However, even the idea that societal decisions should be made primarily by "appealing to facts" and to technical calculations and models for predicting probable consequences, rather than by more inclusive political processes that base risk management actions and policies more on understanding and respect for the perceptions, concerns, value judgments, and participation of interested community members, has provoked a powerful backlash (Healy, 2001). Increasingly, opponents of QRA portray it as part of the problem, and even as an instrument by which a technocratic elite uses the language of science to justify unjustifiable actions, rather than as a promising way to make more effective societal decisions in the presence of risk, uncertainty, and complexity. These doubts are sharpened if different risk assessors (e.g., funded by industry vs. regulators) reach vastly different conclusions (Ruden, 2005), especially if their risk assessments appear to be driven by assumptions or calculations that lack objective validity but tend to promote particular political agendas or points of view.

Example: Use of Incorrect Modeling Assumptions in Antimicrobial Risk Assessment

In the preceding example of a QRA for enrofloxacin, the original calculation of 24,000 cases of compromised responses to antibiotics per year assumed that the fraction of patients with fluoroquinolone-resistant infections (estimated as 19%) times the fraction of infections caused by eating poultry (estimated as 72%) gives the fraction of patients with a compromised response caused by eating poultry (due to fluoroquinolone use in poultry): $0.19 \times 0.72 = 0.137$. This crucial assumption is not valid (Cox, 2005b). As a simple counterexample, suppose that the fraction of all infections caused by poultry were indeed 72%, with the rest caused by something else (e.g., contaminated water), and that all and only the 28% of infections caused by the latter source are resistant. Then this procedure would misestimate the fraction of resistant infections caused by eating poultry as (72% of infections caused by eating poultry) × (28% of infections resistant) = 20% of resistant infections caused by eating poultry. But, by hypothesis, the correct value for the fraction of resistant infections caused by eating poultry is zero, not 20%. The key formula assumed in this risk assessment is mistaken.

Clearly, risk assessments that use such incorrect formulas to calculate risks should engender distrust among those who receive the results (or who are forced to live by them, such as farmers denied the continued use of enrofloxacin). The field of risk analysis currently has no professional license or other mechanisms for assuring the competence and validity of QRA calculations. Risk assessments are sometimes performed by self-styled experts lacking training or competence in risk analysis, or by experts in other disciplines who attempt to calculate risks using formulas of their own devising (such as the one above). Therefore, skepticism and critical thinking are crucial in evaluating and deciding whether to use QRA results.

A second key assumption in this example is that reducing enrofloxacin use would necessarily reduce the fluoroquinolone resistance in bacteria (*Campylobacter*) from food animals, thus benefiting human health. This assumption is contradicted by practical experience (e.g., DANMAP, 2004), which shows that reducing enrofloxacin use has not reduced fluoroquinolone resistance in *Campylobacter* isolates from pigs or human patients. Clearly, QRA estimates and methods will and should lose credibility if they promise benefits that are not actually achieved when their prescriptions are followed.

These and other difficulties (e.g., the use of hypothetical numbers as if they were data, conflation of guesses with facts, and so forth) are not unique to QRA, but also plague other areas of quantitative analysis intended to inform and improve policy making (Best, 2001). Perhaps the best defense is to make sure that all risk assessment calculations are explicit, transparent, well documented, and open to public inspection, so that errors can be detected and corrected and new information can be introduced as it becomes available. Unfortunately, some popular alternatives to QRA, including those that use qualitative judgments and consensus in place of explicit calculations, replace transparent calculations with judgments and guesses that cannot necessarily be checked and independently reproduced, removing the opportunity to identify and correct errors or conduct sensitivity analyses.

Example: Use of Unvalidated Assumptions in a QRA for BSE ("Mad Cow" Disease)

In December of 2002, the Canadian Food Inspection Agency (CFIA) issued a document entitled *Risk Assessment on Bovine Spongiform Encephalopathy in Cattle in Canada*. (As of this writing in mid-2008, this document is available online at www.inspection.gc.ca/english/sci/ahra/bseris/bserise.shtml.) This QRA estimated the fractions of at-risk cattle, previously imported from countries with BSE, that might have arrived infected with BSE, been slaughtered, rendered, entered the cattle feed chain, and resulted in infection of other cattle. The executive summary explained the motivation and results of this QRA as follows:

> In order to evaluate the risk for BSE [bovine spongiform encephalitis, or "mad cow disease"] in Canada, the Canadian Food Inspection Agency (CFIA) has carried out a risk assessment on BSE in cattle in Canada.... The Government of Canada is committed to safeguarding the Canadian food supply and preventing the entry and establishment of foreign diseases such as BSE, and Canada has committed significant resources to this end.... One case of BSE was diagnosed in Canada in 1993, in a cow imported from the United Kingdom.... The estimated probability of at least one infection of BSE occurring prior to 1997 was 7.3×10^{-3} and *therefore the likelihood of establishment of BSE in Canada was negligible.* The risk was even further reduced by the mitigating measures in place since 1997.... In conclusion, the measures applied prior to the 1997 Feed Ban (import policies, disease control measures, detection system on-farm and at slaughter plants) combined with Canadian feed production and feeding practices, *were effective in preventing the entry of BSE and its subsequent amplification through the feed system.* (Emphases added)

Supported by these comforting calculations, the CFIA decided that there was no need for Canadian meat processors to undertake the unpopular and expensive step of removing especially risky material (tissue and organs) from cattle prior to rendering: "Given the controls in place for BSE (e.g. import policies, the Feed Ban) and the lack of evidence that BSE is present in native Canadian cattle, the CFIA does not exclude specific risk material from rendering."

The following year, a new case of BSE was found in Alberta. A second BSE-infected cow, imported from Alberta into the United States, was found in the state of Washington. From 2003 to 2008, more than 12 cases of BSE were confirmed in Canadian cattle, mostly from Alberta; these included cattle born in 2000, well after the 1997 Feed Ban was supposed to have taken effect (www.cdc.gov/ncidod/dvrd/bse/). Clearly, the CFIA's QRA had been overly optimistic in reassuring policy makers that "the measures applied prior to the 1997 Feed Ban (import policies, disease control measures, detection system on-farm and at slaughter plants) combined with Canadian feed production and feeding practices, were effective in preventing the entry of BSE and its subsequent amplification through the feed system."

What went wrong? Arguably, QRA in this case led to overly optimistic conclusions and policy recommendations (as became clear after the fact) due to a combination of (a) the use of unvalidated risk modeling assumptions by CFIA's risk assessors, (b) the presentation of results in ways that did not emphasize their contingent nature or uncertainty about the validity of their premises (as in the strong statement "The estimated probability of at least one infection of BSE occurring prior to 1997 was 7.3×10^{-3} and therefore the likelihood of establishment of BSE in Canada was negligible"), and (c) the willingness of policy makers to embrace reassuring conclusions based on these unvalidated modeling assumptions. Insisting on empirical evidence of the model's predictive validity (such as a demonstration that it could successfully explain surveillance results in multiple countries) before accepting its conclusions as a guide to risk management policy might have helped to prevent the overreliance on results whose validity was contingent on that of a largely unproved and hypothetical model. (Ironically, as of 2008, the United States continues to import cattle at risk of BSE from Canada, arguing that the BSE risk to the United States from doing so is "negligible." Perhaps, when and if this policy leads to a sustained outbreak of BSE in the United States, similar to the one in Alberta – as it seems to this author that it well might do – then U.S. policy makers will again learn that speculative risk models are only speculative.) Chapter 3 discusses further how to conduct QRAs and present results and uncertainties in ways that help to avoid unsound and misleading conclusions. Chapter 7 offers some technical recommendations for avoiding confirmation biases in risk modeling.

Such examples illustrate untrustworthy QRAs. Their results have been used to support risk management policies (e.g., banning beneficial animal antibiotics, or treating the risk of BSE in Canada as negligible) different from those that might have been based on more realistic QRAs. Initial skepticism about QRA models and results is well justified in such cases. Failures of QRA-based predictions teach that critical thinking about QRA models and results is essential before they are accepted and trusted for use in risk management decisions. Unvalidated risk modeling assumptions should not be used as a basis for policy making.

Yet the fact that QRA has sometimes been done badly does not mean that it cannot be done well. Chapters 2 and 3 of this book present basic principles of QRA that can help to avoid the types of errors in the preceding examples. Chapters 4 and 5 present easy but wrong methods to avoid. Chapters 6–13 develop and illustrate methods for using available data and knowledge, with all their gaps and imperfections, to improve QRAs for traditionally hard-to-model systems – especially, systems with unknown or uncertain causal mechanisms (Chapters 7–11), complex interactions among many variables (Chapters 6 and 13), delayed and dynamic responses to exposures (Chapters 11–13), and initially unknown and possibly nonlinear or stochastic relations between inputs (such as exposures or actions) and outputs (the probable consequences of the inputs) (Chapters 6–9). A motivating theme for these chapters is that QRA can greatly improve risk management decisions, even for such challenging applications, provided that appropriate methods are used.

Reconciling or choosing between the very different perspectives held by QRA advocates and opponents may be impossible on the basis of a priori arguments alone. QRA advocates are apt to resort to the conceptual tools and frameworks of QRA itself to justify its value. Those already skeptical of such methods are not likely to be persuaded by them. However, a useful common ground from which to evaluate QRA, compared to alternatives such as expert judgment-based decision making or direct application of the Precautionary Principle without formal risk assessment modeling, may be an examination of the real-world performance of QRA compared to that of proposed alternatives.

Toward Less Analytic, More Pluralistic Risk Management

Much of traditional decision analysis and QRA is based on a clear separation of facts or *beliefs* about the consequences of actions from *values* assigned to different consequences as well as on a clear distinction between preferences for *consequences* and preferences for *actions*. Each component is analyzed separately, with beliefs about and preferences for consequences ultimately determining the preferences for actions (e.g., via expected utility calculations). Decision analysis is admittedly very artificial and cognitive; most people do not naturally draw such clear distinctions and follow such a rigid discipline in evaluating and choosing among decision options in their daily lives (Loewenstein et al., 2008). Yet it can be well worth doing so when decisions have momentous consequences, such as when public health, safety, the environment, or well-being are at risk. Important decisions in engineering and business can often benefit tremendously from careful analysis (Russo and Schoemaker, 1989), although visceral decision making ("gut feel") may be easier and more satisfactory for many routine decisions, and decision aids may be inadequate to clarify some personally momentous decisions.

Rather than embracing these distinctions and separate analyses of these components, critics of QRA often encourage community members and other stakeholders to express holistic preferences directly for *actions* (e.g., banning animal antibiotics, GMOs, industrial emissions, use of DDT, and so forth). From this perspective, a key role of the public is seen as being to *express concerns and demand specific actions*, rather than simply expressing clear preferences for consequences and trusting and empowering regulators, aided by technical experts (and motivated by appropriate accountability and incentives), to figure out how best to achieve them. Instead of holding regulators accountable for *results,* the public is invited to hold them accountable for taking specific *actions,* with a minimal delay to assess likely consequences and to compare them to those from alternative actions often being seen as a virtue rather than as folly, no matter what a value of information (VOI) calculation might show.

In this view, risk assessors, technical experts, and scientists are just some of the many stakeholders in an essentially political decision process, rather than being seen as holding privileged information based on objective truths ("facts and data") about the probable consequences of alternative actions. The traditional concept of rational choice as the choice that is most consistent with achieving desired consequences is replaced by a more pluralistic concept in which different groups (e.g., scientists, policy makers, and the public) are seen as having different, yet equally legitimate, "forms of rationality" (Garvin et al., 2001). Even asking whether a form of rationality has legitimacy (a social and political construct), rather than whether it works well, reflects a shift away from an instrumental view of rationality in risk analysis (i.e., one that asks, "Is it successful in bringing about desired ends?"). Rather than being driven by a narrow focus on *outcomes,* marked by questions such as "Will this action achieve its stated goals with high confidence?" and "Is there a more effective way to achieve these goals?", this more pluralistic perspective typically focuses more on *process* and asks questions such as "Can the different stakeholders trust each other?" and "Can new types of knowledge brokering help different groups to work together?" (e.g., Choi et al., 2005). (An anonymous reviewer noted that, in some areas, "The trend towards concern-driven decision making has gone beyond the use of expert judgment as an alternative to QRA, to the use of public judgment as a replacement for both, with expert knowledge being used only ′on demand′ [to support particular points of view]. There are examples in Europe of where this kind of 'relativism' has taken hold, including the current decision process over what to do with Britain's nuclear waste. The high moral ground has been captured by the concern-driven lobby.")

Still, a valuable role remains for techniques that focus on achieving desired results. The traditional concept of rationality, focusing on identifying actions that tend to yield preferred consequences, delivers insights that often significantly improve upon unguided individual decision making (Shafir and LeBoeuf, 2002), even though it may not easily extend to situations where social interactions are the main determinants of outcomes (Colman, 2003).

Alternatives to QRA in Recent Policy Making: Some Practical Examples

Misgivings such as those outlined above have led to some noteworthy recent departures from QRA in important recent public policy decisions potentially affecting human health and safety. For example,

- In 2005, the U.S. Department of Agriculture (USDA) decided to immediately resume imports of cattle from Canada despite the unknown prevalence of BSE-infected cattle in Alberta. The USDA argued that its qualitative judgments that the risk to the United States was "very low" and that existing overlapping safeguards effectively "precluded" BSE from entering the United States (recent history to the contrary notwithstanding) should suffice for such a risk management decision, with no need or obligation to define "very low" in quantitative terms or to further address the quantitative risks created by resuming imports. [Later, however, the USDA did apply quantitative risk models and calculations similar to those that led the CFIA in 2002 to conclude that the BSE risk in Canada was "negligible." In a news release dated September 14, 2007 (USDA Release No. 0247.07), the USDA, too, concluded that "The assessment found that the risk of BSE establishment in the United States as a result of the imports [of cattle from Canada] announced today and those announced in January 2005 is negligible." In light of this conclusion, the USDA decided to allow the riskiest cattle (those over 30 months old) to be imported into the United States from Canada. However, other countries, such as South Korea, and the World Organisation for Animal Health (OIE), did not agree with the "negligible risk" designation. They assigned the United States the same BSE risk level as Canada ("controlled," but not "negligible") – the same as many other BSE-affected countries, including the United Kingdom (www.evira.fi/portal/en/animals_and_health/current_issues/?id=1084; www.thebeefsite.com/news/23227/uk-steps-up-to-controlled-bse-risk-status).]
- In 2003, a World Health Organization report (WHO, 2003) stated that traditional QRA, as developed for chemical and microbial hazards, is "inadequate" for assessing the biological risks associated with animal antibiotics and the

emergence of resistance, largely because of the inherent complexity and adaptive nature of bacterial systems (as opposed, presumably, to the inherent complexity and adaptive nature of other biological systems treated in traditional risk assessment modeling for chemical carcinogens and environmental hazards). WHO advocated relying instead on qualitative, precautionary judgments and on expert perceptions about the relative "importance" of antibiotics in human medicine – ideally, reached through expert consultations and consensus processes – as the primary bases for risk assessment and risk management.

- Similarly, in 2004, the U.S. Food and Drug Administration's Center for Veterinary Medicine (CVM) finalized a Guidance Document describing a qualitative risk-rating procedure as an alternative to QRA for assessing animal antibiotics as objects of regulatory concern (FDA-CVM, 2003). This guidance sheds many of the restrictive logical coherence requirements important in QRA. It does not assess or compare the human health effects of alternative risk management actions, does not prevent arbitrarily small or zero risk from being assigned the highest possible qualitative risk rating if subjective concerns call for it, avoids potentially burdensome data requirements (e.g., considering correlations among uncertain inputs) needed to reach correct quantitative conclusions, and does not insist that its qualitative ratings accurately reflect the true relative sizes of risks (Cox, 2006b). Rather, it bases its results and recommendations solely on the qualitatively expressed extent of concerns (e.g., *High*, *Medium*, or *Low*) about the release and exposure potentials for antibiotic-resistant bacteria and about the perceived "importance" of relevant antibiotics in human medicine. The guidance emphasizes pragmatic ease of use and flexibility in expressing and documenting concerns specifically about resistance. (Larger impacts on health risks that are not of current regulatory concern, such as the effects of removing antibiotics on increasing bacteria that are not antibiotic-resistant, are not considered.) The question of how the resulting risk management recommendations will change the frequency and severity of human health effects – the *sine qua non* for QRA – is not addressed.
- More generally, as discussed in Chapter 4, simple qualitative rating and ranking methods (typically organized as "risk matrices") have been developed and applied to the risk management of a wide range of important hazards. Risk matrices have been applied to terrorism risks, highway construction project risks, bridge and airport safety, office building risk analysis, climate change risk management, and enterprise risk management (ERM). National and international standards have stimulated the adoption of risk matrices by many organizations and risk consultants. Unfortunately, these methods can lead to very poor risk management decisions and ineffective resource allocations, as discussed further in Chapter 4.
- Activists, regulators, and scientists have advanced various versions of "Precautionary Principles" to justify intervening (or, sometimes, to justify maintaining the status quo) when the probable consequences caused by alternative available actions are uncertain. These principles are usually invoked in the presence of uncertainty to support the conclusion that a particular action should (or should

not) be taken, without the need for further causal analysis or quantification of the risks from following (or rejecting) this recommendation. For example, as Klinke and Renn (2002) state,

> If uncertainty plays a large role, in particular, indeterminacy or lack of knowledge, the risk-based approach becomes counterproductive. Judging the relative severity of risks on the basis of uncertain parameters does not make much sense. Under these circumstances, management strategies belonging to the precautionary management style are required. The precautionary approach has been the basis for much of the European environmental and health protection legislation and regulation.

An advocate of QRA would presumably challenge whether "judging the relative severity of risks on the basis of uncertain parameters does not make much sense," arguing that much of decision and risk analysis has been developed to do exactly that, enabling effective choices (tending to produce preferred outcomes) even – or especially – among prospects posing uncertain risks. In addition, he or she might point out that QRA is fully consistent with taking precautions in the absence of complete information, but that the real question is how to use available information to determine which actions truly are "precautionary," or at least undominated. Otherwise, the consequences produced by actions intended as "precautionary" may turn out to be unexpectedly adverse (Pugh, 2002).

Concern-Driven Risk Management

A common theme in the preceding examples is the principle that risk management interventions should be based on the qualitative *levels of concern* felt by responsible regulatory authorities, exercising their expert judgment, and not necessarily be based on quantitative analyses of their likely human health consequences (i.e., on the outcome of a risk assessment). In this view, regulatory agencies should be adequately sensitive to the concerns of their constituents and stakeholders, so that public concerns are translated quickly and effectively into corresponding agency concerns and regulatory actions. This leads to advice such as "Awareness and understanding of public concerns must be the basis of an effective risk management strategy" – whether or not these concerns align with calculations or actuarial reality (Frewer, 2004).

According to this principle, which we call the principle of *concern-driven risk management*, if a regulatory agency is sufficiently concerned about risks from current human behaviors (e.g., a particular animal antibiotic use), then it should be able to ban or restrict those behaviors without having to quantify the probable effects of doing so and without having to compare them to the probable effects of alternative actions. A strong version of concern-driven risk management is that the agency should be allowed – and perhaps encouraged – to take prompt actions based on its concerns, even if they have no transparent or objective basis in data and calculations and even if the actions taken will not necessarily produce beneficial consequences (or even if they cause severe adverse human health consequences that the agency is not concerned about, e.g., by increasing the quantity of antibiotic-susceptible pathogens reaching consumers in meat). The rationale is that the regulatory *concerns* motivating interventions, rather than the likely human health *consequences* of regulatory actions (and of other available alternatives), should drive regulatory risk management actions.

Conversely, if a responsible agency (e.g., the USDA) and its major stakeholders and experts do *not* feel much concern about a hazard (such as BSE risks from imported cattle), perhaps believing that existing regulations are sufficient to assure that "it can't happen here," then the principle of concern-driven decision making implies that the agency should be allowed to take actions (such as authorizing resumed imports of cattle from BSE-positive countries) without any need to quantify the risks of doing so. Again, it is the level of concern felt, rather than the likely quantitative health consequences of different actions, that is taken as the most appropriate guide to risk management action.

Potential Political Advantages of Concern-Driven Regulatory Risk Management

Concern-driven decision making has proved politically and legally viable for several important regulatory risk management problems in recent years in Europe and the United States, as evidenced by the above examples. Why? What about it allows it to successfully compete with – or even outcompete – QRA? QRA's main strength is that it offers to help identify actions that will cause the most desirable probable consequences, from among the available alternatives being evaluated. What does concern-driven risk management offer to rival this claim?

The study of human choice behavior and decision making reveals that real people care about much more than just the probable consequences of actions, as posited by idealized models of rational decision making (*homo economicus*). Most people also care about the *intent* behind an action, the perceived *fairness* of decisions, the perceived *equity* of opportunities and outcomes embedded in a decision process and its results, and whether the selected act respects norms of *reciprocity* and *fairness* in its allocation of gains and losses (Ohmura and Yamagishi, 2005; Sanchez and Cuesta, 2005; Camerer and Fehr, 2006). Many people would rather select, trust, support, and follow leaders or public servants who conspicuously share their concerns and plainly intend to address them, rather than to entrust the management of life, health, and safety issues to dispassionate calculators of theoretically "optimal" courses of action.

Concern-driven risk management has the potential advantages that it is often easy to explain (by listing the concerns that it intends to address), simple, actionable, and addressed specifically to topics that worry and upset constituents. Rather than engaging in complex and assumption-laden hypothetical calculations about alternative potential futures, the concern-driven risk manager can simply say, "This threat looks serious to me. I think we should do something about it now, and here is what I think we should do." Such a conspicuously responsive, action-oriented, well-intended approach may be far more politically attractive than the careful fact-gathering, analysis, and deliberation associated with QRA. It mirrors the physiological reflex response to an unpleasant stimulus – no higher-level cognition is needed!

Less benignly, concern-driven risk management may encourage taking politically popular actions that feel right, even if they do not further the achievement of explicitly stated outcomes, such as reducing illness rates. For example, recent brain imaging studies (Fehr and Rockenbach, 2004) show that people derive psychological satisfaction from punishing others whom they perceive to have violated social norms – so much so, indeed, that they are often willing to incur some cost or harm to themselves in order to punish such others (Knoch et al., 2006). Theoretical models in evolutionary biology help to explain how such *altruistic punishment* might arise and persist (Bowles and Gintis, 2004). Concern-driven risk management is commonly applied in situations where its recommended actions are seen as punishing or harming parties, especially, large corporations or organizations, whose actions or existence are perceived or portrayed as violating social norms. For example, the "Polluter Pays" catchphrase is often deployed in the same contexts as the "Precautionary Principle" (Hayes, 2005), viz., when cause-and-effect relations have not actually been established but action is being urged anyway. In such settings, requiring companies to address the expressed concerns of members of the public through expensive actions may feel like the right thing to do, even if those actions are truly not likely to produce health benefits or risk reductions. Although punishment is seldom offered as an explicit rationale for risk management decisions, concern-driven risk management permits it, as well as other emotional and psychological motives (drivers of concern) distinct from quantified cause-and-effect relations and the pursuit of explicitly stated goals.

Concerns about real or perceived risks are not uniformly distributed among demographic and political groups. Demographic attributes (e.g., sex and race) are significantly associated with the extent of concerns about health risks from environmental hazards (e.g., Johnson, 2002). For example, in many studies, women are significantly *less* likely than men (especially, white men) to express high trust in public health experts and *more* likely to be concerned about environmental hazards (Flynn et al., 1994), except in neighborhoods that truly are affected by multiple environmental stresses (Greenberg and Schneider, 1995). They are also more likely to believe that apparently random cancer clusters are not truly due to chance (Siegrist et al., 2001). Low trust in science and technology to protect public health, as well as low trust in neighborhood (e.g., local) officials to do so, is more common among younger people who have spent relatively little time in their current neighborhoods (and, ironically, who do not engage in many personal public health practices) (Greenberg and Williams, 1999). These and other systematic differences between the groups in society who are most and least concerned about suggested hazards provide a possible basis for a "politics of concern" that addresses itself primarily to those groups that are most likely to feel concern, even if other members of society do not.

In summary, a strong form of concern-driven risk management that maps levels of concern directly to specific recommended actions to address those concerns, without quantitative analyses of likely consequences or comparisons to alternatives, may have several emotional and political advantages over QRA. These include the advantages of clarity, responsiveness, and emotional satisfaction that worries are being promptly addressed. Those who articulate and implement concern-driven risk management policies may thereby display desirable leadership characteristics such as good intent, responsiveness to constituent concerns, and willingness to take action that attract political support. If the recommended actions are coupled to perceptions that those who bear the costs of addressing concerns are those who ought to do so, then the concern-driven actions may feel right and become popular even if their effectiveness is uncertain or unlikely. Such factors may help to explain the growing popularity of concern-driven risk management as an alternative to QRA. However, its effectiveness in producing desired results remains to be assessed.

How Effective Is Judgment-Based Risk Management?

Concern-driven risk management usually relies on the judgments of those in positions of power, responsibility, or authority (e.g., expert committees or staff teams appointed by regulatory agencies and public health organizations, sometimes with members of the public) to decide on appropriate risk management actions based on levels of concern (usually qualitatively expressed), thus obviating the need for formal QRA. These judgments may be informed by technical analyses and by the advice of subject matter experts, some of whom may apply QRA or other quantitative models and methods to help form their opinions. But ultimately, the risk management decision taken depends on the informed judgments of those in charge.

Example: Expert Judgment vs. QRA for Animal Antibiotics

Not infrequently, the results of expert consultations and expert judgment conflict directly with the results of simple QRA models. For example, WHO (2003) used expert judgment to conclude that

> Therefore, the expert workshop recommends that an expert clinical medical group appointed by WHO defines which antimicrobials are considered critically important [potentially subject to maximally restricted use] in humans.... [A] list of critically important classes of antimicrobials should include: the fluoroquinolones and 3rd generation cephalosporins for *Salmonella* spp. and other *Enterobacteriaceae*; the fluoroquinolones and macrolides for *Campylobacter* spp. ; and glycopeptides, oxazolidinones and streptogramins for Gram positive bacteria such as *Enterococcus* spp.

By contrast, QRA models suggest that fluoroquinolones and macrolides for *Campylobacter* spp. and streptogramins for *Enterococcus* spp. pose at most only minimal quantitative risk to human health: less than 1 quality-adjusted life-year (QALY) lost per year for the entire population of the United States; and less than

0.001 QALYs lost per year from the use of the streptogramin drug virginiamycin (Cox, 2006b). A draft QRA for virginiamycin completed by the FDA-CVM (2004) likewise suggests that quantitative human health risks are at most relatively small, even under some worst-case assumptions. Thus, expert judgment may yield quite different results from QRA models, making it important to decide how to choose between them when they differ.

This raises the following key empirical question: *How good are human judgments, including expert judgments, as guides to effective risk management actions?* If the proposed alternative to formal quantitative QRA calculations is less formal and less quantitative expert judgment – as explicitly recommended by WHO/FAO (2003) and FDA-CVM (2003), for example – then it is important to understand how well such judgment performs in identifying risk management actions that produce desired consequences. Similar critical evaluations have already been conducted for stakeholder-based decisions, with generally reassuring results (e.g., Beierle, 2002), but not yet for concern-driven risk management decisions implemented through expert judgment and consensus.

Performance of Individual Judgment vs. Simple Quantitative Models

The performance of human judgment on a variety of prediction and choice tasks under uncertainty has been studied extensively for several decades, leading to some striking, but very well-supported, conclusions. In general, *the performance of individual judgments and judgment-based decisions compares poorly to the performance of even very simple (e.g., linear) quantitative models* on a variety of prediction and decision tasks, including identifying causal explanations or diagnoses for observations (Ceci and Bjork, 2000), predicting what will probably happen next in many situations, and predicting the probable consequences of alternative actions or interventions. Hundreds of studies have confirmed this pattern, not only in contrived psychological experiments, but also for real judgments and decisions with important consequences in business (Russo and Schoemaker, 1989), medicine (Bornstein and Emler, 2001), social work, academics, and other fields (Plous, 1993; Meehl, 1954, 1993). In almost all cases, human judgment performs relatively poorly, compared to even simple quantitative models. Indeed, this effect is so strong that just fitting a simple quantitative model *to one's own judgments* typically yields predictions that outperform them (Russo and Schoemaker, 1989)!

In psychology, the situation has been described as follows:

> The accuracy of judgments made in a methodical way from just a few relevant pieces of information is usually equal or superior to those of experts who combine a wide array of information using unaided human judgment.... Indeed, to date there is no replicable counterexample to this empirical generalization. This superiority of SPRs [statistical prediction rules] over clinical judgment has been attributed to two complementary sources: the desirable mathematical properties of SPRs and the cognitive limitations and biases of human judgment.... We are also strongly biased in favor of our prior beliefs and are adept at

> constructing post hoc explanations (Ruscio, 1998b) that sacrifice historical truth for narrative truth. All of these biases, as well as many others ... can contribute to a poor use of information, especially relative to SPRs. Available evidence suggests that unaided human judgment cannot compete with a more mechanical process that involves a comparatively simple combination of a small handful of relevant variables (Swets et al., 2000). This conclusion has been supported in a tremendous number of disciplines.... When provided with identical information, SPRs tend to achieve greater empirical accuracy than do professionals. This remains true when one provides professionals with information not available to the SPR, and even when one provides the results of the SPR itself, in which case professionals identify too many "exceptions" to the rule.... Meehl (1986) attributed what he refers to as irrational adherence to an inferior decision-making procedure to several sources. Many individuals′ unwavering belief in the efficacy of their own judgment or in the importance of their preferred theoretical identification (as contrasted, for example, with an atheoretical SPR) is a potent stumbling block.... It is noteworthy that clinical judgments – and even less so holistic claims – are rarely tolerated when large sums of money are at stake. For example, when making decisions involving loans, insurance rates, or gambling odds, actuarial decision making is the norm. (Ruscio, 2003)

Much of the explanation for the relatively poor performance of nonquantitative and judgment-based methods can be grouped into the following three areas:

1. *Individual judgments are sensitive to logically irrelevant details of how information is presented.* For example, decisions may be affected by whether information is presented in rows vs. in columns, in bar charts vs. in pie charts, and in graphs vs. in tables (e.g., Elting et al., 1999); by giving probabilities of "survival" instead of logically equivalent probabilities of "mortality"; by presenting cost information before vs. after other information (Jenkins-Smith and Kunreuther, 2001); by disaggregating columns within a table or sequences of repeated choices; and by including vs. excluding information on inferior alternatives that will not be selected but that may change the relative salience of attributes for other, better alternatives (Stewart et al., 2003; Schwartz and Chapman, 1999, for medical decisions). In some circumstances, individuals seek out and use logically irrelevant information to make decisions (Bastardi and Shafir, 1998). Many human judgments tend to overemphasize the importance of human actions, compared to other events and conditions, in bringing about undesirable consequences (Morris et al., 1999). Such preoccupation with whom to blame for adverse consequences, rather than how best to prevent or mitigate them, may have served well in furthering the evolution of cooperation in small hunter-gatherer societies (Bowles and Gintis, 2004) yet has limited value in improving current risk management decisions.
2. *Individual judgments are often insufficiently sensitive to relevant information.* This can be due to *confirmation bias*, in which information is sought to support or confirm already formed opinions, while disconfirming evidence is not sought and is disregarded or underweighted when presented (Russo and Schoemaker, 1989; Plous, 1993). It also arises when decision makers pay attention to only a few attributes of a complex decision problem in deciding what to do, even though these factors do not suffice to predict outcomes or to distinguish between good and bad choices for achieving desired ends (ibid.) (In

the examples previously discussed, considering only antibiotic-resistant bacteria while neglecting antibiotic-susceptible bacteria may illustrate such overly narrow framing, leading to importantly incomplete assessments of how risk management actions affect human health.) Human judgments also often suffer from *overconfidence*, with uncertainty around best judgments or guesses being systematically underestimated (Hammond et al., 1999).

3. *Information is often combined and used ineffectively.* For example, some information may be overweighted (e.g., based on vividness), while other information that could be more valuable in improving predictions and decisions is disregarded or underweighted (Russo and Schoemaker, 1989). Decision makers often consider components of portfolios independently (sometimes called *narrow framing*), neglecting the portfolio in which they are embedded. Similarly, they may inappropriately evaluate in isolation choices made within sequential plans, repeated choices, or sequences of opportunities (Benartzi and Thaler, 1999). Decisions based on the presentation of information can differ significantly from decisions based on information acquired through direct experience (Hertwig et al., 2004).

Table 1.1 summarizes some well-studied cognitive *heuristics and biases* (in the famous term of Tversky and Kahneman, 1974) that can strongly affect how information is processed and used in human judgment and decision making (Plous, 1993; Russo and Schoemaker, 1989; Hammond et al., 1999). These heuristics and biases affect real decisions having important (e.g., medical) consequences (Bornstein and Emler, 2001). Formal quantitative risk assessment and decision analysis can help to identify and overcome such heuristics and biases in situations where they might otherwise adversely affect decision making. (As usual in this book, poor choices are defined as ones that are dominated by other available choices that were not made. Heuristics and biases can lead to poor choices in this sense. Formal analysis can promote better choices, in many practical situations, as documented at length in the preceding references.)

Table 1.2 presents glimpses from the young field of neuroeconomics, which offers some possible biological explanations for why we make risky choices as we do. In brief, neuroeconomics suggests that the brain integrates results from multiple "modules" (brain processes and structures) in making decisions involving risk. Both cognitive (slow, deliberate, conscious, controlled) and affective-emotional (quick, reflexive, unconscious, uncontrolled) reactions are crucial for the effective evaluation of risky options and decision making in risky situations. Decisions with different characteristics (e.g., immediate vs. delayed outcomes, known vs. unknown probabilities, fair vs. unfair offers, isolated vs. social decisions, anticipated vs. surprising outcomes, and so forth) activate different brain systems differently, thus leading to choices that are inconsistent with standard models of coherent decision making. Similarly, reframing decisions in logically equivalent ways that trigger these subsystems to different extents can lead to different decisions. Such insights can potentially explain many of the well-documented ways in which real decisions

Table 1.1 Some heuristics and biases in individual judgments

Heuristic/Bias	Explanation
Anchoring	Judged probabilities, frequencies, or values remain too close to initial values (perhaps based on irrelevant cues), despite new information.
Availability	Background information is underweighted compared to readily retrieved information.
• Primacy and recency	• Initial events and/or most recent events are most easily recalled and hence receive excessive weight.
• Saliency	• Salient information is overweighted compared to other information.
• Status quo bias	• Historical experience is more vivid and is weighted more heavily than hypothetical alternatives.
Base rate neglect	Base rates are neglected or underweighted in many situations.
Overconfidence, miscalibration	Assessed probabilities are not accurate. (For example, events judged to be impossible in some experiments happen about 20% of the time, while events judged to be certain happen only about 80% of the time.)
Credulity/superstition	Evidence that supports patterns and causal explanations for coincidences is too readily accepted.
Confirmation bias	Evidence that supports previously formed hypotheses (especially about likely causes or diagnoses) is overweighted compared to other evidence.
Conjunction fallacy	A conjunction of events (e.g., patient was exposed *and* exposure caused observed symptoms) is considered *more* probable than the individual components of the conjunction. This violates probability theory.
Conservatism	Sample information is often underweighted.
Focal effects	Recalled or stated values are binned into approximate categories.
Disjunction ("irrational prudence")	A decision maker who will take the same action whether or not an event occurs may still prefer to wait for the uncertainty to be resolved.
Framing	The presentation of data (e.g., in ascending or descending order) affects judgments of likelihoods or estimates for uncertain events or quantities.
Gambler's Fallacy	Belief that random fluctuations will tend to occur to cancel out previous unusual patterns and restore the representativeness of the sample.
Hindsight bias	Belief that whatever happened was inevitable or was predictable in advance.
Illusion of control	Subjects perceive that their skill can affect the outcomes of chance events.
Law of Small Numbers	Subjects assume that small samples are representative of the populations from which they are drawn. Since representativeness is not sensitive to sample size, they tend to gather too little data and to overgeneralize.
Regression to the mean	Many subjects expect that observed patterns will continue (i.e., that observations are representative) rather than expecting unusual fluctuations to be followed by a return to more usual levels ("regression to the mean").
Representativeness	$\Pr(X \mid E)$ is estimated by how "representative" evidence E is of explanation X, ignoring or being inadequately sensitive to base rates for $\Pr(X)$ and $\Pr(E)$ and violating the probability rule $\Pr(X \mid E) = \Pr(E \mid X)\Pr(X)/\Pr(E)$.

References: Plous (1993), Russo and Schoemaker (1989), Hammond et al. 1999.

Table 1.2 Selected findings from neuroeconomics studies of risk

Key findings and interpretations (emphases in bold added)	References
"**Neuroeconomics has primarily challenged the standard economic assumption that decision making is a unitary process** – a simple matter of integrated and coherent utility maximization – suggesting instead that **it is driven by the interaction between automatic and controlled processes**." Such interacting brain processes affect decision making under risk and uncertainty, intertemporal choice, and social decision making.	Loewenstein G, Rick S, Cohen JD. Neuroeconomics. *Ann Rev Psych.* 2008;59:647–72.
"Here we use functional magnetic resonance imaging [fMRI] to show that **neural activity in several brain regions** – particularly the ventral striatum, medial prefrontal cortex and posterior cingulate cortex **– tracks the revealed subjective value of delayed monetary rewards**. This similarity provides unambiguous evidence **that the subjective value of potential rewards is explicitly represented in the human brain**."	Kable JW, Glimcher PW. The neural correlates of subjective value during intertemporal choice. *Nat Neurosci.* 2007 Dec; 10(12):1625–33.
"To a first approximation, decision making is a process of optimization in which **the decision maker tries to maximize the desirability of the outcomes resulting from chosen actions.** Estimates of desirability are referred to as utilities or value functions, and they must be continually revised through experience according to the discrepancies between the predicted and obtained rewards. . . . Similar to the findings from other brain areas, such as cingulate cortex and basal ganglia**, activity in the dorsolateral prefrontal cortex often signals the value of expected reward and actual outcome**."	Lee D, Seo H. Mechanisms of reinforcement learning and decision making in the primate dorsolateral prefrontal cortex. *Ann NY Acad Sci.* 2007 May; 1104: 108–22.
"It is known that humans vary widely in their performance in decision-making tasks. Here, we used a simple four-armed bandit task in which subjects are almost evenly split into two groups on the basis of their performance: those who do learn to favor choice of the optimal action and those who do not. . . . **While the learners showed robust prediction error signals in both the ventral and dorsal striatum during learning, the nonlearner group showed a marked absence of such signals**. Moreover, the magnitude of prediction error signals in a region of dorsal striatum correlated significantly with a measure of behavioral performance across all subjects."	Schönberg T, Daw ND, Joel D, O'Doherty JP. Reinforcement learning signals in the human striatum distinguish learners from nonlearners during reward-based decision making. *J Neurosci.* 2007 Nov 21; 27(47):12860–7.
"Reinforcement learning models that focus on the striatum and dopamine can predict the choices of animals and people. Representations of reward expectation and of reward prediction errors that are pertinent to decision making . . . are also found in prefrontal and cingulate cortex." [In addition to expected reward], "**uncertainty in the estimate of the reward expectation, the value of information that might be gained by taking a course of action and the cost of an action** all influence [how] decisions are made through prefrontal and cingulate cortex."	Rushworth MF, Behrens TE. Choice, uncertainty and value in prefrontal and cingulate cortex. *Nat Neurosci.* 2008 Apr; 11(4):389–97.

Table 1.2 (continued)

Key findings and interpretations (emphases in bold added)	References
"Expectation of a high reward can motivate a subject to go for an action despite a large cost, a decision that is influenced by dopamine in the anterior cingulate cortex. **Uncertainty of action outcomes can promote risk taking and exploratory choices**, in which norepinephrine and the orbitofrontal cortex appear to be involved."	Doya K. Modulators of decision making. *Nat Neurosci*. 2008 Apr; 11(4):410–6.
"Individual differences in sensitivity to delays and uncertainty may contribute to impulsivity and risk taking. **Learning and choice with delayed and uncertain reinforcement are related but in some cases dissociable processes**."	Cardinal RN. Neural systems implicated in delayed and probabilistic reinforcement. *Neural Netw.* 2006 Oct; 19(8): 1277–301.
"Here, we demonstrate by functional magnetic resonance imaging (fMRI) that **individuals′ preferences for risk (uncertainty with known probabilities) and ambiguity (uncertainty with unknown probabilities)** predict brain activation associated with decision making. ... [D]ecision making under ambiguity does not represent a special, more complex case of risky decision making; instead, these two forms of uncertainty **are supported by distinct [brain] mechanisms**."	Huettel SA, Stowe CJ, Gordon EM, Warner BT, Platt ML. Neural signatures of economic preferences for risk and ambiguity. *Neuron*. 2006 Mar 2; 49(5):765–75.
"Results indicate that **deciding advantageously in a decision-making task with explicit and stable rules is linked to applying calculative strategies**. In contrast, **individuals who decide intuitively prefer risky or disadvantageous choices** in the Game of Dice Task. . . . The results support the view that **calculative processes and strategies may improve decision-making under explicit risk conditions**."	Brand M, Heinze K, Labudda K, Markowitsch HJ. The role of strategies in deciding advantage-ously in ambiguous and risky situations. *Cogn Process.* 2008 Jan 30.
"Using [fMRI] and a guessing task in two large cohorts, we are able to **confirm ventral striatal responses coding both reward probability and magnitude during anticipation, permitting the local computation of expected value (EV)**. However, the ventral striatum only represented the gain-related part of EV (EV+). At reward delivery, the **same area shows a reward probability and magnitude-dependent prediction error signal,** best modeled as the difference between actual outcome and EV+. In contrast, **loss-related expected value (EV–) and the associated prediction error was represented in the amygdala.** Thus, the ventral striatum and the amygdala distinctively process the value of a prediction and subsequently compute a prediction error for gains and losses, respectively."	Yacubian J, Gläscher J, Schroeder K, Sommer T, Braus DF, Büchel C. Dissociable systems for gain- and loss-related value predictions and errors of prediction in the human brain. *J Neurosci.* 2006 Sep 13; 26(37):9530–7.

Table 1.2 (continued)

Key findings and interpretations (emphases in bold added)	References
"In [fMRI] research, nucleus accumbens (NAcc) activation spontaneously increases before financial risk taking. As anticipation of diverse rewards can increase NAcc activation, **even incidental reward cues may influence financial risk taking**. Using event-related [fMRI], we predicted and found that anticipation of viewing rewarding stimuli (erotic pictures for 15 heterosexual men) increased financial risk taking, and that this effect was partially mediated by increases in NAcc activation."	Knutson B, Wimmer GE, Kuhnen CM, Winkielman P. Nucleus accumbens activation mediates the influence of reward cues on financial risk taking. *Neuroreport.* 2008 Mar 26; 19(5):509–13.
"We used [fMRI] of Ultimatum Game players to investigate neural substrates of cognitive and emotional processes involved in economic decision-making. In this game, two players split a sum of money; one player proposes a division and the other can accept or reject this. We scanned players as they responded to fair and unfair proposals. **Unfair offers elicited activity in brain areas related to both emotion (anterior insula) and cognition (dorsolateral prefrontal cortex). Further, significantly heightened activity in anterior insula for rejected unfair offers suggests an important role for emotions in decision-making.**"	Sanfey AG, Rilling JK, Aronson JA, Nystrom LE, Cohen JD. The neural basis of economic decision-making in the Ultimatum Game. *Science.* 2003 Jun 13; 300(5626):1755–8.
"Humans restrain self-interest with moral and social values. They are the only species known to exhibit reciprocal fairness, which implies the punishment of other individuals′ unfair behaviors, even if it hurts the punisher′s economic self-interest. Here we show that **disruption of the right, but not the left, dorsolateral prefrontal cortex (DLPFC)** by low-frequency repetitive transcranial magnetic stimulation **substantially reduces subjects′ willingness to reject their partners′ intentionally unfair offers** [in an Ultimatum Game], which suggests that subjects are less able to resist the economic temptation to accept these offers. Importantly, however, subjects still judge such offers as very unfair, which indicates that the right DLPFC plays a key role in the implementation of fairness-related behaviors."	Knoch D, Pascual-Leone A, Meyer K, Treyer V, Fehr E. Diminishing reciprocal fairness by disrupting the right prefrontal cortex. *Science.* 2006 Nov 3; 314(5800):829–32.

under risk depart from the prescriptions of normative models of rational decision making (Loewenstein et al., 2008).

Psychologists have also discovered that many lay and expert judgments about causality and risk are strongly biased by prior beliefs (White et al., 1995; Hagmayer and Waldmann, 2002) and by envisioned causal mechanisms (Ahn and Bailenson, 1996; Tangen and Allan, 2004). These biases can lead to severe underweighting of empirical evidence and excessive resistance to new evidence and data (Plous, 1993), unawareness of the sources of beliefs and the influence of preconceptions (Fugelsang and Thompson, 2003), and maladaptive causal inferences and ineffective decisions (Elstein, 1999, for individuals; Jones and Roelofsma, 2000, for teams). Again, even relatively simple quantitative models often provide more useful and reliable insights and conclusions (e.g., more accurate predictions on a variety of tasks) than judgment (Grove et al., 2000).

Performance of Consensus Judgments vs. Simple Quantitative Models

Critics of QRA who call for judgment-based processes usually propose that *teams* of experts (usually selected or appointed by the relevant agency or authority) should make the required judgments. Thus, the question arises of how well these team-based expert decision processes can be expected to perform.

Group decision processes have been studied extensively over the past several decades. Like individuals, groups (including teams of highly trained experts) tend to consider too few alternatives and to collect and use too little information (Plous, 1993; Russo and Schoemaker, 1989), due in large part to confirmation bias (Jonas et al., 2005; Schulz-Hardt et al., 2000) but also to poor sharing and use of individual information within the group (Jones and Roelofsma, 2000; Lam and Schaubroeck, 2000). Unlike individuals, groups can also be subject to the strategic misrepresentation of knowledge and beliefs by group members (Steinel and De Dreu, 2004). Empirically, groups often make suboptimal judgments and decisions that are then strongly defended against conflicting information and divergent points of view (most notoriously, in "group think"), especially when group norms encourage consensus (Postmes et al., 2001).

Example: Resistance of Expert Judgments to Contradictory Data

In 1997, building on three decades of concerns and speculations, but without firm empirical evidence (NAS, 1999), Danish researchers warned that "A general increase in antimicrobial resistance among pathogenic bacteria is causing concern worldwide that the widespread use of antimicrobial agents in animal production may promote the development of resistant bacteria or resistance genes that can be transferred to bacteria that cause disease in humans" (Wegener et al., 1997). In 1998, these concerns were addressed via an application of the Precautionary Principle: Five animal antibiotics used as feed additives were banned in Denmark and the European Union. No QRA showing that this was likely to produce desirable human health consequences was used or considered necessary (Pugh, 2002).

Immediately after the ban, animal illness and mortality rates surged. Although some animal illnesses (e.g., necrotic enteritis in chickens) were brought back under control over the next five years by the increased use of therapeutic antibiotics and other countermeasures (VLA, 2004), other changes have persisted. As of 2005, pig mortality rates in Denmark remained significantly elevated (by about 25%) and productivity remained significantly depressed. Human illness rates for foodborne illnesses such as campylobacteriosis also continued to increase in parts of Europe, including Denmark (Patrick et al., 2004; see also ElAmin, 2006), although they fell significantly over the same time period in the United States. Resistance rates to antibiotics such as erythromycin and streptomycin among human isolates sharply *increased* (by several hundred percent between 1997 and 1999), while pig illness and production costs also increased (Hayes and Jensen, 2003). Outside observers reported that many of these problems persisted in Denmark as of 2005. Simple QRA models predict and explain these observed outcomes if withdrawing animal antibiotics increases rates of animal illnesses (Cox, 2006). Although inferences about causality are complex and uncertain, and a full understanding of the historical record would admittedly require more extensive analysis and discussion, in part because comprehensive programs to reduce foodborne illnesses were being implemented in parts of Europe at the same time as the bans (e.g., Wegener et al., 2003), the belief that the bans would have human health benefits is not well supported by the data.

Yet, despite this poor empirical record for human (and animal) health, many of the experts who advocated the bans continue to see them as desirable and successful. Instead of focusing on the increased rates of animal mortality, animal and human illnesses, and human resistance, they have emphasized statistics (such as decreasing the prevalence of resistance in isolates from *healthy* animals and people) that do not measure the deterioration in animal and human health, but that are more consonant with the original belief that the bans should be implemented. As late as 2003 and 2004, original advocates of animal antibiotic bans were still writing that "The data shows that although the levels of resistance in animals and food, and consequently in humans, has been markedly reduced after the termination of use, the effects on animal health and productivity have been very minor" (Wegener, 2003) and that "Several European countries have demonstrated that restricting the use of antimicrobial agents in food animals can be followed by a decrease in antimicrobial resistance in humans without compromising animal health or significantly increasing the cost of production" (Angulo et al., 2004). Although such beliefs contrast with historical data showing sharply *increased* resistance rates in isolates from humans and severely compromised animal health and increased animal mortality and productions costs (e.g., Hayes and Jensen, 2003), they help to defend and reinforce the original beliefs of the expert groups involved in this concern-driven risk management intervention (e.g., WHO, 2003).

Example: Ignoring Disconfirming Data About BSE Prevalence

In 2005, the Canadian Food Inspection Agency (CFIA, 2005) explained the policy context for its BSE risk assessments as follows:

> As North American Free Trade Agreement (NAFTA) partners, Canada, the U.S. and Mexico have applied independent but harmonized transmissible spongiform encephalopathy (TSE) risk-management strategies that are aligned with the OIE. . . . The continuing goal is that of a unified position on TSE issues in North America.

They urged that Canada and the United States should be classified as having the same BSE risks: "[T]he structure of the cattle industries in both the U.S. and Canada are virtually identical and share the same BSE risk factors" and

> Internationally, Canada and the U.S. are also viewed as having the same BSE risk, both before and after the detection of BSE in 2003. . . . The equivalency of risk status between Canada and the U.S. is . . . [a]lso the basis of the U.S. BSE risk analysis, "*Evaluation of the Potential for Bovine Spongiform Encephalopathy in the United States*" and Canada's BSE risk analysis "*Risk Assessment on Bovine Spongiform Encephalopathy in Cattle in Canada.*"

(These were risk assessments used by the CFIA and USDA *before* the onset of BSE cases in 2003 to argue that Canada's BSE risks were "negligible.") Fortifying its position against possible further disconfirming data (which soon materialized), the CFIA emphasized that "It is important to note that the detection of some additional BSE-positive animals does not change the risk profile for either Canada or the U.S." – a curious proposition from a QRA perspective, since risk estimates are typically sensitive to new occurrences, especially when the initial number of observed occurrences is small. This enabled the CFIA to leave its policy position unchanged as additional cases of BSE were found in Canada, including cases among cattle born long after the 1997 feed ban. Indeed, as of 2008, the United States and Canada were assigned the same risk category ("controlled," but not "negligible") by the OIE. The policy goal of identical risk classifications for the United States and Canada was achieved.

What is striking about these qualitative arguments is that they ignore quantitative data showing that Canada has a BSE prevalence rate many times greater than that in the United States. By 2008, Canada had confirmed over a dozen BSE cases (mostly of the typical type found in the U.K.) among about 200,000 high-risk Canadian-born cattle inspected. The United States had confirmed only two cases of BSE (both atypical H-strain, possibly spontaneous and probably of noninfectious origin) among about 800,000 U.S.-born cattle inspected. Thus, a crude calculation suggests that the prevalence rate of all BSE types in Canadian cattle might be about $(12/200{,}000)/(2/800{,}000) = 24$ times higher than in the United States. [In 2008, the U.S. Centers for Disease Control and Prevention (CDC) posted results of a refined calculation that accounted for additional relevant details, such as age-specific risks. It concluded that "the true prevalence of BSE in Canada is 90% likely to be between 18-fold and 48-fold higher than the previously published best estimate of the prevalence of BSE in the United States."] Moreover, the CDC noted, "The BSE strain responsible for most of the BSE cases in Canada is the same classic or typical strain linked to the outbreak in the United Kingdom. This typical strain has not yet been identified in any U.S.-born cattle" (www.cdc.gov/ncidod/dvrd/bse/).

In summary, the available quantitative data have long revealed that Canada has a much higher prevalence rate of BSE cases than the United States (which, as of early 2008, had found no typical BSE in U.S.-born cattle). Policy statements to the contrary simply ignore this large difference.

In general, group decision making and judgment often converge prematurely to incorrect or suboptimal consensus judgments, and participants then waste substantial time and effort talking themselves into increased confidence and support for their conclusions, even though such confidence-boosting does not lead to improved decisions (Russo and Shoemaker, 1989). Although various techniques have been developed to avoid premature consensus and to improve group decisions (ibid.; Lam and Schaubroech, 2000), many groups (including groups of experts) perform no better than – and sometimes considerably worse than – their individual members in a variety of prediction, judgment, and decision tasks (Plous, 1993).

Group dynamics, strategic, and psychological explanations have often been advanced for the limitations of group decision making. However, there are also mathematical reasons to expect that group consensus decisions will often be poorer than decisions based on QRA models. For example, consider a group decision process in which (a) individuals are first invited to express their individual opinions or judgments, (b) these are aggregated somehow (e.g., by averaging, majority rule, or some other group decision rule) to create a proposed group judgment, and (c) the individual members of the group are repeatedly invited to either agree with or propose changes to the group's current judgment (e.g., a best estimate of an uncertain quantity, or a recommendation for what action to take) and/or to modify their own judgments. Suppose that we wish group decisions to satisfy the following two properties:

Unanimity implies consensus: If each member of the group agrees with the group's current judgment, the adjustment process terminates, with the unanimously agreed-to judgment becoming the group's final judgment. In other words, unanimous agreement ends discussion and produces a consensus judgment.

Effective information pooling: If the group reaches a consensus judgment, then it is consistent with the facts known by the individuals (i.e., the consensus judgment should not be contradicted by the information of the individual members).

A basic limitation of many group decision processes and consensus judgment is that these two conditions are not necessarily compatible. If a group decision process stops when consensus is reached, then it may make poor use of individual information.

Example: Consensus Decision Making Can Waste Valuable Individual Information

Suppose that each of three experts, A, B, and C, offers an opinion about whether an animal antibiotic poses a high risk to human health. Before gathering any specific information, the experts agree that human health risk should be called "high" if and only if at least one of the following three conditions is satisfied: (1) use results in the release of resistant organisms; (2) these resistant organisms can cause exposures and resistant infections in human patients; and (3) such resistant infections lead to compromised treatments or treatment failures for human patients. The experts agree that an *uncertain risk* should be considered "high" if the probability that it is high exceeds 50%. Before collecting any specific data, all experts agree that each of conditions (1)–(3) independently has a prior probability of 0.5. Now, suppose that expert A collects information that reveals that (1) does not hold; expert B discovers that (2) does not hold; and expert C learns that (3) does not hold. No expert knows what specific information the others have collected, and no expert has any other relevant information (other than these prior statistics and collected information). Then each individual expert must assess the probability that the true risk is "high" as $(1 - 0.5 \times 0.5) = 75\%$ (i.e., it is high unless both conditions that the expert has not collected information on fail to hold). Hence, each expert concludes that the uncertain risk should be treated as "high" (specifically, 75%). They agree on this unanimously. If each expert announces this conclusion, then the process will terminate with the consensus judgment that the risk from this antibiotic use is 75%, or "high" (depending on whether a quantitative or qualitative expression of judgments is used), even though this is inconsistent with the facts known to the individual experts: that none of (1)–(3) holds, so the true risk is zero. Therefore, in this example, *consensus has not led to the effective pooling of individual information.*

As this example illustrates, consensus-driven decision-making procedures that examine only *whether* experts agree, or *what* number they agree on, but not *why* (by using an explicit model of how the different facts known to individual members interconnect) may waste valuable individual information – even if there is no deliberate intent to hold back or to distort individual information. Processes such as Delphi and modified Delphi procedures, which share reasons and knowledge, rather than only numbers and conclusions, can help to avoid such limitations.

Other simple mathematical models of group decision making reach similar conclusions. For example, Berend and Sapir (2002) consider the probability that a group of decision makers can choose correctly between two alternatives, based on the advice of its members and partial information about their relative expertise (modeled as the probability of being correct). The probabilities of the individual decision makers being correct are modeled as independent random variables, with only their relative ranks being (at least partly) known to the group. The main result is that, for any member of a broad family of distributions of expertise, the "expert rule" that accepts the advice of the best expert, disregarding all others, "is far more likely to be optimal than the majority rule, especially as the deciding body becomes larger." Of course, such simple models and results are artificial, but they do suggest that the performance of group vs. individual (or subgroup) decisions may be worth studying and validating empirically before assuming that more inclusive decision processes are necessarily better.

In summary, a substantial body of empirical research, as well as theoretical modeling, indicates that judgments arrived at by expert groups do not necessarily outperform individual judgments – and, a fortiori, often perform less well than explicit quantitative models. In light of these findings, proposals to replace quantitative risk modeling with group decision processes for making judgments about risks and decisions about risk management face a significant burden of proof to show that the recommended group processes perform better than (or as well as) more quantitative modeling.

How Effective Can QRA Be?

That some common alternatives to QRA do not always perform well provides no guarantee that QRA will be useful (or better than alternatives) in any particular application. Indeed, after several decades of enthusiastic development and application, a mix of QRA methods has been proposed; not all methods are equally good. As discussed in Chapters 4 and 5, some methods now in widespread use for both qualitative and quantitative risk assessment can be *worse than useless*, meaning that they can recommend decisions that are, on average, inferior to those that would result from simply making decisions at random. Opponents and skeptics of quantitative modeling methods also suspect that they are most speculative and least trustworthy precisely when they are most needed: in situations with uncertain, complex, and nonlinear relations between controllable inputs and resulting outputs or consequences. Most of this book is about QRA methods for such situations.

It is fair to challenge QRA advocates to explain exactly how their methods can be applied in these challenging situations and why the results should be believed. Responding constructively to this challenge is the main purpose of the rest of this book. Chapters 2 and 3 introduce QRA methods widely used in engineering and health risk analysis, respectively. Chapters 4 and 5 warn against popular methods of qualitative and quantitative risk analysis, respectively, that do not necessarily contribute to improved risk management decisions (and that may be worse than useless). Chapters 6 through 11 explain and illustrate several practical methods for quantifying risks in uncertain, nonlinear, and complex systems. Two traditionally challenging application areas, antimicrobial risk assessment and cancer risk assessment, are used to illustrate many of the ideas and methods.

Chapter 6 discusses how to use information theory and data mining to discover and quantify nonlinear causal relations in large data sets with many variables, some of which may be causally irrelevant or may confound true cause-and-effect relations. Chapter 7 proposes that these and other statistical methods can help to overcome preconception and confirmation biases. Chapters 8 and 9 show how to use available (but realistically incomplete and imperfect) data to bound uncertainties about predicted risks. Chapters 10 and 11 consider QRA when several alternative sets of assumptions are consistent with available data. They consider how to use resulting *sets* of risk estimates to support inferences and decisions.

Chapters 12 through 16 present several applications and case studies, ranging from risk modeling for complex diseases [lung cancer and chronic obstructive pulmonary disease (COPD)] to systems in which complexity arises from the interactions of multiple decision makers. These include evaluating the risk management options for reducing economic risks resulting from "mad cow" disease in imported cattle, efforts to improve upon terrorist risk assessments by incorporating optimization models of attacker behavior, and techniques for making telecommunications networks resilient to deliberate attacks.

Summary and Conclusions

This chapter has discussed the goals and some limitations of quantitative risk assessment (QRA). It has introduced one of the chief competitors to QRA in the ecology of competing ideas that shape current national risk management policies: a concept of *concern-driven risk management* that generalizes and extends features of the Precautionary Principle. It is typified by the use of expert judgment and consensus decision making or group decision making, rather than QRA, to decide when concerns about a perceived risk should be considered great enough to warrant specified risk management interventions, even if cause-and-effect relations between these actions and human health consequences are not yet well understood and quantified. Such decisions are made without relying on QRA or a formal quantitative assessment of the probable consequences that will be caused by the recommended actions compared to alternatives, including the status quo.

Concern-driven risk management has recently been recommended by many critics of QRA (e.g., WHO, 2003). It has been implemented in bans of animal antibiotics in Europe (Pugh, 2002) and in the USDA's 2007 decision to allow older cattle (those over 30 months old, which have the greatest risk of BSE) to be imported into the United States. It may have several potential political and psychological advantages over QRA. However, it is not clear that concern-driven risk management performs better than (or as well as) QRA in identifying risk management interventions that successfully protect human health or achieve other desired consequences. To the contrary, there are strong empirical and theoretical reasons to expect that judgment-based risk management in response to concerns, conducted without formal QRA, may lead to worse decisions and outcomes than would more quantitative models and methods. Therefore, those who advocate replacing QRA with alternatives, such as group deliberation and decision processes not driven by quantitative modeling, face a substantial burden of proof to show that their recommended alternatives truly outperform QRA.

Concerns are important and should be addressed in applied risk analysis. Indeed, the main business of risk analysis is to provide trustworthy answers to questions such as "How bad is it?", "What might go wrong?", "What are the consequences if something does go wrong?", and "What should we do about it?" However, concern over a current situation, while perhaps motivating a formal quantitative risk assessment, should not substitute for it for purposes of assessing the probable consequences of alternative risk management interventions. The leap from "This situation raises concerns" to "Let's address them by taking some specific action, *X*" can be improved upon by QRA. QRA replaces "Let's take action *X*" with "Let's assess the probable consequences of *alternative possible actions or interventions*, and then implement the one(s) with the most desirable probable consequences."

QRA documents facts and assumptions, clearly articulates value judgments, and causally links current actions to desired consequences. Not following these steps risks taking actions that will not produce intended results and makes it more difficult to diagnose failures and to learn from experience. The following chapters consider methods for achieving the QRA ideals – rational action based on well-articulated, explicit analyses of probable consequences – in managing the risks of complex (possibly nonlinear) and uncertain systems.

Chapter 2
Introduction to Engineering Risk Analysis

Overview of Risk Analysis for Engineered Systems

Can contemporary organizations and societies design, build, and operate complex engineering systems safely and reliably for long periods? Being able to do so is crucial if nuclear power is to be a viable option, if infrastructure such as advanced transportation systems and energy distribution networks is to be trustworthy, if minerals and petroleum are to be discovered and extracted safely, and if hazardous manufacturing and chemical storage facilities are to be located in convenient proximity to transportation hubs and population centers. This chapter, which is an update and extension of Bier and Cox (2007), discusses methods for quantifying the extent to which complex engineering systems can be designed and operated safely.

Opinions about the answer are divided. One school of thought, sometimes called *Normal Accident Theory* after the book that articulated it (Perrow, 1984), holds that engineered systems with high "interactive complexity" (presenting unexpected and surprising sequences of events that are difficult to detect or comprehend at first) and "tight coupling" of interdependent components or subsystems (so that changes propagate quickly among them) are inherently unpredictable and uncontrollable by human operators. Resulting accidents and catastrophic failures in such high-risk technological systems are seen as inevitable and unavoidable: in this sense, they are "normal." In this pessimistic view, adding redundancy to complex systems to reduce accidents makes them even more complex and prone to unpredictable failures. Case studies of accidents and near-accidents at chemical plants, nuclear reactors, airports, and other complex industrial facilities well illustrate Normal Accident Theory.

A different view, popularized in the catchphrase "Failure is not an option" (made famous by Oscar-nominated actor Ed Harris playing NASA Flight Director Gene Kranz in the movie *Apollo 13*), is that complex engineering systems *can* be built and operated safety by sufficiently disciplined, creative, well-organized, and well-trained teams and organizations. Sociologists, psychologists, and other researchers have sought common features of "high-reliability organizations" (HROs), meaning organizations with significantly fewer accidents and failures than normally expected. They have proposed lists such as preoccupation with failure, reluctance

L.A. Cox, Jr., *Risk Analysis of Complex and Uncertain Systems*,
International Series in Operations Research & Management Science 129,
DOI 10.1007/978-0-387-89014-2_2,

to (over-)simplify interpretations or to prematurely reject unexpected interpretations of observations, sensitivity to operations, commitment to resilience, and appropriate deference to expertise (as opposed to rank, seniority, or power) (Weick and Sutcliffe, 2001). Such habits can help to create the vigilant "mindfulness" needed to operate safely and to catch and correct potential problems before they cascade out of control. Routinely safe operations on aircraft carriers and in other high-stress, risky, and complex environments vividly illustrate that well-trained teams in well-designed organizations and environments can manage risks successfully. Similar principles might be applied in different settings, such as operating rooms and intensive care units. Eliminating "mindlessness" (e.g., blind rule following or deference) in the implementation of federally and locally funded programs to reduce infant mortality and preterm birth has been proposed as a way to reduce the frequency and severity of poor outcomes (Issel and Narasimha, 2007).

Probabilistic risk assessment (PRA) of engineered systems gives engineers and risk managers practical tools to understand, predict, and manage risks for a variety of complex engineered systems. It identifies how systems might fail, the likely (and not-so-likely) potential adverse consequences of failures, and how best to prevent failures and mitigate adverse consequences while meeting other goals, such as the continued productive operation of a hazardous facility. PRA methods include probability modeling techniques (both analytic and simulation-based) for quantifying engineering risks, typically expressed as the probabilities of adverse events and as the frequencies and severities of their adverse consequences over a stated period of time. PRA also includes optimization methods from operations research and safety and reliability engineering that can identify cost-effective ways to improve safety and reliability while satisfying other constraints (e.g., on system cost, weight, or performance).

Examples of complex engineering systems to which PRA has been successfully applied include nuclear power plants (beginning with the Reactor Safety Study in 1975, and continuing to the present day); the space shuttle (both before and especially after the Challenger disaster); dam and reservoir planning and operations; highway, bridge, and transportation infrastructure; emergency planning; liquefied natural gas (LNG) terminals and storage facilities; other hazardous chemical plants and operations; and electric power generation and distribution planning. The common elements in such systems is that they all involve (1) a *designed system* intended to withstand different levels of stress, with the option of incorporating different levels of backup and fail-safe design, (2) a *system operator/risk manager* faced with decisions about how to inspect, maintain, and use the system (e.g., when to launch, when to shut down, and generally what level of precaution to adopt), and (3) an uncertain *environment* that generates stresses and adverse conditions that the system should ideally be able to withstand. Uncertainties from the environment may involve random events, such as equipment failures or unexpectedly high or stressful transient loads (as in the case of the Tacoma Narrows bridge collapse); natural disasters such as earthquakes, floods, or hurricanes; terrorist attacks; or operator errors, perhaps arising from miscommunication, miscoordination, or misunderstanding of systems behavior among those running it. Unexpected behaviors of interacting software modules or other subsystems may also cause a system to fail, even if each component performs as it was designed to (Leveson, 2004).

Example: Unreliable Communication with Reliable Components

Setting: Suppose that two people call each other on their cell phones at the same time, so that each receives a busy signal. If each caller can reattempt a call at the beginning of each new time period (say, every 20 seconds), then what retry strategy minimizes the average time to connect?

Solution: If each caller tries to call with probability p at the start of a period, then the probability that they conflict again is p^2, the probability that neither calls is $(1 - p)^2$, and the probability that they connect is $2p(1 - p)$, which is maximized for $p = 0.5$. Thus, each should call with probability 0.5 at the start of each period, until they connect. The probability of connecting at the start of a period is then $2^*(0.5)^*(1 - 0.5) = 0.5$ and the expected number of periods until connection is established is therefore $1/0.5 = 2$. Thus, even if all parts of the system work perfectly and the two callers behave optimally (given what they know), the effective availability of a connection to the callers is less than 100%. This illustrates a *coordination failure* in the use of the system. Of course, such coordination failures are usually only minor annoyances... unless the safety of a system depends on being able to establish contact promptly when something goes wrong!

A system's designer and its operator/risk manager usually both want to make decisions so that the system operates as planned, given engineering and cost constraints and the uncertain environment. Of course, the decisions and trade-offs faced by the operator/risk manager typically reflect the decisions made by the system designer. PRA can help to quantify the trade-offs between cost and safety at the design stage and can help to identify policies and schedules for cost-effective inspection and testing, preventive maintenance, spare parts provisioning, redundancy allocation, and replacement of working parts to keep complex systems operating as intended throughout their design lives.

Example: Optimal Number of Redundant Components

Setting: Suppose that an airplane can have one, two, or four engines. Each engine independently has a probability $1 - p$ of failing during the course of a mission. (Equivalently, it has probability p of surviving). A plane fails (crashes) if more than half of its engines fail.

Problem: What number of engines should a plane have, to maximize the probability of completing its mission?

Solution: A plane with one engine has success probability p. A plane with two engines has success probability $1 - (1 - p)^2$, the probability that both engines do not fail. Since $1 - (1 - p)^2 = 1 - (1 - 2p + p^2) = p(2 - p)$, this success probability is greater than p if and only if $2 - p > 1$ and $p > 0$; in other words, for $0 < p < 1$. Thus, *a twin-engine plane is at least as likely to survive as a single-engine plane*, with equality only if $p = 1$ or $p = 0$. For a four-engine plane, the probability of success is one minus the probability of losing more than two engines: $1 - [\Pr(\text{lose 4 engines}) + \Pr(\text{lose 3 engines})] = 1 - [(1 - p)^4 + 3p(1 - p)^3] = 1 - (1 - p)^3\,[(1 - p) + 3p]$.

This is greater than the success probability for two engines if and only if the following inequalities hold:

$$\begin{aligned}
&1-(1-p)^3[(1-p)+3p]>1-(1-p)^2,\\
&(1-p)^2>(1-p)^3[(1-p)+3p],\\
&1>(1-p)[(1-p)+3p],\\
&1>(1-p)(1+2p),\\
&1>1+p-2p^2,\\
&2p^2>p,\\
&2p>1,\\
&p>0.5.
\end{aligned}$$

Thus, a four-engine plane is more likely to survive than a two-engine plane if and only if the individual engines are more likely than not to survive the mission ($p > 0.5$).

Example: Optimal Scheduling of Risky Inspections

Setting: Suppose that, in the absence of intervention, a component (perhaps an engine in the previous example) of age t has probability $1 - e^{-ht}$ of a defect that will increase the risk of failure when the component is next used. h is called the *hazard rate* for occurrence of the defect, and ht is the *cumulative hazard* accumulated by age t in the absence of intervention. At any time, an expensive inspection may be performed, and, if a defect is present, it will be found and repaired, effectively setting the age of the component back to 0. However, careless inspection may itself introduce an uncorrected defect that would not otherwise have occurred. The probability of this is $p \geq 0$ for each inspection.

Problem: What time between inspections minimizes the expected number of uncorrected defects per unit time?

Solution: If inspections take place every T time units, then each inspection removes hT expected defects and adds p expected defects. The optimal time between inspections makes the marginal "cost" (here meaning loss of reliability) from an inspection – that is, the expected new defects created, p – equal to its marginal benefit (that is, the expected effects removed, hT). Thus, the optimal time between inspections, denoted by T^*, satisfies $hT^* = p$, and so $T^* = p/h$ (for $h > 0$). More frequent inspections than this, with $T < T^*$, are expected to create more problems than they solve ($p > hT$). Less frequent inspections, with $T > T^*$, let the expected costs of not intervening sooner exceed the costs of doing so ($hT > p$).

PRA is usually applied to rare and catastrophic events for which it may be difficult to estimate risks directly due to the lack of empirical data, the possibility of unobserved changes (e.g., deterioration) in the system, and changes in the system's environment or use. Risk assessment can also be applied to predict routine (e.g., occupational accident) risks, although in such cases it may be possible to rely primarily on empirical data, reducing the need for modeling. In general, PRA is used to estimate, predict, and find ways to reduce the risks to facility or system owners, employees, and the public. This chapter focuses on methodological advances in engineering risk analysis, with selected applications (including some applications of PRA methods and insights to fields other than engineering) to illustrate the methodology.

Using Risk Analysis to Improve Decisions

Risk analysis can help to inform design decisions (e.g., trade-offs among safety and performance, cost, etc.) as well as operational decisions (e.g., when to shut down a facility). It can be useful regardless of who makes the decisions – for example, facility owners and operators, regulators, or multiple stakeholders interacting through a participatory risk management and conflict-resolution process. Key technical challenges that PRA must address include: how to predict the probable performance and quantify the behaviors – both probable and improbable – of a complex system, given a design and the operator's decisions, in the face of inadequate data; how to optimize the joint decisions faced by the system designer and owner/operator (which can involve NP-hard combinatorial optimization problems, as well as problems of coordination and communication between different organizations); how to most effectively model interdependencies and uncertainties about the system's current state; the development of cost-effective "screening"-type methods for addressing the myriad possible risks in "open" systems (such as the risk of terrorist attack); and scale-up problems for extremely complex systems, such as infrastructure networks. Also, there is still room to benefit more fully from adaptation of methods developed in other fields, including decision analysis and related fields (such as Bayesian statistics).

Hazard Identification: What Should We Worry About?

Probabilistic risk assessment typically begins by defining a system to be analyzed and identifying undesired outcomes that might occur when it is operated. *Hazard identification* methods have been developed to identify the potential adverse consequences of system operation. Structured qualitative techniques include hazard and operability (HAZOP) studies and failure modes and effects analysis (FMEA), which describes potential failure modes, causes, effects, safeguards, and recommendations for reducing risks.

Fault trees and event trees can be used in a qualitative mode for hazard identification but can also be quantified to estimate the likelihood of adverse events. *Fault tree analysis* (Barlow, 1998) begins with an undesired outcome, called the "top event," and reasons backward to identify which combinations of more basic events (e.g., component failures) could bring about the top event (e.g., failure of the system). The result is a tree that represents those sets of basic events that would be sufficient to cause the top event using "AND" and "OR" logic (and possibly more complicated logic gates as well). The tree generally goes down to the level of basic events whose probabilities can be reliably estimated from experience, judgment, and.or data.

Example: Fault Tree Calculations for Car Accidents at an Intersection

Setting: Suppose that a car accident (the top event) occurs at an intersection if and only if (two cars approach the intersection at the same time from different directions) AND (both cars proceed). The event "both cars proceed" can be further decomposed into a logical subtree, as follows: (both cars proceed) if and only if [(the signal is broken AND both cars proceed) OR (the signal is not broken AND both cars proceed)]. Reliable statistics show that the first event (sometimes called the *imitating event*), namely, "Two cars approach the intersection at the same time from different directions," occurs with an average annual frequency of 100 times per year. The signal is broken on 0.1% of these occasions (independently of traffic) and the conditional probability that both cars will proceed, following the initiating event, is 0.1 if the signal is broken and 0.01 if it is not broken.

Problem: (a) What is the average annual frequency of accidents at the intersection, given these numbers? (b) What fraction of accidents would be prevented if the signal never failed?

Solution: (a) The conditional probability of an accident, given the initiating event, is Pr(signal is broken)*Pr(both cars proceed | signal is broken) + Pr(signal is not broken)*Pr(both cars proceed | signal is not broken) = (0.1%)*(0.1) + (1 − 0.1%)∗(0.01) = 0.0001 + 0.9999*0.01 = 0.0101 (to four significant digits). (Here "|" is read as "given" or "conditioned on.") The average annual frequency of accidents is this conditional probability times the average annual frequency of initiating events: 0.0101*100 = 1.01 accidents per year. (b) The contribution of accidents with a broken signal to the total average annual frequency of accidents is only (0.1%)*(0.1)*100 = 0.01 accidents per year. If the signal were never broken, then the average frequency of accidents per year would still be 100*0.01 = 1 accident per year.

Comments: (a) *Dominant contributors.* In this example, accidents with the traffic signal working constitute a *dominant contributor* to the average annual accident frequency. This means that ignoring other, rarer events (namely, accidents with the signal broken) yields the same calculated risk number (about one expected accident per year), to one significant digit. One way to simplify fault tree calculations is to focus on dominant contributors, neglecting events that are rare enough that they do not change the numerical answer (within some desired level of precision, such as one or two significant digits). (b) *Poisson probabilities.* The calculated risk of about one accident per year can be viewed as the mean of a rare-event (approximately Poisson) process. This allows the probabilities for any number of accidents per year (under current conditions) to be estimated: It is Pr(x accidents in a year) $= \lambda \exp(-\lambda)/x!$ for $x = 0, 1, 2, \ldots$, where λ is the mean number of accidents per year (approximately 1, in this example). For example, the probability of zero accidents at this intersection in a year, if the accident process is Poisson with mean 1 accident per year, is $e^{-1} = 1/2.718 = 0.368$. (c) *Obtaining probabilities for basic events.* If reliable statistics were not available for the probabilities that both cars proceed when the signal is working and when it is broken, they might be estimated from experiments (e.g., using driving simulator results), models of driver behavior, or expert judgment. Uncertainty and sensitivity analyses would then typically be used to determine by how much the calculated risk might change if different plausible estimates or better future information about these inputs were to be used in the analysis. (d) *Recursive deepening of a tree.* Each event in a model, such as "both cars proceed," can potentially be expressed as a subtree consisting of a logical combination of more refined event descriptions, e.g., "(both cars proceed and weather is good) or (both cars proceed and weather is not good)." Infinite recursion is prevented by stopping further decomposition when the current description allows basic event probabilities to be quantified accurately enough to support risk management decisions.

Event tree analysis begins with an "initiating event" and works forward to identify its potential consequences. In essence, an event tree is a decision tree without decision nodes. It shows potential sequences of events, with the probability of each branch leaving an event node (representing the possible resolution of an uncertainty, often modeled as a possible value of a random variable) being conditionally independent of earlier information, given that the branch point (i.e., that event node) has been reached. The frequency of a given event sequence is then just the product of the conditional branch probabilities along that path multiplied by the frequency of the initiating event. Both fault trees and event trees can be represented as logically equivalent influence diagrams. They can be solved by more general-purpose influence diagram algorithms (Barlow, 1998; Bobbio et al., 2001).

Structuring Risk Quantification and Displaying Results: Models for Accident Probabilities and Consequences

A *quantitative risk model* typically consists of a formal mathematical and or simulation model of the system of interest, together with one or more consequence attributes of interest and one or more alternative risk management decisions to be evaluated or decision variables to be optimized. The model is used to predict the probable consequences of alternative decisions. Preferred decisions are those that yield preferred probability distributions (or, more generally, preferred stochastic processes) for the consequences of interest.

Risk modeling typically involves some or all of the following components.

- *System representation* (Barlow, 1998; Smith, 2005). An engineered system is often represented mathematically in one of the following forms: (a) A "black-box" *statistical model* (e.g., a lifetime hazard function quantifying the conditional failure rate of a system for different ages or elapsed times, given that it has not failed so far): (b) component failure rates combined via a *coherent structure function* (such as a fault tree or an event tree) mapping the states of system components to the states of the system. (A coherent structure function must be monotonically increasing, going from a system failure probability of zero if all components work to a system failure probability of one if all components fail.); (c) a stochastic *state-transition* model (e.g., a Markov or semi-Markov model for transitions among working and failed components, representing component failure and repair rates); (d) a discrete-event *simulation model* (Smith, 2005).
- *Environment representation.* Like a system model, a model of the environment may be a statistical black-box model (e.g., a function describing the frequency and intensity of stresses to the system's components), a stochastic process, or a simulation model. Plausible worst-case or bounding scenario analyses are sometimes used when probabilistic descriptions of uncertainty are unavailable or are difficult to obtain. The model of the environment is often incorporated directly into the system model, as with traffic levels and weather conditions in a traffic accident model.
- *Decision-rule representation.* A *decision rule* for managing an engineered system maps observed information about the system into a resulting action or intervention. For example, a component may be replaced based on the observed history of failures and repairs for its components. Optimization methods, including recently developed simulation-optimization techniques (see, for example, Ólafsson and Kim, 2002), can help to identify "good" or "best" decision rules, given a system model, an objective function (e.g., a multiattribute utility function), and a model of the environment. Of course, many decisions in the real world (even when informed by PRA) are made without a formal decision rule, either because the PRA results themselves make the best decision clear or because of the need to address the concerns of multiple stakeholders.

Example: Bug-Counting Models of Software Reliability

An example of a simple black-box risk model for software reliability is a "bug-counting" model in which the (unknown) initial number of bugs in a piece of code is represented by a random variable N with a prior distribution. As the code is tested and debugged, the remaining number of bugs presumably decreases, and the random times between successive bug discoveries stochastically increase. (Relatively sophisticated models also allow for the possibilities that detection and repair are imperfect processes and that debugging activities may introduce new bugs.) The empirical record of bug discoveries can be used to trigger a decision rule such as "If no bugs have been discovered within M tester-hours, then release the software." Simulation optimization can then be used to numerically optimize the parameter M. For analytic alternatives, see Singpurwalla and Wilson (1999) and Wilson and Samaniego (2002).

Example: Risk Management Decision Rules for Dams and Reservoirs

Wurbs (2005) describes the use of decision rules to manage water releases for dams and reservoirs as follows:

> Release decisions depend upon whether or not the flood control storage capacity is exceeded ... federal reservoirs are typically sized to contain at least a 50-year recurrence interval ... flood and, for many projects, design floods greater than the 100-year flood ..., perhaps much greater. A specified set of rules, based on downstream flow rates, are followed as long as sufficient storage capacity is available to handle the flood without having to deal with the water surface rising above the top of the flood control pool. ... For extreme flood events which would exceed the reservoir storage capacity, moderately high damaging discharge rates beginning before the flood control pool is full are considered preferable to waiting until a full reservoir necessitates much higher release rates.

The outputs from quantitative risk models are often summarized as *F–N* curves (also sometimes called exceedance probability curves, or complementary cumulative frequency distributions), showing the expected annual frequency F of fatalities or damages exceeding any given level, N, for $N > 0$. (Technically, as discussed in Chapter 5, such diagrams make sense only for compound Poisson processes, not for more general renewal processes. However, *F–N* curves are often used to summarize the results of PRA calculations, which typically use compound-Poisson approximations to risk in any case.) *F–N* curves are not perfect summaries of the distribution of risk within a population, however – largely because they do not describe individual risks, which may differ substantially. Other risk displays show how risk varies by location, over time, and with other covariates. For example, it is common practice to plot "risk contours" showing risks to individuals at different locations around a potentially hazardous installation or transportation route.

Example: Different Individual Risks for the Same Exceedance Probability Curve

Suppose that three people, 1, 2, and 3, live near two hazardous facilities, A and B. Facility A can have any of three accidents: A small accident that kills individual 1 only; a medium-sized accident that kills individuals 1 and 2; or a large accident that kills individuals 1, 2, and 3. If an accident occurs at facility A, it is equally likely to be small, medium, or large. By contrast, an accident at facility B is equally likely to kill individual 3 only, kill individuals 1 and 2, or kill all three. Accidents at facilities A and B are equally frequent. Then A and B have identical *F-N* curves, since each accident (at either facility) has probability 1/3 of causing one fatality, 1/3 of causing two fatalities, and 1/3 of causing three fatalities. But the individual risks from the two facilities are very different. An accident at facility *A* has a 100% probability of killing individual 1, a 2/3 probability of killing individual 2, and only a 1/3 probability of killing individual 3; but an accident at facility *B* has a 2/3 probability of killing each individual. This difference in the distribution of individual risks is not captured in an *F-N* curve, but could be shown in a risk contour plot if the three individuals are positioned at different locations.

Major technical challenges for developing PRA results include

1. *Constructing and validating models* of the system and its environment. Statistical analysis of accident precursors uses data on "near-misses" to validate and refine model-based predictions (Yi and Bier, 1998; Borgonovo et al., 2000; Phimister et al., 2004). Powerful model-building and model-checking methods have also been developed in the areas of *system identification*, which attempts to identify dynamic system descriptions of input-output relations from observed time course data (see Chapter 11), and *data mining and machine learning*, which seek to learn correct models (or at least subsets of especially plausible models) directly from data (see Chapters 6 and 7).
2. *Calculating, simulating, or estimating probabilities of rare events.* Methods for addressing this challenge, such as importance sampling, adaptive importance sampling, cross-entropy, and Markov chain Monte Carlo (MCMC) methods with carefully designed transition kernels, have advanced significantly in recent years (e.g., Bucklew, 2004; Rubinstein et al., 2004).
3. *Treatment of dependencies* among failure events and system components. Methods for treatment of dependencies presently include common-cause failure analysis (to show dependence in the failure rates of similar components due to a common underlying cause), dependency matrices and event trees (to show the dependence of some systems on "support" systems such as electric power), and external-events analysis (to capture the fact that events such as earthquakes, fires, and floods can affect multiple components of a system).

Quantifying Model Components and Inputs

A model typically expresses risk (e.g., the probability of failure by a certain time) as a function of the performance of model components and or input parameters. These must be quantified from available data, perhaps using a combination of expert judgment and Bayesian statistics (due to the sparseness of directly relevant data). In Bayesian statistics, a prior distribution is updated by conditioning on observed data to yield a posterior probability distribution for the quantities of interest (Lee, 2004). Such methods include hierarchical Bayesian methods (in which partially relevant data are used to help construct the prior distribution) as well as empirical Bayesian methods (in which the actual data for the problem at hand are used to help construct the prior distribution); see Carlin and Louis (2000).

Although Bayesian approaches to quantifying risk models are frequently applied in practice, advances are still being made in numerous areas. These include designing more flexible and tractable models for treating probabilistic dependence in risk models, alternatives to relying on subjective prior distributions (which can be problematic if plausible differences in subjective priors significantly affect risk results), and treatment of model uncertainty.

Modeling Interdependent Inputs and Events

If the state of a system is described by a coherent structure function, and each component independently undergoes stochastic transitions over time (e.g., from "working" to "failed" to "repaired" or "replaced"), then the probability distribution for the system's state (i.e., the probability that it will be working rather than failed at any time) can be obtained relatively easily. Stochastic simulation of the behaviors of the components, or the routine application of combinatorial reliability models and algorithms, such as fault tree analysis or event tree analysis, is practical even for large systems. However, if component behaviors are interdependent (e.g., if each component failure increases the stress on those components that have not yet failed), then it becomes more complex to calculate the risk that the system will have failed by any given time. Simulating interdependent behaviors may be straightforward in principle, but, in practice, it requires specifying how events depend on each other – a potential combinatorial nightmare.

Dependence can also be a problem for uncertainty analysis. In particular, the failure rates (or probabilities) of the various components can be uncertain and statistically dependent on each other, even if their behaviors are conditionally independent given their failure rates. For example, learning that one component had a higher failure rate than expected may cause one to increase estimates of the failure rates of other similar components. The failure to take such dependence into account can result in substantial underestimation of the uncertainty about the overall system failure rate (or probability), and in some cases also underestimation of the mean failure probability of the system (e.g., if the components whose failure probabilities are dependent are functionally in parallel with each other); see Apostolakis and Kaplan (1981), Burmaster and Anderson (1994), and Kraan and Cooke (1997).

Historically, for reasons of computational tractability (among others), dependencies among random variables have often been either ignored, or else treated using unrealistic and simplistic assumptions such as perfect correlation. Fortunately, substantial progress is being made in modeling dependencies among components (and/or in the information about components). Two techniques, copulas and Bayesian networks, have become popular for specifying dependency relations. Bayesian networks are directed acyclic graphs (influence diagrams without decision nodes) in which nodes represent events and directed arcs ("arrows") between nodes show probabilistic dependencies. Each node's value has a conditional probability distribution that depends only on the values of the variables that point into it. (Equally important, absent arrows indicate the conditional independence of each variable from those that do not point into it, given the values of those that do.) Sampling from the conditional distribution of each variable in turn, given the sampled values of its predecessors (after sorting the variables so that each appears only after those that point into it, if any), and repeating many times provides a way to sample from the joint distribution of the variables without having to explicitly specify it. [Such "Gibbs sampling" is a simple form of Markov chain Monte Carlo (MCMC) sampling that is well suited for Bayesian networks. In effect, the joint distribution is factored as a product of marginal distributions (for the input variables, meaning those with no predecessors) and conditional distributions (for all other nodes), thus allowing the potentially large size of a full joint distribution to be tamed by the relative sparseness of significant dependencies among variables in most real-world systems.] Free Windows software for Bayesian inference using Gibbs sampling ("WinBUGS"), called from the free statistical computing environment R, can be obtained by Googling on R2WinBUGS.

The use of copulas (functions that link a multivariate cumulative distribution to its one-dimensional cumulative marginal distributions; see, for example, Nelsen, 1999) has also become increasingly common in both financial and engineering risk analysis. Copulas have been applied, for example, to model dependencies between opinions from different experts (Jouini and Clemen, 1996; Lacke, 1998) and between system failure rates during normal and accident conditions (Yi and Bier, 1998). They are used extensively in financial risk analysis (e.g., in the Gaussian CreditMetrics or Basel II model) to describe correlated credit portfolio risks and interdependent risks of default (Frey et al., 2001).

Of course, copulas are not always the most convenient way to represent dependencies; see Joe (1997) for a compendium of multivariate distributions. Recently, Merrick et al. (2005) used an inverted Wishart distribution to model uncertainty about the dependencies among experts in assessing risks to the Washington State Ferries system while allowing the analyst to "learn about the dependencies between the experts from their responses." This is achieved by asking the experts to provide multiple different assessments of maritime risk under differing circumstances.

Cooke and colleagues (Bedford and Cooke, 2001; Kurowicka and Cooke, 2004) developed a practical method for specifying a joint distribution over n continuous random variables with specified rank correlations, using only $n(n - 1))2$ assessments of conditional correlations. Kurowicka and Cooke (2004) point out that use of continuous multivariate distributions for a Bayesian belief net (a Bayesian network) allows for more tractable Bayesian updating than the commonly used discrete distributions (Lauritzen and Spiegelhalter, 1998).

Example: Analysis of Accident Precursors

Consider a risk analyst attempting to estimate the failure probabilities of critical safety systems in a nuclear power plant in the event of an accident. Fortunately, few if any accidents will have been observed on plants of that type, suggesting the analyst may use data regarding failure probabilities of those systems during routine testing. However, this data will clearly be only partially relevant to the probabilities to be assessed; for example, one might expect that many systems will have higher failure probabilities under accident conditions than during routine testing.

Yi and Bier (1998) show how copulas can be used to represent dependency between the system failure probabilities under normal versus accident conditions. This makes it possible to perform a Bayesian update showing the effect of data collected under normal conditions on the system failure probabilities under accident conditions. Thus, for example, if routine testing showed a particular system to be much less reliable than was previously believed, this information could be used to update the expected failure probability of the system in the event of an accident. However, Yi and Bier's model is not sufficiently general to account for all relevant prior assumptions about dependencies. Thus, further work is needed to enhance ability to model dependencies.

Example: Flight-Crew Alertness

A challenge in modeling flight-crew alertness (Roelen et al., 2003) is that various predictive variables are correlated not only with crew alertness, but also with each other. For example, the crew's workload on a given flight is likely to be a function of both the length of the flight (with longer flights having higher total workload) and how much the crew members rest during the flight (with more rest being associated with a lower workload). However, assessing the combined impact of these variables on crew alertness may be difficult if longer flights also allow more rest time during flight.

Kurowicka and Cooke (2004) develop a continuous Bayesian belief net for this situation to allow airline managers to identify ways to compensate for known causes of poor alertness (such as long flights, or insufficient sleep prior to flight time). By allowing the variables in their model to have continuous distributions (rather than discrete distributions, which are more common in applications of Bayesian belief nets), they were able to achieve a highly parsimonious model requiring the assessment of only eight conditional rank correlations, compared to the many more assessments that would have been required for a discrete model.

Some Alternatives to Subjective Prior Distributions

Unlike classical statistical procedures, Bayesian analysis can be used in situations of sparse data, because subjective judgments and other nonstatistical types of evidence can be used in Bayesian estimation, inference, and decision processes. However, with sparse data, the results of Bayesian analyses are often sensitive to the analyst's choice of prior probabilities for models and parameters. Hence, Bayesian methods can be more subjective and less readily accepted when data are sparse.

Maximum-entropy distributions have sometimes been proposed to help solve this problem. They use whatever information is available about the uncertain quantity of interest (e.g., mean, median, or mean and variance) to constrain the assumed distribution for that quantity but presuppose as little additional information as possible beyond that, to avoid inadvertently assuming more than is actually known. A maximum-entropy distribution is defined to be the least informative distribution (in a precise technical sense) that satisfies the specified constraints (Jaynes, 2003). The resulting distribution can then be used either as a prior distribution for Bayesian analysis (if additional data become available) or as a partially informative distribution without updating. For example, Meeuwissen and Bedford (1997) use maximum entropy to identify the minimally informative distribution with a given set of rank correlation coefficients, using a piecewise constant numerical approximation (a so-called chessboard distribution).

However, maximum entropy and related approaches (such as "noninformative prior" distributions) lead to significant problems even in some relatively simple examples. For example, if all we know about a random variable X is that it is bounded by 0 and 1, then a maximum-entropy distribution for it would be uniform between these limits. Of course, exactly the same reasoning presumably applies to X^2, but X and X^2 cannot both be uniformly distributed between 0 and 1. Such lack of invariance to transformations of variables (e.g., from half-life to decay rate) means that maximum-entropy distributions may depend on essentially arbitrary choices of scale, or of how to represent the same physical situation. In addition, the maximum-entropy distribution can be difficult to compute in some cases (especially when quite a bit is known about the quantity of interest, so that the maximum-entropy distribution must satisfy numerous constraints).

Such limitations have raised interest in "robust" Bayesian methods and other bounding approaches. Robust Bayesian methods (Rios Insua and Ruggeri, 2000) update an entire class, family, or set (usually convex) of prior distributions with observed data, rather than just a single prior distribution. If the class is chosen carefully, the computational effort required to update all distributions in the class need not be substantially greater than for a single distribution. If all (or most) prior distributions in a suitably broad class give similar results, this can lead to greatly improved confidence in the results of the analysis.

In a similar spirit, probability bounds analysis (Ferson and Donald, 1998) propagates uncertainties (rather than choosing a prior distribution for Bayesian updating). The analyst specifies bounds on the cumulative distribution functions of the various input parameters to a model, rather than selecting specific cumulative distributions. These bounds are then propagated through the model. The uncertainty propagation process, which again can be quite computationally efficient, yields valid bounds on the cumulative distribution function for the final result of the model (e.g., a risk level). This approach can take into account not only uncertainty about the probability distributions of the model inputs, but also uncertainty about their correlations and dependence structure. This is valuable, because correlations will often be more difficult to assess accurately than marginal distributions, and correlations of 1 or −1 among the input variables do not necessarily produce the most extreme possible distributions for the output variable(s) of interest; see, for example, Ferson and Hajagos (2006).

Example: Effects of Exposure to Contaminated Soil

Ecological and environmental risk models frequently involve a high degree of uncertainty, because some important parameters in the model may not be readily measurable. Consider the problem of attempting to estimate the effect of soil contamination on predator species (Hope, 1999), which may be exposed to contamination both directly (through ingestion of soil) and indirectly (by ingestion of a variety of prey species). Estimating the exposure to the predator species requires estimating the concentration of the contaminant in the flesh of all prey species, some of which may themselves be predators. This requires estimating the overall food and water intake and diet composition for each relevant species, as well as the uptake of the contaminant. Good data or expert opinion may be available for some parameters, but for others (such as the fraction of a particular predator's diet made up of a particular prey species), experts may feel uncomfortable assessing an informative probability distribution and may prefer simply to state, for example, that the fraction must be between 0 and 1. Standard practice would either press the experts to provide informative distributions, or simply assume a uniform distribution between 0 and 1, but this may not always conform to the experts' judgments. Correlations between the fractions of the diet made up of differing foods can also obviously be difficult to estimate reliably.

Regan et al. (2002) compare a traditional two-dimensional Monte Carlo analysis of this problem to the results obtained using probability bounds. Even using bounds of 0 and 1 for some parameters, the qualitative conclusions of the analysis (e.g., that the predator species of interest was "potentially at risk" from exposure to soil contamination) remained essentially unchanged between the two-dimensional Monte Carlo analysis and the probability bounds analysis. Thus, bounding analysis can help support a particular decision if it shows that the qualitative results and recommendations resulting from the analysis are not highly sensitive to the specific choices of probability distributions used in the simulation.

The use of subjective prior probabilities and judgment-based probability models can also be simplified or avoided in many situations where probability theory provides the required forms of distributions and/or useful bounds on the probable values of uncertain quantities. Table 2.1 summarizes some important classes of situations where probability theory prescribes distributions and bounds. [Table 2.1 assumes familiarity with the various distributions mentioned, such as Poisson, Weibull, exponential, gamma, Gumbel, normal, and lognormal. See Ross (1996) and the hyperlinks in the table for technical details of these distributions and topics. Googling on the distribution names and italicized topics in Table 2.1 will provide a host of web resources and authoritative references, even if these specific links become obsolete.]

Many of these results can be applied even when the correct probability distributions are unknown or are only partly known, perhaps from statistical sampling or

Table 2.1 Selected asymptotic distributions and bounds from probability theory

Situation	Key results and references	Examples
Random occurrences, with the numbers of occurrences in disjoint time intervals being statistically independent count random variables, and with the expected number of occurrences (counts) in any interval being proportional to its length.	The random number of occurrences in an interval has a Poisson probability distribution with mean (and variance) λt, where t = length of interval and λ = mean number of occurrences per unit time (the "intensity" of the arrival process). If each occurrence has a random consequence, and consequences are independent identically distributed (i.i.d.) random variables, the process is a *compound Poisson* distribution process (Ross, 1996). If λ depends on a linear combination of explanatory variables, a *Poisson regression model* results.	Annual numbers of fires in a city, sporadic food poisoning cases, auto thefts, bank robberies, car accidents on a stretch of highway, etc. Number of typos per page, defects per square foot of material, per mile of pipe, etc.
Waiting times between events in a Poisson process.	The random time between consecutive events in a Poisson process with intensity λ is an *exponential distribution* with mean $1/\lambda$. The random time for k consecutive events to occur has a *gamma distribution* (Ross, 1996).	Time-to-failure in a process where failure occurs after k "hits" (or other event arrivals).
Rare events, independent or weakly dependent numbers of occurrences in disjoint time intervals.	The number of events in a time interval is approximately Poisson or compound Poisson. Bounds are available for approximation error (Barbour et al., 1995; Pekoz, 2006).	Reliability systems; sequences with positive values tending to be clumped in time.
Smallest value among a large number N of i.i.d. nonnegative random variables, e.g., failure times of N independent components, or first time for one of N competing processes to reach a certain stage.	• For large N, the smallest value has approximately a *Weibull distribution*. • If the N variables are unbounded (instead of being non-negative or, more generally, bounded from below), then the distribution of the smallest value approaches a *Gumbel distribution for minima* [an Extreme Value Type I distribution, having hazard function $h(x) = e^x$ when location = 0 and scale = 1]. www.itl.nist.gov/div898/handbook/apr/section1/apr163.htm www.itl.nist.gov/div898/handbook/eda/section3/eda366g.htm mathworld.wolfram.com/ExtremeValueDistribution.html	Failure time of a system with N components in series or with N competing sources or causes of failure.

Table 2.1 (continued)

Situation	Key results and references	Examples
Maximum value in a large set or sequence of N i.i.d. random variables.	The largest value has a probability distribution that approaches a *Gumbel distribution for maxima* for large N, provided that the variables have finite moments and unbounded tails that decrease at least as fast as an exponential (e.g., Normal distributions). More generally, maxima and minima in many situations have approximately a *generalized extreme value* (GEV) distribution. [This is a three-parameter family that includes Weibull (or reversed Weibull), Gumbel, and Fréchet distributions as special cases.] www.itl.nist.gov/div898/handbook/eda/section3/eda366g.htm http://rss.acs.unt.edu/Rdoc/library/evd/html/00Index.html	Maximum floods, rainfalls, traffic loads, insurance claims, bank deposits/withdrawals. Failure time of a reliability system with N components in parallel.
Sum of N i.i.d. random variables with finite mean μ and variance σ^2. *Sum* of N random variables with finite means and variances, *not necessarily i.i.d.*, satisfying the Lindeberg condition (each variance is small compared to the sum of all the variances).	The sum approaches a *Normal distribution*, with mean $N\mu$ and variance $N\sigma^2$ for large N. (This is a *central limit theorem*. Explicit bounds are available for the rate of convergence and closeness of the approximation.) For large N, the sum has approximately a *Normal distribution*, with mean = sum of the N means and variance = sum of the N variances. (This is the Lindeberg-Feller central limit theorem.) http://mathworld.wolfram.com/LindebergCondition.html http://mathworld.wolfram.com/Lindeberg-FellerCentralLimitTheorem.html	Total losses from N consecutive insurance claims; total damage accumulated in a population of N units exposed to a source of random damage; total time to use up N spare parts having i.i.d. random lifetimes.
Product of many positive well-behaved (square-integrable) i.i.d. random variables. Roughly, this may be interpreted as a product of many individually "small" random factors.	The product is asymptotically *lognormal*. Distributions of chemicals, particles, or microbes in the environment, and other *exposure variables*, are often lognormally distributed. Corrosion, diffusion or migration of ions, crack growth, and other material and chemical processes often lead to lognormal degradation processes and *failure times*. http://stat.ethz.ch/~stahel/lognormal/bioscience.pdf http://www.itl.nist.gov/div898/handbook/apr/section1/apr164.htm	Probability that all of many conditions hold; survival times for cancer patients; failure times of processes where further degradation occurs at a rate proportional to current degradation.

Table 2.1 (continued)

Situation	Key results and references	Examples
Any *monotone graph property P*, i.e., any graph property (such as connectivity) that does not depend on the labeling of graph vertices and is not destroyed by adding edges to a graph. *Percolation processes*, in which an effect propagates from site to neighboring site in a lattice if they are linked (and there is probability p that any two neighboring sites are linked).	The probability that property P holds in a random graph with N vertices and with edge probability p is close to 0 for p below some threshold value and is close to 1 for p above a second, larger threshold value. For large N, the two threshold values approach each other, forming a "sharp transition threshold," t, such that that the graph almost surely has property P if $p > t + \varepsilon$ and almost surely does not have property P if $p < t - \varepsilon$, where ε approaches 0 for large N. The system is said to undergo a *phase transition* for property P at the sharp transition threshold, t (Friedgut, 2005). Sharp transition thresholds also hold for *percolation processes* and for *random geometric graphs* (in which nodes are uniformly distributed in a spatial region and any two nodes within a specified distance d of each other are linked). http://www.stanford.edu/~ashishg/papers/gnrn.pdf	Spreading of forest fires, infectious diseases, invasive species, etc. have been modeled as percolation processes. Phase transitions from unreliable to reliable wireless networks as power and/or density of wireless nodes in an area increases. http://portal.acm.org/citation.cfm?id=970847
X is a possibly unknown nonnegative random variable with mean $E(X)$.	$\Pr[X \geq k^*E(X)] < 1/k$, for $k \geq 1$. This is *Markov's inequality*. It implies Chebyshev's inequality (next). http://mathworld.wolfram.com/MarkovsInequality.html	Random lifetime with unknown probability distribution, mean estimated from data.
X is a possibly unknown random variable with finite mean $E(X)$ and finite variance σ^2.	For any $k > 0$, $\Pr(\lvert X - E(X)\rvert \geq k\sigma) \leq 1/k^2$. Interpretively, values of a random variable are unlikely to be many standard deviations away from the mean. This is *Chebyshev's inequality* (Ross, 1996; http://en.wikipedia.org/wiki/Chebyshev's_inequality).	Probability of gain or loss exceeding a given limit. Financial risk analysis, banking risks.
Martingales: $X_0, X_1, \ldots$ is a sequence of random variables with finite means and bounded increments such that $E(X_{t+1} \mid X_0, \ldots, X_t) = X_t$ and $\lvert X_k - X_{k-1}\rvert \leq a_k$ almost surely. The variables need not be i.i.d.	$\Pr(\lvert X_t - X_0\rvert \geq d) \leq 2\exp(-d^2/[2(a_1{}^2 + a_2{}^2 + \ldots + a_t{}^2)]$. This is a form of *Azuma's inequality* for martingales. For example, if each step size $\lvert X_k - X_{k-1}\rvert \leq 1$, then the probability that the process will have increased by $\geq d$ after t steps satisfies $\Pr(X_t \geq X_0 + d) \leq \exp(-d^2/2t)$. Interpretively, the probability of a "large deviation" (of size d or more) for cumulative winnings in a sequence of fair gambles is exponentially small (Ross, 1996; http://en.wikipedia.org/wiki/Azuma's_inequality).	Probability of gain or loss exceeding a given limit. Machine learning with limited error probabilities; fault detection and change detection in martingale processes.

Table 2.1 (continued)

Situation	Key results and references	Examples
Binomial trials with unknown success probability, *p on each trial.*	If data consist of N trials with no successes, then there is approximately 95% confidence that p does not exceed $3/N$. This is the *rule of 3* (Chen and McGee, 2008; www.sinica.edu.tw/~jds/JDS-401.pdf).	Upper bound for probability of a never-observed event.
Coherent structure reliability system, with unknown structure and unknown independent component failure rates.	If each component has a failure rate (or hazard function) that is *increasing on average* [i.e., $H(t)/t$ is increasing in t, where $H(t)$ = cumulative failure rate], then so does the whole system. Many other similar results and bounds are available (Barlow, 1998).	Reliability system with uncertain structure and/or component failure rates.

simulation modeling that provides estimates of means and variances. For example, the sums, maxima, and minima of repeated random samples from most distributions encountered in practice have asymptotic distributions (as the number of samples becomes large) that do not depend on the specific distribution being sampled. Thus, it is unnecessary to know the underlying "parent distribution" to quantify the distribution of these statistics, all of which are of interest in various risk analysis applications. Similarly, a variety of inequalities quantify how unlikely it is that a value sampled from a distribution will fall far from its expected value. Again, these bounds do not require detailed knowledge of the parent distribution. As a result, empirical data that give only limited information about a risky process may still be adequate to obtain useful quantitative bounds on risks of interest.

Example: The "Rule of Three" for Negative Evidence

Setting: People sometimes worry about events that *might* happen in theory, even though they *have not* (yet) happened in practice. How reassuring should one consider such "negative evidence" (i.e., the absence of occurrences of a feared event, despite past opportunities for occurrence), bearing in mind the adage that "Absence of proof [of a hazard] is not proof of absence"? This can be an important topic when new technologies or poorly understood systems are involved, ranging from the Large Hadron Collider particle accelerator at CERN, which some feared might destroy the world by producing micro black holes, to the systems of interlocking safeguards that countries establish to try to protect against diseases such as bovine spongiform encephalitis (BSE, or "mad cow" disease). We will use the latter example to illustrate how negative evidence (i.e., the observation that a feared event has not yet been observed) can be used to bound risk.

Problem: Supposed that a country concerned about the possibility that its domestic cattle might be infected with BSE tests 1,000,000 randomly selected cattle and finds no cases. How confident can one be, based on this data, that the true prevalence proportion of BSE in the sampled population is not large? Assume that how BSE originates and spreads among cattle is not understood well enough to simulate or model with high confidence and that the effectiveness of any safeguards against BSE is not yet known. Thus, we want an upper-bound risk estimate based on the empirical "negative evidence" of no observed cases among a million animals tested, since calculations based on a well-validated understanding of the BSE disease process are not available.

Solution: A useful nonparametric confidence bound is based on the following "rule of 3" (Chen and McGee, 2008): If an event that has the same probability p (which may be unknown) of occurring on each trial has not occurred in any of N independent trials (e.g., in a simple random sample of size N), then, with at least 95% confidence, its occurrence probability on each trial satisfies $p \leq 3/N$. Thus, the unknown prevalence proportion of detectable BSE in this example would satisfy $p \leq 3/1{,}000{,}000 = 0.000003$. This bound does not require or assume any specific prior distribution for p, or any knowledge of the (probably complex) processes by which BSE might enter the country and spread domestically.

Example: A Sharp Transition in a Symmetric Multistage Model of Carcinogenesis

Setting: This example illustrates how probability laws can be used to model complex processes such as cancer, even if the molecular-level details of causal pathways are unknown. As a simplified illustration, consider the following *symmetric multistage model* of carcinogenesis. A cell line gradually accumulates transformations (e.g., somatically heritable mutations) from a set of K possible transformations. Transformations occur randomly and independently over time. The K transformations arrive according to independent Poisson processes, with (at least approximately) equal intensities, given by λ average occurrences per unit time. (Transformations with occurrence rates much less than this common value are not rate-limiting and thus may be disregarded.) Once any of the K transformations has occurred, we assume that it is permanent and irreversible. If a specific transformation occurs more than once, the occurrences after the first one are wasted, i.e., the cell genotype has already acquired that transformation and does not reach malignancy any faster if it occurs again. The cell line survives for a finite lifetime of duration T. If all K distinct transformations occur before time T, then the cell line becomes malignant.

Problem: Under these conditions, what is the probability that the cell line will become malignant before death at time T? If it does become malignant before time T, then what can be said about the (random) time at which the first malignant cell is formed?

Solution: The somewhat surprising answer is that, for sufficiently large K, there is a "sharp transition" time such that the first malignant cell is very unlikely to be formed much sooner or much later than that time. In other words, a nearly deterministic occurrence time for the first malignant cell emerges simply as a consequence of there being many stages in this simple stochastic transition model.

Result: In this completely symmetric multistage model, there is a "sharp transition" time $T^* \approx (1/\lambda)[(\ln(K) + \gamma]$, where λ is the expected number of transformations events per unit time, i.e., their average occurrence rate, and γ = Euler's constant = 0.57721.... In particular, the expected time until the first malignant cell is formed is T^*; moreover, the coefficient of variation of the actual (random) time of formation of the first malignant cell (i.e., the ratio of its standard deviation to T^*) approaches 0 for large K.

Proof: The expected *number* of transformation occurrences, including wasted (i.e., repeated) ones, until a malignant cell is formed (i.e., until all K transformations have occurred at least once) is given by the harmonic sum: $E(n^*) = K(1 + 1/2 + 1/3 + \ldots + 1/K) \approx K[(\ln(K) + \gamma]$, where n^* denotes the random number of the transformation occurrence event at which all K transformations are first completed and γ is Euler's constant, $\gamma = 0.57721\ldots$. This follows from previously known results for the "Coupon Collector's Problem" with equal probabilities (e.g., Ross, 1996, p. 414; Motwani and Raghavan, 1995) or for the maximum of K independent exponential random variables (e.g., Nelson, 1995, p. 173). [Intuitively, this result is motivated by the fact that any of the K transformations can occur first and be nonredundant, after which the probability that the next transformation is nonredundant drops to $(K-1)/K$, then to $(K-2)/K, \ldots$, and finally, for the last transformation, to $1/K$.] The expected *time* until a malignant cell is formed is therefore $T^* = E(t^*) = E(n^*)/(K\lambda) \approx (1/\lambda)[(\ln(K) + \gamma]$, where T^* denotes the random time at which all K transformations are first completed, and $K\lambda$ is the rate at which transformation events arrive (since each of the K types independently arrives at rate λ). This proves part (a) of the theorem. The fact that the probability distribution of n^* has a sharp concentration around $E(n^*)$ is proved in Motwani and Raghavan (1995). Given this key result, hold n^* fixed. The time until n^* transformations (including redundant ones) have occurred has a gamma distribution with mean $n^*/(K\lambda)$ and variance $n^*/(K^2\lambda^2)$, by standard results for waiting times in Poisson arrival processes and for the mean and variance of the gamma distribution (e.g., Ross, 1996, p. 18). The ratio of the standard deviation to the mean of this waiting time is therefore $(n^*)^{-1/2} \approx [K((\ln(K) + \gamma)]^{-1/2}$, which goes to 0 as K increases.

Discussion: An interesting, and perhaps unexpected, aspect of this result is that it establishes a form of nearly deterministic behavior for a stochastic system: If the sharp transition time T^* is smaller than the death time T, then formation of a malignant cell by time T is almost certain; otherwise, it is very unlikely. (This qualitative behavior is typical of what is sometimes called a 0–1 law in stochastic processes.)

If K is not large enough to guarantee a sharp transition at time T^*, then the qualitative behavior can be generalized as follows: For any $\varepsilon > 0$, no matter how small, there is an interval of times $[T^-, T^+]$ such that the probability of a malignant cell being formed before T^- or after T^+ is less than ϵ. The cumulative probability distribution for the occurrence time of the first malignant cell increases from almost 0 to almost 1 over this interval. As K increases, the width of this interval shrinks toward zero, with T^- and T^+ approaching a common value, T^*.

Realistic models of carcinogenesis are more complex than this example (see, for example, Chapters 11 and 12), but this simplified illustration shows that sometimes the behaviors of complex stochastic systems can be described well by phase transitions and probability laws, even if the details of the systems (such as which specific events occur along different causal pathways leading to cancer) are unknown. (Chapter 16 describes a similar phase-transition result for the ability of telecommunications networks to recover from deliberate coordinated attacks at multiple locations.)

Dealing with Model Uncertainty: Bayesian Model Averaging (BMA) and Alternatives

Copulas and maximum-entropy methods are mainly used to deal with uncertainties about the *parameters* and *input distributions* for particular models. However, *model uncertainties* about (a) which variables to include in a model when many potential predictors (including some possibly irrelevant ones) have been measured and (b) the most appropriate *functional form* for a model – or, more generally, how to calculate or predict a model's outputs from its inputs – are even more important in practice than input and parameter uncertainties, in applications ranging from dose-response models in toxicology to the reliability modeling of complex systems. Some researchers have suggested assessing a probability distribution over multiple plausible models by evaluating the consistency of the various models with the observed data (in much the same way as the likelihood function in Bayesian updating evaluates the consistency of various parameter values with observed data) and determining how much weight to put on each model based on its consistency with the data. Failing to consider model uncertainties can lead to spuriously narrow statistical confidence intervals for parameter estimates and to spuriously high confidence in model-based predictions (Hoeting et al., 1999).

However, it is frequently not reasonable to attempt to estimate the probability that a given model is "correct," because, as Box (1979) pointed out, "All models are wrong, some models are useful." For example, it seems highly implausible that any of the current models for estimating the probability of human error on a given task is close to being "correct" (because all are gross oversimplifications of the real world), nor can the current models be considered a collectively exhaustive set of possible models of human error. Bayesian updating of probability distributions over such partial subspaces of possible models may not always work well in practice. Some models may be intentionally conservative (e.g., for regulatory and/or screening purposes) or intentionally simplified (e.g., for computational tractability, or to yield qualitative insights). That such models may be inconsistent with observed data does not necessarily invalidate their use for their intended purposes.

Finally, of course, more complex models, with larger numbers of parameters, will often fit the observed data well in many situations (subject to the possible limitations of overfitting), but may not always be preferable, if only for reasons of parsimony and/or generalizability. Thus, standard approaches for dealing with uncertainty probabilistically are often not well suited for handling model uncertainty. Bayesian model averaging (BMA) (see Chapter 7) was motivated largely by these challenges. BMA avoids basing all of one's conclusions on any single model if multiple models are about equally plausible. It avoids giving high weight to models that are excessively complex if simpler ones give comparably good (or better) descriptions of the data, as measured by the likelihood of the data given a model. BMA generally performs reasonably well in practice, e.g., as evaluated by its ability to give well-calibrated uncertainty interval estimates for uncertain outputs, taking into account model uncertainty (Hoeting et al., 1999; Raftery and Zheng, 2003).

An alternative that avoids assigning probabilities to individual models, "comprehensive uncertainty evaluation" (Brown, 1999), involves subjectively adjusting the probability distributions resulting from a particular model to try to take into account known weaknesses of the model (such as conservatisms, or risks that are not adequately modeled). This is consistent with subjective utility theory and avoids some of the theoretical conundrums associated with assigning probabilities to models. Brown has applied this method (for example, to support regulatory decision making for nuclear power plants), but it has not yet seen widespread application by other analysts in practice.

In many applications of Bayesian analysis to situations involving model uncertainties, the input parameters are assumed to be known, and the model results are used to update the prior distribution over model outputs (see, for example, Chick, 1997). However, observing the output of a model could also cause one to revise the prior distribution over model inputs if the true values of the model outputs were known reasonably well (e.g., from empirical data). Thus, for example, Bayesian analysis could be used to estimate which values for the rate of disease progression are most consistent with the observed data on disease prevalence and severity (Andradóttir and Bier, 2000).

Risk Characterization

The output of a PRA to support risk management decision making is a characterization of the risk for each decision option being evaluated. Occasionally, the decision task is to identify an optimal risk management policy from a large set of possibilities, rather than to explicitly characterize the risks for each of a small number of alternatives. Then, simulation-optimization algorithms or special-purpose techniques such as Markov decision processes or stochastic optimal control theory may be required (see Tables 2.2 and 2.3). However, explicit comparison of risks from a few options is more usual, and is the main focus of this section.

"Risk" is usually defined in engineering risk assessments and PRA as the frequency and severity of losses arising from operation of the designed system in its uncertain environment, including a specification of losses (i.e., which adverse consequences matter, and to whom). An effective display of risk shows how it is affected by different actions (e.g., different risk management decisions) and allows "drill-down" to view the risks to particular subpopulations, as well as the contributions of various different causes to the overall level of risk. For example, seeing how risk curves shift when risk-reducing measures are implemented would help managers identify the most effective measures. Uncertainty and sensitivity analysis are also essential to risk characterization, because they support estimates of the value of information.

Engineering vs. Financial Characterizations of "Risk": Why Risk Is Not Variance

The variance (or standard deviation) of the return on investment is widely used as a measure of risk in financial risk analysis, where mean-variance analysis is applied to calculate "efficient" frontiers and undominated portfolios, defined as those having maximum expected return for a given variance. Why, then, do health, safety, environmental, and reliability risk analysts insist on defining risk more flexibly, as being determined by probabilities *and* consequences, rather than simply by variances (or, for that matter, semivariances, value-at-risk, or modern coherent risk mea-

Table 2.2 Some decision modeling frameworks

Framework	Act	State	Consequence	Optimization algorithms
Normal form • $\Pr(s)$ = state probability model • $u(c)$ = utility model $c \varepsilon C$, a set of possible consequences	a = act = choice of *controlled input* to system, or of a *decision rule* mapping observations to controlled input values	s = state (random or uncontrolled inputs) $s \varepsilon S$, a set of possible states	• $c(a, s)$ = deterministic consequence model • $\Pr(c \mid a, s)$ = stochastic consequence model • $a \varepsilon A$, a set of feasible acts	Choose a from A to maximize expected utility (EU): $\max_{a \in A} \Sigma_c u(c)p(c \mid a)$ s.t. $p(c \mid a) = \Sigma_s \Pr(c \mid a, s)\Pr(s)$
Utility table	a = row	s = column	expected utility for each cell (i.e., for each row-and-column pair)	Eliminate dominated rows, then choose act to maximize EU
Decision tree	Choice of an act at each decision node	Outcome at each chance node	Utilities at tips of tree	Backward dynamic programming
Influence diagrams and Bayesian networks	Choice of an act at each decision node	Outcome at each chance node	Value at value node	• Gibbs sampling • Bucket elimination • Graph algorithms (e.g., arc reversal)
Markov decision process (MDP); semi-Markov decision process	Choice of act in each state	Transition time and next state	Reward per unit time in states and/or at transitions; transition rates among states	• Value iteration • Policy iteration • Neurodynamic programming • Linear programming

Table 2.2 (continued)

Framework	Act	State	Consequence	Optimization algorithms
Partially observable MDP	Decision rule mapping observations to acts	Underlying state of process	Function of uncertain state and choice of acts, and/or transitions	Similar to MDP (A POMDP is an MDP with probabilities of states for its state.)
Stochastic optimal control and robust control of uncertain systems	Decision rule (maps observed history to choice of controlled input values)	Probability of next state, given history to date (and inputs)	Function of state and/or control trajectory	Stochastic dynamic programming; adaptive optimization, robust control, reinforcement learning algorithms
Simulation model	Controllable inputs	States of model components	Function of component states and/or input sequence	Simulation-optimization heuristics and algorithms
Unknown model	Sequence of acts		Sequence of rewards	Adaptive learning, online, and minimal-regret algorithms

Table 2.3 Selected literature on decision optimization frameworks and algorithms

- Decision trees:
  - Game Trees for Decision Analysis – Shenoy (1996), http://citeseer.ist.psu.edu/shenoy96game.html
- Influence diagrams and Bayesian networks:
  - Sampling Methods for Action Selection in Influence Diagrams – Ortiz, Kaelbling (2000), http://citeseer.ist.psu.edu/ortiz00sampling.html
  - A Forward Monte Carlo Method for Solving Influence Diagrams. – Charnes, Shenoy (2000), http://citeseer.ist.psu.edu/charnes00forward.html
  - A Simple Method to Evaluate Influence Diagrams – Xiang, Ye (2001), http://citeseer.ist.psu.edu/ye01simple.html
  - Learning Bayesian Networks with R, http://www.ci.tuwien.ac.at/Conferences/DSC-2003/Proceedings/BottcherDethlefsen.pdf;
    see also http://www.cs.ubc.ca/~murphyk/Software/bnsoft.html
- Markov decision processes (MDPs) and partially observable MDPs (POMDPs):
  - Reinforcement Learning for Factored Markov Decision Processes – Sallans (2002), http://citeseer.ist.psu.edu/sallans02reinforcement.html
  - Symbolic Dynamic Programming for First-Order MDPs – Boutilier, Reiter, Price (2001), http://citeseer.ist.psu.edu/boutilier01symbolic.html
  - Speeding Up the Convergence of Value Iteration in POMDPs – Zhang, Zhang (2001), http://citeseer.ist.psu.edu/zhang01speeding.html
  - Solving POMDP by On-Policy Linear Approximate Learning Algorithm – He (1999), http://citeseer.ist.psu.edu/335710.html
- Optimal and robust control and reinforcement learning for uncertain and nonlinear systems:
  - Feedback Control Methodologies for Nonlinear Systems – Beeler, Tran, Banks (2000), http://citeseer.ist.psu.edu/Beeler 00feedback.html (for deterministic nonlinear systems)
  - An Overview of Industrial Model Predictive Control Technology – Qin, Badgwell (1997), http://citeseer.ist.psu.edu/qin97overview.html
  - http://citeseer.ist.psu.edu/kaelbling96reinforcement.html
  - http://citeseer.ist.psu.edu/sutton98reinforcement.html
  - http://www.princeton.edu/~noahw/palgrave2.pdf (introduces robust control)
- Simulation-optimization:
  - A Survey of Simulation Optimization Techniques and Procedures – Swisher, Jacobson et al. (2000), http://citeseer.ist.psu.edu/517471.html
  - Simulation Optimization of Stochastic Systems with Integer Variables by Sequential Linearization – Abspoel et al. (2000), http://citeseer.ist.psu.edu/516176.html
  - Simulation Optimization: Methods and Applications – Carson, Maria (1997), http://citeseer.ist.psu.edu/carson97simulation.html
  - http://opttek.com/simulation.html (overview and link to commercial software)
- Minimal-regret, online, and adaptive learning algorithms:
  - Minimizing Regret: The General Case – Rustichini (1999), http://citeseer.ist.psu.edu/rustichini98minimizing.html
  - Adaptive Strategies and Regret Minimization in Arbitrarily Varying Markov Environments – Mannor, Shimkin (2001), http://citeseer.ist.psu.edu/467490.html
  - Nearly Optimal Exploration-Exploitation Decision Thresholds – Dimitrakakis (2006), http://citeseer.ist.psu.edu/dimitrakakis06nearly.html
  - Combinatorial Online Optimization in Real Time – Grötschel, Krumke, Rambau (2001), http://citeseer.ist.psu.edu/448491.html; see also http://citeseer.ist.psu.edu/foster97regret.html

sures developed in financial risk theory)? This section suggests a partial answer by providing a simple proof that mean-variance decision making violates the principle that a rational decision maker should prefer higher to lower probabilities of receiving a fixed gain, all else being equal. Indeed, simply hypothesizing a continuous increasing indifference curve for mean-variance combinations at the origin is enough to imply that a decision maker must find unacceptable some prospects that offer a positive probability of gain and zero probability of loss. Unlike some previous analyses of the limitations of variance as a risk metric, this section does not require the additional framework of von Neumann-Morgenstern utility theory.

Incompatibility of Two Suggested Principles for Financial Risk Analysis

Two plausible principles for managing financial investment risks are the following:

1. *Rule 1: Make dominating choices.* Other things being equal, given a choice between a smaller probability of gain and a larger probability of gain, a decision maker should always choose the larger probability of gain. For example, given a choice between winning \$100 with probability 0.1 and winning \$100 with probability 0.2, rational decision makers who prefer more dollars to fewer should choose the option that gives a 0.2 probability of winning the \$100.
2. *Rule 2: Seek mean-variance efficiency (higher variance requires higher mean return).* Given a choice among risky prospects, an investor should require more expected return to accept a prospect with more variance than to accept a prospect with less variance. For example, a 0.2 chance of winning \$100 (else nothing) has a higher variance than a 0.1 chance of winning \$100, but it also has a higher mean.

Rule 1 is implied by the decision-analytic principle of first-order stochastic dominance (Sheldon and Sproule, 1997): Prospects that give higher probabilities of preferred outcomes (and lower probabilities of less preferred outcomes) should be preferred. Rule 2 provides the basis for many current efficient portfolio and mathematical optimization (e.g., quadratic programming) approaches to optimal investment (http://en.wikipedia.org/wiki/Modern_portfolio_theory). Although theorists have noted that some risk-averse decision makers may prefer some mean-preserving increases in variance (ibid.), the idea that volatility in returns, as measured by variance or standard deviation, is generally undesirable to risk-averse investors, and that it should be avoided or compensated by higher expected returns, is still widely taught and practiced.

However, *Rules 1 and 2 are incompatible* in general. Simply hypothesizing that a decision maker has continuous upward-sloping indifference curves for mean-variance combinations (so that increasing the variance in the random return from an investment prospect or portfolio requires increasing its mean return in order to leave the investor equally well off) violates Rule 1 for some simple prospects, as demonstrated next.

Following the literature on mean-variance decision making, suppose that a decision maker has positively sloped continuous indifference curves in mean-variance space (e.g., Wong, 2006). To any mean-variance pair (m, v) (a point in the mean-variance space) there corresponds a *certainty equivalent*: namely, the point at which the indifference curve through (m, v) reaches the horizontal (mean) axis. The indifference curve through the origin (0, 0) separates *acceptable risks* (those with positive certainty equivalents, lying below and to the right of the curve, if return is desirable) from *unacceptable risks* (those with negative certainty equivalents, lying above and to the left of it). To make an unacceptable risk acceptable in this framework, one must either increase its mean return or reduce its variance. (A risk-neutral decision maker who cares only about means and not about variances would have vertical indifference curves, but we will focus on the case, implied by Rule 2, of positively sloped indifference curves.)

The hypothesis that upward-sloping mean-variance indifference curves exist has some surprising consequences.

Theorem 1 *If the indifference curve through the origin slopes upward, then the decision maker finds unacceptable some prospects with positive expected values and no possibility of loss.*

Proof The proof is constructive. Let the slope of the indifference curve through the origin be s at the origin. By hypothesis, $0 < s\ \infty$. Now, consider a Bernoulli random variable $X(p)$ that gives a positive return of $2s$ with probability p (the "win probability") and no return ($0) with probability $(1 - p)$. For a given value of p between 0 and 1, inclusive, $X(p)$ has mean $2ps$ and variance $4s^2p(1 - p)$ (since it is a scaled version of a Bernoulli random variable). Therefore, as p ranges from 0 to 1, $X(p)$ traces out a parabola in mean-variance space, with variance = 0 at $p = 0$ and at $p = 1$, and with a positive maximum variance of s^2 at $p = 0.5$ (see Fig. 2.1). A line from (0, 0) to the point on this parabola corresponding to a particular value of p has slope $4s^2p(1 - p)/2ps = 2s(1 - p)$. As p approaches 0, this slope approaches $2s$. Hence, the parabola traced out by $X(p)$ as p ranges from 0 to 1 starts above and to the left of the indifference curve through the origin (since it is constructed in such a way as to have twice the slope of the indifference curve at the origin), but it ends below and to the right of the indifference curve [since it is constructed to pass through the point $(2s, 0)$ when $p = 1$]. Therefore, the parabola must intersect the indifference curve somewhere above and to the right of the origin (since it starts above it and ends below it). Let p^* denote the value of the win probability for this intersection point. Then the decision maker prefers (0, 0) to all prospects $X(p)$ with $p < p^*$ since, by construction, these are unacceptable (i.e., above and to the left of the indifference curve through the origin). Hence, the decision maker finds unacceptable all such prospects giving probability p of $2s$ (else $0) for $p < p^*$ even though he or she has positive win probabilities and even though none of them offers the possibility of a loss.

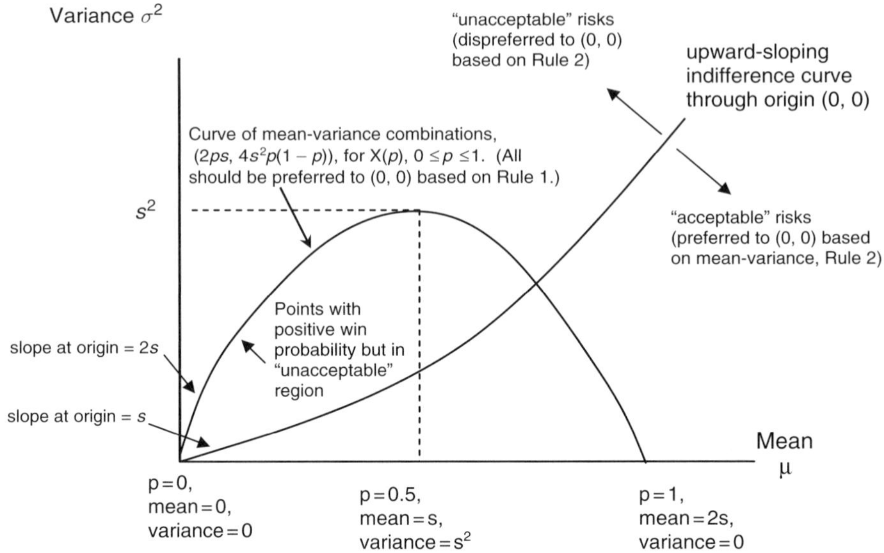

Fig. 2.1 Geometry of inconsistency between Rules 1 and 2

The proof of Theorem 1 implies that if the indifference curve through the origin has positive slope, then the decision maker prefers some prospects that give zero probability of winning a positive amount (namely, $2s$) to other prospects that give a positive probability of winning the positive amount (and otherwise nothing). Such a decision maker prefers the status quo or "nothing ventured, nothing gained" point (0, 0) to the possibility of winning a positive amount without any possibility of a loss, violating Rule 1. In this sense, Rules 1 and 2 are incompatible.

More generally, other parabolas can easily be constructed that intersect indifference curves twice, once for the ascending (positively sloped) portion of the parabola and once for its descending (negatively sloped) portion (Borch, 1969). In any such construction, the rightmost intersection represents a stochastically dominant prospect (which should be preferred, by Rule 1) compared to the leftmost intersection. That both points lie on the same indifference curve violates Rule 1.

In summary, although students of elementary finance are often taught that "risk" should be characterized by the variance or standard deviation of returns around an expected value, students of health, safety, environmental, and reliability risk analysis are usually taught instead that "risk" is determined by the probabilities of different consequences. Theorem 1 shows why the second approach (considering different specific consequences, such as \$0 and \$2s, and their probabilities) can be preferable to considering only means and variances.

The finding that variance is problematic as a measure of risk has a history at least several decades old in the financial and decision sciences literatures. A common critique in the theoretical decision analysis and financial economics literatures is that mean-variance analysis is compatible with von Neumann-Morgenstern expected utility theory only under restrictive conditions (e.g., if all risky prospects have normal or location-scale distributions and utility functions are quadratic, implying that less money is preferred to more, for some amounts) (Markowitz, 1959; Baron, 1977). Mean-variance dominance and stochastic dominance relations for location-scale distributions do not coincide in general (Wong, 2006). Indeed, *expected utility theory is inconsistent with all possible moment-based preference models* (in which preferences are determined by mean, variance, skewness, kurtosis, etc.) for many utility functions (Brockett and Kahane, 1992). Variance is also inconsistent with proposed normative axioms for "coherent" financial risk measures (nonnegativity; homogeneity and subadditivity, which together imply that deterministic outcomes have zero risk; and shift-invariance, which implies that adding a constant to a random variable does not change its risk) (Pedersen and Satchell, 1999). Empirical studies since the 1960s have demonstrated that real decision makers pay attention to more than mean and variance in their choices among risky prospects (Jia et al., 1999).

Thus, Theorem 1 is consistent with a long line of previous research. However, in contrast to much previous work, it demonstrates a conflict between Rules 1 and 2 making only minimal assumptions (in particular, not requiring the framework of von Neumann-Morgenstern expected utility theory or other sets of normative axioms for risk measures) and using only elementary mathematics. It may therefore be useful for understanding why specifying the variances and expected returns from alternative investment choices (or other actions) does not adequately characterize risk or identify the choice with the most desirable probability distribution of consequences.

In fairness, it should be noted that financial risk analysts have developed much more sophisticated and satisfactory measures of risk than variance and that characterizing risk by frequency and severity is not problem-free (see Chapter 5). A recent triumph of financial risk theory has been the definition and analysis of *coherent risk measures* (Artzner et al., 1999). These provide formulas for assigning numbers to risky prospects so that normative axioms are satisfied, such as that risk remains unchanged if the same predictable constant is added to or subtracted from all possible consequences of a prospect; and that comparisons of risk should be logically consistent with each other over time. Financial risk theorists have shown that various sets of normative axioms imply intuitively pleasing quantitative representations of risk, such as that the "risk" of a financial prospect is its minimum (worst-case) expected net present value (ENPV), when ENPVs are calculated for each of a set of mutually consistent probability measures (Riedel, 2004). Older proposed measures of financial risk, including variance and Value-at-Risk (VaR), which reflects the probability of losing at least a specified amount, do not satisfy these axioms (Artzner et al., 1999; Pedersen and Satchell, 1999). Although coherent risk measures provide a substantial advance in methodologies for characterizing financial risks, they do not apply to other risks that cannot be traded, valued, or diversified away via financial markets. The characterization of health, safety, environmental, and reliability risks in terms of probabilities or frequencies of different consequences, having different magnitudes or severities, is still the norm. Chapter 5 discusses further the use and limitations of frequency in risk characterization.

Challenges in Communicating the Results of PRAs

Risk communication (including both presenting the results of risk analyses to stakeholders, decision makers, and other audiences, and listening to and actively eliciting their concerns so that they can be addressed in the risk analyses in the first place) facilitates the effective participation and interaction of technical experts, stakeholders, and decision makers in risk management decisions and deliberations. There is an extensive body of literature on risk communication (including guidelines, survey results, experiments, and web resources on risk communication).

Even more than PRA, risk communication is still an art rather than a science, but one that can be informed and improved by theory, experience, and experiments. Current challenges in risk communication include dealing with framing effects, communicating highly technical results to decision makers who may not be intimately familiar with some of the methods used in the risk analysis, and building trust among affected stakeholders and members of the public more generally. Adding to the difficulty is the fact that the communication and presentation styles that are most effective in accurately expressing the technical content of risk assessment findings may not always be those that invite and elicit public understanding, participation, and interaction.

Cullen and Frey (1999) discuss the distinctions between state-of-knowledge uncertainty and population variability (sometimes referred to simply as uncertainty and variability, respectively). State-of-knowledge uncertainty typically reflects uncertainties that affect all of the units being studied (e.g., can certain standardized industrial systems fail in a particular way? Can a certain chemical cause particular health effects?). These uncertainties could be reduced through further research. Variability refers to variations among the elements being studied (often assumed to be due to randomness in production processes, phenotypes, etc.). For example, differences in how different individuals in the population would respond to a chemical being studied, or strengths of different samples of a material, would reflect variability. Variability is often taken to be essentially irreducible through further study.

With the development and increased popularity of so-called second-order Monte Carlo analysis for quantifying uncertainty about risks, it is now common practice to distinguish between uncertainty and variability. This increases the value of the risk analysis for decision making, because different policy options may be appropriate for dealing with uncertainty rather than variability. For example, in situations of high population variability but low state-of-knowledge uncertainty, such as airbag effectiveness (Thompson, 2002), it may make sense to target risk-reducing efforts at those facilities or members of the population with the highest estimated risks (in this case, children and small adults). By contrast, situations of low variability but high uncertainty would tend to suggest that further research may be desirable before undertaking costly risk reduction actions. However, the widespread use of second-order Monte Carlo simulation does increase the challenges of effectively communicating ever more sophisticated and sometimes abstruse risk analysis methods and results to decision makers and members of the public in a way that clearly supports improved decision making (Bier, 2001a).

Of course, technically accurate risk communication by itself is not sufficient to achieve other key goals of risk communication, such as changing people's behavior (Blaine and Powell, 2001), gaining their trust in the results of the analysis, or even giving them the information they need to make improved decisions. Rather, effective and persuasive communication about risks generally requires a concerted effort to build trust, gain and maintain credibility and legitimacy, and summarize relevant information simply and clearly (Bier, 2001b). Brevity, clarity, focus, candor, the use of cogent examples, and avoiding negative stereotypes of risk communicators may be crucial for communicating technical risks to nonspecialist audiences in a way that ensures the message is heard and absorbed rather than tuned out or dismissed (e.g., Byrd and Cothern, 2000). As discussed in Chapter 1, audience members generally respond not only (and sometimes not primarily) to technical information about risks, but also to message framing, the source of the information, and the emotional style and assumed motives of the presenter in assessing the credibility of risk communication messages (Chartier and Gabler, 2001).

Methods for Risk Management Decision Making

Formal methods of decision analysis and optimization for uncertain systems have been extensively developed in operations research and systems engineering and applied to both the design and the operation of complex engineering and industrial systems. Table 2.2 sketches some of the best-known frameworks for decision making when a decision maker's choice of act is related only probabilistically to resulting consequences.

Although some of the methods and algorithms mentioned in Table 2.1 are quite sophisticated, most share a simple common structure. The risk manager must choose from a set of feasible *controllable inputs* that influence a system's behavior. There are other facts and inputs (sometimes thought of as being selected by "Nature" or "Chance," and referred to as the *state* of the world) that cannot be directly selected by the risk manager but that also influence the system's behavior. The risk manager's acts and the state of the world together determine probabilities for different *consequences* (and for the system's next state, in systems dynamics and optimal control formulations of decision problems). Finally, a *utility function* represents preferences for different consequences (or time streams of consequences) produced by the system.

Various optimization algorithms and heuristics can be used to identify optimal (i.e., expected utility-maximizing) or approximately optimal *acts* (i.e., values of controllable inputs), given available information, or to identify optimal or approximately optimal *decision rules* (also called *policies*) that prescribe what acts to take based on the information available when decisions are made. Optimization algorithms for solving decision problems are constantly being refined and improved by ongoing research. Thus, it is worth Googling the topics and solution methods in Table 2.2 (leftmost and rightmost columns, respectively) before selecting a framework and solution methods for a particular problem. As of this writing (in 2008), the links in Table 2.3 provide points of entry to the technical literature and solution algorithms.

An important principle that cuts across many solution techniques for complex decision models is that *adaptive random sampling of potential solutions* is computationally tractable and finds "good" (optimal or nearly optimal) solutions to many problems that are too hard (e.g., too computationally complex) to solve using exact methods. Monte Carlo methods and related meta-heuristics (such as genetic algorithms, simulated annealing, Tabu Search, or particle swarm optimization) can estimate and optimize the expected utility of different acts or decision rules even for large and complex stochastic systems. Much as the mean value of a variable in a population can be estimated accurately from a random sample, regardless of the uncertainties and complexities of processes that created the population distribution of the variable, so the expected utility of a decision rule, policy, or act (followed by future optimized acts, in dynamic settings) can often be estimated accurately using optimization algorithms that incorporate random sampling and adaptive improvement components.

Example: A Bounded-Regret Strategy for Replacing Unreliable Equipment

Setting: Suppose that a piece of machinery (such as a crane) that is being used in a major construction project breaks down frequently. The construction project must continue for 1,000 more days in order to meet a key deadline; after 1,000 more days, all activity (and further costs) on this effort will stop. Use of the current unreliable equipment costs \$1,000 per day in maintenance, repair, insurance, and overtime costs. The unreliable equipment will eventually break down completely; if this happens before the end of the project, it must then be replaced. The cost of replacement is \$1,200,000, and a new machine is highly reliable, costing \$0 per day, after it has been purchased, for the remaining duration of the project. Suppose that the current machine will last for an unknown number, T, of additional days before breaking down completely, where T is an unknown integer. Assume that not enough is known about the machine's remaining lifetime (e.g., from historical experience, accelerated life testing, reliability modeling, etc.) to assess a credible, well-calibrated probability distribution for T.

Problem: Assuming that the probability distribution for the remaining lifetime T is unknown, devise a decision rule for when to replace the machine (if at all) that is guaranteed to cost no more than twice as much as the least cost that could be achieved if T were known. (For simplicity, ignore discounting.)

Solution: If the lifetime T of the current machine were known, then the total cost of replacing the machine after $t < T$ days would be \$1,000t + \$1,200,000 if $T < 1{,}000$ days, in which case the optimal decision would be to replace the current machine immediately (set $t = 0$) and the minimized cost would be \$1,200,000; otherwise, if $T > 1{,}000$ days, then the unreliable machine should be used for the rest of the project's duration, for a total cost of \$1,000,000. Now consider the following myopic decision rule: *Wait until the machine fails completely, and then replace it.* The worst (most expensive) case is that the machine fails on day $T = 999$, in which case the total cost is \$999,000 + \$1,200,000 = \$2,199,000. This is less than twice the optimal cost if T were known, \$1,200,000 (for immediate replacement). If $T \geq$ 1,000 days, then the myopic decision rule yields the same optimal decision as if T were known. If $T <$ 1,000 days, then the myopic rule has less than twice the cost of the optimal decision if T were known.

Discussion: Although this example is trivial, it illustrates that analysis of decision rules is possible for some problems, even if uncertain quantities cannot be characterized by probability distributions. A number of nontrivial results on "online" decision and optimization procedures show that, in many sequential decision problems, it is possible to do almost as well on average, over the long run, using cleverly designed decision rules, as if the uncertain quantities (such as T in this example) were known.

Despite these advances in methods for decision analysis and optimization under uncertainty, in practice, formal decision analysis is seldom applied directly to make important risk management decisions. In part, this is because different participants may have different utility functions (which may be their own private information), different trade-offs among goals (e.g., minimizing average risk versus reducing inequities in the distribution of risks), and different tolerances for risk. In such cases, consensus utilities may not exist, and risk management decision making requires not only analysis and deliberation (Stern and Fineberg, 1996), but also negotiation and compromise.

Even when decision analysis is not directly applied, however, its conceptual framework is still useful for organizing analysis and deliberation (Apostolakis and Pickett, 1998), separating beliefs from preferences, and identifying and resolving relevant conflicts and/or uncertainties about facts and values. Byrd and Cothern (2000) and Cox (2001) further discuss individual and group decision-making processes and frameworks for risk management decision making.

Methods of Risk Management to Avoid

Well-informed and effective risk management (i.e., risk management that is likely to produce the desired consequences) requires considering *all* of the most important impacts – good and bad – that an intervention is likely to create. Unfortunately, many risk assessments exhibit a form of tunnel vision, focusing on one or a few narrowly defined issues (such as quantifying the reduction in risk that would be caused by contemplated actions) while ignoring other, possibly more important, ones, such as the risks that proposed risk management interventions might inadvertently *create* (Dowell and Hendershot, 1997; Bier, 1997). This represents a breakdown in sound risk assessment and risk management. Rational risk management requires considering and comparing the *total* consequences of the risk management decision options being evaluated. Risk characterization should therefore provide risk managers with a balanced accounting of the adverse effects that a risk management intervention might *cause*, as well as of those that it might *prevent*.

Risk management recommendations that are based primarily on protecting the status quo or on beliefs about what might constitute "precautionary" actions should also be avoided if they do not explicitly identify and compare the probable consequences of alternative decision options. Decision analysis teaches that it is more effective to use quantitative information about the *probable consequences* of alternative interventions to eliminate dominated options, and to choose the best among those that remain. Heal and Kriström (2002) have argued on theoretical grounds that precautionary measures might make sense in situations where harm is irreversible, but their argument is based on, and consistent with, utility theory and real options theory.

Game-Theory Models for Risk Management Decision Making

Game theory has long been viewed by risk analysts as being of little relevance for practical risk management decision making. Several recent developments have started to change that view. These include not only increased interest in terrorism, homeland security, and critical infrastructure protection (which can be viewed as games between an attacker and a defender), but also increased interest in risk-informed regulation (which can be viewed as a game between a regulator and a regulated firm). As a result of such developments, game theory is becoming an important research tool in a variety of application areas related to risk.

Hausken (2002) has applied game theory to study the allocation of resources to ensuring component (and hence system) reliability in situations where different agents are responsible for the reliability of different components. In this situation, system reliability is viewed as a "public good." For example, agents responsible for the reliability of a component in a parallel system or subsystem might "free-ride" on investments in the reliability of other components in that system – e.g., postponing needed reliability enhancements in the hopes that some other agent will implement such improvements instead.

Recent work on reliability optimization (e.g., Levitin et al., 2001; Levitin and Lisnianski, 2003) attempts to identify cost-effective risk reduction strategies; for example, by optimizing physical separation of components that are functionally in parallel with each other, or by allocating physical protection to various hierarchies of a system (e.g., whether to harden the system as a whole, or individual components). However, the "threat" against which systems are to be hardened is generally taken to be static in this work.

Game-Theory Models for Security and Infrastructure Protection

Following September 11, 2001, there has been increasing interest in security, including the protection of public and commercial buildings, water supply systems, and computer systems and software. Numerous researchers and practitioners have proposed the use of risk analysis in one form or another for homeland security (e.g., Paté-Cornell and Guikema, 2002; Garrick et al., 2004), especially for critical infrastructure (Haimes et al., 1998; Ezell et al., 2001; Apostolakis and Lemon, 2005). Most of this work is not formally game-theoretic. For instance, Paté-Cornell and Guikema discuss the need for periodic updating of the model and its input to account for the dynamic nature of counterterrorism but do not attempt to anticipate the effects of defensive investments on attacker strategies. Protection from intentional sabotage or terrorism differs from many other areas of risk management, because sabotage protection involves an intelligent adversary that can adapt in response to protective measures. Thus, reducing the vulnerability of some systems may cause adversaries to shift their attacks to other systems that have not yet been "hardened" to the same degree. Risk management in this context can be modeled as a game against an adversary or, conversely, as a game between defenders, because security investment by one defender can have either positive or negative externalities on the threats faced by other defenders (Kunreuther and Heal, 2003).

There is a large body of work on applications of game theory to security, much of it by economists (e.g., Frey and Luechinger, 2003; Arce et al., 2001; Enders and Sandler, 2004; Keohane and Zeckhauser, 2003; Lakdawalla and Zanjani, 2005). Much of this work is intended to inform policy-level decisions, e.g., by clarifying the relative merits of public versus private funding of defensive investments, or deterrence versus other protective measures. Recently, efforts have begun to focus more on operational risk management decisions, such as deciding how much defensive investment to allocate to particular assets (e.g., O'Hanlon et al., 2002), and have more of a risk analysis flavor (e.g., taking the success probabilities of potential attacks into account); see, for example, Bier et al. (2005) and Woo (2002).

Game-Theory Models of Risk-Informed Regulation

In health, safety, and environmental regulation, regulated parties often know more than regulators about the operations and risks of facilities. As a result, regulators may wish to provide incentives to encourage regulated parties to accurately disclose unfavorable information about their risks. Such situations can be modeled as games of asymmetric information between regulators and regulated parties. More widespread use of risk analysis results in regulatory decision making has the potential to both reduce risk and decrease compliance cost, by increasing management flexibility in determining how to achieve acceptable levels of safety (Bier and Jang, 1999). However, this approach has been slow to be adopted in practice, in part because of the inability of regulators to directly and accurately measure risk (Chinander et al., 1998) and because companies may have incentives not to disclose unfavorable risk information to regulators and or not to collect such information in the first place (Wagner, 1997).

Game-theoretic work in environmental economics to date (e.g., Heyes, 2000; Livernois and McKenna, 1999) has emphasized applications such as pollution monitoring, in which a regulator can (with some effort) determine a firm's level of performance essentially with certainty, and firm performance can reasonably be modeled as binary (e.g., compliant with pollution-control regulations or not). Lin (2004) considers *risk-informed regulation*, in which regulators may not be certain to detect high risk levels even with substantial effort, and continuous risk levels may be more relevant than binary compliance status. Lin shows conditions under which it is still optimal (more efficient than traditional direct-monitoring regulation) for regulators to offer a loosened standard to firms that voluntarily disclose their risk levels.

Conclusions

This chapter has surveyed methods and concepts for PRA and decision making in engineered systems. Although the modeling of uncertain systems has been tremendously enabled by recent advances (such as Bayesian belief networks, with dependencies among inputs expressed via copulas), PRA still poses many challenges. Technical challenges remain in how best to construct useful (and at least approximately valid) models of systems and their environments from engineering knowledge and data, and in identifying optimal or near-optimal risk management policies. Communicating the results effectively and using them to guide improved decision making by multiple parties (e.g., teams of stakeholders) also poses practical questions that go beyond the framework of single-person decision theory. If the environment in which a system operates includes intelligent adversaries, then insights from novel methods (e.g., game-theoretic principles) may be needed to ensure that risk reduction strategies are effective and cost-effective (see Chapters 14 and 16). These challenges are likely to stimulate further advances in both the theory and practice of decision sciences for engineering risk analysis.

Chapter 3
Introduction to Health Risk Analysis

Introduction

This chapter, which is an update of Cox (2007), introduces methods of quantitative risk assessment (QRA) for public health risks. As noted in the Preface, public health risk analysis often falls in the intersection of politics, business, law, economics, and science and technology, as stakeholders with different interests seek to use QRA for their own ends. Public health risk analysis deals with decisions about which potential risk management *interventions* (usually including the status quo or "do-nothing" option) should be implemented to maintain or increase the safety of complex social, economic, and technological systems, such as the food supply network or industrial emissions control systems. The best course of action is often hotly disputed. For example, should emissions of gases or particles from a facility be further restricted even if permitting the current levels has unquestioned benefits in industry or agriculture? Should cell phone use in cars be banned? ("Public health" is often extended to include such issues of public safety.) Should cattle be imported from countries that have low levels of diseases such as BSE? Should antibiotics used in human medicine be prohibited from uses in food animals, even if doing so will lead to more sick animals (and perhaps more sick people), in order to preserve the effectiveness of the antibiotics in treating human patients? QRA seeks to inform policy-making deliberation and debate in such controversial cases by clarifying the probable consequences of alternative decisions.

Health risk analysis consists of a set of methods, principles, and high-level procedures for using scientific data (e.g., biological and epidemiological knowledge) to assess and compare the probable human health consequences of different exposures to hazards (i.e., sources of risk); to assess the likely changes in exposures and risks arising from alternative risk management interventions; and to evaluate and choose among alternative risk management interventions based on their probable health consequences. The goal is usually to identify and select actions or interventions that will cause relatively desirable (e.g., stochastically undominated) probability distributions of human health consequences in the affected population. Health risk analysis is often divided into the overlapping stages of *risk assessment, risk management,* and *risk communication,* organized as an iterative process. Table 3.1 summarizes several traditionally defined steps in this process.

L.A. Cox, Jr., *Risk Analysis of Complex and Uncertain Systems*,
International Series in Operations Research & Management Science 129,
DOI 10.1007/978-0-387-89014-2_3,

Table 3.1 Traditional steps in health risk analysis

Step	Purpose and description	Relevant information and techniques
Hazard identification	Identify potential sources of harm or loss. These sources are called *hazards*.Hazard identification identifies the possible adverse health effects of activities or exposures and the possible causes of the observed adverse effects.	• Human data: epidemiology, clinical, and public health statistics; surveillance data. • Animal tests and bioassays. • *In vitro* tests. • Structure-activity patterns, molecular modeling, pattern recognition, and statistical classification techniques.
Exposure assessment	Quantify the number of people receiving various levels or intensities of exposure to a hazard over time. Relevant exposure metrics may depend on dose-response relations.	• Environmental fate and transport models, possibly summed over multiple media (paths) and sources. • Studies of human activity patterns • Biological monitoring of exposed individuals and receptors.
Quantitative exposure-response and dose-response modeling	Quantify the magnitude of risk created by exposure of a target to a hazard. Characterize the probable frequency and severity of adverse health outcomes or losses caused by exposure to the hazard.	A quantitative risk assessment (QRA) runs multiple exposure scenarios through *dose-responsemodels* to predict likely health impacts. Statistical, simulation, or biomathematical models of biological processes are used to quantify dose-response relations.
Risk characterization and uncertainty analysis	Combine estimated probabilities and severities of adverse consequences (harm), with indications of uncertainty or confidence, to create an overall summary and presentation of risk.	Monte Carlo simulation calculates risks by sampling many scenarios. Risk profiles, probability distributions, and trade-off and sensitivity analyses display risk, uncertainty, and variability.
Risk communication	Deals with how to present risk information to stakeholders. Considers how different types of recipients perceive risks and internalize/act on messages about them, in deciding what messages to send via what media.	Psychological theories and models and behavioral/experimental findings on risk perception and effective risk communication.
Risk management decision making	Decide what actions to take to control risks and hazards – i.e., accept, ban, abate, monitor, further research, reduce, transfer, share, mitigate, or compensate.	Risk-cost-benefit analysis, formal decision analysis for groups and individuals, risk quantification and comparison.

The *hazard identification* step deals with how to establish cause-and-effect relations from data. *Exposure assessment* quantifies the changes in exposures caused by alternative interventions, while dose-response modeling (or exposure-response modeling) quantifies the causal relation between changes in exposures and the probable resulting changes in adverse consequences. Finally, *risk characterization* integrates the preceding components to predict the probable changes in health caused by risk management actions that change exposures.

Health risk assessment uses available facts, data, and models to estimate the health risks to individuals, to an entire population, and to selected subpopulations (e.g., infants, the elderly, immunocompromised patients, and so forth) caused by hazardous exposures, and the decisions and activities that create those exposures. The health risks of sporadic illnesses due to exposure to chemicals, radiation, bacteria, or other hazards are measured quantitatively by the changes in the *frequencies and severities* of adverse health effects caused by the exposures.

Quantitative Definition of Health Risk

For sporadic illnesses (as opposed to epidemics), individual and population health risks can be defined as follows:

- The *individual risk* of sporadic illnesses (or accidents, injuries, or other adverse outcomes) caused by an exposure can be represented by the *frequency and severity of additional adverse health effects per capita-year caused by that exposure.* It can often be tabulated or plotted as the expected number of cases per capita-year in each severity category – e.g., mild, moderate, severe, or fatal, as defined in Buzby et al. (1996) based on illness-days and mortality. To avoid having to carefully define, describe, and compare the severities of different illnesses, one can simply use days of illness per year for each category of illness (e.g., mild, moderate, or severe) to summarize morbidity impacts, perhaps broken down by different age groups or other population subgroups. Alternatively, the loss due to increased mortality and morbidity can be expressed in terms of quality-adjusted life-years (QALYs), which can serve as a single summary measure of severity if the required preference-independence conditions justifying QALYs are accepted (Hazen, 2003; Miyamoto, 1999). Individual risk is then given by the joint probability distribution of the number of cases per capita per year and the associated severities (i.e., QALYs lost per case).
- *Population risks* are the sum of individual risks over all person-years in the population. They can be expressed as *numbers of additional adverse health effects per year* (of each type or severity category) occurring in the population. Population risks can also be further characterized by identifying subpopulations with especially high individual risks.

Technical note: Use of expected values. Using the expected number of events per year to quantify risk is justified for sporadic illnesses that occur independently, or with only weak statistical dependence, in large populations, when the Poisson approximation (Janson, 1994) or the compound Poisson approximation (Barbour et al., 1995) holds. The expected number of cases per year then determines the full probability distribution of the number of illnesses per year, to a close approximation (made precise in the above references). Moreover, the Poisson probability distribution is stochastically increasing in its mean; thus, larger numbers of expected cases correspond to less preferred distributions for *all* decision makers who prefer fewer cases per year to more. The formulas *individual risk* = *expected number of additional illnesses per year* × *expected QALYs lost per illness* and *population risk* = *sum of individual risks* are useful for sporadic illnesses, although they must be generalized for other types of risks, e.g., to allow for risk aversion (Cox, 2001).

The main goals of risk assessment are to produce information to improve risk management decisions by *identifying and quantifying valid cause-effect relations-between alternative risk management decisions and their probable total human health consequences*, and by *identifying decisions that make preferred outcomes more likely.* Health risk assessments typically use explicit – and, if possible, validated – analytic models (e.g., statistical, biomathematical, or simulation models) of causal relations between actions and their probable health effects. In general, quantitative risk assessment applies specialized models and methods to quantify likely exposures and the frequencies and severities of their resulting consequences.

Example: Statistical and Causal Risk Relations May Have Opposite Signs

As illustrated by the following (perhaps counterintuitive) example, *there is no necessary relation between statistical exposure-risk associations and the change in risk that would be caused by changing exposure.* As a simple counterexample, consider a hypothetical population in which 100% of men and 0% of women are exposed (i.e., *exposure* = 1 for men, *exposure* = 0 for women). (Perhaps thinking of exposure as drinking beer while eating fried chicken and watching Monday night football will aid intuition.) Suppose that *risk* = 0 for all women, *risk* = 100% for unexposed men, and *risk*= 10% for exposed men. In this example, exposure *reduces* risk, yet the statistical association between them is *positive*. The *statistical* relation between exposure and risk in this population is

$$risk = 0.1 \times exposure.$$

That is, when *exposure* = 1, *risk* = 10% (for exposed men), and when *exposure*= 0, *risk* = 0 (for unexposed women). Yet the *causal* effect of reducing *exposure* is to *increase* risk in the population, by shifting men from the lower-risk exposed group to the higher-risk unexposed group. The causal relation between *exposure* and *risk* in this population is thus

$$risk = 1 - 0.9 \times exposure \text{ for men; } risk = 0 \text{ for women.}$$

Thus, the statistical and causal relationships between *exposure* and *risk* have opposite signs.

As discussed further in Chapter 5, fitting a simple reduced-form statistical model (such as $risk = 0.1exposure$) to data does *not* in general allow one to correctly predict the effects of changing the independent variables on resulting changes in the dependent variable (Shipley, 2000; Freedman, 2004). [This example is motivated by empirical relations found in a real data set collected by the CDC (Friedman et al., 2000) for the foodborne bacterial pathogen *Campylobacter*. Men *do* appear to have greater susceptibility to campylobacteriosis than women; they *do* appear to have greater exposure to risk factors such as eating undercooked meat in restaurants and swimming in untreated water; and exposure to chicken (e.g., buying and handling raw chicken, preparing and eating chicken at home, etc.) *does* appear to reduce the risk of campylobacteriosis, for both sexes. The above counterexample exaggerates these empirical patterns to extremes to provide a simple illustration of the disconnect between statistical and causal relations.]

A Bayesian Network Framework for Health Risk Assessment

To support effective risk management decisions, human health risk assessments must characterize known or suspected potential causal relations between risk management actions (including the status quo or "do-nothing" option), on the one hand, and probable resulting human health consequences on the other. Actions typically affect exposures to sources of risk (i.e., hazards), while consequences typically include changes in the frequency or severity of resulting illnesses or deaths in affected populations. Hazard identification identifies causal relations (possibly including causal paths) leading from risk management actions to their human health consequences. Hazard identification often precedes any plan to develop a risk management strategy, as effective risk management is often impossible if causal relations are not understood.

Figure 3.1 outlines a causal graph (Shipley, 2000; Greenland and Brumback, 2002; Ellis and Wong, 2008) for assessing risks to humans from changes in exposures to hazards. In this template, risk management *actions* can change *exposures* of individuals to potentially harmful agents (the hazards). Changes in exposures, in turn, change expected *illness rates* and hence adverse *health consequences* (e.g., illness-days or early deaths per capita-year) in susceptible members of the exposed population. If desired, different human health consequences can be aggregated into a single summary measure, such as quality-adjusted life-years (QALYs) if the required preference conditions hold (Hazen, 2003), but this is optional. The effects of such changes on the number of QALYs lost per year in the population can be mediated by individual behaviors or attributes (e.g., immune status, age, gender, diet, and other covariates that affect susceptibility to infections). These covariates may also influence each other (indicated by the brackets [] around them in Fig. 3.1. For example, an AIDS patient may have food consumption and

Fig. 3.1 A causal graph for health risk analysis

Act → ΔExposure → ΔIllnesses → Δ**Consequences** →ΔQALYs

↑ ↑ ↑

[Behavior Susceptibility Treatment] = **type** of case

preparation behaviors and medical treatments that differ from those of a non-AIDS patient. Risk assessment helps to identify risk management options (acts) that decrease adverse health consequences, taking into account the distribution of covariates in the population.

Technical note: Influence diagram interpretation. Figure 3.1 can be interpreted as a Bayesian belief network or causal graph model (Greenland and Brumback, 2002; Chang and Tian, 2002). In this framework, each variable to which arrows point is interpreted as a random variable with a conditional probability distribution that is completely determined by the values of the variables that point into it. Because this diagram has a decision node ("act") and a value node ("ΔQALYs"), it is an example of an *influence diagram* (Owens et al., 1997). Important details are represented only implicitly, by conditional probability distributions. Algorithms to identify possible causal graph structures from data (and hence to test whether hypothesized causal theories are consistent with data) have been developed (e.g., Tsamardinos et al., 2003; Wood 2006; Silander and Myllymäki, 2006; Ellis and Wong, 2008) but are not yet routinely applied in risk assessment. Such causal graph models are useful because effective algorithms have been created to (a) quantify the conditional probability distributions of any subset of their variables, given observed values of the rest; and (b) solve for the act that maximizes expected utility (once a utility function has been defined for outcomes such as ΔQALYs) (see Crowley, 2004, www.cs.ubc.ca/~crowley/academia/papers/aiproj.pdf).

Each choice of a risk management act in Fig. 3.1 generates a corresponding random number of incremental illness cases ("responses") caused or prevented each year in each severity class of consequences (e.g., mild, moderate, severe, fatal) in the population (and in each subpopulation, if there are several). The expected health consequences of this change can be calculated from the following three submodels, which are common to most risk assessments:

- An *exposure model* (the "act → Δexposure" link in Fig. 3.1) that quantifies the units of exposure received per unit time for individual risks.
- A *dose-response* or *exposure-response* model (the "Δexposure → Δillnesses" link in Fig. 3.1) that quantifies the probability of illness, or the expected number of cases, at each given severity level, per unit of exposure. In general, this relation may depend on the individual's "type" (i.e., on the combination of covariate values that influence risk for that individual), as well as on the dose (units of exposure) received.
- A *health consequence model* (the "Δillnesses → Δconsequence" link in Fig. 3.1) quantifying the conditional probabilities of different health outcomes (e.g., survival vs. fatality, or number of QALYs lost) from each case. These outcome

probabilities may depend on factors such as physician prescription behavior or hospital infection-control standards, for example.

These three submodels determine the expected illnesses and QALYs lost per year in each severity class for each act. Multiple exposure pathways and at-risk populations (perhaps including groups receiving different medical treatments) can be included to quantify the *total* human health impact of different acts. Summing health impacts over all distinct combinations of hazards, exposure routes, and target populations (each corresponding to an instance of Fig. 3.1) gives the total probable change in human health consequences for the act.

Technical note: Monte Carlo simulation. If there are too many combinations of hazards, exposure routes, and target populations for explicit summation over all of them to be practical, then Monte Carlo simulation can be used to obtain accurate numerical approximations of the average risk (and the distribution of health effects). For example, suppose that risk is given by $f(x_1,x_2,\ldots, x_n) = f(\boldsymbol{x})$ and that one can sample from the joint probability density function (PDF) of the x_i, $\Pr(x_1,x_2,\ldots, x_n) = \Pr(x_1)\Pr(x_2 | x_1)\ldots \Pr(x_n | x_1, \ldots, x_{n-1})$. Then Markov chain Monte Carlo (MCMC) simulation techniques such as Gibbs sampling (Andrieu et al., 2003; Lange, 2003) can be used to generate random samples from the joint PDF of $\boldsymbol{x}$. Taking a simple arithmetic average of the values of $f(\boldsymbol{x})$ obtained for a sufficiently large random sample of $\boldsymbol{x}$ values will give an accurate estimate of the true average risk $E_{\Pr(\boldsymbol{x})}[f(\boldsymbol{x})]$ implied by $f(\boldsymbol{x})$ and $\Pr(\boldsymbol{x})$. Commercial risk analysis software tools such as Analytica^{TM}, @RISK^{TM}, and Crystal Ball^{TM} include Monte Carlo simulation routines that can generate estimated means, confidence bands, and entire estimated probability distributions for $f(\boldsymbol{x})$. Vose (2000) provides a basic introduction to Monte Carlo simulation in spreadsheet models for microbial risk assessment, and Cassin et al. (1998) discuss how to use Monte Carlo simulation for tasks such as priority setting and risk management.

The conceptual framework in Fig. 3.1 can be implemented with greater or lesser degrees of sophistication. Perhaps the simplest approach is to generate point estimates for each risk management act and exposure pathway for each of the following:

- *Exposure factor* = units of exposure received per capita per year;
- *Dose-response* factor = expected cases of illness per unit of exposure;
- *Health consequence factor* = expected QALYs lost (or illness-days created, etc.) per case of illness. (Alternatively, a vector of expected numbers of different health outcomes can be estimated; e.g., mild, moderate, severe, and fatal outcomes per case.)

In this case, each submodel (corresponding to a horizontal arrow in Fig. 3.1) is represented by a single number. One can then multiply these numbers together, and multiply by the number of people affected, for each causal path and each risk management action. (Note that causal paths need not be restricted to different exposure paths, but may encompass all three links.) Summing the results over all causal paths provides an estimate of the total human health impact per year for each action. A more refined calculation can be made by considering how these factors might change over time and then summing over time periods (perhaps with discounting).

At the other end of the spectrum, Fig. 3.1 can be applied to risk estimation using conditional probability algorithms developed for Bayesian networks and causal graphs (Chang and Tian, 2002). In this case, *hazard identification* can be thought of as identifying instances of Fig. 3.1 that are consistent with available data. Statistical methods are available to test whether specified causal graph models are indeed consistent with data (Greenland and Brumback, 2002; Shipley, 2000), and practical algorithms have been developed to identify potential causal graph models from multivariate data (Aliferis et al., 2003; Tsamardinos et al., 2003; Wood, 2006; Silander and Myllymäki, 2006; Ellis and Wong, 2008). The remaining steps in the risk assessment process can then be interpreted as quantifying and applying the resulting Bayesian network. Multiplying exposure, dose-response, and consequence factors generalizes to combining arbitrary probability distributions for inputs, and conditional probability relations, via Monte Carlo simulation (Andrieu et al., 2003) to derive joint probability distributions of outputs.

Bayesian network methods, combined with objective statistical tests for potential causality, such as conditional independence tests (Shipley, 2000; Greenland and Brumback, 2002), appear promising for providing more effective, data-driven risk assessments while also allowing for the use of expert judgment when necessary.

Hazard Identification

Risk assessment begins with *hazard identification,* the process of specifying the scope of the assessment and summarizing the available empirical evidence that exposure to a specific "hazard" causes specified adverse health effects in exposed individuals or populations. Thus, hazard identification can serve to

1. *Rapidly screen potential hazards* by identifying whether available data support the hypothesis that the hazard might cause specific adverse health effects (possibly using formal statistical methods of causal analysis; e.g., Shipley, 2000, Ellis and Wong, 2008) .
2. *Identify causal relations between specific hazards and specific adverse human health effects.*
3. *Identify risk factors, behaviors, and exposure conditions that increase risks to specific exposed populations* (e.g., the old, the young, the immunocompromised, etc.).
4. *Summarize empirical evidence both for and against the hypothesis that exposures to specific hazards cause specific adverse human health effects* (Patton, 1993) .

In reality, of course, *joint causation* is common; i.e., observed adverse consequences are often due to a combination of a hazardous agent, activities resulting in exposures to that agent, failure to undertake protective actions, and possibly other confounding factors (such as decreased immunity in a subpopulation). In general, *any* event or condition that hastens the occurrence of an adverse effect or increases its likelihood can be viewed as a contributing "cause" of the effect; for more on the philosophical definition and ambiguities of "causation," see Williamson (2005). Thus, "the cause" of an adverse health effect is often not uniquely defined. Nonetheless, for purposes of risk management, it is often adequate to predict the effects of alternative risk management interventions on the rates of adverse events of different severities. Hazard identification helps to identify such interventions (typically, reductions in exposure).

Table 3.2 outlines steps for forming and testing causal hypotheses about exposure-response relations using epidemiological data. As more of these steps are completed, the empirical support increases for a causal relation between exposure and risk. Most statistical methods in epidemiological risk analysis focus on steps 1–3, i.e., identifying nonrandom associations, and then eliminating potential biases and confounders as likely explanations. These steps can often be carried out using observational data without experimental controls using the *refutationist approach* (Maclure, 1990, 1991), namely, by systematically enumerating possible competing explanations for the observed data, and eliminating each of those potential explanations (if possible) using statistical tests on the available data.

Many epidemiologists have recognized that, to draw valid causal inferences, it is necessary to refute competing (noncausal) hypothesized explanations for observed exposure-response associations (Maclure, 1990, 1991). Table 3.3 summarizes common competing explanations (mainly, confounding and/or sampling, information, or modeling biases) and some suggested statistical methods to refute them (Cox, 2001, Chapter 3).

Table 3.2 Steps to establish a causal exposure-risk relation

a. *Identify a statistically significant exposure-response association,* e.g., using case-control, prospective cohort, or other cross-sectional or longitudinal epidemiological data.
b. *Eliminate confounding* as a possible explanation of the association by accounting for factors such as lifestyle, age, or exposure to other hazards, e.g., using conditional independence tests (Grimes and Shulz, 2002; Feldman, 1998; Greenland and Morgenstern, 2001).
c. *Eliminate biases in sampling, information collection, and modeling choices* as possible explanations for the association (Choi and Noseworthy, 1992; Deeks et al., 2003).
d. *Test and confirm hypothesized causal and conditional independence relations*, for example, by showing that the response is *not* conditionally independent of the hypothesized exposure that causes it, given other variables (Shipley, 2000; Friedman and Goldszmidt, 2006; Frey et al., 2003).
e. *Confirm the efficacy of interventions*, e.g., by experimental manipulations (Ellis and Wong, 2008) and/or intervention and change point analyses of time-series data (e.g., Swanson et al., 2001; Green, 1995).
f. *Identify and elucidate causal mechanism(s)*, identified from experimental data and/or from generally accepted principles.

Table 3.3 Potential noncausal explanations for associations and some statistical methods to overcome them

Potential noncausal explanations	**Statistical methods to refute potential noncausal explanations (See http://cran.r-project.org/ for more on statistical methods and R software.)**
Modeling biases	
Variable selection bias (includes selection of covariates in model)	Bootstrap variable selection, Bayesian model averaging (BMA), and cross-validation for variable selection (Wang et al., 2004).
Omitted explanatory variables (including omitted confounders)	Include potential confounders in an explicit Bayesian network or causal graph model; test for unobserved latent confounders.
Variable coding bias (Coding of variables may affect risk estimates. See Streiner, 2002; Brenner and Loomis, 1994.)	Don't unnecessarily discretize continuous variables (Royston et al.,2005; Gustafson and Le, 2002; Ragland, 1992). Use automated variable-coding methods such as classification trees (see Chapter 6).
Aggregation bias/Simpson's paradox	Test hypothesized causal relations at multiple levels of aggregation, down to individual-level data.
Multiple testing/comparisons bias	Adjust p-values (Romano and Wolf, 2005).
Choice of exposure and dose metrics.	Use multiple exposure indicators as explanatory variables (e.g., concentration and time. Don't combine them.).
Choice of response/effect metrics	Use survival functions and transition rates among health states.
Model form selection bias; uncertainty about correct model	Use flexible nonparametric models (e.g., kernel smoothers, wavelets) and BMA for multiple models. Report model diagnostics and sensitivities of results to model forms (Greenland, 1989).
Missing data (Little and Rubin, 1987; Horton and Kleinman, 2007)	Use data augmentation, EM, multiple imputation (Harrell, 2007), or Markov chain Monte Carlo (MCMC) algorithms (Schafer, 1997).
Measurement and misclassification errors in explanatory variables	Use Bayesian measurement error models; treat unknown true values as missing data (Schafer, 1997; Ibrahim et al., 2005); use bias-correction formulas, regression-calibration, instrumental variables, simulation-extrapolation (SIMEX) corrections (Carroll et al., 2006), or Bayesian smoothing and regression splines (Berry et al., 2002).
Unmodeled heterogeneity in individual response parameters	Use latent variable models, finite mixture distribution models, or frailty models of interindividual variability (http://cran.r-project.org/).
Biases in interpreting and reporting results	Report results (e.g., posterior PDFs) *conditioned* on data, models, and statistical methods. Show sensitivities.

Table 3.3 (continued)

Potential noncausal explanations	Statistical methods to refute potential noncausal explanations (See http://cran.r-project.org/ for more on statistical methods and R software.)
Sample selection biases	
Sample selection (sample does not represent population)	Randomly sample *all* cohort members if possible.
Data set selection bias (i.e., selection of studies may affect results)	Conduct meta-analysis of sensitivity of conclusions to studies. Use causal graph models to integrate diverse data sets.
Health status confounding, hospital admission/referral bias	If possible, use prospective cohort design and population-based cases and controls (Choi and Noseworthy, 1992).
Selective attrition/survival (e.g., if exposure affects attrition rates) Differential follow-up loss	Use a well-specified cohort. "Include non-surviving subjects in the study through proxy interviews" (Choi and Noseworthy, 1992). Compare counterfactual survival curves.
Detection/surveillance bias	Match cases to controls (or exposed to unexposed subjects) based on cause of admission.
Membership bias (e.g., lifestyle bias, socioeconomic history)	• In cohort studies, use multiple comparison cohorts. • Hard to control in case-control studies.
Self-selection bias; response/volunteer bias	Achieve a response rate of at least 80% by repeated efforts. Compare respondents with sample of nonrespondents.
Information collection biases	
Intra-interviewer bias	Blind interviewers to study hypotheses, subject classifications.
Interinterviewer bias	Use same interviewer for study and comparison groups.
Questionnaire bias	Mask study goals with dummy questions; avoid leading questions/response options.
Diagnostic suspicion bias Exposure suspicion bias	Hard to prevent in case-control studies. In cohort studies, make diagnosis and exposure assessments blind to each other.

Example: Some Traditional Criteria for Causality Fail to Refute Other Explanations

Several traditional criteria (most often identified with Sir Austin Bradford Hill) that are now widely used in epidemiology to evaluate evidence for causality actually are not necessary, are not sufficient, and are not always useful as indicators of a possible causal relation. To illustrate how traditional criteria can be irrelevant to establishing evidence of causation, suppose that an analyst tries to use them to assess the evidence for this hypothesis:

$$A \rightarrow B,$$

interpreted as "*A* causes *B*," or as "increasing *A* will increase *B*," where *A* and *B* represent the exposure and response variables, respectively, in a large population. Suppose that the correct causal relations (unknown to the analyst) are

$$A \leftarrow C \rightarrow B,$$

meaning "changes in *C* cause changes in *A*, and changes in *C* also cause changes in *B*, but directly changing either of *A* or *B* does not change the other." Finally, suppose that changes in *C* propagate quickly to change *A*, and more slowly to change *B*. Then the observed association between *A* and *B* (due to confounder *C*) can satisfy all of the following traditional criteria:

- *Strength* of association (For example, there may be up to 100% correlation between levels of *A* and levels of *B*, if both are directly proportional to *C*.)
- *Consistency* (Different observers using different methods in different places can observe that *A* is positively associated with *B*.)
- *Specificity* (High levels of *B* may be found only in cases with high levels of *A*. If *C* is not known or understood by the investigator, hence not considered as an explanation, then such specificity may be misconstrued as supporting the hypothesis that high levels of response *B* are specifically caused by high levels of exposure *A*.)
- *Temporality* (Changes in *B* follow changes in *A*.)
- *Biological gradient* (if both *A* and *B* are increasing functions of *C*, and hence *B* is an increasing function of *A*)
- *Plausibility* (if the investigator thinks that $A \rightarrow B$ is a plausible hypothesis)
- *Coherence*, *experiment*, and *analogy* (if the investigator thinks that $A \rightarrow B$ is a plausible hypothesis because of an analogy with some other – as it turns out, irrelevant – *in vitro* results)

In other words, all of these traditional criteria for concluding that there is evidence that "*A* causes *B*" can be thoroughly satisfied, even if *A*, in fact, does not cause *B*.

This example used just one competing explanation (a hidden confounder, *C*) that explains away all of the "evidence" for causation between *A* and *B* considered by these criteria. Other competing explanations (e.g., model misspecification bias or multiple testing bias) can also explain away such "evidence," even if there are no confounders. Hence, better methods are needed for identifying potential valid causal exposure-response relations from epidemiological data. Chapters 6 and 8 discuss more recent methods that are far more useful than traditional criteria for identifying possible causal relations.

As stated by Savitz et al. (1990), "Biases that challenge a causal interpretation can always be hypothesized.... It is essential to go beyond enumerating scenarios of bias by clearly distinguishing the improbable from the probable and the important from the unimportant." Fortunately, well-developed statistical methods and algorithms are now available to identify significant statistical associations from data on the spatial and temporal associations between exposures and health effects (Mather et al., 2004) and to screen them for potential causality based on the above criteria.

Technical note: Statistical tests for assessing potential causality. Over the past 40 years, intuitive criteria for causality used in epidemiology (such as the Bradford Hill considerations, emphasizing the strength, consistency, biological gradient, coherence, etc. of an association) have been improved upon by more objective statistical approaches. For example, an approach based on information theory (see Chapter 6) proposes that, roughly speaking, a data set provides evidence that exposure variable X is a *potential cause* of response variable Y if and only if X is (a) INFORMATIVE about Y, i.e., the mutual information between X and Y, denoted by $I(X; Y)$ and measured in bits (see Chapter 6), is positive in the data set (this allows for nonlinear and even nonmonotonic relations); (b) UNCONFOUNDED: X provides information about Y that cannot be removed by conditioning on other variables, i.e., $I(X; Y \mid Z) > 0$ for all subsets of variables Z; (c) PREDICTIVE: Past values of X are informative about future values of Y, even after conditioning on past values of Y. (This generalizes the concept of *Granger causality* for time series, e.g., Guatama and Van Hulle, 2003.) (d) CAUSALLY ORDERED: Y is conditionally independent of the parents of X, given X, i.e., $I(P\ ; Y \mid X) = 0$, for any parent or ancestor P of X (Ellis and Wong, 2008). These criteria yield practical algorithms (e.g., Bayesian Lab^{TM}, Tsamardinos et al., 2003; Wood et al., 2006; Silander and Myllymäki, 2006) for detecting potential causation in cohort, case-control, and time-series data sets, even if the functional relations involved are nonmonotonic. (Causation may be present even if these conditions are not satisfied, but then the data do not provide evidence of it.) Formal tests for statistically significant associations between the timing of one event (e.g., introduction or cessation of exposures) and subsequent changes in a series of measurements (e.g., human illness rates in a surveillance program) can be based on *intervention analysis* and *change point analyses* (Green, 1995) for time series. These methods for testing for potential causality are entering common biostatistical and risk analysis practice only slowly, but appear to be very promising (Shipley, 2000).

Exposure Assessment

For environmental risk assessment, U.S. EPA experts have stated that "Questions raised in the exposure analysis concern the likely sources of the pollutant ... its concentration at the source, its pathways (air, water, food) from the source to target populations, and actual levels impacting target organisms" (Patton, 1993).

Similarly, the U.S. FDA has defined exposure assessment as "a component of a risk assessment that characterizes the source and magnitude of human exposure to the pathogen." The magnitude of human exposure, also called the dose, is defined as "the amount or number of a pathogen that is ingested or interacts with an organism (host)" (http://www.foodsafety.gov/~dms/lmriskgl.html).

Exposure assessment seeks to identify exposed subpopulations at risk from exposures to hazards, and to identify conditions leading to high-risk exposures. It also describes the extent of exposures (frequency and magnitude of individual exposure in the population in relation to susceptibility and covariates) and uses models to predict how risk management decision options will probably affect exposures. A successful exposure assessment should describe the frequency distribution of exposures received by members of exposed populations and subpopulations, and should show how these distributions change for different risk management decisions. The descriptions should contain enough detail to discriminate among different exposure distributions that would cause significantly different health outcomes. This information is used, together with dose-response information, in risk characterization.

The shape of the frequency distribution of exposures relative to the dose-response relation (e.g., how frequent are exposures that are likely to cause illness?) drives quantitative risk. It is common for exposures to be very uncertain, especially if they depend on unmeasured and/or highly variable processes. The exposure assessment influence diagram may then look like this:

$$\begin{array}{l}\text{Act} \rightarrow \text{ exposures} \rightarrow \text{illnesses} \leftarrow \text{individual covariates}\\ \qquad\qquad\quad\downarrow \\ \text{measured exposure surrogates.}\end{array}$$

For example, available data may consist of surrogate measurements (e.g., contaminant levels in exposure pathways) rather than direct measurements at the point of exposure. True exposures then play the role of *latent variables* in causal modeling, i.e., they affect observed outcomes but are not observed themselves. Appropriate statistical techniques for causal diagrams with latent variables (e.g., Shipley, 2000, for linear models; Pearl, 2002, and Hartemink et al., 2001, for more general Bayesian network models) can be applied to the above diagram with surrogate measurements of exposure for data. Software such as WinBUGS helps to automate the required computations for inference with missing data and unobserved or surrogate variables.

Exposure models describe the transport and distribution of hazardous materials through different media and pathways (e.g., air, foods, drinking water) leading from their source(s) to members of the exposed population. In addition, exposure models may consider the distribution over time of human populations among locations and activities that result in exposures. Simulation models of transport and behavioral processes, often developed using discrete-event simulation software, are used to estimate frequency distributions of population exposures from assumptions about, or submodels of, the more detailed microprocesses involved.

Example: Simulation of Exposures to Pathogens in Chicken Meat

The World Health Organisation (WHO) has described a process simulation model of human exposures to the foodborne pathogen *Salmonella* as follows:

> The exposure assessment of *Salmonella* in broiler chickens mimics the movement of *Salmonella*-contaminated chickens through the food chain, commencing at the point of completion of the slaughter process. For each iteration of the model, a chicken carcass was randomly allocated an infection status and those carcasses identified as contaminated were randomly assigned a number of *Salmonella* organisms. From this point until consumption, changes in the size of the *Salmonella* population on each contaminated chicken were modeled using equations for growth and death. The growth of *Salmonella* was predicted using random inputs for storage time at retail stores, transport time, storage time in homes, and the temperatures the carcass was exposed to during each of these periods. Death of *Salmonella* during cooking was predicted using random inputs describing the probability that a carcass was not adequately cooked, the proportion of *Salmonella* organisms attached to areas of the carcass that were protected from heat, the temperature of exposure of protected bacteria, and the time for which such exposure occurs. The number of *Salmonella* consumed were then derived using a random input defining the weight of chicken meat consumed, and the numbers of *Salmonella* cells in meat as defined from the various growth and death processes. Finally, in the risk characterization, the probability of illness was derived by combining the number of organisms ingested (from the exposure assessment) with information on the dose-response relationship (hazard characterization). (www.who.int/foodsafety/publications/micro/Salmonella/en/)

The results of the Monte Carlo simulation exposure modeling are presented as (a) an estimated 2% prevalence of contaminated chicken servings, and (b) the following conditional frequency distribution for the dose (CFUs) per serving from contaminated servings:

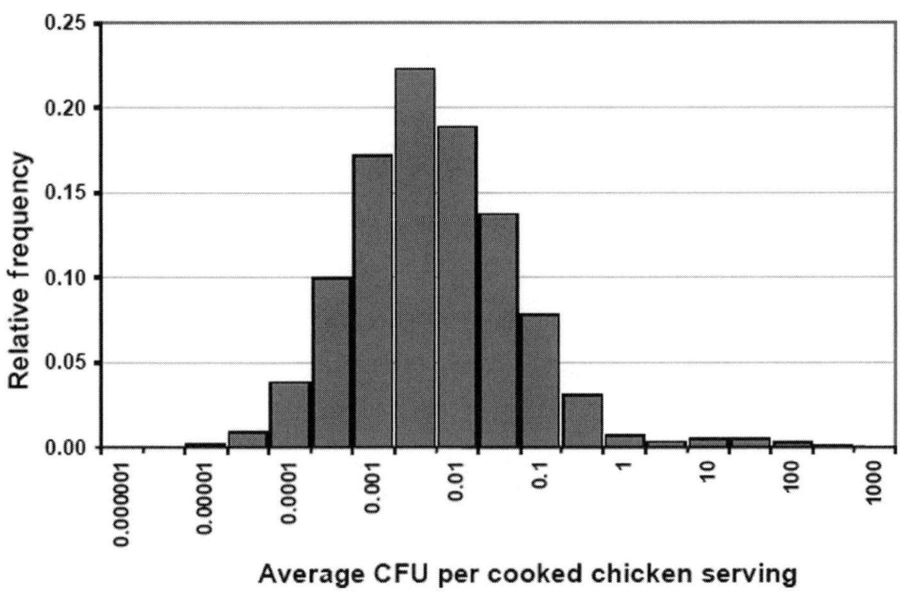

Source: http://www.who.int/foodsafety/publications/micro/*Salmonella*/en/

This frequency distribution shows how large an exposure a person is likely to receive from a serving of contaminated, undercooked broiler chicken. This is the main output of the exposure assessment and the main input to the dose-response model for calculating illness risk per serving.

Example: Mixture Distributions and Unknown Dose-Response Models

Exposure-response relations in a population can often be estimated even when the shapes of dose-response relations are unknown, by decomposing the risk as follows:

$$\Pr(\text{illness}|\text{exposure} = x) = \Sigma_r \Pr(\text{illness}|\text{exposure} = x$$
$$\& \text{ response type} = r)^* \Pr(\text{response type} = r),$$

where "response type" is an unobserved (latent) variable summarizing all of the missing information needed to predict the probability of illness from a known level of exposure. (For example, if each individual has an unknown threshold number of bacteria that must be ingested in one meal to cause illness, then r would be that threshold number. If there is a continuum of response "types," the above sum is replaced by an integral.) An important development in mathematical statistics is the recognition that the uncertain quantities Pr(response type = r) can be interpreted as *statistical coefficients* to be estimated directly from data on the aggregate number of responses observed in populations for different exposure conditions, while the conditional response probabilities that are paired with these coefficients, Pr(illness | exposure = x, type = r), can be estimated simultaneously from the same data (provided that technical identifiability conditions are met. These are automatically satisfied by many large families of distributions.). The required statistical technology is that of *finite mixture distribution models* if the number of types is finite, or continuous mixture models if types are continuous. Well-developed computational Bayesian algorithms can be applied to estimate the number of components in the mixture (i.e., the number of statistically significantly different "types") and the corresponding coefficients and conditional response probabilities (see, e.g., Richardson and Green, 1997). Note that in this construction, the definition of the exposure variable x can be any measured quantity that can be paired with corresponding illness rates. All unobserved details of subsequent processing, handling, preparation, etc. are then absorbed into the latent "type" variable, r. Missing values and errors in measured values of x can also be handled within the computational Bayesian framework (e.g., using the data augmentation algorithm; Schafer, 1997) to allow the conditional distributions of outputs given observed data to be quantified, even when other data are missing. There is thus great flexibility to use all available data (via conditioning) but without requiring the use of unavailable data.

Dose-Response Modeling

Dose-response models quantify the conditional probability of illness caused by each level of exposure; thus, the term *exposure-response model* is also appropriate. Figure 3.2 shows an example of a dose-response model developed for *Listeriamonocytogenes* in ready-to-eat foods. A specific parametric dose-response model was assumed (an exponential model) and fit to epidemiological data for immunocompromised ("High risk") and non-immunocompromised ("Normal") subpopulations. The dark solid curve in Fig. 3.2 is the estimated dose-response model for the "Normal risk" subpopulation. The dashed line above and to the left of it is the dose-response model for the "High risk" subpopulation. The lighter gray curves indicate estimated statistical confidence bands around these best-estimate curves – an upper confidence band for each (corresponding to the upper end of the 95% confidence interval estimated for the parameter of the exponential dose-response model), and a lower 95% confidence band for the rightmost (Normal) dose-response model.

As in Fig. 3.2, it is often necessary to fit separate dose-response models to "normal" and "susceptible" subpopulations within the general population to account for interindividual variability in dose-response relations. While more than two gradations of susceptibility can potentially be modeled using finite mixture distributions, distinguishing between only two levels or response "types" in the population, i.e., susceptible and normal, often suffices to explain most of the variability in the data. If different degrees or severities of illness are distinguished, ranging from mild through severe to fatal (Buzby et al., 1996), then a *health consequence model* describing the conditional probabilities of different levels or severities of health outcomes, given

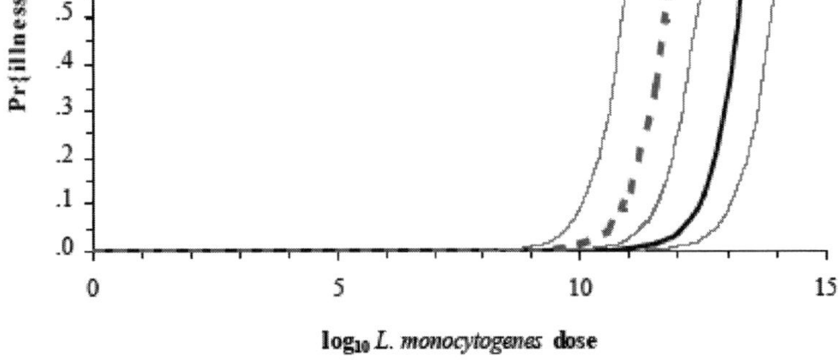

Fig. 3.2 Example dose-response function for *Listeriamonocytogenes*
Source: FAO/WHO, 2001. http://www.who.int/foodsafety/publications/micro/en/may2001.pdf

that illness occurs, is needed to augment the conditional probability of illness as a function of exposure. In general, risk characterization requires describing the severities as well as the frequencies of adverse health outcomes caused by exposures.

In practice, biologically motivated parametric dose-response models are the most common, and usually the best justified, models in widespread use. They are typically fit to data by a combination of maximum-likelihood estimation (MLE) for point estimates and computationally intensive resampling techniques (e.g., bootstrapping algorithms) for confidence intervals, simultaneous confidence bands around the dose-response curve, and joint confidence regions for model parameters (e.g., Haas et al., 1999, Chapter 7, cf. p. 293).

Example: Apparent Thresholds in Cancer Dose-Response Data

Table 3.4 illustrates some of the challenges of selecting useful exposure or dose metrics and dose-response models for predicting cancer risks. It shows several apparent carcinogenic thresholds in experimental data for mice exposed via inhalation to isoprene. Liver and lung adenomas and carcinomas are only significantly elevated at concentrations of 140 ppm and above. Similarly, histiosarcomas exhibit an apparent response threshold between 140 and 280 ppm. These thresholds are specific to concentration, rather than to cumulative exposure. For example, doubling concentration (from 70 to 140 ppm between exposure groups 3 and 5) increases liver adenomas from 0.29 to 0.44, whereas doubling weeks of exposure (from 40 to 80 between exposure groups 3 and 4) does not increase the risk significantly at any site, and even appears to reduce it. Similarly, quadrupling exposure concentration from 70 to 280 ppm while quartering exposure duration from 80 to 20 weeks unambiguously

Table 3.4 Results of a stop-exposure experiment for isoprene in male B6C3F1 mice

Group	ppm	Weeks	Hr/ day	Liver adenomas	Lung adenomas	Other adenomas	Liver carcinomas	Lung carcinomas	Histio- sarcomas
1	0	0	8	0.22	0.22	0.14	0.18	0	0
2	10	80	8	0.24	0.32	0.12	0.12	0.02	0.04
3	70	40	8	0.29	0.16	**0.30***	0.22	0	0.04
4	70	80	8	0.30	0.08	0.18	0.18	0.04	0.04
5	140	40	8	**0.44***	0.20	**0.28***	0.20	0.02	0.02
6	280	20	8	0.36	0.32	**0.36***	0.24	0.06	**0.16***
7	2,200	80	4	**0.42***	0.30	**0.56***	0.30	0.06	**0.14***
8	2,200	40	8	**0.57***	**0.59***	**0.65***	**0.37***	0.06	**0.14***

Explanation: Columns 2–4 summarize the exposures defining each dose group. The remaining columns show the fraction of animals in each dose group that were found to have each tumor type at necropsy.

Source: Cox LA Jr, Bird MG, Griffis L. Isoprene cancer risk and the time pattern of dose administration. *Toxicology*. 1996 Oct 28;113(1–3):263–72.

$*$ Tumor incidence rates in bold and marked with an asterisk are significantly greater than in the control group ($p < 0.05$ by Fisher's Exact Test).

increases tumor risk (compare exposure groups 4 and 6), even though cumulative exposures are identical.

Apparent carcinogenic thresholds or threshold-like nonlinearities can also exist for components of the exposure's duration. For example, at higher concentrations, doubling the hours per day of exposure while halving the weeks of exposure increases the risk of adenomas and liver carcinomas (compare exposure groups 10 and 11), even though the cumulative exposures remain identical.

Such threshold-like behavior for exposure concentrations and durations occurs for many chemical carcinogens. It requires explanations that go beyond the usual multistage stochastic transition dose-response models with linear-in-dose transition rates (see Chapter 11), since such models predict that risk changes smoothly with the concentration and duration of internal doses received by target tissues or cell populations. Physiologically based pharmacokinetic (PBPK) modeling shows that these internal dose attributes, in turn, are typically smooth (and, at low doses, approximately linear) functions of the concentration and duration of administered dose over the range of experimental data, even though the induction, depletion, or saturation of enzyme-mediated processes can introduce important nonlinearities at very high administered doses.

Chapter 2 introduced the concept of *phase transitions with sharp transition thresholds* as one possibility for how threshold-like nonlinearities can occur at low doses (even without nonlinearities in the PBPK component). Smooth, possibly linear no-threshold dose-response relations at the level of individual cells can, in principle, lead to the types of sharp transitions and threshold-type behaviors observed in some empirical dose-response relations, such as those in Table 3.4. Better understanding the apparent thresholds and other nonlinearities in pharmacodynamics – the study of how doses of agents affect cell populations and other physiological systems – is one of the current grand challenges of quantitative dose-response modeling.

Example: Best-Fitting Parametric Models May Not Fit Adequately

Figure 3.3 for *Salmonella* feeding trial data shows that even the best-fitting model in a certain class of parametric models (here, the approximate Beta-Poisson dose-response family, widely used in microbial risk assessment) may not adequately describe the observed data. The parametric family of models is then said to be *mis-specified* for the data, i.e., it is not appropriate for describing the empirical relation. In this example, the approximate Beta-Poisson model family is inappropriate for the data because even the best-fitting curve in the family dramatically underpredicts low-dose risks.

If the correct dose-response model is unknown and several models all provide adequate fits to the available data, multiple plausible models may be used to carry out the rest of the assessment. In this case, the analysis can be organized and presented as a *model uncertainty decision tree* in which different modeling choices correspond to different branches in the tree. The results of the risk analysis at the end

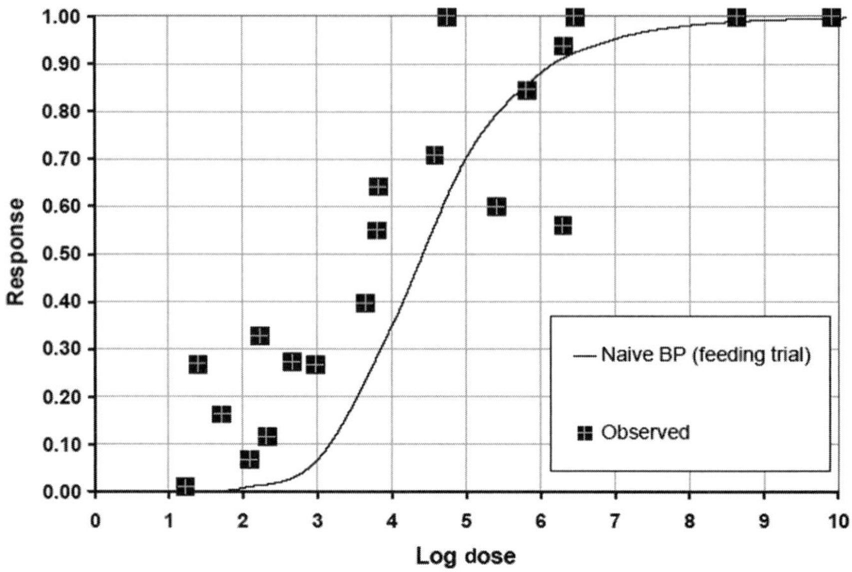

Fig. 3.3 The best-fitting Beta-Poisson model underpredicts low-dose risks
Source: WHO/FAO, 2002 (Naïve BP = approximate Binomial Poisson)

of each branch are contingent on the assumptions and modeling choices that lead to it. Different branches may be weighted by the relative strength of the evidence supporting them (Kang et al., 2000). Bayesian model averaging provides a similar way to combine predictions from multiple models (Viallefont et al., 2001; Keiding and Budtz-Jorgensen, 2004). Model uncertainty decision trees can also be used to present and analyze uncertainties due to choices of dose metrics, response definitions, and other modeling decisions, as well as choices of particular dose-response models.

Uncertainty about the illness probabilities caused by a given dose is often dominated by uncertainty about the most appropriate dose-response model (sometimes called structural uncertainty). A decision tree presentation of alternative modeling choices and the resulting predicted risks (or even a simple plot of different plausible dose-response curves) can express much of the relevant uncertainty with a minimal amount of statistical sophistication. Other important computational methods and algorithms for uncertainty analysis include

- *Monte Carlo uncertainty analysis* using commercial software products such as Analytica™, @RISK™, Crystal Ball™ (Vose, 2000). For more on uncertainty and sensitivity analysis software, see the descriptions at the product web sites.
- *Bayesian uncertainty analysis* for model parameters and predictions (e.g., based on the WinBUGS software for inference with missing data).
- *Bootstrapping and other resampling techniques* for estimating joint confidence regions for model parameters and predictions.

- *Model cross-validation* techniques for estimating the accuracy and prediction error characteristics of model predictions from performance on multiple subsets of data.

These methods are discussed in general computational statistics texts and for dose-response modeling in risk analysis texts such as Haas et al. (1999), Vose (2000), Cox (2001).

Risk and Uncertainty Characterization for Risk Management

Risk characterization is the ultimate output of a risk assessment. It integrates hazard identification, exposure assessment, and dose-response information to determine the probable frequency and severity of adverse health effects in a population caused by exposures to a hazard. Characterizing the change in risk for different risk management interventions helps decision makers choose among them. Risk characterization also includes the characterization of current *uncertainty about risk*. This allows the value of gathering additional information to be assessed as part of risk management deliberation and decision making, based on the potential value of such information (VOI) in enabling risk managers to make choices that are more likely to result in desired consequences (Yokota and Thompson, 2004).

Given the results from

(a) Exposure assessment (i.e., the conditional probability distribution of exposures, for each act),
(b) Exposure-response or dose-response modeling (i.e., the conditional probability of illness for each exposure pattern), and
(c) Consequence modeling (the conditional probability distribution of adverse consequences given illness),

the risk characterization step calculates, for each act being assessed, the resulting probability distributions for adverse consequences. This can be done by summing or integrating expressions such as Pr(consequence $= c \mid$ illness)Pr(illness $\mid$ exposure $= x$)*Pr*(exposure $= x \mid$ act) over all exposure levels x, to obtain the probability for each consequence, c.

Example: Risk Characterization Outputs

Figure 3.4 shows one of the risk characterization outputs from a risk assessment of *Listeria monocytogenes* (FAO/WHO, 2001). The solid curve shows the median estimate of the mortalities per year caused among the elderly subpopulation by *L. monocytogenes* in deli meats, for different maximum allowed storage times. The

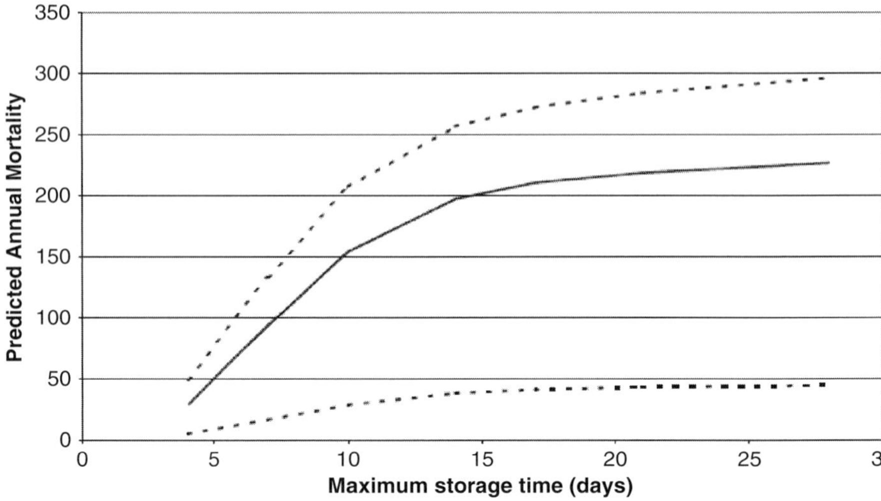

Fig. 3.4 Predicted annual mortality in the elderly subpopulation attributable to deli meats as a function of maximum storage time
Source: FAO/WHO, 2001, http://www.cfsan.fda.gov/~dms/lmr2-6.html

dotted curves represent the 5th and 95th percentiles of the uncertainty distribution, as assessed by Monte Carlo uncertainty analysis.

This display shows how predicted risks in this subpopulation vary with the effects of different potential interventions that would limit the maximum storage times allowed for deli meats. Similar curves can be shown for the effects of such interventions for other foods or groups of foods (e.g., dairy products, produce, seafood products, etc.) and for other subpopulations and the U.S. population as a whole.

Risk management is often viewed as a decision process that takes scientific information obtained from risk assessment as an input, along with value judgments and policy goals and constraints, and that recommends choices of risk management actions as its output. Risk management options typically include risk acceptance, prevention or avoidance (e.g., by reduction of exposures), mitigation of consequences (e.g., by appropriate clinical screening, diagnosis, and prescription procedures), transfer (e.g., health insurance), or compensation.

A successful risk analysis shows the estimated changes in the frequencies and magnitudes of adverse human heath consequences resulting from different risk management decision options. (Of course, if hazard identification and risk management reveal that the risk from the status quo is so small that no risk management action is needed, risk analysis may stop there. A full risk analysis is usually carried out only when a risk management intervention is being contemplated.) Risk analysis uses probability distributions, confidence intervals, and other displays to show uncertainties about the human health consequences of different decisions. It identifies a subset of one or more decision options leading to preferred (e.g., stochastically undominated) probability distributions of health risks and other outcomes.

A successfully completed risk analysis should allow a risk manager to answer the following questions for each risk management decision alternative being evaluated or compared:

- *What probable change in human health risk would result from each risk management intervention?* If the risk management decision option or action being assessed is implemented, how will the probable adverse human health effects [e.g., the expected numbers of mild, moderate, severe, and fatal illnesses per year; the expected numbers of illness-days and, if desired, quality-adjusted life-years (QALYs) lost per year] change in the whole population and in subpopulations with distinct risks?
- *How certain is the change in human health risk that would be caused by each risk management action?* Instead of a single value, i.e., a "point estimate" of risk, uncertain risks are characterized by intervals or probability distributions indicating how closely the change in human health risk caused by a proposed risk management intervention can be predicted. There are several technical options for expressing uncertainty around point estimates (e.g., plausible upper and lower bounds, confidence limits, coefficients of variation, tolerance intervals, prediction intervals, Bayesian posterior probability intervals and distributions, etc.) The essential information to provide about uncertainty in any risk assessment is how large or how small the true risks might be, consistent with the data and with the specified assumptions of the risk assessment. Point estimates that are "best" with respect to various technical statistical criteria will typically fall between these extremes.

Technical note: Statistical point estimates and interval estimates. Many criteria have been used to define and identify the "best" point estimates in risk models, e.g., maximum-likelihood estimates (MLE), maximum a posteriori (MAP) Bayesian estimates, maximum entropy, minimum description length, least squares, minimum absolute deviation, and minimum expected loss (for various loss functions) (see Cox, 2001, for a survey of methods for risk analysts). While these criteria have led to useful theory and algorithms for estimating the parameters of risk models, *none* of them is satisfactory as the sole output from a risk assessment. *It is essential to provide intervals or probability distributions around any point estimate of risk* to inform the users of a risk assessment about the full range of risks that might be caused by a risk management intervention. This principle applies to qualitative and fuzzy risk ratings as well. If a point estimate of a risk is "High," then some indication must be given of how certain this value is and of how compatible the frequency and severity components of the risk are with other qualitative labels, such as "Low." A risk assessment that produces a single overall value for risk with no indication of uncertainty should be avoided.

- *What are the key drivers of risks and uncertainties for each option?* The analysis should make clear to the user the main reasons *why* the estimated risk from each decision option is as high or low as it is. Are the results driven mainly by predicted exposure levels, by the responses of sensitive subpopulations, by genetic or epidemiological data that establish tight constraints on the plausible values, or by other factors? Sensitivity analyses that plot how estimated risks would change as input assumptions and estimates vary within plausible ranges (e.g., within a few standard deviations of their median values) can help to identify the combinations of input values that drive the main conclusions and the extent to which these could be changed without changing the comparison of different risk management interventions.

- *Which risk management interventions are undominated?* One risk management intervention *dominates* another if it produces smaller probabilities of exceeding any specified level of adverse consequences per year. For example, if two different interventions lead to different expected numbers of sporadic salmonellosis cases per year (with the actual number being a Poisson random variable), and if the probable health consequences per case (e.g., the distribution of the number of days of illness of given severity) is the same for each intervention, then the one giving the smaller expected number of illnesses per year dominates the other. Scientific risk assessment can, at most, identify undominated risk management alternatives for risk managers to further assess and choose among.

Conclusions

This chapter has discussed how quantitative risk assessment (QRA) can collect, organize, and present information to support improved risk management decision making about public health risks. A successful risk analysis estimates the causal relations between decisions and probable resulting exposures, and between exposures and their probable total human health consequences. To guide rational decision making, a risk analysis should yield evaluations and comparisons of proposed risk management *actions and interventions*, not simply descriptions of the current situation. It should show the estimated changes in frequencies and magnitudes (and uncertainties) of human heath consequences resulting from different proposed risk management decisions. It is important to identify an adequate range of risk management options to assure that dominant alternatives are not overlooked. For each option, the total health consequences are found by summing the impacts of proposed actions on human exposures over all relevant pathways that contribute significantly to the outcome. Applying an exposure-response model to the changed exposures for different decisions then yields the estimated risks associated with them.

A well-conducted risk analysis enables its recipients to participate more effectively in risk management deliberations and to communicate questions and concerns more clearly and concisely than would otherwise be possible. It does so by providing them with the relevant information needed to determine the probable consequences of proposed actions and by showing how sensitive these predicted consequences are to specific remaining uncertainties.

However, not all risk analyses are well conducted. Unsound risk analysis techniques that do *not* necessarily identify valid causal relations or recommend undominated actions are in common use. Their use is still increasing, stimulated in part by regulatory requirements and enthusiastic consultants, together with widespread desire and demand for the benefits of QRA that are not always tightly tied to a clear understanding of the methods needed to achieve them. Chapters 4 and 5 consider in more detail approaches to risk analysis that may do more harm than good, degrading rather than improving the effectiveness of risk management decisions. The current popularity of such methods poses a potential threat to the long-term reputation and value of the field of risk analysis. To overcome this threat, better methods of risk analysis must be used. Possibilities for doing better are explored in detail after Chapters 4 and 5.

Part II
Avoiding Bad Risk Analysis

Chapter 4
Limitations of Risk Assessment Using Risk Matrices

This chapter focuses on the use and abuse of risk matrices – tables mapping "frequency" and "severity" ratings to corresponding risk priority levels. Such matrices have become very popular in applications as diverse as terrorism risk analysis, highway construction project management, office building risk analysis, climate change risk management, and enterprise risk management (ERM). Their use is now so widespread in important applications that it is worth devoting an entire chapter to understanding the main concepts – and limitations – of risk matrices as a framework for practical risk analysis.

National and international standards (e.g., Military Standard 882C and AS/NZS 4360:1999) have stimulated the adoption of risk matrices by many organizations and risk consultants. However, little research rigorously validates their performance in actually improving risk management decisions. This chapter examines some mathematical properties of risk matrices and shows that they have the following limitations:

- *Poor resolution.* Typical risk matrices can correctly and unambiguously compare only a small fraction (e.g., less than 10%) of randomly selected pairs of hazards. They can assign identical ratings to quantitatively very different risks ("range compression").
- *Errors in comparative rankings.* Risk matrices can mistakenly assign higher qualitative ratings to quantitatively smaller risks. For risks with negatively correlated frequencies and severities, they can be "worse than useless," leading to worse-than-random decisions.
- *Suboptimal resource allocation.* The effective allocation of resources to risk-reducing countermeasures cannot, in general, be based on the categories provided by risk matrices.
- *Ambiguous inputs and outputs.* Categorizations of relative severity cannot necessarily be made objectively – independent of subjective risk attitudes – for uncertain consequences. Inputs to risk matrices (e.g., frequency and severity categorizations) and resulting outputs (i.e., risk ratings) require subjective interpretations, and different users may obtain opposite ratings of the same quantitative risks. These limitations suggest that risk matrices should be used with caution, if at all, and only with careful explanations of embedded judgments.

L.A. Cox, Jr., *Risk Analysis of Complex and Uncertain Systems*, International Series in Operations Research & Management Science 129, DOI 10.1007/978-0-387-89014-2_4,

These limitations of risk matrices illustrate some important pitfalls of attempting to do risk analysis without using appropriate QRA concepts and techniques. The QRA frameworks and methods discussed in previous chapters can avoid these limitations, although they, too, can be misapplied and can lead to poor decision recommendations if they are used carelessly (see Chapter 5).

Introductory Concepts and Examples

A *risk matrix* is a table that has several categories of "probability," "likelihood," or "frequency" for its rows (or columns) and several categories of "severity," "impact," or "consequences" for its columns (or rows, respectively). It associates a recommended level of risk, urgency, priority, or management action with each row-column pair, i.e., with each cell. Table 4.1 shows an example of a standard 5×5 risk matrix developed by the Federal Highway Administration for assessing risks and setting priorities in addressing issues as diverse as unexpected geotechnical problems at bridge piers and the unwillingness of land owners to sell land near critical road junctions.

The green, yellow, and red cells indicate low, medium, and high or urgent risk levels based on ratings of probability (vertical axis) and impact (horizontal axis) ranging from "VL" (very low) to "VH" (very high).

Table 4.2 shows a similar example of a 5×5 risk matrix from a 2007 Federal Aviation Administration (FAA) Advisory Circular (AC) introducing the concept of a safety management system for airport operators. The accompanying explanation states that "hazards are ranked according to the severity and the likelihood of their risk, which is illustrated by where they fall on the risk matrix. Hazards with high risk receive higher priority for treatment and mitigation." Many similar examples can be found for regulatory agencies, regulated industries, and public- and private-sector organizations. Training courses and software tools, such as MITRE's *Risk Matrix* tool for program risk management (MITRE, 1999–2007), help to automate risk matrix creation, application, and documentation.

Table 4.1 Standard 5×5 risk matrix for the Federal Highway Administration

Probability\impact	VL	L	M	H	VH
VH	Green	Yellow	Red	Red	Red
H	Green	Yellow	Red	Red	Red
M	Green	Green	Yellow	Red	Red
L	Green	Green	Yellow	Red	Red
VL	Green	Green	Green	Yellow	Red

Source: Federal Highway Administration, 2006, http://international.fhwa.dot.gov/riskassess/images/figure_12.htm.
For color version, please see the online version at www.springerlink.com

Table 4.2 Example of a predictive risk matrix for the Federal Aviation Administration

Severity / Likelihood	No Safety Effect	Minor	Major	Hazardous	Catastrophic
Frequent					
Probable					
Remote					
Extremely Remote					
Extremely Improbable					

HIGH RISK
MEDIUM RISK
LOW RISK

Source: Federal Aviation Administration, 2007, www.faa.gov/airports_airtraffic/airports/resources/advisory_circulars/media/150-5200-37/150_5200_37.doc.
For color version, please see the online version at www.springerlink.com

As mentioned above, the use of such risk matrices to set priorities and guide resource allocations has also been recommended in national and international standards. It has spread through many areas of applied risk management consulting and practice, including enterprise risk management (ERM) and corporate governance (partly under the influence of the Sarbane Oxley Act and international standards such as AUS/NZ 4360:1999), highway construction project risk management (Table 4.1), airport safety (Table 4.2), homeland security, and risk assessment of potential threats to office buildings, ranging from hurricanes to terrorist attacks (Renfroe and Smith, 2007).

Risk matrices have been widely praised and adopted as simple, effective approaches to risk management. They provide a clear framework for the systematic review of individual risks and portfolios of risks; convenient documentation for the rationale of risk rankings and priority setting; relatively simple-appearing inputs and outputs, often with attractively colored grids; opportunities for many stakeholders to participate in customizing category definitions and action levels; and opportunities for consultants to train different parts of organizations on "risk culture" concepts at different levels of detail, from simply positioning different hazards within a predefined matrix to helping thought leaders try to define risk categories and express "risk appetite" preferences in the color-coding of the cells. As many risk matrix practitioners and advocates have pointed out, constructing, using, and socializing risk matrices within an organization requires no special expertise in quantitative risk assessment methods or data analysis.

Yet despite these advantages and their wide acceptance and use, there has been very little rigorous empirical or theoretical study of how well risk matrices succeed in actually leading to improved risk management decisions. Very little prior technical literature specifically addresses the logical and mathematical limitations of risk matrices (but see Cox et al., 2005). Risk matrices are different enough from other topics (such as multivariate classification, clustering, and learning with correct classes provided as training data) to require separate investigation of their properties, in part because "risk" is not a measured attribute, but is derived from frequency and severity inputs through a priori specified formulas such as *risk* = *frequency*severity*. This chapter explores fundamental mathematical and logical limitations of risk matrices as sources of information for risk management decision making and priority setting.

A Normative Decision-Analytic Framework

Many decision makers and consultants believe that while risk matrices may be only rough approximate tools for risk analysis, they are very useful for distinguishing qualitatively between the most urgent and least urgent risks in many settings and are certainly much better than nothing, e.g., than purely random decision making. This section examines these beliefs from the standpoint of optimal statistical decision making in a simple framework for which it is possible to obtain exact results.

The simplest possible risk matrix is a 2×2 table that results from dichotomizing each of the two axes, referred to here as "probability" and "consequence." (Many other axes such as "frequency" and "severity" or "likelihood" and "magnitude" are also used, but changing the names does not affect the logic.) Table 4.3 shows such a matrix. Now, consider using it to categorize quantitative risks. For simplicity, suppose that the two attributes, *probability* and *consequence,* have quantitative values between 0 and 1, inclusive (where 0 = minimal or zero adverse consequence and 1 = maximum adverse consequence). Define the quantitative risk for any (*probability*, *consequence*) pair to be their product, *risk* = *probability* × *consequence*, as advocated in many risk matrix methodology documents. The risk matrix designer can choose where to draw the boundaries between Low and High values on each axis. Let the boundary between Low and High consequence correspond to a numerical value x between 0 and 1; and let the boundary between Low and High probability correspond to a value y between 0 and 1.

Table 4.3 A 2 × 2 risk matrix

Consequence Probability	Low	High
High	Medium	High
Low	Low	Medium

For color version, please see the online version at www.springerlink.com

To assess the performance of the risk matrix in supporting effective risk management decisions, consider the following specific decision problem. The decision maker must choose which of two risks, A and B, to eliminate. (She can only afford to eliminate one of them.) The quantitative values of *probability* and *consequence* are a priori independently and uniformly distributed between 0 and 1 for each of A and B. The only information that the decision maker has is knowledge of which cell of the risk matrix each risk falls in. (Thus, the risk matrix provides statistical information about the true but unknown quantitative risk; it is a lossy information channel.) *How well can the information provided by the risk matrix be used to identify the quantitatively greater risk?* Equivalently, how well can the categorizations of quantitative risks provided by the matrix be used to identify the decision that maximizes expected utility (minimizes expected loss)?

The answer depends on how the risk matrix is designed and on the joint probability distribution of *probability* and *consequence* values. In general, the two risks can be ranked with no error if one risk falls in the High (red) cell in the upper right of Table 4.1 and the other falls in the Low (green) cell in the lower left (since every risk in the High cell is quantitatively as well as qualitatively greater than any risk in the Low cell). The probability of this event is $2^*(1 - x)(1 - y)xy$. This symmetric function is maximized by choosing $x = y = 0.5$. (Otherwise, if the two risks have the same qualitative rating, then there is no way to choose among them based on the risk matrix, and we can assume that there is a fifty-fifty chance of making the right choice, i.e., a 50% error probability. If one of the two ratings is Medium and the other is not, then the error probability from choosing the risk with the higher rating is positive, since some points in the cell with the higher qualitative rating have smaller quantitative risk values than some points in the cell with the lower qualitative rating; see Lemma 1 in the next section.)

The probability that two risks can be unambiguously ranked (i.e., with zero error probability) using the risk matrix with $x = y = 0.5$ is $(1/2)^*(1/4) = 0.125$ (i.e., it is the probability that one of them falls in one cell of the "High/Low" diagonal and the other falls in the other cell of that diagonal). The probability that the two risks cannot be compared using the matrix with better than random accuracy (50% error probability) is the probability that both risks receive the same qualitative rating; this is $0.375 = (1/4)^*[(1/2) + (1/4) + (1/2) + (1/4)]$ (considering the four cells clockwise, starting with the upper left). The probability that the two risks can be compared using the matrix with error probability greater than zero but less than 50% is $1 - 0.125 - 0.375 = 0.5$.

Next, suppose that the risk matrix is constructed with $x = y = 0.5$ but that it is applied in decision settings where the joint probability distribution of *probability* and *consequence* is uncertain. Now, how well the matrix can identify which of two risks is greater depends completely on the joint probability distribution of the (*probability*, *consequence*) pairs. For example, if the *probability* and *consequence* values are uniformly distributed along the diagonal from (0, 0) to (1, 1), then there is a 50% probability that the two risks can be classified with zero error probability (if one of them is in the High cell and the other is in the Low cell); otherwise, the error probability is 50% (if both are in the same cell). Thus, under these very favorable conditions of perfect positive correlation, the error probability is 0.5*0.5 = 0.25. Conversely, if the *probability* and *consequence* values are perfectly negatively correlated and are concentrated along the diagonal from (0, 1) to (1, 0), then all risks will be assigned a risk rating of "Medium" (although their numerical values range from 0 at the ends of the upper left to lower right diagonal to 0.25 in the middle), and the risk matrix will provide no useful information for discriminating between greater and lesser risks. Under these less favorable conditions, the decision maker using the risk matrix can do no better than random decision making, and the error probability increases to 50%.

Finally, if the *probability* and *consequence* values are negatively correlated and concentrated along the line *probability* = 0.75 – *consequence* (for *consequence* values between 0 and 0.75), then all points on this line in the Medium cells (i.e., for *consequence* values between 0 and 0.25 or between 0.5 and 0.75) have *smaller* quantitative risks than any points in the Low cell (i.e., for *consequence* values between 0.25 and 0.5). For example, the pair (0.1, 0.65) would be classified as a Medium risk (although its quantitative risk value is 0.1*0.65 = 0.065), while the pair (0.37, 0.38) would be classified as a Low risk, even though its quantitative risk value is more than twice as great, 0.37*0.38 ≈ 0.14. (More generally, such counterexamples can be constructed by noting that each iso-risk contour *probability***consequence* = *constant* is convex, so that a straight line passing through the two points where such a contour intersects the edges of a cell of the matrix will lie above the contour within the cell but below it outside the cell.)

For this unfavorable joint distribution of (*probability*, *consequence*) pairs, the information provided by the risk matrix is *worse than useless* (Cox and Popken, 2007) in the sense that whenever it discriminates between two risks (by labeling one Medium and the other Low), it reverses the correct (quantitative) risk ranking by assigning the higher qualitative risk category to the quantitatively smaller risk. Thus, a decision maker who uses the risk matrix to make decisions would have a lower expected utility in this case than one who ignores the risk matrix information and makes decisions randomly, e.g., by tossing a fair coin. [Similar examples can be constructed for the High risk cell in the upper right corner of Table 4.3. For example, the (*probability*, *consequence*) pair (0.6, 0.6) is rated as High and the pair (0.48, 1) is rated as Medium, even though the latter has a higher quantitative risk (0.48) than the former (0.36).]

The question of how risk matrices ideally should be constructed to improve risk management decisions has no simple answer, both because risk matrices are typically used as only one component in informing eventual risk management decisions and also because their performance depends on the joint distribution of the two attributes *probability* and *consequence*, as illustrated in the above examples. Since risk matrices are commonly used when quantitative data are limited or unavailable, this joint distribution is typically unknown or very uncertain. This knowledge gap implies that the actual performance of a risk matrix, and whether it is helpful, no better than random, or worse than useless, may be unknown. It also prevents the easy application of traditional decision-analytic, statistical, artificial intelligence, and engineering methods for similar problems (e.g., for optimal classification and for discretization of multivariate relations) that require the joint distribution of the attributes as an input.

However, the simplest case of a 2 × 2 risk matrix does suggest two important related conclusions. First, it is not necessarily true that risk matrices provide qualitatively useful information for setting risk priorities and for identifying risks that are high enough to worry about and risks that are low enough to be neglected or postponed. (As just discussed, the information they provide can be worse than useless when probability and consequence are negatively correlated.) Second, the use of a risk matrix to categorize risks is not always better than – or even as good as – purely random decision making. Thus, the common assumption that risk matrices, although imprecise, do some good in helping to focus attention on the most serious problems and in screening out less serious problems is not necessarily justified. Although risk matrices can indeed be very useful if the probability and consequence values are positively correlated, they can be worse than useless when these values are negatively correlated. Unfortunately, negative correlation may be common in practice, e.g., when the risks of concern include a mix of low-probability, high-consequence and high-probability, low-consequence events.

Although this section has been restricted to 2 × 2 risk matrices, the nature of the counterexamples in which the optimal statistical decision is to ignore risk matrix information (e.g., examples with joint distributions of probability-consequence pairs concentrated on negatively sloped lines that intersect with convex iso-risk contours where they cross cell boundaries) implies that simply changing the position or number of grid lines cannot eliminate the problem. A similar construction can be carried out no matter how many cells a matrix has and no matter where the cell boundaries are located. Generalizing the decision problem to that of selecting a subset of risks to remediate, from among a larger set of many risks (rather than only deciding which of two risks is greater), also does not change the main conclusion. For some joint distributions of probability and consequence values, normative decision theory would require *not* using the qualitative risk rating information provided by a risk matrix, as it reverses the correct (quantitative) risk ratings that would be obtained using perfect information.

What can be salvaged? Several directions for advancing the research on risk matrices appear promising. One is to consider applications in which there are sufficient data to draw some inferences about the statistical distribution of (*probability*, *consequence*) pairs. If data are sufficiently plentiful, then statistical and artificial intelligence tools such as classification trees (Chen et al., 2006), rough sets

(Dreiseitl et al., 1999), and vector quantization (Lloyd et al., 2007) can potentially be applied to help design risk matrices that give efficient or optimal (according to various criteria) discrete approximations to the quantitative distribution of risks. In such data-rich settings, it might be possible to use risk matrices when they are useful (e.g., if probability and consequence are strongly positively correlated) and to avoid them when they are not (e.g., if probability and consequence are strongly negatively correlated).

A different approach is to consider normative properties or axioms that risk matrix designers might ideally want their matrices to satisfy, and then to identify whether such matrices exist (and, if so, whether they are unique). This normative axiomatic approach, explored in the following section, can be used even when sufficient data are not available to estimate the joint distribution of probability and consequence values.

Logical Compatibility of Risk Matrices with Quantitative Risks

What does a risk matrix mean? One natural intuitive interpretation is that it provides a rough discrete (ordered categorical) approximation to a more detailed – but not readily available – underlying quantitative relation. At least in principle, the underlying relation is described by a risk formula such as one of the following:

$$risk = probability \times consequence \text{ [or } frequency \times severity \text{ or } likelihood \times impact \text{ or } threat \times (vulnerability \times consequence), \text{ etc.]}.$$

(We will use "frequency" or "probability" and "severity" or "consequence" as the default names of the two axes, and "risk" as the name for their product, but the analysis applies to any similar mathematical structure, regardless of the names.) For example, it might be supposed that the division of the probability axis into five ordered qualitative categories (e.g., from very rare to almost certain) corresponds roughly to a partitioning of a quantitative probability axis into the intervals [0, 0.2), [0.2, 0.4), [04, 0.6), [0.6, 0.8), and [0.8, 1] (where square brackets indicate that the corresponding endpoint is included in an interval and parentheses indicate that it is not). Similarly, the five ordered categories for the severity axis might naturally be interpreted as corresponding to numerical intervals, [0, 0.2), [0.2, 0.4), [04, 0.6), [0.6, 0.8), and [0.8, 1], on a quantitative value scale (e.g., a von Neumann-Morgenstern utility scale) normalized to run from 0 to 1, where 0 = no adverse impact, 1 = worst possible adverse outcome considered, and values between 0 and 1 represent adverse impacts or consequences with values intermediate between no adverse impact and worst possible adverse impact.

However, such an intuitive interpretation of the risk matrix as an approximation to an underlying quantitative model can only be sustained if the risk matrix satisfies certain constraints. To be most useful, a risk matrix should, at a minimum, discriminate reliably between very high and very low risks, so that it can be used as an effective screening tool to focus risk management attention and resources. This requirement can be expressed more formally as the following principle of *weak consistency* between the ordered categorization of risks provided by the matrix and the ranking of risks by an underlying quantitative formula, such as one of those above.

Definition of Weak Consistency

A risk matrix with more than one "color" (level of risk priority) for its cells satisfies weak consistency *with a quantitative risk interpretation if points in its top risk category represent higher quantitative risks than points in its bottom category.*

Here "quantitative risk" is defined as the product of a point's coordinates when the axes are interpreted quantitatively, e.g., *frequency* × *severity*. If weak consistency holds, then all risks in the top qualitative category are quantitatively larger than all risks in the lowest qualitative category. In this case, the risk matrix can discriminate reliably between at least some risks, even though it does not require quantifying the *probability* and *consequence* attributes. It may then serve as a useful screening tool, which is one of the main practical uses of risk matrices. But if weak consistency does not hold, then risks that are screened out as being relatively small according to the matrix may, in fact, be larger than some of those that the matrix classifies as top priority, thus leading to a misallocation of risk management resources. It is therefore desirable to construct risk matrices that satisfy weak consistency, if possible.

Weak consistency is not an arbitrary axiom. It is implied by the hypothesis that *some* quantitative interpretation of the risk categories in a matrix exists, at least in principle (i.e., that there is some underlying quantitative risk scale such that the consecutive ordinal risk categories of the matrix correspond, at least approximately, to consecutive intervals on the quantitative scale), even if this scale is unknown, imprecise, or undefined in practice. If it does not hold, then a risk matrix does not mean what many users might expect it to mean, i.e., that risks rated in the top category (red) are larger than those rated in the bottom category (green). Thus, transparency of interpretation provides another incentive for designing risk matrices to satisfy weak consistency.

Discussion of Weak Consistency

More generally, a risk matrix partitions alternatives (typically representing different threats, hazards, risk reduction or investment opportunities, risk management actions, etc.) into distinct categories corresponding to the different priority levels or "colors" of the matrix cells. Weak consistency implies that this partitioning assigns the highest qualitative level (e.g., red) to the alternatives that actually do have higher quantitative risk values than those assigned the lowest qualitative level (e.g., green). If weak consistency holds, the qualitative classification given by the matrix is, in this sense, at least roughly consistent with what a quantitative analysis would show. Red

Table 4.4 A 5 × 5 matrix compatible with *risk = probability × consequence*

Prob\Consequence	0–0.2	0.2–0.4	0.4–0.6	0.6–0.8	0.8–1
0.8–1	Green	Green	Yellow	Red	Red
0.6–0.8	Green	Green	Yellow	Yellow	Red
0.4–0.6	Green	Green	Green	Yellow	Yellow
0.2–0.4	Green	Green	Green	Green	Green
0–0.2	Green	Green	Green	Green	Green

For color version, please see the online version at www.springerlink.com

cells do represent unambiguously higher risks than green cells, where we use "red" to denote the highest urgency level (that of the upper rightmost cell, if the matrix axes are oriented to represent increasing probability or frequency on one axis and increasing severity of consequences on the other) and we use "green" to denote the lowest urgency level (that of the lowest leftmost cell in such a table). This provides a logical basis for screening risks into "larger" (red) and "smaller" (green) categories.

Table 4.4 shows an assignment of risk levels that satisfies weak consistency for a 5 × 5 matrix in which the rows and columns are interpreted as equal partitions of two numerical scales, each normalized to run from 0 to 1. Any point in a red cell has a quantitative value (calculated as the product of the horizontal and vertical coordinates) of at least 0.48, while no point in any green cell has a value greater than 0.40.

Logical Implications of Weak Consistency

Weak consistency is more restrictive than might be expected. For example, neither of the colorings in Tables 4.1 and 4.2 satisfies weak consistency (see Lemma 2). Indeed, weak consistency implies some important constraints on possible colorings of risk matrices.

Lemma 1. *If a risk matrix satisfies weak consistency, then no red cell can share an edge with a green cell.*

Proof. Suppose that, to the contrary, a red cell and a green cell do share an edge. The iso-risk contour (i.e., the locus of all frequency-severity combinations having the same value of the product *frequency* × *severity*) passing through the midpoint of the common edge is a curve with negative slope. (It is a segment of a rectangular hyperbola, running from northwest to southeast.) Thus, it divides both cells into regions above and below this contour curve. Points that lie above this contour in the green cell have higher quantitative risk values than points lying below it in the red cell, contradicting weak consistency. Therefore, in a risk matrix satisfying weak consistency, red and green cells cannot share an edge. QED.

Comment: It is sufficient for this proof that iso-risk contours exist and have negative slopes. Thus, risk could be any smooth increasing function of frequency and severity (or whatever attributes the two axes of the matrix represent), not necessarily their product. However, the product of the coordinates is often used in practice in discussions of the concept of quantitative risk that accompany risk matrices, and we will use it as the default definition for quantitative risk in numerical examples.

Lemma 2. *If a risk matrix satisfies weak consistency and has at least two colors ("green" in the lower left cell and "red" in the upper right cell, and if the axes are oriented to show increasing frequency and severity), then no red cell can occur in the left column or in the bottom row of the risk matrix.*

Proof. Contours for all sufficiently small risk values (namely, values of all risk contours below and to the left of the one passing through the upper right corner of the lower leftmost cell) pass through all cells in the leftmost column and in the bottom row of a risk matrix. If any of these cells is red, then all points below one of these contours in the red cell will have lower quantitative risk levels than points above it in the green lower leftmost cell of the table. This would contradict weak consistency; thus, no such red cell can exist. QED.

An implication of Lemmas 1 and 2 is that any risk matrix that satisfies weak consistency and that does not assign identical priorities to all cells must have at least three colors: e.g., red for the upper rightmost cell; green for the lower leftmost cell; and at least one other color (i.e., priority rating), which we will call yellow, to separate the red and green cells.

The Betweenness Axiom: Motivation and Implications

The hypothesis that a risk matrix provides an approximate qualitative representation of underlying quantitative risks also implies that arbitrarily small increases in frequency and severity should not create discontinuous jumps in risk categorization from lowest priority ("green") to top priority ("red") without going through any intermediate levels ("yellow"). (Notice that this condition is violated in Tables 4.1, 4.2, and 4.3 but holds in Table 4.4.) Indeed, if the successive risk categories in a risk matrix represent (at least approximately) successive intervals on some underlying quantitative risk scale, then continuously increasing quantitative risk from 0 to 1 should cause the corresponding qualitative rating to pass through increasingly severe categorical values. A weaker condition is that the qualitative risk should pass through at least one intermediate value between green and red as the quantitative risk increases continuously from 0 to 1. Otherwise, a risk matrix does not mean what users might intuitively expect: that intermediate risk categories describe risks between the highest (red) and lowest (green) ones. These considerations motivate the following axiom.

Definition of betweenness. A risk matrix satisfies the axiom of betweenness *if every positively sloped line segment that lies in a green cell at its lower (left) end and in a red cell at its upper (right) end passes through at least one intermediate cell (meaning one that is neither green nor red) between them.*

Comment: Tables 4.1 and 4.2 both have red cells in row 2 and violate betweenness, i.e., in each an arbitrarily small increase in frequency and severity can cause a risk to be reclassified as red instead of green, without going through yellow. A 2×2 table such as Table 4.3 lacks sufficient resolution to allow betweenness, since there are no cells between the green lower left cell and the red upper right cell. Thus, betweenness can only be required for 3×3 and larger risk matrices.

Only some risk matrices satisfy both weak consistency and betweenness. Among all 3×3 matrices having more than one color, only one coloring of the cells satisfies both axioms. Using our conventional coloring scheme (green for lowest risk, red for highest risk, yellow for intermediate risk), this is the matrix with red in the upper right cell, green throughout the left column and bottom row, and yellow in all other cells.

Consistent Coloring

The final normative axiom considered in this chapter is motivated by the idea that equal quantitative risks should ideally have the same qualitative risk rating (color). Although this condition is impossible to achieve exactly in a discrete risk matrix, for the reason shown in the proof of Lemma 1 (essentially, horizontal and vertical grid lines cannot reproduce negatively sloped iso-risk contours), one rough approximation might be to enforce it for at least the two most extreme risk categories, red and green, while accepting some inconsistencies for intermediate colors. Accordingly, we will consider a requirement that all cells that contain red contours (meaning iso-risk contours that pass through other red cells) should themselves be red, unless the low resolution of the risk matrix causes them also to contain green contours. (A cell that contains both red and green contours has insufficient resolution to separate top-priority and bottom-priority risks and will not be required a priori to have either color.) Conversely, cells that contain green contours but no red ones should themselves be green. This motivates the following axiom of *consistent coloring*.

Definition of Consistent Coloring. 1. A cell is red if it contains points with quantitative risks at least as high as those in other red cells (and does not contain points with quantitative risk as small as those in any green cell). 2. A cell is colored green if it contains some points with risks at least as small as those in other green cells (and does not contain points with quantitative risks as high as those in any red cell). 3. A cell is colored an intermediate color (neither red nor green) only if either (a) it lies between a red cell and a green cell or (b) it contains points with quantitative risks higher than those in some red cells and also points with quantitative risks lower than those in some green cells.

Intuitively, one might think of an iso-risk contour as being colored green if it passes through one or more green cells but not through any red cells, as being colored red if it passes through one or more red cells but not through any green cells, and as being colored yellow (or some other intermediate color) if it passes through both red and green cells (or through neither red nor green cells). Then the consistent coloring principle implies that any cell that contains green contours but no red contours must itself be green, while any cell that contains red contours but no green ones must itself be red. This is admittedly only one possibility for trying to capture the intuitive idea that all sufficiently high risks should have the same color ("red") and all sufficiently low risks should have the same color ("green"). Other normative axioms could perhaps be formulated, but this chapter will only use the three already defined.

Implications of the Three Axioms

Theorem 1. *In a risk matrix satisfying weak consistency, betweenness, and consistent coloring: (a) All cells in the leftmost column and in the bottom row are green (lowest-priority); and (b) all cells in the second column from the left and in the second row from the bottom are non-red.*

Proof. Appendix A.

Corollary. A 3×3 or a 4×4 risk matrix satisfying weak consistency, betweenness, and consistent coloring (and having more than one color) has a unique coloring, as follows. The left column and bottom row are green, the top right cell (for a 3×3 matrix) or the four top right cells (for a 4×4 matrix) are red, and all other cells are yellow.

Proof. Theorem 1 implies that the left column and bottom row are green. Assuming that the upper right cell is red (since there is more than one color and this is the most severe cell), consistent coloring implies that the two cells in a 4×4 matrix that share edges with it must also be red and that the cell that both of these share edges with (diagonally below and to the left of the upper right cell) must also be red. Betweenness then implies that all other cells in a 3×3 or 4×4 matrix must be yellow. QED.

This result shows that it is possible to construct 3×3 and 4×4 matrices (although not 2×2 matrices) satisfying all three of the normative axioms proposed in this section. There is only one way to do so, however: Any other colorings violate one or more of the axioms. For larger matrices, there is greater flexibility, as illustrated next.

Example: The Two Possible Colorings of a Standard 5×5 Risk Matrix

Table 4.5 shows two possible colorings of a 5×5 risk matrix that are consistent with the axioms of weak consistency, betweenness, and consistent coloring and also with a fully quantitative interpretation of the two axes, whose product gives a quantitative measure of risk (e.g., *risk* = *frequency* $\times$ *severity, expected utility* = *success*

Table 4.5 The two possible colorings of a standard 5 × 5 risk matrix

	0–0.2	0.2–0.4	0.4–0.6	0.6–0.8	0.8–1
0.8–1	0.18, 1	0.21, 0.86	Yellow	Red	Red
0.6–0.8	Green	0.24, 0.75	Yellow	Yellow	Red
0.4–0.6	Green	0.36, 0.5	0.42, 0.42	Yellow	Yellow
0.2–0.4	Green	Green	0.5, 0.36	0.75, 0.24	0.86, 0.21
0–0.2	Green	Green	Green	Green	1, 0.18
	0–0.2	0.2–0.4	0.4–0.6	0.6–0.8	0.8–1
0.8–1	0.18, 1	0.21, 0.86	Green	Yellow	Red
0.6–0.8	Green	0.24, 0.75	Green	Yellow	Yellow
0.4–0.6	Green	0.36, 0.5	0.42, 0.42	Green	Green
0.2–0.4	Green	Green	0.5, 0.36	0.75, 0.24	0.86, 0.21
0–0.2	Green	Green	Green	Green	1, 0.18

For color version, please see the online version at www.springerlink.com

probability × utility of success, reduction in perceived risk = perceived reduction in expected annual frequency of adverse events× perceived average severity per event, and so forth). The axes are normalized to run from (0, 0) at the lower left corner of the matrix to (1, 1) at the upper right corner, and the grid lines partition the axes into equal quantitative intervals.

In these tables, a "green contour" (with numerical value of 0.18) extends from the upper left cell to the lower right cell of the matrix (both of which are green, by Theorem 1), passing through a total of nine cells. (The table shows pairs of numbers in these cells that have 0.18 as their product. All cells containing this contour are green, as are all cells below and to the left of it, by consistent coloring.) The upper right cell is defined to be red (top risk priority). The cell to its left and the cell below it each contain points with higher quantitative risks than those of points in this top-priority cell's lower left corner; therefore, they must also be red (by consistent coloring) unless adjacent green cells make them yellow. The other yellow cells are implied by betweenness.

Risk Matrices with Too Many Colors Give Spurious Resolution

The foregoing analysis implies that for a 5 × 5 risk matrix to be consistent with a fully quantitative interpretation, as in Table 4.5, it must have exactly three colors. This is violated in many practical applications. For example, Table 4.6 shows a default risk matrix used in some commercial risk management software tools designed to help support risk analysis standards and recommendations. Such a four-color matrix is inconsistent with the assumption that the colors represent relative sizes of underlying quantitative risks as in Table 4.5. For example, if the horizontal and vertical axes of Table 4.6 are interpreted quantitatively as in Table 4.5, then

Table 4.6 Default 5 × 5 risk matrix used in a risk management software system

Likelihood\Consequence	Insignificant	Minor	Moderate	Major	Catastrophic
Almost certain	Blue	Orange	Red	Red	Red
Likely	Light Green	Blue	Orange	Red	Red
Possible	Light Green	Blue	Blue	Orange	Red
Unlikely	Green	Light Green	Blue	Blue	Orange
Rare	Green	Green	Light Green	Light Green	Blue

Source: Adapted from www.incom.com.au/risk.asp?ID=471.
For color version, please see the online version at www.springerlink.com

Table 4.6 assigns a higher rating to (0.81, 0.21) than to (0.79, 0.39), even though the former has a product of 0.17 and the latter a product of 0.31.

Example: A 4 × 4 Matrix for Project Risk Analysis

The use of risk matrices for risk analysis of projects has been described as follows by the California Division of the Federal Highway Administration:

> Risk is computed as the probability of occurrence multiplied by the consequence of the outcome. Probability is between 0 [minimal] and 1 [certain]. Consequence is expressed in terms of dollars, features, or schedule. Multiplying probability of occurrence and consequence [impact analysis] together gives a risk assessment value between 0 [no risk] and 1 [definite and catastrophic]. . . . Below is an example of the matrix used for such an evaluation. The numbers are the order in which the risks are to be considered. Anything that is in the box labeled "1" is the highest priority.

	Likely 0.7–1.0	Probable 0.4 to 0.7	Improbable 0.0 to 0.4	Impossible 0
Catastrophic 0.9 to 1.0	1	3	6	
Critical 0.7 to 0.9	2	4	8	
Marginal 0.4 to 0.7	5	7	10	
Negligible 0 to 0.4	9	11	12	

Source: California Department of Transportation, 2007, www.fhwa.dot.gov/cadiv/segb/views/document/Sections/Section3/3_9_4.htm.

Table 4.7 presents this risk matrix with its horizontal and vertical axes exchanged and oriented to be increasing, consistent with the conventions in previous examples.

The matrix has 13 priority levels as possible outputs, far greater than the three levels needed for a quantitative risk interpretation consistent with our axioms. The excess levels make it inconsistent with a coherent quantitative interpretation. For example, it assigns a priority rating of 8 to a quantitative risk of 0.42 (from a probability = 0.65 of a loss of relative severity 0.65 on a scale from 0 = no loss to

Table 4.7 Example risk matrix for airport projects

Consequence \ Probability	Negligible 0 to 0.4	Marginal 0.4 to 0.7	Critical 0.7 to 0.9	Catastrophic 0.9 to 1.0
Likely 0.7–1.0	9	5	2	1
Probable 0.4 to 0.7	11	8	4	3
Improbable 0.0 to 0.4	12	10	7	6
Impossible 0				

1 = worst catastrophic loss considered), but it assigns a much higher priority rating of 3 to a lower quantitative risk of 0.37 (probability = 0.41, consequence = 0.91). (Recall that outputs levels in the cells are numbered so that 1 = top priority.) Similarly, a loss of 0.6 with probability 1 receives a lower priority level than a quantitative loss of 0.8 with probability 0.5 (5 vs. 4), even though the former has a quantitative risk greater than the latter (0.6 vs. 0.4). A priority level of 12 is assigned to a probability 0.33 of consequence 0.33, but a priority level of only 6 is assigned to a numerically identical risk consisting of a probability 0.11 of consequence 0.99. Thus, as expected, the priority ratings implied by the 13 distinct priority levels in this matrix do not successfully represent the relative sizes of these quantitative risks. (That the qualitative ratings reverse the quantitative ratings in such examples cannot be justified by risk aversion, since the consequence axis is explicitly assumed to have been already transformed, scaled, or defined in such a way that the product of the two coordinate axes, probability and consequence, is the measure of quantitative risk that the qualitative matrix attempts to represent.)

The upper left cell of the risk matrix in Table 4.7 illustrates *range compression*: Discrete categorization lumps together very dissimilar risks, such as an adverse consequence of severity 0 occurring with probability 1 and an adverse consequence of severity 0.39 occurring with probability 1.

The two possible 5 × 5 risk matrices in Table 4.5 have very limited resolution. They assign a green rating to all risks less than 0.24, and a red rating to all risks greater than 0.64 (on a scale normalized to run from 0 to 1). Attempts to use more colors or risk-rating levels to improve resolution, as in the preceding example, necessarily create more ranking-reversal errors, in which quantitatively smaller risks are assigned qualitatively higher rating levels than some quantitatively larger risks.

As a rough measure of the degree to which these limitations might affect practical work, suppose that the cases being classified by a risk matrix have their two components independently and uniformly distributed between 0 and 1. Then the probability that a randomly selected pair of points can be correctly and unambiguously rank-ordered by a matrix such as the one in Table 4.4 (i.e., the probability that one point falls in a red cell and the other in a green cell) would be only (3/25 red fraction)*(17/25 green fraction) = 8.2%. Thus, over 90% of the time, the matrix will not be able to rank-order the two points correctly with certainty.

Risk Ratings Do Not Necessarily Support Good Resource Allocation Decisions

How well can the information provided by a risk matrix guide risk management resource allocation decisions? This section examines some limitations that hold even if the risk matrix provides qualitative ratings that perfectly represent the underlying quantitative risks.

Example: Priorities Based on Risk Matrices Violate Translation Invariance

Suppose that a risk manager can afford to eliminate all but one of the following three risks: (A) Lose \$95 with certainty. (B) Lose \$75 with certainty. (C) Lose \$95 with probability 50% (else lose nothing). Which one should she keep to minimize risk (here defined as expected loss)? According to the priority ranking in Table 4.7 (and interpreting the normalized consequence axis running from 0 to 1 as corresponding dollar losses running from \$0 to \$100), the answer is C. (This has the lowest rating, 3, compared to ratings of 1 for A and 2 for B. Recall that in Table 4.7, lower numbers in the cell indicate higher priority.)

Now suppose that all potential losses are reduced by \$15, so that the new alternatives are as follows: (A′) Lose \$80 with certainty. (B′) Lose \$60 with certainty. (C′) Lose \$80 with probability 50% (else lose nothing). According to Table 4.7, one should now choose to keep B′ (rating = 5, compared to ratings of 2 and 4 for A′ and B′, respectively). Thus, simply reducing the potential loss by the same amount for all three risks changes the prescribed priority ordering among them. This violates the principle of translation invariance for coherent risk measures (Artzner et al., 1999). Moreover, keeping B′ instead of C′ is inconsistent with minimizing risk (defined as expected loss in this example). Thus, the risk matrix in Table 4.7 does not necessarily support effective risk management decision making.

Similarly in Table 4.6, if a risk manager can eliminate exactly two out of four risks, corresponding the four lower left cells in the table, and if ties are broken at random, then the probability that the risk in the second column and the bottom row will be eliminated is 1/3 (since the risk in the higher-rated cell to its northeast will certainly be selected, followed by any one of the remaining three tied risks). Translating all consequences one cell to the right (by adding the same incremental consequence value to each of them) increases the probability to 1/2 (since this alternative will now tie with one other for second place). But a second translation by one step to the right reduces the selection probability to zero (since now the two blue cells in the second row dominate the two cells in the first row). Finally, one more rightward shift of the four alternatives increases the probability that this one will be selected to 1/2 again.

In Table 4.5, if only one of four risks in the four upper left cells [e.g., with respective (*probability*, *consequence*) values of (0.9, 0.1), (0.9, 0.3), (0.7, 0.1), and

(0.7, 0.3)] can be selected to eliminate, and if ties are broken at random, then the probability that the numerically greatest of these risks, namely, (0.9, 0.3), would be selected for elimination is only 1/4. Translating all four consequences rightward by the same amount, 0.4, would increase this selection probability to 1. Translating them further rightward by an additional 0.2 would reduce the selection probability to 1/3 (since the three red cells would then be tied). Thus, the probability of assigning top priority to the numerically greatest risk does not satisfy translation invariance. (This same pattern also occurs for successive rightward translations of the four lower left cells in Table 4.1.)

Example: Priority Ranking Does Not Necessarily Support Good Decisions

Setting: A risk manager has identified the following three risk reduction opportunities:

- Act A reduces risk from 100 to 80. It costs \$30.
- Act B reduces risk from 50 to 10. It costs \$40.
- Act C reduces risk from 25 to 0. It costs \$20.

(This example can also be constructed so that all three acts start from the same base level of risk, say 50, and A, B, and C reduce risk by 20, 40, and 25, respectively. Using different base levels allows for the possibility that the different options A, B, and C being compared protect different subpopulations.) The risk manager's goal is to purchase the largest possible total risk reduction for the available budget.

To assist risk management decision making, suppose that a risk matrix is used to categorize opportunities A, B, and C. Resources will then be allocated first to the top-rated alternatives, working down the priority order provided by the risk matrix until no further opportunities can be funded.

Problem: How should a risk matrix categorize A, B, and C to support the goal of achieving the largest risk reduction from the allocation of limited funds?

Solution: The answer depends on the budget. For a budget of \$40, the largest feasible risk reduction is achieved by funding B, so the best priority order puts B first. If the budget is \$50, then funding A and C achieves the greatest risk reduction, so B should be ranked last. At \$60, the best investment is to fund B and C, so now A should be ranked last. In short, *no categorization or rank ordering of A, B, and C optimizes resource allocation independent of the budget.* No possible priority order (or partial order, if some ratings are tied) is optimal for budgets of both \$49 and \$50. This illustrates a limitation on the type of output information – ordered categorical classification – provided to decision makers by risk matrices. In general, such information is not sufficient to support the effective allocation of risk-reducing resources, because solutions to such resource allocation optimization problems cannot generally be expressed as priority lists or categories that should be funded from the top down until no further items can be afforded (Bertsimas and Nino-Mora, 1996).

Thus, the input information going into a risk matrix (ordinal ratings of event frequencies and severities) is simply not sufficient to optimize risk management resource allocations, or even to avoid very poor allocations, as in the above example. Calculating optimal risk management resource allocations requires quantitative information beyond what a risk matrix provides, e.g., about budget constraints and about interactions among countermeasures. In general, risk rankings calculated from frequency and severity do not suffice to guide effective risk management resource allocation decisions.

Categorization of Uncertain Consequences Is Inherently Subjective

To use a risk matrix, it is necessary to be able to categorize the alternatives being compared in the cells of the matrix. However, decision analysis principles imply that there is no objective way to categorize severity ratings for events with uncertain consequences. Subjective risk attitudes play an essential (but seldom articulated) role in categorizing severity for such events. Thus, the information in a risk matrix represents a mixture of factual (probability and consequence) information about the risk and (usually unstated) psychological information about the risk attitude of the person or people performing the risk categorization. Since the risk attitudes of the builders are seldom documented, it can be impossible to determine how consequence severity classifications should be changed when someone else views or uses the matrix.

Example: Severity Ratings Depend on Subjective Risk Attitudes

For a decision maker with an exponential utility function, the certainty-equivalent (CE) value of a prospect with normally distributed consequences is $\mathrm{CE}(X) = E(X) - k^*\mathrm{Var}(X)$, where k is a parameter reflecting subjective risk aversion ($k = 2 \times$ coefficient of risk aversion), $E(X)$ is the mean of prospect X, $\mathrm{Var}(X)$ is its variance, and $\mathrm{CE}(X)$ is its certainty-equivalent value (i.e., the deterministic value that is considered equivalent in value to the uncertain prospect) (Infanger, 2006, p. 208). Consider three prospects, A, B, and C, with identical probabilities or frequencies and normally distributed consequences (on some outcome scale) with respective means of 1, 2, and 3 and respective variances of 0, 1, and 2. Their certainty equivalents are

$$\begin{aligned}\mathrm{CE(A)} &= 1,\\ \mathrm{CE(B)} &= 2 - k,\\ \mathrm{CE(C)} &= 3 - 2k.\end{aligned}$$

For a risk-neutral decision maker (for whom $k = 0$), the ordering of the prospects from largest to smallest certainty-equivalent value is therefore C > B > A. For a risk-averse decision maker with $k = 1$, all three prospects have the same certainty-equivalent value of 1. For a more risk-averse decision maker with $k = 2$, the ordering of the prospects is A > B > C. Thus, *the certainty equivalents of the severities of the prospects are oppositely ordered by decision makers with different degrees of risk aversion.* There is no objectively correct ordering of prospect severity certainty equivalents independent of subjective attitudes toward risk. But risk matrices typically do not specify or record the risk attitudes of those who use them. Users with different risk attitudes might have opposite orderings, as in this example. Neither is objectively (independent of subjective risk attitude) more correct than the other. As a result, there is no objective way to classify the relative severities of such prospects with uncertain consequences.

Example: Pragmatic Limitations of Guidance from Standards

In practice, various standards provide written guidance on how to classify severities for use in risk matrices. For example, Table 4.8 shows the severity ratings suggested in a 1998 General Accounting Office report on *Combating Terrorism*, based on the widely cited Military Standard 882C (https://crc.army.mil/guidance/system_safety/882C.pdf). As that standard notes,

> These hazard severity categories provide guidance to a wide variety of programs. However, adaptation to a particular program is generally required to provide a mutual understanding ... as to the meaning of the terms used in the category definitions. The adaptation must define what constitutes system loss, major or minor system or environmental damage, and severe and minor injury and occupational illness.

Even with these caveats, the guidance in Table 4.8 does not resolve the type of ambiguity in the previous example. For example, it offers no guidance on how to rate a consequence that is zero with probability 90% but catastrophic otherwise (perhaps depending on wind direction or crowding of a facility or of evacuation routes at the time of a terrorist attack). Moreover, it introduces other ambiguities. For example, how should one rate the severity of a consequence that consists of one death and one severe injury compared to that of a consequence of 0 deaths but 50 severe injuries? The answer is not obvious from Table 4.8.

The discrete qualitative categories provided in guidance such as Table 4.8 are also inconsistent with the continuous quantitative nature of many physical hazards. For example, should a condition that causes "negligible" environmental damage on each occurrence (e.g., leaking 1 ounce of jet fuel per occurrence) but that causes a high frequency of these small events (e.g., averaging 5 events per hour) truly have a lower

Table 4.8 Severity levels of undesired event consequences for combating terrorism

Severity level	Characteristics
I Catastrophic	Death, system loss, or severe environmental damage
II Critical	Severe injury, severe occupational illness, major system or environmental damage
III Marginal	Minor injury, minor occupational illness, or minor system or environmental damage
IV Negligible	Less than minor injury, occupational illness, or less than minor system or environmental damage

Source: GAO, 1998.

severity rating than a second condition that causes more damage per occurrence (e.g., leaking 10 pounds of jet fuel per occurrence) but that causes less frequent occurrences (e.g., once per week)? (Both would be assigned the highest-possible frequency rating by Military Standard 882C.) If so, then the risk matrix analysis could give a lower priority to eliminating a threat of leaking 52.5 pounds per week (= 5 ounces per hour × 24 hours/day × 7 days per week) than to eliminating a threat of leaking only 10 pounds per week, due to the greater "severity" of 10 pounds than 1 ounce and the equal "frequency" rating of common events (an example of range compression). In such cases, the idea of rating severity independently from frequency appears flawed.

Focusing on applying qualitative rating criteria, rather than on more quantitative comparisons of risks, can create irrational risk management priorities. The following example illustrates how the uncritical application of risk matrix guidance might promote misperceptions and misrankings of the relative risks of different strategic investment opportunities.

Example: Inappropriate Risk Ratings in Enterprise Risk Management (ERM)

Suppose that a company must choose between the following two risky investment strategies for responding to major and pervasive uncertainties, such as climate change risks:

- Strategy A has probability 0.001 of leading to a small growth rate that barely meets shareholder expectations; otherwise (probability 99.9%), shareholder value and growth will increase by a negligible amount (e.g., <0.00001%), disappointing shareholders and failing to meet their expectations.
- Strategy B has probability 50% of causing rapid and sustained growth that greatly exceeds growth shareholder expectations; otherwise (e.g., if the outcome of a crucial R&D project is unsuccessful), shareholder value and growth will not grow (growth rate = 0%).

Which strategy, A or B, better matches a responsible company's preferences (or "risk appetite") for risky strategic investments?

Common sense might suggest that Strategy B is obviously better than Strategy A, as it offers a 50% probability of greatly exceeding expectations instead of a 0.1% probability of barely meeting them, with no significant difference in the downside risk. However, the uncritical application of risk matrices suggested as examples for enterprise risk management (ERM) systems could rate B as riskier than A. For example, Australia published a risk management "guide for business and government ... [that] is consistent with the Australian and New Zealand Standard for Risk Management, AS/NZS 4360:2004, which is widely used in the public and private sectors to guide strategic, operational and other forms of risk management. The Guide describes how the routine application of the Standard can be extended to include the risks generated by climate change impacts" (Australian Government,

2006). The illustrative risk matrix and category definitions for a commercial business (Tables 10–12 of the Guide) could be used to assign a "Medium" risk priority to Strategy A but a "High" risk to Strategy B, making B appear to be less attractive than A. [For A, the likelihood of the adverse consequence, 99.9%, is classified as "almost certain." The consequence is described as "Growth would be achieved but it would fail to meet expectations," which is classified as a "minor" consequence. The risk matrix example in Fig. 12 of the Guide categorizes the likelihood-consequence pair (*almost certain, minor consequence*) as a "Medium" risk. For B, the likelihood of the adverse consequence is classified as "likely," the consequence is described as "There would be no growth," and this is classified as a "moderate" consequence. The combination (*likely, moderate consequence*) is categorized as a "High" risk.] Thus, a tight focus on implementing the discrete categorization criteria in the guidance could distract attention from the fact that most shareholders would gladly trade a negligible increase in adverse consequences for a large increase in the probability of a much better outcome. In the terminology of multicriteria decision making, the discrete categorization of consequences and probabilities inherent in risk matrices can produce noncompensatory decision rules that do not reflect the risk trade-off preferences of real decision makers and stakeholders.

Quantitative risk assessment was developed in part to help prevent the types of paradoxes illustrated in these examples. Even if the quantities in the fuel-leaking example were quite uncertain (e.g., an average of 1–10 ounces every few minutes in the first case and 0–100 pounds every few months in the second), a rough quantitative calculation would reveal that the first threat is much more severe than the second. Similarly, even a rough quantitative comparison of Strategies A and B in the enterprise risk management example would show that B is much more attractive than A. By contrast, qualitative or semiquantitative risk assessments based on ordered categories do not necessarily prevent rating reversals and misallocations of resources, as in these examples – and may even unintentionally encourage them, by directing the risk management effort and attention away from the key quantitative comparisons involved and toward the (often inherently subjective) task of categorizing frequency and severity components.

Discussion and Conclusions

This chapter has shown that quantitative and semiquantitative risk matrices have limited ability to correctly reproduce the risk ratings implied by quantitative models, especially if the two components of risk (e.g., frequency and severity) are negatively correlated with each other. Moreover, effective risk management decisions cannot, in general, be based on *any* scheme for mapping categorical (or ordered-categorical) ratings of frequency and severity into recommended risk management decisions or priorities. Instead, optimal resource allocation typically depends crucially on other quantitative information, such as the costs of different countermeasures, the risk reductions that they achieve, budget constraints, and possible interactions among risks or countermeasures (such as when fixing a leak protects against multiple subsequent adverse events).

Categorizing severity may require inherently subjective judgments (e.g., reflecting the rater's personal degree of risk aversion, if severity is modeled as a random variable) and/or arbitrary decisions about how far to aggregate multiple small and frequent events into fewer and less frequent but more severe events. The need for such judgments, and the potential for inconsistencies in how they are made by different people, implies that there may be no objectively correct way to fill out a risk matrix.

Conversely, the meaning of a risk matrix may be far from transparent, despite its simple appearance. In general, there is no unique way to interpret the comparisons in a risk matrix that does not require explanations – seldom or never provided in practice – about the risk attitude and subjective judgments used by those who constructed it. In particular, if some consequence severities are random variables with sufficiently large variances, then there may be no guarantee that risks that receive higher risk ratings in a risk matrix are actually greater than risks that receive lower ratings.

In summary, the results and examples in this chapter show that risk matrices do not necessarily support good (e.g., better-than-random) risk management decisions and effective allocations of limited management attention and resources. Yet the use of risk matrices appears to be too widespread (and convenient) to make cessation of use an attractive option. Therefore, research is urgently needed to better characterize conditions under which they are most likely to be helpful or harmful in risk management decision making (e.g., when frequencies and severities are positively or negatively correlated, respectively) and that develops methods for designing them to maximize potential decision benefits and limit potential harm from using them. One promising research direction may be to focus on placing the grid lines in a risk matrix to minimize the maximum loss from misclassified risks.

Most of the difficulties identified in this chapter do not apply to quantitative risk assessment (QRA) methods that treat frequency and severity as continuous, rather than categorical, variables. However, even using continuous quantitative descriptors of risk and of its components – namely, conditional probabilities of exposures given actions, and of frequencies and severities of adverse consequences given exposures – does not guarantee that QRA results will be sound, coherent, or useful. The following chapter examines some limitations of quantitative risk concepts and ways to overcome them.

Appendix A: Proof of Theorem 1

By definition, the lower left cell is green. Consistent coloring implies that any contour must be green if it lies below and to the left of the one passing through the upper right corner of this lower left cell [i.e., the contour through the points

(0.04, 1), (0.2, 0.2), (1 0.04) in the numerical example in Table 4.5], since (a) it passes through the lower left cell (which is green by definition) and (b) none of the cells it passes through is red (by Lemma 2). By construction, such a green contour passes through all cells in the leftmost column and in the bottom row.

Now, consider the cell directly above the lower left cell [i.e., the cell containing the point (0.1, 0.3) in Table 4.5]. Suppose that, contrary to the claimed result, this cell is not green. It cannot be red, by Lemma 2. For it to be an intermediate color (not green), it must contain at least one red contour (by color consistency and the fact that a green contour passes through it. This cell cannot be "between" red and green cells, since it is on an edge of the matrix, so it cannot acquire an intermediate color that way.). This red color cannot come from the cell above it in the leftmost column (which is non-red, by Lemma 2), nor from any cell in the bottom row (again by Lemma 2). Since contours are downward-sloping, the only remaining possibility is that the cell to its right [the cell containing (0.3, 0.3) in Table 4.5] must be red. But this would violate betweenness [at the point (0.2, 0.2) in Table 4.5]. Therefore, the assumption that the cell directly above the lower left cell is not green leads to a contradiction. Hence, it must be green. By a symmetrical argument, the cell directly to the right of the lower left cell [the cell containing (0.3, 0.1) in Table 4.5] must also be green.

Next, suppose that the third cell in the leftmost column [the one containing (0.1, 0.5) in Table 4.5] is not green. Since green contours pass through it (as it is in the leftmost column), it can only be non-green if some red contour also passes through it (by color consistency and the fact that it is an edge cell). This red contour could not come from a red cell below it in the leftmost column, or in the bottom row (by Lemma 2), nor from the cell directly to its southeast [containing (0.3, 0.3) in Table 4.5] (since if that were red, it would violate Lemma 1 and betweenness for the cells so far proved to be green). The only remaining possibility is that the cell to its right [the one containing (0.3, 0.5) in Table 4.5] is red. But this would violate betweenness [with the second cell in the leftmost column, the cell containing (0.1, 0.3) in Table 4.5, which we have proved above must be green]. Hence, the assumption that the third cell in the leftmost column is not green implies a contradiction. So, it must be green. Symmetrically, the third cell in the bottom row must be green. This construction (showing that a cell directly above a green cell in the first column, with only non-red cells to its southeast, must itself be green) can be iterated for all remaining cells in the leftmost column, thus establishing that they all must be green; symmetrically, all remaining cells in the bottom row must be green. This proves part (a). Part (b) is then an immediate consequence of part (a) and Lemma 2. QED.

Comment: This proof does not depend on the number of rows or columns in the table. Therefore, its conclusion (that the leftmost column and bottom row consist entirely of green cells) holds for risk matrices of any size, under the stated conditions of weak consistency, betweenness, and consistent coloring.

Chapter 5
Limitations of Quantitative Risk Assessment Using Aggregate Exposure and Risk Models

Chapter 4 showed that risk matrices can assign small risks to high-risk categories and larger risks to lower-risk categories, defeating the intent of the classification system. Do other methods necessarily do better? This chapter shows that careless use of quantitative risk assessment concepts can also lead to worse-than-useless risk comparisons and recommendations. This happens if causal drivers of risk (such as age-specific failure rates, detailed exposures, or individual dose-response relations) are ignored in favor of potentially meaningless aggregate quantities (such as "average annual frequency," "aggregate exposure," or "population exposure-response ratio," respectively). A lesson from Chapter 4 was that risk matrices cannot correctly compare some risks. The main lesson from this chapter is milder. Care must be taken in using quantitative risk concepts to make sure that they correctly represent causal relations among actions, exposures, and probable consequences. Otherwise, they may give rise to meaningless or misleading numbers and predictions.

This chapter develops the following main ideas:

- "Frequencies" of adverse events are not well defined when the times between them have any of a number of common probability distributions. When frequency is not well defined, other, more sophisticated probability methods and models (such as point-process descriptions of failure processes) should be used instead.
- "Exposure" cannot be summarized adequately by a single number if it has multiple constituents that affect risk in different ways. Trying to use a single aggregate exposure metric to summarize a list of different exposures may lead to nonsensical, or worse-than-useless, results. This is no surprise: It simply states that one cannot, in general, avoid the need to consider relevant details in quantifying risks.
- Causal relations among explanatory variables (such as exposures) and their probable adverse consequences (such as illnesses or system failures) cannot, in general, be estimated or described by aggregate statistical relations (e.g., regression equations) linking observed levels of explanatory variables to observed levels of effect variables. Causal relations are not the same as statistical relations.
- Risk scores or priority ratings are inconsistent with effective risk management in many real-world applications with correlated uncertainties about the effects of alternative risk management actions.

L.A. Cox, Jr., *Risk Analysis of Complex and Uncertain Systems*, International Series in Operations Research & Management Science 129, DOI 10.1007/978-0-387-89014-2_5,

The common theme in all these cases is that for quantitative risk assessment (QRA) methods to work well, it is essential not to try to short-circuit the QRA process by ignoring essential details. Random times between events may have to be described by probability distributions, not single frequency numbers. Exposures to complex mixtures may have to be described in terms of the different components in the mixture, rather than by using a single aggregate number that ignores this complexity. Causal relations may have to be quantified by considering how changing some variables (e.g., inputs to a system) will change others (e.g., outputs from the system; see Chapters 2 and 3), not simply by fitting aggregate statistical relations to observed input-output pairs. Effective risk management often requires optimizing decisions, and the results cannot be expressed by priority rankings.

In short, a certain amount of detailed work is usually required to quantify valid causal relations between controllable inputs or decisions and resulting probable outputs or consequences. This work must be done to obtain sound and useful risk models. Attempting to build risk models without doing it – that is, without identifying, quantifying, and validating these essential causal relations – is a recipe for producing meaningless results and, in some cases, worse-than-useless recommendations. Chapters 6–11 address practical methods for developing the required causal models from data. This chapter focuses on the limitations of methods that try to quantify risk without doing the required detailed modeling.

What Is Frequency?

As explained in Chapter 2, frequency is one of the concepts often used to simplify and explain risk, using visual aids such as *F-N* or "exceedance probability" curves, showing magnitudes of loss (or other adverse consequences) on the horizontal axis and the average annual frequency of exceeding each loss level on the vertical axis. Students of risk analysis are often taught that "Risk is frequency times consequence," or, more generally, that risk is determined by the frequency *and* severity (magnitude of loss) of adverse consequences. But is it? This section reviews the concepts of frequency as the average annual occurrence rate and as the reciprocal of mean time-to-failure (MTTF) or mean time between failures (MTBF) in a renewal process. It points out that if two risks [represented as two (*frequency*, *severity*) pairs for adverse consequences] have identical values for *severity* but different values of *frequency*, then it is not necessarily true that the one with the smaller value of *frequency* is preferable – and this is true no matter how *frequency* is defined.

In general, there is not necessarily an increasing relation between the reciprocal of the mean time until an event occurs, its long-run average occurrences per year, and other criteria, such as the probability or expected number of times that it will happen over a specific interval of interest, such as the design life of a system. *Risk depends on more than the frequency and severity of consequences.* It also depends on other information about the probability distribution for the time of a risk event that can become lost in simple measures of event "frequency." More flexible descriptions of risky processes, such as point-process models of hazardous processes, can avoid these limitations.

An Example: Comparing Two Risks

Which of the following two risks is preferable?

- Risk A: *frequency* of risk event = 0.01 expected occurrences per year; *severity* of adverse consequences per occurrence = 1 (on some consequence scale, e.g., 1 occupational fatality per occurrence or 1 QALY lost per occurrence)
- Risk B: *frequency* of risk event = 0.02 expected occurrences per year; *severity* = 1 (the same as for A)

Many engineering risk analyses assume that such questions are meaningful and that the correct answer for this example is that Risk A is (obviously) preferable to Risk B, since A has a smaller frequency and the same adverse consequences per occurrence as B. However, the truth is that descriptions of risk in terms of the frequency and severity of adverse consequences, as in descriptions A and B above, do not contain enough information to enable a clear decision.

Event Frequencies in Renewal Processes

For simplicity, consider a renewal process in which a specified risk event is of interest (e.g., burn-out of a light bulb, loss of coolant accident at a nuclear power plant, etc.). We will refer to the risk event generically as a *failure*. The process starts in a good or operating state and eventually the specified risk event (failure) may occur. We care about how long it takes for this to happen. It is usual to model the time until failure occurs either as a random variable with some probability distribution function, or as an average annual occurrence frequency, measured in units of expected failures per year.

To define frequency in this setting, we can envision the process as immediately starting over from its original state as soon as a failure occurs. Thus, the light bulb is imagined to be replaced immediately with a new one (having the same failure time distribution as the original) as soon as it burns out, or the plant is promptly brought back to its initial operating state as soon as a failure occurs, and so forth. While such immediate renewals may not occur in physical reality, they define a conceptual renewal process in which the long-run average number of failures (and renewals) per unit time is the reciprocal of the mean time between failures (MTBF) (Ross, 1996, Proposition 3.3.1, p. 102). This provides one definition of the "frequency" of failures.

It is common practice in engineering risk assessment to identify the frequencies of different events, along with the severities of their consequences, and to prepare curves showing the cumulative average annual frequencies of events exceeding certain magnitudes. Such curves have often been suggested as useful for defining, displaying, and comparing the risks of various systems (e.g., Kaplan and Garrick, 1981; Thompson, 1988).

Example: Average Annual Frequency for Exponentially Distributed Lifetimes

Problem: Suppose that the random times between successive renewals (i.e., failures) have an exponential distribution with mean $\mu = 66.67$ years. (a) What is the average annual frequency of the renewal event? (b) How many renewal failures are expected to occur in 50 years, if that is the design life of the system? (c) What is the probability of no failures within 50 years?

Solution: (a) The average annual frequency is $1/\mu = 1/(66.67 \text{ years}) = 0.015$ expected occurrences per year. (b) The expected number of failures in 50 years is $0.015{*}50 = 0.75$ expected failures. (c) The probability of no failures in 50 years is given by the survivor function: $\exp(-50/\mu) = 0.47$.

Exponentially distributed interoccurrence times imply that the actual (random) number of failures in any interval $[0, t]$ is Poisson distributed with mean (and variance) μt. The expected failure frequency is simply μ expected failures per unit time. The Poisson distributions are stochastically increasing in μt (Ross, 1996, Chapter 9); therefore, a decision maker who prefers fewer failures by any given time should also prefer smaller frequencies.

The "Frequency" Concept for Nonexponential Failure Times

If interoccurrence times are not exponentially distributed, then the expected number of failures per unit time depends on the length of the time window used to compute the average. In general, different failure time distributions cannot be ranked usefully for decision-making purposes according to the reciprocals of their means (assuming that these exist); i.e., this definition of frequency is no longer adequate for comparing processes.

Example: Average Annual Frequency for Uniformly Distributed Lifetimes

Problem: Suppose that the random times between successive renewals (i.e., failures) have a uniform distribution with mean $\mu = 50$ years. (a) What is the average annual frequency of the renewal event? (b) How many renewal events are expected to occur in 50 years, if that is the design life of the system?

Solution: (a) The average annual frequency is $1/\mu = 1/(50 \text{ years}) = 0.02$ expected occurrences per year. (b) The expected number of occurrences in the interval $[0, t]$ for $0 \leq t \leq 1$ if the interoccurrence time is $U[0, 1]$, denoted by $M(t)$, may be found from the renewal equation (stating that the expected number of occurrences in $[0, t]$, if the first occurrence is at time s, is $1 + M(t - s)$, i.e., it is 1 + expected number after the first one):

$$M(t) = \int_{0 \leq s \leq t} [1 + M(t - s)]ds = t + F(t),$$

where $F(t) = \int_{0 \leq s \leq t} M(t-s)ds = \int_{0 \leq s \leq t} M(s)ds$. Using the initial condition $F(0) = 0$ yields $F(t) = \exp(t) - t - 1$, whence $M(t) = dF(t)/dt = \exp(t) - 1$ for any $0 \leq t \leq 1$ (Ross, 1996, Exercise 3.7). Rescaling the time axis to run from 0 to 100 years gives the solution:

$$M(t) = e^{0.01t} - 1 \text{ for } 0 \leq t \leq 100 \text{ years.}$$

For $t = 50$, $M(t) = \exp(0.5) - 1 \approx 0.65$ occurrences in 50 years.

Conflicts Among Different Criteria for Comparing Failure Time Distributions

Table 5.1 compares the average annual frequencies and the cumulative expected numbers of failures at different time points (1 year, 50 years, and 100 years) for three different processes. The first two have the exponential and uniform failure time distributions, respectively, discussed in the two preceding examples. The third has a discrete failure time distribution with equal probabilities of immediate failure (probability 0.5) or no failure for 100 years (0.5 probability). (These two probability masses may be positioned at some very small number greater than zero and another number slightly greater than 100 to avoid ambiguities about the timing of failures compared to the start and end of the system's design life.) The expected number of immediate failures, denoted by N, before the process exits to its long-duration

Table 5.1 Comparison of three processes with different failure time distributions

	Exponential time to failure, μ = 66.67 years	Uniform time to failure, $U[0, 100]$ μ = 50 years	Bernoulli time to failure, 0 or 100, μ = 50 years
Average annual frequency	0.015	0.02	0.02
Expected accidents by 1 year	0.015	0.01	1
Expected accidents by 50 years	0.750	0.65	1
Expected accidents by 100 years	1.500	1.73	1
Probability of $\geq$1 accident in 50 years	0.530	0.50	0.50
Probability of $\geq$1 accident in 1 year	0.015	0.01	0.50

(100-year) failure-free state can be calculated from the following equation, which says that there is a probability 0.5 of proceeding immediately to the 100-year failure-free lifetime (0 immediate failures) and a complementary probability of 0.5 of suffering an immediate failure and starting the process over:

$$N = 0.5^{*}0 + 0.5^{*}(1 + N).$$

Solving reveals that $N = 1$, as shown in Table 5.1.

The entries in Table 5.1 illustrate that *there is not necessarily an increasing relation between event frequencies and the expected number of failures.* Processes with lower frequencies of failure do not necessarily have smaller expected numbers of failure (or smaller probabilities of failure) over any time interval of interest, such as the design life of a system. For example, the exponential failure time process with $\mu = 66.67$ years has the smallest average annual frequency of failures (0.015 compared to 0.02) but has a higher expected number of failures by both 1 year and 50 years than the process with the uniformly distributed $U[0, 100]$ failure time. For longer design lives, however, the exponential failure time process eventually has fewer expected failures than the uniform failure time process, and the Bernoulli failure time process has the lowest expected number of failures (one).

More generally, *there is not necessarily an increasing relation between the long-run average frequency of a failure event and the expected number of failures over the design life of a system.* For example, a Bernoulli process that puts probability p on immediate failure and probability $(1 - p)$ on a greater number T of years prior to failure has the following frequency (still defined as the reciprocal of the mean time between failures):

$$\text{frequency} = 1/[p^{*}0 + (1 - p)T] = 1/[(1 - p)T].$$

The expected number of immediate failures is given by

$$N = (1 - p)^{*}0 + p^{*}(1 + N), \text{ implying that } N = p/(1 - p).$$

Since the frequency involves T and the formula for N depends only on p, we can adjust them independently. For example, with $T = 1{,}000$ and $p = 0.9$, the frequency would be $1/[(1 - 0.9)^{*}1{,}000] = 0.01$ and the expected number of immediate failures would be $N = 0.9/0.1 = 9$. Clearly, the lower frequency of 0.01 does *not* imply a lower expected number of failures over the life of the system (or over any shorter interval starting at $t = 0$) than the distributions in Table 5.1.

Do These Distinctions Really Matter?

It is natural to wonder whether the conflicts among different criteria illustrated in Table 5.1 arise in real systems. The Bernoulli and uniform distributions of failure times are admittedly contrived examples, not common in real systems. The Bernoulli distribution of failure times with point masses at times 0 and $T > 0$ is a highly idealized simplification of more realistic "bathtub-shaped" hazard functions for which high risks (hazard rates) of failure occur only early or late in the life of a system. The uniform distribution over a finite interval is a simplification of "hill-shaped" lifetime distributions (e.g., approximately normal or lognormal) in which most of the risk of failure is concentrated in an interval around the mean failure time, with failure risks sufficiently far from the mean being negligible by comparison. Would less idealized examples produce similar results? The answer is yes. We show next that *no satisfactory definition of "frequency" is possible* for a very wide variety of hazard functions, including realistic ones.

The most general concept of "frequency" that we will consider is that it provides a formula or algorithm for assigning *numbers* (i.e., frequencies) to *probability distributions* for time-to-failure so that each probability distribution receives a unique corresponding frequency number. [Probability distributions for failure times can be represented in multiple mutually equivalent ways, such as by cumulative distribution functions, decumulative distribution functions, survivor functions, hazard functions, cumulative hazard functions, mean residual life functions, and so forth. These different representations contain the same information, and it is easy to convert among them (Thompson, 1988; Cox, 2001). Hence, we will simply refer to "probability distributions" or "lifetime distributions," without caring about which specific representation is used. A specific definition of frequency maps all of these mutually equivalent representations in any case to the same single number, the frequency of failure.] The reciprocal of mean time-to-failure (when it exists) is an example of one possible definition of frequency. Expected-failures-per-unit time is another. Still others might be devised. Table 5.1 suggests that different definitions rank at least some probability distributions differently.

Suppose that a risk manager prefers one lifetime distribution to another if a system is more likely to survive to complete its mission when its lifetime (or time-to-failure) has the preferred distribution instead of the other one. Call a particular definition of frequency *satisfactory* (for purposes of comparing lifetime distributions and choosing among them) if it always assigns smaller numbers (i.e., lower "frequencies" of failure) to preferred lifetime distributions. That is, a satisfactory definition of frequency assigns numbers to lifetime distributions so that preferred distributions always receive smaller numbers than dispreferred distributions. We now show that no definition of frequency that depends only on lifetime distributions (and not on broader knowledge about the risk management decision problems that require choosing between lifetime distributions) can possibly be satisfactory in general.

Theorem 1 *No satisfactory definition of "frequency" exists for comparing or choosing among lifetime distributions.*

Proof The proof is by contradiction. Suppose that, to the contrary, a satisfactory definition of "frequency" does exist. Consider two different lifetime distributions, F and G, such that F is positive on the interval [4, 8] (else 0) and G is positive on the interval [2, 10] (else 0). [It may aid intuition to think of F and G as being uniform (or Bernoulli, if the condition of being nonzero throughout the interval is dropped) distributions with these endpoints, but no such restriction is required: F and G may be quite arbitrary distributions within their respective intervals.] Now, a decision maker who requires a system to survive for three time units to complete its mission will always prefer F to G, since F guarantees success and G does not. Thus, a satisfactory definition of frequency must assign F a lower frequency than G in order to give this decision maker the comparative information needed to choose correctly between them. Now consider a second decision maker who requires a system to survive for nine time units to complete its mission. The second decision maker will always prefer G to F, since G gives a higher success probability than F. Thus, a satisfactory definition of frequency should assign G a lower frequency than F in order to give this decision maker the information needed to choose correctly between them. But this contradicts the requirement for the first case. Therefore, no definition of frequency can exist that correctly orders F and G in both cases.

More generally, many common parametric families of lifetime distributions – including the Weibull, gamma, log-gamma, extreme value, and lognormal families that are often encountered in practical applications – contain members that cannot be compared by stochastic dominance or hazard rate orderings (Ross, 1996). The hazard functions for a pair of such noncomparable distributions intersect, so that one is preferred for some situations (e.g., for some specific required length of a mission) but not for others. This is more often the rule than the exception. Only in very special cases can the possibility of intersecting hazard functions be completely avoided. For example, the family of exponential distributions – that is, of distributions with constant hazard rates – is completely ordered by first-order stochastic dominance (ibid.), and for this family, the reciprocal of mean time-to-failure is a satisfactory definition of frequency.

The limitation in Theorem 1 is also relevant to qualitative risk-rating systems and risk matrices that require users to categorize "frequency" using ordered categorical scales such as *High*, *Medium*, and *Low*. As shown in the proof, the correct ordinal ranking of lifetime distributions to be used in choosing among them cannot, in general, be determined solely from the lifetime distributions themselves.

Summary of Limitations of the "Frequency" Concept

This section has illustrated two main points. The first is that for a given severity of adverse consequences, *a lower frequency of a risk event is not necessarily preferable to a higher frequency.* The reason is that a frequency number does not address how risk changes over time. Therefore, frequency-severity diagrams with axes such as "severity" (horizontal axis) and "average annual frequency" (or, for the more common cumulative format, "average annual exceedance frequency") (vertical axis) are not necessarily adequate for summarizing and comparing event rates.

The second point is that *there is, in general, no way to define frequency that is satisfactory,* in the sense that all risk managers who prefer longer to shorter survival times (or longer to shorter times between failures, in the renewal process setting) should necessarily prefer distributions with lower frequencies of failure to distributions with higher frequencies of failure. No possible definition of "frequency" can have this property in general, or even in the cases most often encountered in practice (e.g., for the families of Weibull or lognormal lifetime distributions). Intuitively, the reason is that which of two distributions should be preferred depends in general on system requirements (e.g., how long the system must function to accomplish its mission) other than the lifetime distribution itself. Thus, in general, any definition of frequency that is derived solely from the lifetime distribution – as all common definitions are, including those illustrated in Table 5.1 – does not provide sufficient information to determine which of two lifetime distributions is preferable. There are exceptions, such as when all distributions being compared are exponential (or, more generally, when all distributions belong to a one-parameter family totally ordered by first-order stochastic dominance). But when more general and realistic models are considered that allow for bathtub-shaped, hill-shaped, or other realistic hazard functions, then no satisfactory definition of frequency is possible.

To experts in probability and quantitative risk modeling, these caveats will, of course, not come as a surprise. Excellent texts and monographs (e.g., Thompson, 1988) have developed renewal process models and other point-process models for reliability and risk analysis that fully address the timing issues that are missing from simple frequency-severity descriptions. Probability theorists have rigorously extended the intuitive idea of frequency as the mean number of events per unit time to allow for time-varying intensities (Korolyook's theorem), and have also extended stochastic ordering relations for random variables (such as one decumulative frequency-severity curve or *F-N* curve lying above/to the right of another) to allow comparisons of renewal processes (Ross, op cit., p. 412). Thus, models and tools for describing time-varying risks are available and the somewhat ambiguous concept of frequency need not be used.

Nonetheless, many students are still introduced to quantitative risk assessment with sayings such as "Risk is probability times consequence," "Risk is probability *and* consequence," "Risk is likelihood and severity of consequences," or "Risk is frequency times consequence." Students who press for details – asking, "Probability (or likelihood or frequency) over what time interval?" – may be told about frequency-severity diagrams and average annual frequencies of consequences exceeding given severity levels. The caveats in this section may then prove useful.

Limitations of Aggregate Exposure Metrics

This section turns from the quantification of risk using "frequency" to the quantification of exposure using aggregate exposure metrics. These are numerical summaries of exposures to complex mixtures that are based on sums or weighted averages of component exposures. They are widely used in risk assessments of complex mixtures such as asbestos-associated dusts and fibers. Allowed exposure levels based on total particle or fiber counts and estimated ambient concentrations of such mixtures may be used to make costly risk management decisions intended to protect human health and to remediate hazardous environments. However, in general, *aggregate exposure information alone may be inherently unable to guide rational risk management decisions* when the components of the mixture differ significantly in potency and when the percentage compositions of the mixture's exposures differ significantly across locations. Under these conditions, which are common in practice, *aggregate exposure metrics may be "worse than useless,"* meaning that risk management decisions based on them are less effective than decisions that ignore the aggregate exposure information and select risk management actions at random.

The potential practical significance of these results is illustrated by a practical example where applying an aggregate unit risk factor (from the EPA's IRIS database) to aggregate exposure metrics produces average risk estimates about 25 times greater – and of uncertain predictive validity – compared to risk estimates based on specific components of the mixture that have been hypothesized to pose risks of human lung cancer and mesothelioma.

Use of Aggregate Exposure Metrics in Risk Assessment

Many regulatory risk assessments for known and suspected carcinogens use linear nonthreshold exposure-response models of the form

$$\textit{excess risk caused by exposure} = K \times \textit{exposure},$$

where K is the potency of exposure, i.e., the increase in risk per unit of exposure, for an adverse human health effect. Typical units for expressing excess risk are excess numbers of illnesses, deaths, QALYs lost, etc. per person-year in an exposed population. In symbols, the linear model is $r = Kx$, where x = exposure and r is the excess risk (e.g., in units of additional illnesses per person-year) when the exposure level is x.

When the exposure variable x represents a *mixture* of components with different potencies, fitting the simple linear model to data consisting of (x, r) pairs raises the possibility of *aggregation errors* in risk estimation, arising from the fact that the same value of the exposure metric, x, can represent different mixtures having different corresponding risks. It is important to consider how such aggregation errors can affect the quality of risk management decisions.

Aggregate Exposure Information May Not Support Improved Decisions

Suppose that there are n types or categories of particles in a mixture, having different potencies. If exposure is summarized by some aggregate exposure metric, consisting of a sum or weighted sum of the n components, then it may be impossible to use this aggregate exposure information to approximate the risk management decisions that would be made if the components were known. In other words, aggregate metrics summarizing exposures may not provide enough information to make good decisions. Even the best risk management decision based on a sum or weighted sum of components may be much less effective than the best decision based on disaggregated information. Indeed, when used with simple intuitive risk management decision rules such as "Clean up sites having sufficiently high exposures" (where "sufficiently high" refers to an action threshold set by regulators or other decision makers), aggregate exposure estimates may lead to decisions that are less effective than simply choosing actions at random.

Example: How Aggregate Exposure Information Can Be Worse Than Useless

Suppose that each of several geographic sites has its own mixture of two types of hazardous dust particles in its air. Type 1 particles have relative potency 1; type 2 particles have relative potency 8; and thus the risk from exposure to a mixture (x_1, x_2) of particles of types 1 and 2, respectively, at any site is $r = x_1 + 8x_2$. A risk manager measures the sum of type 1 and type 2 particles at each site (their sum, $x = x_1 + x_2$, is called "exposure"). She decides which sites to clean by giving priority to those with the highest levels of exposure. We will call the aggregate exposure measure, x, *worse than useless* if using it leads to worse decisions (e.g., less risk removed) than ignoring it and randomly selecting which sites to clean up.

For example, suppose that the characteristics of four sites are as follows:

Table 5.2 Higher aggregate exposures have lower risks

Site	x_1 = exposure to type 1 particles	x_2 = exposure to type 2 particles	Aggregate exposure $x = x_1 + x_2$	Risk $= x_1 + 8x_2$
A	0	3	3	24
B	2	2	4	18
C	4	1	5	12
D	6	0	6	6

In this hypothetical example, higher aggregate exposures correspond to lower risks. Thus, any decision rule of the form "Clean up a site if and only if its aggregate exposure level is sufficiently high" or "Clean up the sites with the highest aggregate exposure levels" will clean up lower-risk sites instead of higher-risk sites. Indeed, if we can afford to clean only one site, then picking one at random results in an expected risk removed of (24 + 18 + 12 + 6)/4 = 15, but using aggregate exposure to pick the site with the highest aggregate exposure (i.e., site D) to clean removes a risk of only 6. Thus, using the aggregate exposure data to allocate resources using any decision rule that assigns higher priority to sites with greater aggregate exposures is "worse than useless."

In this example, the aggregate exposure information x could be used to manage risks effectively by changing the decision rule to give priority to sites with the lowest aggregate exposure levels. But such rules perform poorly if the risk function is $r = x_1 + 0.08x_2$ instead of $r = x_1 + 8x_2$.

In general, as shown next, an effective decision rule using aggregate exposure data can be designed *only* by using information about the joint distribution of disaggregate exposures [such as (x_1, x_2) instead of only $x = x_1 + x_2$ in the above example]. But this takes away much of the point of using aggregate exposure data, such as total particle counts, when more detailed disaggregated exposure data are judged to be too expensive or otherwise impractical to obtain. Information about disaggregated exposure data may be essential to making use of aggregate exposure data to improve risk management decision making.

Theorem 2 *(Aggregate exposure information alone is inadequate for risk management decision making.). From aggregate exposure information alone (if the aggregation is nontrivial, meaning that different detailed exposure patterns creating different risks may have the same value for aggregate exposure), it is impossible to determine which of two sites has the higher risk.*

Proof To avoid nontriviality, there must be at least two components of an exposure vector $\boldsymbol{x}$ having different potencies (i.e., risk coefficients k_1 and k_2 with $k_1 \neq k_2$); thus, we will prove the result for the simplest case of two dimensions. Consider two exposure components, x_1 and x_2, with exposure-related risks given by $r = k_1x_1 + k_2x_2 = \boldsymbol{kx}$, $k_1 \neq k_2$. Here, $\boldsymbol{k} = (k_1, k_2)$ is a vector of component potencies. Let $\boldsymbol{a}$ be a vector of positive aggregation weights (to avoid triviality), with $x = \boldsymbol{ax} = a_1x_1 + a_2x_2$ being the aggregate exposure metric for exposure vector $\boldsymbol{x} = (x_1, x_2)$. Consider two sites with aggregate exposures x and y, respectively, where $x > y$. Then two possible values of $\boldsymbol{x}$ that are consistent with x are $\boldsymbol{x}^1 = (0, x/a_2)$, corresponding to risk level k_2x/a_2, and $\boldsymbol{x}^2 = (x/a_1, 0)$, corresponding to risk level k_1x/a_1. (Superscripts here indicate different values of the $\boldsymbol{x}$ vector, not exponents.) Similarly, two possible values of $\boldsymbol{y}$ that are consistent with y are $\boldsymbol{y}^1 = (0, y/a_2)$, corresponding to risk level k_2y/a_2, and $\boldsymbol{y}^2 = (x/a_1, 0)$, corresponding to risk level k_1y/a_1. Now, choose $\boldsymbol{k} = (0, 1)$; then $\boldsymbol{x}^1$ has risk level $x/a_2 > 0$ and $\boldsymbol{x}^2$ has risk level 0, while $\boldsymbol{y}^1$ has risk level $y/a_2 > 0$ and $\boldsymbol{y}^2$ has risk level 0. Thus, the site with aggregate exposure x has a risk level higher than that of the site with aggregate exposure y if the detailed exposure vectors are x^1 and y^2, but this risk ranking is reversed if the detailed exposure vectors are x^2 and y^1. Since either choice is compatible with the aggregate exposure levels of x and y, it is impossible to determine from these aggregate exposure levels alone which site has the higher risk. QED.

Note: Although choosing $\boldsymbol{k} = (0, 1)$ simplifies the proof, a similar argument can be constructed by choosing $\boldsymbol{k} = (1, 0)$, or any other $\boldsymbol{k} \neq \boldsymbol{a}$. Geometrically, if the lines $\boldsymbol{kx} = constant$ have a different slope from the lines $\boldsymbol{ax} = constant$, as must be true when $\boldsymbol{k} \neq \boldsymbol{a}$, then there will always exist vectors $\boldsymbol{x}$ and $\boldsymbol{y}$ for which $\boldsymbol{ax} > \boldsymbol{ay}$ but $\boldsymbol{kx} < \boldsymbol{ky}$; this holds for any number of dimensions ≥ 2. By varying the choice of $\boldsymbol{k} \geq \boldsymbol{0}$, the risk rankings of the extreme points $\boldsymbol{x}^1$, $\boldsymbol{x}^2$, $\boldsymbol{y}^1$, and $\boldsymbol{y}^2$ can be reversed according to the pattern shown above for the special case $\boldsymbol{k} = (0, 1)$.

More generally, let A denote a set of alternative possible *actions* (e.g., $A = \{clean, do\ not\ clean\}$) and let d be a *decision rule* that maps available information for a site where a decision must be made into a corresponding decision (i.e., a choice from A). Thus, if the information available about a site is denoted by I, then $d(I) \in A$ denotes the action in A specified for that site by decision rule d. We may define information I provided to decision rule d to be *worse than useless* if $d(I)$ is dominated (in the sense of first-order stochastic dominance) by a random choice rule that ignores I and randomly selects an action from A.

Theorem 3 *(Worse-than-useless decisions based on aggregate exposure metrics). Let $d(x)$ be a decision rule that determines which sites to clean up based on their levels of aggregate exposures, x, where x is a sum or weighted sum (with nonnegative weights) of the individual components. Assume that $d(x)$ is a nondecreasing function of x, with $d(x) = 1$ (clean up) for all sufficiently large values of x and $d(x) = 0$ (do not clean up) for all smaller values of x. If the purpose of risk management is to clean up sites that currently have the largest levels of risk, then $d(x)$ is worse than useless for some distributions of detailed exposures among sites.*

Proof Let t be a threshold such that $d(x) = 1$ if $x \geq t$, else $d(x) = 0$. The hyperplane $\boldsymbol{ax} = t$ separates sites that will be cleaned $[d(x) = 1]$ from those that won't $[d(x) = 0]$ according to decision rule $d(x)$. In two dimensions, two extreme points on this separating hyperplane are $\boldsymbol{x}^1 = (t/a_1, 0)$ and $\boldsymbol{x}^2 = (0, t/a_2)$, corresponding to aggregate exposure level t and to risk levels $r_1 = k_1t/a_1$ and $r_2 = k_2t/a_2$, respectively. (For mixture exposures with more than two exposures, set all but two components equal to zero for purposes of this proof.) Since $(k_1, k_2) \neq (a_1, a_2)$, one of r_1 and r_2 exceeds the other; without loss of generality, number the components so that $r_2 > r_1$. Now construct the two additional exposure vectors $\boldsymbol{x}^4 = (r_2/k_1, 0)$ with risk level r_2 and exposure level $(a_1k_2t)/(a_2k_1)$ and $\boldsymbol{x}^3 = (0, r_1/k_2)$ with risk level r_1 and exposure level $(a_2k_1t)/(a_1k_2)$. Since $\boldsymbol{x}^1$ and $\boldsymbol{x}^2$ both have aggregate exposure $\boldsymbol{x} = t$, both will be cleaned by decision rule $d(x)$. If $(a_2k_1)/(a_1k_2) > 1$, then $\boldsymbol{x}^3$ will also be cleaned [since its aggregate exposure $(a_2k_1t)/(a_1k_2)$ will exceed t] and $\boldsymbol{x}^4$ will not [since its aggregate exposure level $(a_1k_2t)/(a_2k_1)$ will be less than t], even though the risk level for $\boldsymbol{x}^4$ exceeds the risk level for $\boldsymbol{x}^3$. So, if the goal is to clean sites with the largest risk levels, then either it is a mistake to clean sites with exposure vector $\boldsymbol{x}^3$ or it is a mistake not to clean sites with exposure vector $\boldsymbol{x}^4$. On the other hand, if $(a_2k_1)/(a_1k_2) < 1$, then $\boldsymbol{x}^3$ will not be cleaned, even though it has the same risk level, r_1, as $\boldsymbol{x}^1$, which is cleaned. Again, it is either a mistake (with respect to the goal of cleaning sites with the highest risks) to clean $\boldsymbol{x}^1$, or a mistake not to clean $\boldsymbol{x}^3$. By nontriviality, $(a_2k_1)/(a_1k_2) \neq 1$; thus, one of the two preceding cases must hold. In either case, giving a sufficiently large majority of sites (e.g., all of them) exposure vectors for which $d(x)$ produces mistaken decisions makes $d(x)$ worse than useless. QED.

Multicollinearity and Aggregate Exposure Data

It is well known that strong correlations among the components of an exposure vector can make it difficult or impossible to draw valid conclusions about the effects of the individual components on risk. They can also make it impossible to estimate parameters from one set of studies with similar mixtures and apply them to predict risk correctly for a very different mixture of the same components.

Example: Multicollinearity Can Prevent Effective Extrapolation of Risk

The data in Table 5.3 are perfectly described by each of the following risk-vs.-exposure regression models: (1) $r = x = 0.1(x_1 + x_2)$; (2) $r = 0.2x_1$; (3) $r = 0.2x_2$.

If the aggregate exposure-response model $r = Kx = K(x_1 + x_2)$ is fit to these data, it will give a perfect fit ($R^2 = 1$) with $K = 0.1$. However, if this same model, $r = 0.1x$, is then applied to predict risks for a new set of exposure data with $x_1 = 0$ and x_2 having values of 2, 4, 6, and 8, it may give perfect predictions (if the true exposure-response relation is $r = 0.1x$) or completely false predictions (e.g., if the true exposure-response relation is $r = 0.2x_1$). There is no way to determine in advance whether the model $r = 0.1x$ has *predictive* validity for new situations, although it clearly has *descriptive* validity for the data in Table 5.3.

Table 5.3 Multicollinearity can prevent valid risk predictions from aggregate exposure data

Site	x_1 = exposure to type 1 particles	x_2 = exposure to type 2 particles	Aggregate exposure $x = x_1 + x_2$	Risk $r = 0.1x = 0.2x_1 = 0.2x_2$
A	1	1	2	0.2
B	2	2	4	0.4
C	3	3	6	0.6
D	4	4	8	0.8

A Practical Example: Different Predictions of Asbestos Risks at El Dorado Hills, CA

The theoretical limitations of the aggregate exposure metrics for managing and predicting risks discussed so far can have important real-world consequences for risk assessments of heterogeneous mixtures such as asbestos-associated dusts. Cox and Popken (2007) examine alternative risk estimates for 27 dust exposure scenarios studied by EPA Region IX in the El Dorado Hills region of California (Ladd, 2005). The different assessments were based on alternative assumptions about which components of the dust mixtures pose risks of lung cancer and mesothelioma, and their potencies. The scenario-specific risks were not significantly different from zero for most scenarios under one set of assumptions [the Berman and Crump, 2003 (BC) assumption set], but were unacceptably high (e.g., >1E-6) in many cases under the EPA assumption set (EPA).

The EPA exposure index is concerned with the airborne concentration of phase contrast microscopy equivalent *(PCME) Fibers*, defined as asbestos fibers with dimensions length $> 5\ \mu\text{m}$, $0.25\ \mu\text{m} \leq$ width $\leq 3\ \mu\text{m}$, and aspect ratio (length/width) > 3 (similar to the definition for PCME fibers provided in ISO, 1995). In contrast, the BC exposure index is concerned only with the airborne concentration of *Long Protocol Structures*, defined as asbestos fibers with dimensions *length* $> 10\ \mu\text{m}$, *width* $< 0.40\ \mu\text{m}$. Let

- x_1 = concentration (fibers/ml) of asbestos particles that are PCME Fibers, but not Long Protocol Structures,
- x_2 = concentration (fibers/ml) of asbestos particles that are both Long Protocol Structures and PCME Fibers.

(Particles that are neither PCME Fibers nor Long Protocol Structures are not of interest, since neither set of assumptions considers exposures to such particles to pose a risk to human health. The other logical possibility, particles that are Long Protocol Structures, but not PCME Fibers, can occur in principle for Long Protocol Structures having width $< 0.25\ \mu\text{m}$. However, in practice, these particles are seldom detected. Although phase contrast light microscopy may not resolve fibers below 0.2 microns in width, these very thin fibers are counted in the Berman Crump model. The asbestos exposures associated with epidemiological studies are typically indices of exposure, rather than absolute counts, because of this limit of resolution of the light microscope.)

The EPA exposure index (in units of fibers per milliliter) can then be written as

$$E_{\text{EPA}} = x_1 + x_2 \text{ (aggregate exposure metric with component weights } a_1 = a_2 = 1\text{)},$$

while the BC exposure index is

$$E_{\text{BC}} = x_2 \qquad (a_1 = 0,\ a_2 = 1).$$

Both models assume that the risk of mortality due to asbestos exposure is of the form

$$risk = \textit{unit risk factor} \times \textit{lifetime average daily exposure},$$

where *lifetime average daily exposure* reflects the proportion of a lifetime that an individual is exposed to asbestos, as defined by the given exposure index. The *unit risk factor* is computed as a function $f(K_L, K_M)$ with inputs K_L and K_M being dose-response coefficients for lung cancer and mesothelioma, respectively. This function is based on a Life Table analysis that accounts for competing risks of mortality at different ages (Berman and Crump, 2003, Appendix E). It assumes a linear function for lung cancer mortality risk as a function of cumulative exposure, a cubic function for mesothelioma mortality risk as a function of cumulative exposure (fibers per milliliter $\times$ years), and a 10-year lag for both types of risk.

The EPA assumptions provide much higher risk estimates than the BC assumptions: overall, the BC risks average only 4.1% of the EPA risks, based on simulations detailed in Cox and Popken (2007). In other geographic regions, however, the BC assumptions do not necessarily provide lower risks than the EPA's. For example, Berman (2006) provide risk computations for four asbestos studies, including El Dorado Hills. Risks were computed using both the BC and EPA assumptions. In each case except El Dorado Hills, the BC assumptions produced higher risks. The computed risk ratio for El Dorado Hills using BC vs. EPA exposure indices was 0.04, very similar to the simulation-based results (0.041).

The very different risk estimates obtained using the BC and EPA exposure metrics reflect the fact that the *particle size distribution* at El Dorado Hills (unlike other sites) contains very few Long Protocol Structures. This difference explains why the BC assumptions give approximately 25-fold lower risk estimates (corresponding to the estimated factor of 0.04) than the EPA risk estimates. Thus, in some practical applications, as well as in mathematical theory, simply applying unit risk factors estimated for one set of mixtures to a new, very different, mixture of the same components may produce risk estimates that are not known to have predictive validity and that may differ significantly from risk estimates based more specifically on the new mixture. In such cases, it may be essential, as Theorems 2 and 3 suggest, to use disaggregate exposure information and unit risk factors to obtain reliable risk estimates for the new mixture.

Summary of Limitations of Risk Assessments Based on Aggregate Exposure Metrics

Total exposures (e.g., particle or fiber counts, in our example) and other aggregate exposure metrics may not provide any useful information about where risks are highest and where remediation money should be spent when some of the components included in the total are irrelevant to human health risk, or, more generally, when different types of particles have very different potencies. Indeed, total exposures (or other aggregate exposure metrics) may be "worse than useless" (meaning, worse than random selection) for purposes of guiding effective risk management decisions. The only way to determine whether this is the case is to use disaggregate (component-specific) exposure and potency information. Doing so may reveal very different risks from those estimated using aggregate exposure data and aggregate potency estimates.

In practice, the differences among risk estimates obtained using different exposure metrics suggest that the discrepancy between risk estimates based on aggregate vs. disaggregate data can be quantitatively significant (e.g., an average of about 25-fold in the El Dorado Hills asbestos duct example). This phenomenon applies well beyond the example context of counting particles or fibers. It holds in contexts as diverse as risk assessment of mixtures of volatile organic compounds (VOCs) and/or particulate matter in air or drinking water; mixtures of different hazardous substances at hazardous waste sites; and mixtures of bacteria in antimicrobial risk assessments. For example, the following section criticizes use of the aggregate linear exposure-response model

$$Risk = K \times exposure$$

for the human health effects of animal antibiotics, in part because the exposure metrics consider only *resistant* bacteria in food servings, whereas most health effects are caused by *susceptible* bacteria. This may be viewed as a special case of the framework discussed in this section, with $\boldsymbol{x}$ = (resistant bacteria per serving, susceptible bacteria per serving), $\boldsymbol{a} = (1, 0)$, and $\boldsymbol{k}$ proportional to $(1, 1)$. The aggregate exposure metric can drive risk management decisions that harm human health (e.g., by reducing resistant bacteria but increasing susceptible bacteria, thus increasing average illness-days per capita-year and per serving ingested in exposed populations). This is a consequence of Theorem 2, with available risk management actions reinterpreted as $A = \{$*ban animal antibiotics*, *do not ban animal antibiotics*$\}$ and the decision rule $d(x)$ reinterpreted as: ban animal antibiotics if exposure to resistant bacteria in foods is sufficiently high. Theorem 3 implies that such decision rules can be worse than useless, resulting in increased rather than reduced harm to human health, because of the risk-relevant information lost in passing from detailed exposures to aggregate exposures.

Limitations of Aggregate Exposure-Response Models: An Antimicrobial Risk Assessment Case Study

Bartholomew et al. (2005) propose a framework for antimicrobial risk assessment that "is based on a linear relationship between the pounds of chicken consumed containing fluoroquinolone-resistant *Campylobacter* and the annual number of fluoroquinolone-resistant cases of campylobacteriosis in the U.S. population resulting from poultry." This relationship can be expressed in symbols as

$$risk = K \times exposure, \tag{5.1}$$

where

- *risk* refers to the expected number of cases of resistant campylobacteriosis per year in the United States caused by eating chicken (or, more generally, "resulting from poultry"),
- *exposure* refers to the pounds of chicken consumed containing fluoroquinolone-resistant *Campylobacter,* and
- *K* is interpreted by Bartholomew et al. as "the population-based dose-response parameter linking the two."

This framework provides an instructive example of key conceptual errors and pitfalls that can lead QRA practitioners to produce meaningless numbers and incorrect inferences if they attempt to use aggregate statistical relations and ratios instead of developing valid causal models (see Chapters 6 and 7). This section uses Equation (5.1) and its application to foodborne risks by the FDA Center for Veterinary Medicine (CVM), described by Bartholomew et al., as a case study of important errors that practitioners of QRA should avoid.

For practical applications, the authors state that "The linear relationship of the CVM risk assessment allows one to predict the effect of reduction in the prevalence of resistance among *Campylobacter*-contaminated chicken carcasses on the number of human cases of resistant campylobacteriosis." They assert that "Confirmation of the algebraically derived linear relationship as a surrogate for a complete farm-to-fork model came from the Danish researchers, Rosenquist et al." and conclude that "From this relationship, it is clear that regulating the prevalence of resistance in *Campylobacter* in poultry will control the prevalence of resistant human cases from poultry. Furthermore, CVM believes that this same basic linear relationship may be sufficiently general to model other pathogens responsible for food-borne illness."

However, as illustrated next, the framework (5.1) incorrectly treats a descriptive statistical equation as if it were a predictive causal one. Hence, it is not a valid framework for supporting risk management decision making that is intended to cause desired outcomes. In particular, while it is certainly possible to fit simple linear regression models such as (5.1) to (*exposure*, *risk*) data points, doing so does *not* in general produce a relationship that correctly predicts how *changes* in *exposure* will change *risk* – the question of greatest interest for risk management.

Statistical vs. Causal Relations

The following thought experiments reveal several potential limitations of the linear model (5.1) for predicting how changes in exposure will change risk.

Example: Significant Positive *K* for Statistically Independent Risk and Exposure

First, consider the performance of model (5.1) under the null hypothesis that the *exposure* and *risk* variables are unrelated. To be concrete, suppose we generate 1,000 pairs of statistically independent values for *exposure* and *risk*, each value being sampled independently from the unit uniform distribution $U[0, 1]$. Then fitting the proposed linear model $risk = K \times exposure$ to these data will produce a highly statistically "significant" positive K (with a value of K greater than 0.7), even though, by construction, the *risk* and *exposure* values are statistically independent. This is because the linear model is *misspecified* when the null hypothesis is true. The correct statistical regression relation is: $E(risk \mid exposure) = 0.5 + 0 \times exposure$. But since model (5.1) has no intercept term, it is unable to provide a correct fit, instead indicating a significant positive relation where none exists. More generally, without postulating anything about the statistical dependence (or lack of it) between the *exposure* and *risk* variables, it is clear that model (5.1) implies a positive K whenever *exposure* and *risk* are positive random variables – even if their values are actually unrelated or are negatively related. [As a practical matter, real-world data often do exhibit a significant *negative* relationship between exposure to chicken and risk of campylobacteriosis (e.g., Friedman et al., 2000; Table 1 of Effler et al., 2001; Table 2 of Phillips et al., 2004). Model (5.1) is intrinsically unable to describe such relationships.]

Example: A Positive K Does Not Imply That Risk Increases with Exposure

The most actionable, policy-relevant implication of model (5.1) is that the value of K shows how much a unit reduction in exposure will reduce risk. That is the interpretation that Bartholomew et al. suggest by stating that "From this relationship, it is clear that regulating the prevalence of resistance in *Campylobacter* in poultry will control the prevalence of resistant human cases from poultry." But this *causal* interpretation of a merely *statistical* relationship is unjustified. Recall from the first example in Chapter 3 that *there is no necessary relation between the value of K in a statistical model (e.g., 0.1 in the example) and the change in risk that would be caused by changing exposure.* The *statistical* relation between exposure and risk in a population can be positive (e.g., $risk = 0.1 \times exposure$) even if the *causal* effect of *exposure* is to *reduce risk* in the population (e.g., by shifting men from a higher-risk unexposed group to a lower-risk exposed group).

In statistical terminology, model (5.1) is a reduced-form equation, not a structural (i.e., causal) equation (Shipley, 2000). The belief that "The linear relationship of the CVM risk assessment allows one to predict the effect of reduction in the prevalence of resistance among *Campylobacter*-contaminated chicken carcasses on the number of human cases of resistant campylobacteriosis" reflects this confusion between structural and reduced-form models. In general, fitting a simple reduced-form model to data does *not* allow one to correctly predict the effects of changing the independent variables on resulting changes in the dependent variable (Shipley, 2000; Freedman, 2004).

Example: Statistical Relations Do Not Predict Effects of Changes

To show why Equation (5.1) does not allow one to predict the effects on *risk* of changing *exposure*, consider the following system of structural equations:

$$risk = age - exposure, \tag{5.1a}$$

$$exposure = (1/3) \times age. \tag{5.1b}$$

These are explicitly intended as structural equations representing causal relations. *Risk* increases in proportion to increases in *age*, decreases in proportion to *exposure*, and *exposure* increases in proportion to *age*. Rewriting (5.1b) as the (non-causal, although algebraically equivalent) equation *age* = 3 × *exposure* and substituting it into (5.a) gives the reduced-form model

$$risk = (3 \times exposure) - exposure = 2 \times exposure. \tag{5.1c}$$

The reduced-form model (5.1c) is of the form *risk* = *K* × *exposure*, with *K* = 2. However, this is only a *statistical* relation. It would be perfectly valid for statistical inference applications such as predicting the magnitude of *risk* in a population from measured values of *exposure*, but it is not valid for predicting how a *change* in *exposure* will affect *risk* – namely, in this case, by decreasing it in direct proportion to the increase in *exposure*, as shown in Equation (5.1a). In short, while the statistical value of *K* based on measured historical levels of *risk* and *exposure* would be (*risk/exposure*) = 2, the corresponding causal value should be (Δ*risk/*Δ*exposure*) = –1. In this sense, the descriptive statistical ratio *K* = 2 has no relevance for the predictive causal ratio (Δ*risk*/Δ*exposure*) = –1. They do not even have the sane sign.

These examples illustrate the crucial distinction between aggregate statistical models for a population and causal models for predicting how changes in exposures will change risks in the population. In general, the interpretation of the parameter *K* in the reduced-form statistical equation (5.1a) as "a population-based dose-response parameter" is not valid. Rather, *K* is simply a ratio of two historical aggregate quantities, interpreted by Bartholomew et al. as *exposure* and *risk*, but this ratio does not necessarily reflect (or predict) anything about how future changes in *exposure* (e.g., due to a ban on enrofloxacin) will affect future values of *risk*. Yet this is what risk managers need to know.

Prevalence vs. Microbial Load as Exposure Metrics

Equation (5.1) uses the *prevalence* of fluoroquinolone-resistant *Campylobacter* ("pounds of chicken consumed containing fluoroquinolone-resistant *Campylobacter*") as a surrogate for exposure. But prevalence (i.e., whether contamination is present) is, in general, *not* an adequate exposure surrogate for predicting risks that depend on *how much* contamination is present. To take an extreme counterexample for simplicity, multiplying all microbial loads of *Campylobacter* on chicken servings by 1,000,000 would presumably greatly increase risk but would leave prevalence completely unchanged.

Bartholomew et al. state that "Confirmation of the algebraically derived linear relationship as a surrogate for a complete farm-to-fork model came from the

Danish researchers, Rosenquist et al." However, a fuller reading reveals that the Rosenquist et al. paper refutes, rather than confirms, the assumption of a linear relationship between relevant exposures and risk. Rosenquist et al. (and common sense) predict that if one doubles the proportion of contaminated birds *while leaving microbial load per bird constant*, then one would double the expected number of illnesses caused by eating contaminated chicken servings. Bartholomew et al. cite and quote this result without the italicized caveat. But Rosenquist et al. also show that *if microbial load per bird changes*, then risk is no longer proportional to prevalence, and prevalence does not predict risk accurately. For example, they state that "The incidence of campylobacteriosis related to consumption of chicken was reduced significantly by reducing the number of *Campylobacter* on the carcasses, even though such a reduction had almost no influence on the fraction of positive chickens." Rosenquist et al. also note that this finding "demonstrates the need for quantitative detection methods" (e.g., estimating changes in microbial loads) and that "qualitative" surrogates (e.g., prevalence) are *not* adequate for predicting risk when microbial loads change. It is this conclusion that is relevant to a ban on enrofloxacin that affects microbial loads, for example, by increasing airsacculitis-positive flocks (Cox and Popken, 2006b; Russell, 2003). It is a conclusion widely accepted in the field of microbial risk assessment. For example, WHO (2002) noted in its *Salmonella* risk assessment that "Unlike a change in prevalence, a change in concentration of the pathogen does not necessarily have a linear relationship with the risk outcome." It gives an example in which an intervention that reduces risk by 62% has no effect on the prevalence of contamination.

Attribution vs. Causation

The framework proposed by Bartholomew et al. estimates K by taking the ratio of two estimated quantities, as in the following formula:

$$K_{\text{res}} = (\textit{the number of fluoroquinolone-resistant campylobacteriosis cases attributable to poultry})/(\textit{the pounds of fluoroquinolone-resistant Campylobacter-contaminated poultry produced}).$$

In this equation, the phrase "attributable to poultry" plays a special role. It reflects a policy decision about how many cases to *blame* on eating chicken, which may be unrelated to the number of cases *caused* by eating chicken. Modelers or policy makers may use a wide range of values for "attributable" numbers with varying rationales, depending on their interests. For example, Bartholomew et al. state that "The estimate of the proportion of all cases due to chicken (57.4%) was taken from two case-control studies." But neither study quantified the proportion of all cases "due to" (i.e., caused by eating) chicken, or provided data from which this proportion could be calculated. Rather, they provided estimated population attributable fractions (PAFs) for special populations (e.g., young male college students). PAFs are epidemiological measures based on statistical associations (uncorrected for key confounders) having no necessary relation to causality. For example, PAFs may be large and positive even for protective factors, for reasons similar to those illustrated above.

Both studies selected by Bartholomew et al. used data from the early 1980s, prior to a historical reduction in *Campylobacter* loads for broilers in the United States of perhaps 90% or more (e.g., Stern and Robach, 2003). Thus, if the PAFs in these studies were about 57.4% then, they might be closer to 6% now. Moreover, multiple recent studies show that the runoff of human-use antibiotics (including fluoroquinolones) in sewage and water is probably a more important source of resistant *Campylobacter* than was recognized in the early 1980s (e.g., Renew and Huang, 2004; Miao et al., 2004). This presumably should reduce the fraction of cases attributed to poultry (since runoff was ignored in the early studies). Bartholomew et al.'s decision instead to attribute the contributions to risk from these and all other unidentified sources to enrofloxacin used in poultry, even though poultry-use-specific antibiotics explicitly do *not* make any detectable contribution to runoff (ibid.), amounts to a policy decision to allocate cases to a source targeted for regulatory action, even if other unrelated sources actually caused the cases in question.

Similarly, larger and more recent case-control studies that show much smaller or zero (or negative) PAFs for campylobacteriosis and chicken (e.g., Friedman et al., 2004, which reports PAF values of less than 25% for restaurant-prepared chicken, similar to other restaurant-prepared meats, based on data that also show a *negative* PAF for home-prepared chicken) were passed over in favor of the two smaller, earlier studies that gave much higher PAF estimates. Finally, some of the risk attribution calculations cited by Bartholomew et al. appear to be based on nonstandard formulas that do not necessarily give correct results. For example, the quantity "probability that a resistant case of campylobacteriosis is attributable to chicken" was estimated as the product of "probability a case of campylobacteriosis is attributable to chicken" $\times$ "probability that a case is resistant." But this formula is incorrect in general. For example, it implies that if *all* campylobacteriosis cases are caused by chicken (so that "probability a case of campylobacteriosis is attributable to chicken" = "probability that a resistant case of campylobacteriosis is attributable to chicken" = 100%) and if 10% of cases are resistant (so that "probability that a case is resistant" = 10%), then 100% = 100% $\times$ 10%, which is incorrect. In short, the number of cases (both resistant and total) that are deemed to be "attributable to poultry" in the framework presented by Bartholomew et al. appears not to have been derived from data or from generally valid scientific formulas in any clear, objective way that shows that the resulting numbers of cases are actually caused by chicken consumption. Rather, they appear to reflect a policy decision about how to allocate illness cases to sources in the context of preparing to regulate an animal antibiotic that has not been shown in this risk assessment to cause any adverse effects in humans.

Human Harm from Resistant vs. Susceptible Illnesses

The risk management framework (5.1) proposes to quantify risk without considering either the frequency or the magnitude of adverse consequences – traditionally considered the *sine qua non* components of quantitative risk assessment. Indeed, Bartholomew et al. define risk as "the expected number of cases of resistant campylobacteriosis per year in the United States caused by eating chicken," which makes no reference to any incremental harm caused by resistance. Bartholomew et al. state that "CVM assumed that susceptible and resistant *Campylobacter* were equally likely to survive and equally likely to cause illness. Given no survival or virulence differential between susceptible and resistant *Campylobacter*, a similar linear relationship also holds between the ... number of cases with *Campylobacter* illnesses, susceptible and resistant combined, and the total number of pounds of chicken V contaminated with *Campylobacter* produced in a year." Yet, if susceptible and resistant *Campylobacter* are approximately equally virulent and pose similar or identical human health risks, then a ban on enrofloxacin that increases microbial loads of susceptible *Campylobacter* while reducing the microbial loads of resistant *Campylobacter* could presumably be either good or bad for human health, depending on which effect is larger.

In light of these possibilities, one key issue is whether resistance causes harm that would not have occurred without it. A current clinical perspective suggests that the answer is usually probably not (Ang and Nacham, 2003). But this question is not addressed at all in the linear modeling framework (5.1). Therefore, the framework does *not* quantify human health risk in the usual sense of quantifying changes in the frequency or severity of adverse human health effects caused by exposures. Without such information, rational risk management decisions cannot be made based on their probable human health consequences.

A second key question is whether banning enrofloxacin is likely to create more or less human health harm than it prevents. To answer this question, it is helpful to generalize model (5.1) as follows:

$$\begin{aligned} risk = {} & K_1 \times (\textit{exposure to susceptible campylobacter per chicken serving}) \\ & + K_2 \times (\textit{exposure to resistant campylobacter per chicken serving}) \end{aligned} \tag{5.2}$$

and to estimate K_1 and K_2 as the average number of illness-days (or QALYs lost, etc.) per "susceptible" case and per "resistant" case, respectively (if this dichotomy is preserved). The exposure variables now refer to microbial loads [measured in colony forming units (CFUs) ingested per serving] rather than to prevalence. If the effects of an enrofloxacin ban are (a) to eliminate all resistant *Campylobacter* CFUs, replacing them with susceptible CFUs, and (b) to increase by 1% the proportion of airsacculitis-positive flocks (having an average risk-per-serving estimated as at least 10 times greater than for airsacculitis-negative flocks; see Cox and Popken, 2005, and Russell, 2003), then the relative number of cases per year from chicken will increase by a factor of $0.99 \times 1 + 0.01 \times 10 = 1.09$, i.e., by 9%. The number of illness-days per case will decrease if $K_1 < K_2$. Although clinical experience suggests that K_1 and K_2 may be nearly identical (Ang and Nacham, 2003), if a conservative assumption is made instead that $(K_1/K_2) = 0.75$, then it can be calculated that a ban on enrofloxacin is expected to cause hundreds or thousands of illness-days (from increased susceptible cases) for each illness-day that it might prevent. (See Cox and Popken, 2005, for details, data, and sensitivity analyses.) [Since a straight line from the origin to a point on any convex dose-response curve provides a linear approximation that *overestimates* the risk reduction due to reductions in exposure and *underestimates* the increase in risk due to an increase in exposure, this conclusion from the linear model (5.2) remains valid if the true but unknown dose-response relation is nonlinear and convex.]

In summary, while Bartholomew et al. assert that "From this relationship [the linear model (5.1)], it is clear that regulating the prevalence of resistance in *Campylobacter* in poultry will control the prevalence of resistant human cases from poultry," the extended model (5.2) shows that even decreasing the prevalence of resistance in poultry to zero by a change (such as banning enrofloxacin) that increases susceptible cases may greatly increase human health risks from chicken consumption, by several orders of magnitude. Continued surveillance and dynamic modeling of experiences in regions or countries where the use of enrofloxacin in poultry is minimal are likely to prove essential in validating and/or improving default modeling assumptions, especially where the true causal relations involved are uncertain. Bartholomew et al. conclude that "Furthermore, CVM believes that this same basic linear relationship may be sufficiently general to model other pathogens responsible for food-borne illness." We suggest that Equation (5.2) provides a useful further generalization that may be important when the total human health impact of regulations that affect both susceptible and resistant bacteria is of interest.

Summary of Limitations of Aggregate Exposure-Response Model, Risk = K × Exposure

The linear model *risk* = *K* × *exposure* has sometimes been proposed as a useful framework for using aggregate data on *exposure* and *risk* levels to estimate how changes in *exposure* will change *risk*. It is not. Risk assessors who try to use such aggregate statistical relations as causal models are deluded.

The FDA Center for Veterinary Medicine (CVM) and Bartholomew et al. (2005) suggest that once *K* has been estimated from historical data, it can be used to predict how limiting future *exposure* will reduce future *risk*. This interpretation is unjustified. (As a counterexample, suppose that one were to estimate the ratio of car accidents in the United States to quarts of orange juice consumed in the United States, for each of several past years. Although this ratio would be positive, it would not justify a causal inference that restricting orange juice consumption would reduce car accidents!) Despite its appealing simplicity, this proposal confuses a possibly meaningless *descriptive* statistical ratio with a valid *predictive* causal relation. The historical ratio $K = (risk/exposure)$ does not necessarily predict how (if at all) changing *future* exposures will affect *future* risks. Hence, it is not an appropriate guide to current risk management actions.

This section has identified several limitations of the proposed framework, including the omission of the frequency and severity of human health harm in quantifying *risk* and the omission of the microbial load from *exposure*. Moreover, an extended linear model that considers impacts of changing animal antibiotic use on susceptible, as well as on resistant, bacteria indicates that reducing *exposure* might greatly increase *risk*!

Some Limitations of Risk Priority-Scoring Methods

This section examines some intrinsic limitations in the performance of all possible priority-setting rules, evaluated as guides to rational action. Most of the results are well known in decision analysis and financial risk analysis and/or are mathematically straightforward. However, they are of great practical importance for understanding the limitations of risk-scoring methods and for improving risk management. In general, *risk-scoring methods are not appropriate for correlated risks.* Indeed, as we will demonstrate, they are not necessarily better than (or even as good as) the purely random selection of which risk management activities to fund.

More constructively, when risk-reducing opportunities have correlated consequences due to uncertainties about common elements (such as the potencies of chemicals or the effectiveness of interventions), then *optimization methods can achieve greater risk reduction benefits for resources spent than can priority-scoring rules.* In general, the best choice of a subset of risk-reducing activities cannot be expressed by priority scores. Instead, optimization techniques that consider interdependencies among the consequences of different risk-reducing activities are essential. Fortunately, such methods are easy to develop and implement. They can substantially improve the risk reduction return on investments in risk-reducing activities.

Motivating Examples

Many organizations currently rate, rank, or score different hazards (sources of risk) or risk-reducing opportunities at least once a year to identify the currently top-ranked opportunities that will be addressed in the current budget cycle. The use of priority-scoring and rating systems is widespread and is becoming even more prevalent as they are incorporated into national and international standards and regulations. Examples of important applications include the following.

Example: Scoring Information Technology Vulnerabilities

The *Common Vulnerability Scoring System* (CVSS) for rating information technology (IT) system vulnerabilities uses scoring formulas such as the following to help organizations set priorities for investing in security risk reductions:

BaseScore = (0.6*Impact + 0.4*Exploitability-1.5)**f*(Impact),
Impact = 10.41*(1-(1-ConfImpact)(1-IntegImpact)*(1-AvailImpact)),
Exploitability = 20*AccessComplexity*Authentication*AccessVector,
f(Impact) = 0 if Impact = 0; 1.176 otherwise,
AccessComplexity = case AccessComplexity of
high: 0.35,
medium: 0.61,
low: 0.71,
Authentication = case Authentication of
Requires no authentication: 0.704,
Requires single instance of authentication: 0.56,
Requires multiple instances of authentication: 0.45,
AccessVector = case AccessVector of
Requires local access: 0.395,
Local Network accessible: 0.646,
Network accessible: 1
(*Source:* NIST, 2008).

Such a rule base, no matter how complex, can be viewed as an algorithm that maps categorized judgments and descriptions (such as that access complexity is "high" and that local access is required) into corresponding numbers on a standard scale. Higher numbers indicate greater vulnerability and the need for remedial action. Proponents envision that, "As a part of the U.S. government's SCAP (Security Content Automation Protocol) CVSS v2 will be used in standardizing and automating vulnerability management for many millions of computers, eventually rising to hundreds of millions" (http://www.first.org/cvss/).

Example: Scoring Consumer Credit Risks

The practice of rank-ordering consumers based on credit scores is ubiquitous in business today. A recent description states that

> FICO® risk scores rank-order consumers according to the likelihood that their credit obligations will be paid as expected. The recognized industry standard in consumer credit risk assessment, FICO® risk scores play a pivotal role in billions of business decisions each year. ... [They] are widely regarded as essential building blocks for devising successful, precisely targeted marketing, origination and customer management strategies by credit grantors, insurance providers and telecommunications companies.

Examples include BEACON® at Equifax US and Canada, FICO® Risk Score, Classic at TransUnion US, and Experian/Fair Isaac Risk Model at Experian (*source:* www.fairisaac.com/fic/en/product-service/product-index/fico-score/).

Example: Scoring Superfund Sites to Determine Funding Priorities

The State of Connecticut (www.ct.gov/dep/lib/dep/regulations/22a/22a-133f-1.pdf) published a *Superfund Priority Score method*, to be used in determining funding priorities for remediation of Superfund sites. Users must score each of many factors (reflecting exposure potential; groundwater impact; surface water impact; toxicity, persistence, mobility, and quantity of hazardous substances; impact to the environment, including Species of Special Concern; and potential air release and fire hazards) using ordered categories. Each category carries a certain number of points. For example, an area that contains a "rare" species gets a score of 4 on this factor. If it has a "declining or infrequent" species, the score is 3; for a "habitat-limited species," the score is 2. If this factor (species of concern) is not applicable, the score for this factor is zero. The scores for all factors are summed. The resulting total score determines "the priority for funding of remedial action at sites on the SPL" [the State of Connecticut Superfund Priority List].

Example: Priority Scoring of Bioterrorism Agents

MacIntyre et al. (2006) propose a risk priority-scoring system for bioterrorism agents. They describe their approach as follows:

> "Disease impact criteria were as follows: infectivity of the agent (person-to-person transmission potential), case fatality rate, stability in the environment and ease of decontamination, incidence of disease per 100,000 exposed persons in the worst-case release scenario, and reports of genetic modification of the agent for increased virulence.
>
> - Probability of attack criteria was [sic] designated as: global availability and ease of procurement of the agent, ease of weaponization, and historical examples of use of the agent for an attack.
> - Prevention/intervention criteria were categorized as: lack of preventability of the disease (such as by vaccination) and lack of treatability of the disease (such as by antibiotics).
> - For each of the scoring categories, a score of 0 to 2 was assigned for each category A agent as follows: 0 = no, 1 = some/low, and 2 = yes/high. The sum of these scores (of a total possible score of 20) was used to rank priority."

This is similar to the Superfund scoring system, in that categorical ratings for various factors are assigned numerical scores, and the sum of the scores is used to set priorities. In neither case did the authors verify whether *additive independence* conditions hold, which are required in multiattribute value and utility theory to justify additive representations of preferences (Keeney and Raiffa, 1976). For example, an agent with a score of 2 for "lack of preventability of disease" and 0 for "lack of treatability" would have the same sum for these two factors (2 + 0 = 2) as an agent with "lack of preventability of disease" = 0 and "lack of treatability" = 2, or as an agent with "lack of preventability of disease" = 1 and "lack of treatability" = 1. Yet risk managers who can completely prevent a disease ("lack of preventability of disease" = 0) might not care as much about whether it is treatable as they would if the disease could not be prevented. Likewise, in Superfund site scoring, many decision makers might care less about the presence of a declining species near a site that creates no exposure than near a site that creates a large, toxic exposure. Such interactions among factor scores are ignored in purely additive scoring systems.

Example: Threat-Vulnerability-Consequence (TVC) Risk Scores and Risk Matrices

Many organizations use numerical priority-scoring formulas such as *risk* = *threat* × *vulnerability* × *consequence* or *risk* = *threat* × *vulnerability* × *criticality* or *risk* = *threat* × *vulnerability* × *impact*. The Department of Homeland Security, the Department of Defense, and the armed services prioritize antiterrorism risk reduction efforts using such formulas (Jones and Edmonds, 2008; Mitchell and Decker, 2004; http://www.ncjrs.gov/pdffiles1/bja/210680.pdf). The formula *risk* = *threat* × *vulnerability* × *consequence* also provides the conceptual and mathematical basis for the RAMCAP™ (Risk Analysis and Management for Critical Asset Protection) standard and related compliance training and software (www.ramcapplus.com/). Chapter 15 discusses RAMCAP™ in greater detail. Law enforcement officers have been trained to use *risk* = *threat* × *vulnerability* × *impact* scoring systems to set priorities for managing security risks at major special events. Unfortunately, when the components on the right-hand side (e.g., *threat*, *vulnerability*, and *consequence*) are correlated random variables – for example, because attackers are more likely to attack facilities with high *vulnerability* and *consequence*, or because larger storage facilities have higher *vulnerability* and *consequence* than small ones – then the product of their means differs from the mean of their product, and it is not clear what either one has to do with risk. Correct expressions require additional terms to adjust for nonzero covariances, as discussed further in Chapter 15. Similar comments apply to risk matrices.

Priorities for Known Risk Reductions

To enable formal analysis in a reasonably general framework, we define a *priority-setting process* as consisting of the following elements:

1. *A set of items to be ranked or scored.* The items may be hazards, threats, interventions, assets, frequency-severity pairs, threat-vulnerability-consequence triples, threat-vulnerability-consequence-remediation cost quadruples, Superfund sites, construction projects, or other objects. We will refer to them generically as "items," "hazards," "prospects," or "opportunities."
2. *An ordered set of priority scores* that are used to compare hazards. These may be ordered categorical grades, such as "High," "Medium," and "Low"; nonnegative

integers indicating relative priority or ranking; or nonnegative real numbers, representing values of a quantitative priority index such as *risk* = *threat* × *vulnerability* × *consequence*, or *priority index* = *expected benefit of remediation/expected cost of remediation*, where the italicized variables are nonnegative numbers.

3. *A priority-scoring rule.* A scoring rule is a mathematical function (or a procedure or algorithm implementing it) that assigns to each hazard a unique corresponding priority score. (This implies that any two hazards having identical attribute values, or identical joint distributions of attribute values, must have the same priority score.)

The priority-scoring rule determines a *priority order* in which hazards are to be addressed (possibly with some ties). Addressing a hazard is assumed to reduce risk, and hence to be valuable to the decision maker: It increases expected utility. For example, it may stochastically reduce the flow of illnesses, injuries, or fatalities resulting from a hazardous process, activity, or environment.

Although items might have multiple attributes, and value trade-offs might make preferences among them difficult to define clearly in practice, we will assume that the decision maker has perfectly clear, consistent preferences for the consequences of addressing different hazards. For example, suppose that addressing hazard j reduces loss of quality-adjusted life-years (QALYs) by an amount, x_j, defined as the difference between the number of QALYs lost if hazard j is left unaddressed and the number of QALYs lost if hazard j is addressed. Suppose that all QALYs are considered equally *intrinsically valuable*, with twice as many being worth twice as much to the decision maker. More generally, we assume that addressing hazards creates gains on a *measurable value scale* satisfying standard axioms (Dyer and Sarin, 1979) that allow preferences for changes in or differences between situations, from before a hazard is addressed to after it is addressed, to be coherently ranked and compared. Let x_j be the measurable value from addressing hazard j. We assume that the value of addressing a hazard, expressed on such a measurable value scale, depends only on its attributes, and we work directly with the measurable values, rather than the underlying attributes. (The value scale need not be measured in QALYs, but thinking of such a concrete example may aid intuition.) If it costs the same amount to address any hazard, and if the resulting increases in value are known with certainty, then, *for any budget, total benefits are maximized by addressing the hazards in order of their decreasing values,* x_j. This provides one useful model for priority-based risk management decision making.

Priorities for Independent, Normally Distributed Risk Reductions

Next, suppose that the value achieved by addressing hazard j is uncertain. This might happen, for example, if the quantities or potencies of hazardous chemicals stored at different waste sites are uncertain, or if the sizes of exposed populations and their susceptibilities to exposure are not known, or if the effectiveness of interventions in reducing risks is in doubt. To model priority-based risk management decisions with uncertainty about the sizes of risk reduction opportunities, we assume that their values are random variables and that the decision maker is risk-averse. For a risk-averse decision maker with a smooth (twice-differentiable) increasing von Neumann-Morgenstern utility function for the value attribute, the conditions in Table 5.4 are all mutually equivalent, and all imply that the utility function is exponential. If one or more of these conditions is considered normatively compelling, then an exponential utility function should be used to choose among prospects with uncertain values.

Assuming an exponential utility function for the value attribute, the expected utility of any random variable corresponds to its moment-generating function. For example, let X_j represent the uncertain measurable value of addressing hazard j, modeled as a random variable on the value axis. Let $\mathrm{CE}(X_j)$ denote the certainty equivalent of X_j, i.e., the value (such as QALYs saved) received with certainty that would have the same expected utility as (or be indifferent to) random variable X_j. Then if X_j is normally distributed with mean $E(X_j)$ and variance $\mathrm{Var}(X_j)$, it follows (from inspection of the moment-generating function for normal distributions) that its certainty equivalent is

Table 5.4 Equivalent characterizations of exponential utility functions

Let X and Y be any two risky prospects (random variables) measured on the intrinsic value scale. They represent the uncertain values (e.g., QALYs saved) by addressing two different hazards.

- *Strong risk independence:* Adding the same constant to both X and Y leaves their preference ordering unchanged. Thus, if $X + w$ is preferred to $Y + w$ for *some* value of the constant w, then X is preferred to Y for *all* values of w.
- *Risk premium independence:* The decision maker'srisk premium (the amount she is willing to pay to replace a prospect with its expected value) for any risky prospect depends only on the prospect. (Thus, it is independent of background levels of the value attribute.)
- *Certainty-equivalent independence:* If a constant, w, is added to every possible outcome of a prospect X, then thecertainty equivalent of the new prospect thus formed is $\mathrm{CE}(X) + w$, where $\mathrm{CE}(X)$ denotes the certainty equivalent (or "selling price" on the intrinsic value scale) of prospect X. (This is sometimes called the "delta property," due to Pfanzagl, 1959.) Thus, for any constant, w, $\mathrm{CE}(w + X) = \mathrm{CE}(X) + w$.
- *Equal buying and selling prices:* For any prospect X and any constant w, the decision maker is indifferent between $w + \mathrm{CE}(X) - X$ and $w + X - \mathrm{CE}(X)$.
- *No buying-price/selling-price reversals:* The ranking of prospects based on theircertainty equivalents (i.e., "selling prices," e.g., how many QALYs would have to be saved with certainty to offset the loss from abandoning the opportunity to save X QALYs) never disagrees with their ranking based on "buying prices" (e.g., how many QALYs a decision maker would give up with certainty to save X QALYs). [This assumes the decision maker isrisk-averse; otherwise, the linearrisk-neutral utility function $u(x) = x$ would also work.]
- *Exponential utility:* $u(x) = 1 - e^{-kx}$.

References: Dyer and Jia (1998); Hazen and Sounderpandian (1999).

$$\mathrm{CE}(X_j) = E(X_j) - (dk/2)\mathrm{Var}(X_j),$$

where k is the coefficient of risk aversion in the exponential utility function (Infanger, 2006, p. 208).

A set of equally costly risk-reducing measures with independent, normally distributed values can be prioritized in order of decreasing $\mathrm{CE}(X_j)$ values. *For any budget, the total expected utility is maximized by funding risk reduction opportunities in order of decreasing priority* until no more can be purchased. Moreover, even if the risk-reducing measures do not have identical costs, an optimal (expected utility-maximizing, given the budget) policy maximizes the sum of certainty equivalents, subject to the budget constraint. (This follows from the additivity of means and of variances for independent risks. Finding an optimal subset in this case is a well-studied combinatorial optimization problem, the knapsack problem.) Thus, for any two feasible portfolios of risk-reducing measures, the one with the greater sum of certainty equivalents is preferred. Certainty equivalents therefore serve as satisfactory priority indices for identifying optimal risk-reducing investments in this case.

Priority Ratings Yield Poor Risk Management Strategies for Correlated Risks

Priority-based risk management successfully maximizes the risk reduction value (expected utility or certainty-equivalent value of risk-reducing activities) of defensive investments in the special cases discussed in the preceding two sections. However, it fails to do so more generally. Selecting a best portfolio of hazards to address (or of risk-reducing measures to implement) cannot, in general, be accomplished by priority setting if uncertainties about the sizes of risks (or of risk reduction opportunities) are correlated (O'Brien and Sculpher, 2000). Unfortunately, this is the case in many applications of practical interest. No priority rule can recommend the best portfolio (subset) of risk-reducing opportunities when the optimal strategy requires diversifying risk-reducing investments across two or more types of opportunities, or when it requires coordinating correlated risk reductions from opportunities of different types (having different priority scores).

Example: Priority Rules Overlook Opportunities for Risk-Free Gains

A priority-setting rule that rates each uncertain hazard based on its own attributes only, as all the real priority-scoring systems discussed above do, will be unable, in general, to recommend an optimal subset of correlated risk-reducing opportunities. For example, any risk-averse decision maker prefers a single random draw from a normal distribution with mean 1 and variance 1, denoted $N(1, 1)$, to a single draw from normal distribution $N(1, 2)$, having mean 1 but variance 2. Therefore, a scoring rule would assign a higher priority to draws from $N(1, 1)$ than to draws from $N(1, 2)$. But suppose that X and Y are two $N(1, 2)$ random variables that are perfectly negatively correlated, with $Y = 2 - X$. (This might happen, for example, if effects depend only on the sum of X and Y, which has a known value of 2, but the relative contributions of X and Y to their sum are uncertain.) Then, drawing once from X and once from Y [each of which is $N(1, 2)$] would yield a sure gain of 2. Any risk-averse decision maker prefers this sure gain to two draws from $N(1, 1)$. Unfortunately, any priority rule that ignores correlations among opportunities would miss this possibility of constructing a risk-free gain by putting X and Y in the same portfolio, as it would always assign draws from $N(1, 1)$ higher priority than draws from $N(1, 2)$.

This example shows that priority-setting rules can recommend dominated portfolios, such as allocating all resources to risk reductions drawn from $N(1, 1)$ instead of pairing negatively correlated $N(1, 2)$ risk reductions, because *they cannot describe optimal portfolios that depend on correlations* among risk-reducing opportunities, rather than on the attributes of the individual opportunities. The next example shows that priority rules can, in principle, not only recommend dominated decisions, but in some cases can even recommend the worst possible decisions.

Example: Priority Setting Can Recommend the Worst Possible Resource Allocation

Setting: Suppose that an environmental risk manager must decide how to allocate scarce resources to remediate a large number of potentially hazardous sites. There are two main types of sites. Hazards at type A sites arise primarily from relatively long, thin chrysotile asbestos fibers. Hazards at type B sites arise from somewhat shorter and thicker amphibole asbestos fibers. The risk manager is uncertain about their relative potencies but knows that removing *mixtures* of approximately equal parts of the chrysotile and amphibole fibers significantly reduces the risks of lung cancer and mesothelioma in surrounding populations. She believes that the following two hypotheses are plausible, but she is uncertain about their respective probabilities. (This is intended for purposes of a simple illustration only, not as a realistic risk model.)

- H1: The relative risk from a type A site is 0 and the relative risk from a type B site is 2 (compared to the risk from a hypothetical site with equal mixtures of chrysotile and amphibole fibers, which we define as 1). This hypothesis implies that all risk is from amphibole fibers.
- H2: The relative risk from a type A site is 2 and the relative risk from a type B site is 0. This hypothesis implies that all risk is from the chrysotile fibers.

For purposes of illustration only, we assume that only these two hypotheses are considered plausible, although clearly others (especially, that the two types of fiber are equally potent) would be considered in reality.

Problem: If the risk manager can afford to clean $N = 10$ sites, then how should she allocate them between type A and type B sites? Assume that she is risk-averse and that more than 10 sites of each type are available.

Solution: If the risk manager cleans x type A sites and $(N - x)$ type B sites, then the total expected utility from cleaned sites is $pu(N - x) + (1 - p)u(x)$. Here, p denotes the probability that hypothesis H1 is correct, $1 - p$ is the probability that H2 is correct, $N = 10$ is the total number of sites that can be cleaned, and $u(x)$ is the utility of cleaning x sites with relative risk of 2 per site cleaned. For *any* risk-averse (concave) utility function $u(x)$, and for *any* value of p between 0 and 1, Jensen's inequality implies that expected utility is maximized for some x strictly between 0 and N. For example, if $u(x) = x^{0.5}$ and $p = 0.5$, then $x = 5$ maximizes expected utility. The worst possible decision (minimizing expected utility) is to allocate all resources to only one type of site (either type A or type B). Yet this is precisely what a priority system that assigns one type a higher priority than the other must recommend. Hence, in this case, any possible priority order (either giving type A sites precedence over type B sites or vice versa, perhaps depending on whether $p < 0.5$) will recommend a subset of sites yielding lower expected utility than even a randomly selected subset of sites. The best subset (e.g., 5 type A sites and 5 type B sites, if $p = 0.5$) can easily be constructed by optimization if p is known. But even if both p and $u(x)$ are unknown, it is clear that a priority order recommends the worst possible decision.

Example: Priority Setting Ignores Opportunities for Coordinated Defenses

Setting: Suppose that an information security risk manager can purchase either of two types of security upgrades for each of 100 web servers. Type A prevents undetected unauthorized access to a web server, and type B prevents the unauthorized execution of arbitrary code with the privileges of the web server, even if the web server is accessed. (For examples of real-world historical vulnerabilities in an Apache web server, see http://www.first.org/cvss/cvss-guide.html#i1.2.) For simplicity, suppose that installing a type A upgrade reduces the annual incidence of successful attacks via web servers from 0.03 to 0.02 per web-server-year and that installing a type B upgrade reduces it from 0.03 to 0.025. Installing both reduces the average annual rate of successful attacks via these machines from 0.03 to 0.

Problem: If the security risk manager can afford 100 security upgrades (of either type), what investment strategy for reducing the average annual frequency of successful attacks would be recommended based on (a) priority ranking of options A and B and (b) minimization of remaining risk? (Assume that the frequency of attempted attacks remains constant, as hackers only discover the defenses of a web server when they attempt to compromise it.)

Solution: (a) A vulnerability-scoring system could assign top priority to installing a type A upgrade on each of the 100 web servers, since a type A upgrade achieves a larger reduction in the vulnerability score of each server than a type B upgrade. Following this recommendation would leave a residual risk of 0.02*100 = 2 expected successful attacks per year. (b) By contrast, a risk-minimizing budget allocation installs both A and B upgrades on each of 50 machines, leaving 50 machines unprotected. The residual risk is then 0.03*50 = 1.5 expected successful attacks per year, less than that from giving A priority over B.

Comment: In this example, a scoring system that considered the interaction between different vulnerability-reducing activities could give "install A & B" a higher priority for each server than either "install A" or "install B." But most deployed scoring systems do not encourage considering interactions among vulnerabilities or among vulnerability-reducing countermeasures. In many applications, doing so could lead to combinatorial explosion. (For example, the guidance for Common Vulnerability Scoring System 2.0 offers this advice: "SCORING TIP #1: Vulnerability scoring should not take into account any interaction with other vulnerabilities. That is, each vulnerability should be scored independently." http://www.first.org/cvss/cvss-guide.html#i1.2.)

Priority Rules Ignore Aversion to Large-Scale Uncertainties

Setting: A bioterrorism risk manager must choose which of two defensive programs to implement this year: (a) a prevention program (e.g., vaccination) that, *if it works*, will reduce the risk of fatal infection from 10% to 0% for each affected person in the event of a bioterrorism attack with a certain agent; or (b) a treatment program (e.g., stockpiling an antibiotic) that will reduce the risk of mortality from 10% to 5% for each affected individual in the event of such an attack. For simplicity, suppose that program A will prevent either N expected deaths (if it works) or none (if it does not) following an attack and that its success probability is p. Program B prevents $0.5N$ expected deaths with certainty, leaving $0.5N$ remaining expected deaths in the event of an attack.

Problem: (a) For a risk-averse decision maker with utility function $u(x) = 1 - e^{-kx}$, where x is the number of expected deaths prevented, which risk reduction measure, A or B, is preferable? (Express the answer as a function of p, k, and N.) (b) How does this compare to the results of a priority ranking system, for $p = 0.8$ and $k = 1$?

Solution: (a) The expected utility of risk reduction is $pu(N) = p(1 - e^{-kN})$ for program A and $u(0.5N) = 1 - e^{-0.5kN}$ for program B. Program A is preferable to program B if and only if $p(1 - e^{-kN}) > 1 - e^{-0.5kN}$, or, equivalently, $p > (1 - e^{-0.5kN})/(1 - e^{-kN})$. For example, if $kN = 1$, then p must be at least 62.2% to make A preferable to B. If $kN = 10$, then p must be at least 99.3% to make A preferable to B. (b) If the probability that program A will work is $p = 0.8$ and the coefficient of absolute risk aversion is $k = 1$, then A is preferred to B for $N = 1$ or 2, and B is preferred to A for $N \geq 3$. In this case, diversification is not an issue (i.e., either A or B is definitely preferable, depending on the value of N). However, *no priority* *ranking of interventions A and B is best for both $N = 2$ and $N = 3$.* The reason is that a risk-averse decision maker who prefers A to B for small N prefers B to A for larger N. Any priority-scoring system that ranks one of A or B above the other, and that is not sensitive to N, will recommend the less valuable decision for some values of N. In practice, most scoring systems use qualitative or ordered categorical descriptions that are not sensitive to quantitative details such as N. (For example, the Common Vulnerability Scoring System rates "Collateral Damage Potential," which scores "potential for loss of life, physical assets, productivity or revenue," as high if "A successful exploit of this vulnerability may result in catastrophic physical or property damage and loss. Or, there may be a catastrophic loss of revenue or productivity." http://www.first.org/cvss/cvss-guide.html#i1.2. Such a qualitative description does not discriminate between $N = 2$ and $N = 3$.)

Discussion: Precisely analogous examples hold for information security and homeland security applications. Suppose that intervention A reduces the average rate of successful attacks per target (e.g., secure facility or web server) per year from 10% to 0% *if it works*, while intervention B reduces the rate from 10% to 5% with certainty. The probability that A will work (i.e., that an attacker cannot circumvent it) is p. If the choice between A and B affects N similar targets, then, by analogy to the above example, a risk-averse risk manager should prefer A to B for sufficiently small N and B to A for larger values of N. Any priority system that is applied to a small number of targets at a time (possibly only 1, by the target's owner, operator, or security manager) will then consistently recommend A, even though B should be preferred when the complete set of N targets is considered. That scoring systems are blind to the total number of similar targets that they are applied to (i.e., to the scale of application) can lead to excessively high-risk exposures arising from large-scale application of priorities that hold for small numbers of targets but that should be reversed for larger numbers of targets.

Discussion and Conclusions on Risk Priority-Scoring Systems

Applied risk analysis is in a curious state today. Highly effective optimization methods for selecting subsets of risk-reducing investments to maximize the value of risk reductions achieved for a given budget are readily available. They can draw on a rich and deep set of technical methods developed in financial investment risk analysis over the past half-century. Yet these methods are having little or no impact on the management of some of the world's most critical risks. Instead, extremely simplistic priority-setting rules and scoring systems are being widely used to set priorities and allocate resources in important practical risk management applications. Scoring systems are being used in important real-world applications as diverse as Superfund site cleanups, computer and IT security vulnerability assessment, counterterrorism and military asset protection, and risk matrix systems (used in everything from designing and defending federal buildings and facilities, to managing construction project and infrastructure risks, to regulating risks of financial and business enterprises). Yet these risk-scoring systems can perform extremely poorly compared to optimization methods, allocating resources so that much less value-of-risk reduction is achieved than could easily be obtained by other methods (including randomized decision making, in extreme cases).

The requirements that scoring systems must meet before being adopted and recommended in standards are evidently not very stringent. In the applications examined in this chapter, there appears to be no requirement that risk-scoring systems should produce effective risk management decisions (or even that they should not produce the lowest-value decisions possible) before they are standardized for widespread use. In all of the applications mentioned, common elements found in multiple risky systems create correlated vulnerabilities, criticalities, consequences, or threats. Priority lists do not generally produce effective risk management decisions in such settings. Applying investment portfolio optimization principles (such as optimal diversification, consideration of risk aversion, and exploitation of correlations among risk reductions from different activities) can create better portfolios of risk-reducing activities in these situations than any that can be expressed by priority scores.

In summary, risk priority-scoring systems, although widely used (and even required in many current regulations and standards), ignore essential information about correlations among risks. This information typically consists of noting common elements across multiple targets (e.g., common vulnerabilities). These common features induce common, or strongly positively correlated, uncertainties about the effectiveness of different risk-reducing measures. It is easy to use this information, in conjunction with well-known decision analysis and optimization techniques, to develop more valuable risk reduction strategies, for any given risk management budget, than can be expressed by a priority list. Thus, there appears to be abundant opportunity to improve the productivity of current risk-reducing efforts in many important applications using already well-understood optimization methods.

Nothing in this section is intended to be new or surprising to experts in decision and risk analysis. Techniques for optimizing investments in risk-reducing (and/or benefit-producing) interventions have been extensively developed in operations research and management science for decades. What is perhaps startling is that these methods are so little exploited in current risk assessment and risk management systems. Risk priority scores can never do better (and often do much worse) than optimization methods in identifying valuable risk-reducing strategies. Perhaps it is time to stop using risk priority scores to manage correlated risks, recognizing that they often produce simple but wrong answers. Optimization techniques that consider dependencies among risk-reducing interventions for multiple targets should be used instead.

Conclusions

This chapter has illustrated several ways in which QRA can go awry if practitioners use summaries of exposure or risk that neglect essential information (usually due to excessive aggregation). The mathematical results presented (Theorems 1–3) are elementary, but their practical implications are important: Representing risky prospects by summary measures (such as frequency, aggregate exposure, or an aggregate ratio of risk per unit of exposure) can lead to incorrect comparisons and poor risk management decisions. The reason is that such summaries omit relevant information about the *causal relations* between actions (including choices among alternative risky prospects) and the probable consequences that decision makers care about (such as completing or failing to complete a mission, reducing larger vs. reducing smaller risks from exposures to complex mixtures of carcinogens, or reducing vs. increasing human health hazards from foodborne pathogens). As a result, prospects that create importantly different risks are assigned identical summary representations, making an informed choice among them, on the basis of these summaries, impossible. (As explained in Chapter 2, *F-N* curves suffer from a similar limitation, insofar as they ignore information about the distribution of individual risks.)

The key to avoiding these limitations is to make sure that risk models describe the causal relations between alternative risk management acts and the probable consequences that the decision maker cares about. This invites the questions of (a) what specific information should a risk model include? and (b) how can it be used to predict the probable consequences of alternative risk management actions? The following chapters address these questions.

Part III
Principles for Doing Better

Chapter 6
Identifying Nonlinear Causal Relations in Large Data Sets

This chapter discusses data-mining methods for identifying potential causal relations in large data sets, such as clinical, epidemiological, or engineering reliability data sets. The causal relations to be discovered may be completely unknown initially; thus, successfully identifying them from data is sometimes called *knowledge discovery*. This is usually more challenging than merely estimating the parameters of a statistical model that is known or specified a priori. The causal relations may be complex and impossible to summarize using only a few parameters. For example, they may contain nonmonotonic (such as n-shaped or u-shaped) or threshold-like exposure-response relations, or more complicated nonlinearities, that render ineffective traditional statistical data analysis techniques (including factor analysis, principal components analysis, discriminant analysis, multiple linear or logistic regression, and so forth) based on linear and generalized linear modeling.

The possibility of nonlinearities in individual dose-response relations undermines traditional epidemiological criteria and tests for causal relations between exposure and response variables. Nonmonotonic exposure-response relations in a large population may lack aggregate consistency, strength, biological gradient, and other traditional hallmarks of causal relations. For example, a u-shaped or n-shaped curve may exhibit zero correlation between dose and response. Thus, possible nonlinearity requires new ways to detect potentially causal exposure-response relations. Conversely, traditional epidemiological criteria for causality may be satisfied even in the absence of a true causal relation, e.g., due to model misspecification (as discussed in Chapter 5 for "$risk = K \times exposure$") or omitted confounders (discussed further in Chapter 7).

Fortunately, information-theoretic criteria provide a unifying framework for several powerful data-mining techniques – including classification trees, conditional independence tests for potential causal relations, and techniques for learning causal graphs (or Bayesian networks) from data – that lead to practical algorithms for detecting potential causal relations in large data sets. Such techniques, explained in this chapter, can be applied to data from large epidemiological studies, including cohort, case-control, cross-sectional, and time-series data. (Throughout this chapter, we repeatedly use the term "causal graph" rather than "Bayesian network" to emphasize that the main challenge is how to identify causal relations from data,

L.A. Cox, Jr., *Risk Analysis of Complex and Uncertain Systems*, International Series in Operations Research & Management Science 129, DOI 10.1007/978-0-387-89014-2_6,

rather than how to draw inferences from a prespecified network of such relations and a collection of data – the usual emphasis in many applications of Bayesian networks in risk analysis.)

In addition to introducing data mining and causal modeling for epidemiological data, this chapter follows up on the case study at the end of Chapter 5 by showing how to use information-theoretic criteria and data-mining methods to identify nonlinear, potentially causal relations between explanatory variables and campylobacteriosis in a case-control data set. [Recall that campylobacteriosis is a foodborne infectious diarrheal illness that typically lasts several days. In most cases (over 99%), it then spontaneously resolves itself without the need for treatment.] In this data set, hundreds of potential explanatory variables and response variables were recorded for each of over a thousand infected people ("cases") and over a thousand uninfected people ("controls"). The risk assessment challenge is to use these data to identify differences in explanatory variables between cases and controls that explain, predict, or cause the differences in health outcomes, i.e., infection vs. non-infection.

In contrast to previous analyses, the methods in this chapter identify a highly statistically significant nonlinear (*u*-shaped) relation between recent fast food consumption and women's risk of campylobacteriosis. The information-theoretic criteria can also be used to resolve ambiguities and apparent contradictions in causal interpretations due to confounding and redundancy or overlap among variables in data sets. This use is discussed in detail in Chapter 7.

Nonlinear Exposure-Response Relations

The existence of a monotonic exposure-response relation has traditionally been regarded as one indication that a statistical association may be causal (Weed and Gorelic, 1996). However, there is no reason that causal exposure-response relations must necessarily be monotonic, and considerable evidence (e.g., on the reality of hormesis, or *u*-shaped exposure-response relations) in many systems suggests that sometimes they are not. Hence, methods are needed to detect and quantify non-monotonic (e.g., *u*-shaped, *n*-shaped, *N*-shaped, or more complicated) relations in epidemiological data. This statistical challenge is exacerbated by the fact that some nonlinear effects may be relatively small over much of the range of available data, or may involve interactions among multiple variables, so that identifying them in complex data sets requires detecting relatively weak signals among a huge number of possibilities.

As an example, Fig. 6.1 shows a *u*-shaped exposure-response relation identified in data from a food safety case-control study of the common foodborne bacterial illness campylobacteriosis (Friedman et al., 2004). The data were collected by the Centers for Disease Control and Prevention (CDC) as part of a study of risk factors for sporadic cases of campylobacteriosis. As described by Friedman et al. (2000):

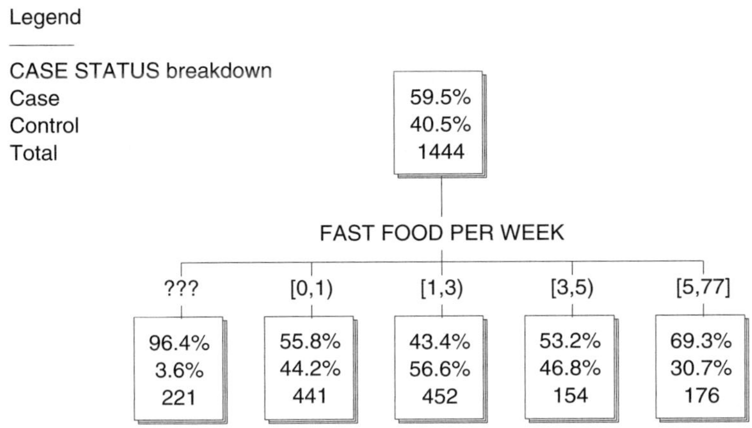

Fig. 6.1 *u*-Shaped relation between fast food consumption and risk of foodborne illness (Campylobacteriosis)

> We enrolled patients with culture-confirmed *Campylobacter* infections from Foodborne Diseases Active Surveillance Network (FoodNet) sites in California, Georgia, Maryland, Minnesota, New York and Oregon. Information about demographics, clinical illness, and exposures occurring within 7 days before diarrhea onset was collected using a standardized questionnaire. By using random-digit dialing, we interviewed one age-group matched, site-matched community control for each patient. . . . From January 1, 1998, to March 1, 1999, 1,463 patients and 1,317 controls were enrolled in the study.

The data set contains one record (with over 800 variables covering demographics, medical information, and recent self-reported food consumption and cooking habits information) for each case and each control. It has previously been analyzed by the CDC and public health researchers (e.g., Kassenborg et al., 2004). The data set was provided to the author as an Excel file by the CDC upon request.

In the classification tree notation of Fig. 6.1 (Lemon et al., 2003), each node in the tree (i.e., each box) indicates the percentages of cases (upper percentage) and controls (lower percentage) for the subpopulation described by that box.

The integer at the bottom of each box indicates the total number of subjects described by it. Among 1,444 subjects, 59.5% were confirmed campylobacteriosis cases and 40.5% were matched controls, as shown in the top node of Fig. 6.1. The variable *FAST FOOD PER WEEK* indicates the number of times that subjects reported eating at a fast food restaurant in the seven days prior to the onset of campylobacteriosis illness. Thus, this small tree displays basic cross-tab information for case status vs. fast food consumption frequency. ("???" denotes missing data and "77" at the right of the tree is a don't know/no answer code.) The methods used to identify the relation in Fig. 6.1 are discussed in the next section.

The classification tree program used to generate Fig. 6.1 (KnowledgeSeeker™, marketed by Angoss Software, 2005) automatically partitions the ordinal variable *FAST FOOD PER WEEK* into the discrete categories shown (i.e., the branches), to create conditional distributions that are statistically significantly different after adjusting for multiple testing bias due to repeated testing with multiple boundary locations. The notation $[x, y)$ on a branch for *FAST FOOD PER WEEK* indicates the interval $x \leq$ *FAST FOOD PER WEEK* $< y$; thus, the branch labeled [0, 1), for example, denotes people who reported eating at fast food restaurants no times in the week prior to illness. These people are at significantly *higher* risk than people who ate one or two meals at fast food restaurants (55.8% vs. 43.4%), but risk then increases for more frequent exposures to fast food, reaching a case rate of 69.3% among subjects reporting five or more fast food meals per week. The KnowledgeSeeker^{TM} program is well suited to this type of analysis as it works with both continuous and discrete variables (including binary and ordered categorical variables) to create highly predictive risk classes without making any specific parametric modeling assumptions (Biggs et al., 1991).

The data display a typical *u*-shaped (or moderately *J*-shaped) relation. Such patterns often are not discovered in standard parametric multivariate modeling (e.g., linear or logistic regression with automatic backward or forward stepwise variable selection) even when they exist, since the *u* shape cannot be expressed by the coefficients in a regression model. (If dummy variables are used to break the domain of the independent variables into downward-sloping and upward-sloping components, then regression methods can be applied successfully, but this requires knowing the correct answer in advance.) The problem is worse when there are hundreds of variables (as in this data set, which has over 800) and many possible interactions: Even highly predictive *u*-shaped relations may be impossible to discover by standard methods in the sea of possible relations.

The following sections propose a solution to the problems of (a) identifying nonlinear relations in large data sets and (b) testing them for potential causality. They illustrate an application to the campylobacteriosis data set, building in part on ideas from classification tree analysis, which is designed to detect high-order interactions in models that need not be linear or monotonic (Lemon et al., 2003). Chapter 7 discusses how classification tree analysis can be used to empirically test hypothesized causal relations, as well as to discover previously unknown ones.

Entropy, Mutual Information, and Conditional Independence

Traditional epidemiology often begins by seeking nonrandom associations between potential explanatory variables and response variables of interest, e.g., by using logistic regression modeling to screen for statistically significant predictors of increased risks of adverse health effects (e.g., Lemon et al., 2003). Fully automated variable selection is a notoriously challenging problem, however. On the one hand, "data dredging" (e.g., using automated variable-selection criteria such as the AIC, BIC, or Mallows criteria included in many standard commercial regression software packages) can easily produce false positives (e.g., Raftery et al., 1997). On the other hand, nonmonotonic relations for predictors having both positive and negative relations with risk over different ranges can easily escape detection by these methods, thus producing false negatives. To help overcome these problems, it is useful to replace more traditional measures of statistical association between variables, such as correlation coefficients or t-tests of regression coefficients, with *mutual information* – a measure that also works for arbitrary nonmonotonic relations and that is nonparametric, or "model-free," thus reducing the problems of multiple testing bias and model selection bias.

Let X and Y be two discrete random variables, e.g., $X =$ level of exposure, $Y =$ level of response. (The following methods can also apply to continuous variables, as in Cover and Thomas, 1991. However, we focus on the discrete case, as this is most useful in conjunction with classification trees.) Uncertainty about any discrete random variable X taking values x_i with corresponding probabilities p_i can be quantified by its *entropy*, defined as

$$H(X) = \text{entropy of } X = -\Sigma_i p_i \log_2 p_i = E[\log_2(1/p_i)] \text{ bits.}$$

$H(X)$ may be interpreted as the *average amount of information* gained when the value of X is learned. [It is also the expected minimum number of binary yes-no questions with equally likely answers, i.e., Pr(yes) = Pr(no) = 0.5, that one would need to have answered about the value of X to uniquely identify its value.]

The *mutual information* between any two random variables X and Y, denoted by $I(Y; X)$, is defined as

$$I(Y; X) = H(Y) - H(Y|X),$$

where $H(Y \mid X) = \Sigma_X \Pr(X = x) H(Y \mid X = x) = E_X[H(Y \mid X)]$ is the conditional entropy of Y given X. For any specific observed value of X, say, x, the conditional entropy of Y given that value of X is

$$H(Y|X = x) = -\Sigma_i \Pr(Y = y_i|X = x) \log_2 \Pr(Y = y_i|X = x).$$

Some intuitively appealing properties of entropy, H, and mutual information, I, include

(a) $I(X ; Y) = I(Y ; X)$, i.e., X and Y provide the same amount of information, or uncertainty reduction, about each other.
(b) $H(Y \mid X) \leq H(Y)$, i.e., conditioning on (or learning the value of) X never increases the expected uncertainty about Y but is expected to decrease it unless they are statistically independent.
(c) $H(X, Y) = H(X) + H(Y \mid X)$, i.e., the entropy of the joint distribution of X and Y, is the entropy of X plus the conditional entropy of Y given X.
(d) $I(X ; Y) > 0$ if $\Pr(Y \mid X = x_i)$ depends on x_i. For example, if the probability distribution of response variable Y depends on the value of exposure variable X, then the mutual information between them is positive. This allows mutual information to be used in screening for possible exposure-response relations.

(e) Let the "causal graph" (also called "Bayesian network") notation $Z \to X \to Y$ indicate that the probability distribution of Y depends on the value of X and that the probability distribution of X depends on the value of Z, but the conditional probability distribution of Y given any specific value of X does not depend on the value of Z. In other words, Y is *conditionally independent* of Z given X. (However, Y is not unconditionally independent of Z, since Z affects Y through X.) Then $I(Y\,;X) \geq I(Y:Z)$, with equality if and only if X is a deterministic, one-to-one function of Z. More generally, in a causal graph, more remote ancestors of a node can never be more informative about it than its direct parents. (In a causal graph, nodes represent variables, and an arrow directed from X to Y indicates that the probability distribution of Y depends on the value of X. Such graphs are required to be acyclic. Each node is conditionally independent of its more remote ancestors, given the values of its parents, i.e., of nodes with arrows pointing into it.)

(f) In a causal graph model $X - Y - Z$ (with the arcs oriented in any directions), more remote ancestors can never be more informative that direct parents. Thus, $I(X\,;Z) \leq I(X\,;Y)$. Moreover, $I(X\,;Z \mid Y) = 0$ (i.e., X and Z are conditionally independent given Y) unless both X and Z point into Y.

For these and other aspects of information theory, see Cover and Thomas (1991).

Classification Trees and Causal Graphs via Information Theory

The above properties suggest that mutual information can be used to help search for potential dose-response relations or exposure-response relations and to identify direct parents of responses in large, multivariate data sets. Two main families of practical data analysis algorithms have exploited this potential: classification tree algorithms and causal graph "learning" algorithms.

A classification tree analysis begins with a specific dependent variable of interest, such as a health response variable in a population, and repeatedly conditions on the "most informative" variables in the data set to calculate its conditional probability distribution, given their values. At any stage in the construction of a tree, each leaf represents a set of values of the variables that have been conditioned on so far. There is a conditional distribution of the values of the dependent variable at each node, given the values of the conditioned-on variables leading to it. At each leaf, the myopically "most informative" variable to condition on next is the one having the *highest mutual information* with the conditional distribution of the response variable at that leaf. (Less myopic, more CPU-intensive procedures seek the subsets of variables that jointly give the greatest reduction in the entropy of the dependent variable, and then condition on combinations of their values. Continuous variables can be discretized into contiguous ranges as part of this search process by taking either the maximum reduction in the entropy of the dependent variable or the maximal increase in the mutual information of all directly related variables as the goal; see Friedman and Goldsmitz, 1996a.) When further conditioning provides no additional useful information about the dependent variable (e.g., as assessed by cross-validation estimation of the true error rate resulting from using the conditional distributions at the current leaf nodes to make predictions about the dependent variable), tree growing stops.

Like classification trees, causal graphs store conditional distributions at each node. However, the conditional distribution at a node is for the variable represented by that node, rather than for some other dependent variable. Instead of there being a single dependent variable, there is usually a set of variables related by statistical dependence and conditional independence relations that are expressed by the directed arcs ("arrows") among the variables (nodes). Moreover, the conditional distribution at any node is conditioned only on the values of its parents, i.e., the variables that point into it. This information may be stored in a *conditional probability table* (CPT) specifying the different conditional probability distributions of that node's variable, for each combination of values of its parents. (Combinations of parent values that lead to the same conditional distribution can be aggregated, e.g., by using *ranges* of values of the variables to create distinct rows in the CPT.) Several computationally practical algorithms for fitting classification trees and causal graphs to large, multivariate data sets are now available (Murphy, 2001; Wood et al., 2006; Silander and Myllymäki, 2006; Ellis and Wong, 2008).

Classification trees and causal graphs are closely related, as follows. Consider an ideal classification tree algorithm, in which X appears in the tree for Y if $I(X ; Y) > 0$ [and $I(X ; Y \mid C) > 0$ even after conditioning on the other variables, C, in the tree]. For any data set with enough observations, with adequate variability in the values of its variables, and with redundant variables eliminated [e.g., by replacing any cluster of redundant variables $A = B = \ldots = C$ with any one of them, or more generally by pruning all variables Z satisfying $I(X ; Z) = H(X)$ for some remaining X], the following properties hold (e.g., Frey et al., 2003):

1. *All* of the parents of a node appear in any mutual information-based classification tree having that node variable as its dependent variable. (This is because, by definition, the conditional distribution of the node depends on the values of its parents.)
2. Once a node's parents (and children, if any) have been included in a classification tree, i.e., conditioned on, no more remote ancestors (or descendants) will enter its classification tree. (By definition, the node's value is conditionally independent of its more remote ancestors, given the values of its parents.)
3. In the causal graph $X \leftarrow Z \rightarrow Y$, the variable Z is called a *confounder* of the statistical relation between X and Y. It explains away an apparent association between them. Including the parents of a health response variable in its classification tree (i.e., conditioning on them) eliminates all variables that are statistically associated with the response variable only due to confounding. [More generally, X is a parent of Y only if there is no subset of variables C such that $I(X ; Y \mid C) = 0$, i.e., only if X provides information about Y that cannot be fully removed by conditioning on any other subset of variables. In the example $X \leftarrow Z \rightarrow Y$, the

tree for Y will include Z but not X. The confounded relation between X and Y is eliminated when the tree conditions on Z.]

4. When a classification tree is grown for a particular node variable, using only its parents as conditioning variables, the leaves of the resulting tree contain the conditional probability table (CPT) information for that node. (The empirical CPT based on the raw data is a maximum-likelihood estimate of the true CPT. It can be used together with a multivariate Dirichlet prior to develop Bayesian posterior estimates of the CPT for purposes of uncertainty analysis; see e.g., Friedman and Goldszmidt, 1996b, and Murphy, 2001.)

By property 1, an automated tree-growing procedure based on conditioning on variables having the highest estimated mutual information with the dependent variable tends to create a tree containing the node's parents. In theory, if the dependent variable is a response variable with no children in the data set, the classification tree should consist only of the parents of that node in a causal graph. In practice, it may also include more remote ancestors (and children and descendants, if there are any) since the empirical joint distribution of the variables among the observed cases may contain sampling variability that causes it to differ from the underlying joint distribution determined by the data-generating process. Property 2 can then be used to prune more remote ancestors (and descendants) by testing whether some variables drop out of the tree when others are conditioned on first. In principle, those that cannot be eliminated in this way are the parents and children of a node. To distinguish among parents and children (which are mutually conditionally independent, given the value of the node variable) for variables that have both, it is necessary to orient the arcs.

Health responses are often known a priori to be possible children of exposure-related variables, but not possible parents. Moreover, earlier observations can usually be causes (parents) of later ones but not consequences (children). These properties help to orient the arrows near exposure and response variables in a causal graph. [More generally, if time-series information is available on variables, as in many longitudinal epidemiological studies, then X is a potential cause of Y only if the history of X up to and including each time t is informative about the future of Y after t, even after conditioning on the past of Y, i.e.,

$I(X^-(t);Y^+(t) \mid Y^-(t)) > 0$, where $X^-(t)$ denotes the set of X values at times $\leq t$, $Y^-(t)$ the set of Y values at times $\leq t$, and $Y^+(t)$ the set of Y values after t. This provides an information-theoretic generalization of the concept of Granger causality for multiple time series (e.g., Guatama and Van Hulle, 2003).]

Property (f) above, which shows that mutual information with a variable Y increases along chains of variables leading to it, helps to orient the remaining arcs. The following *PC algorithm* (Glymour and Cooper, 1999, modified here to use classification trees and mutual information) orients arcs even without such domain-specific knowledge:

1. Grow a classification tree for each node. Create an undirected arc between each node and every node that appears in its tree (and that cannot be forced to drop out after conditioning on other variables).

2. Orient any triple of nodes $X - Y - Z$ as $X \rightarrow Y \leftarrow Z$ if and only if $I(X ; Z \mid Y) > 0$, i.e., if and only if X and Z are dependent when conditioned on Y.
3. Orient any remaining triple $X \rightarrow Y - Z$ as $X \rightarrow Y \rightarrow Z$.
4. Orient any pair $X - Y$ with a directed path through X to Y as $X \rightarrow Y$. [For example, if $X - Y \rightarrow Z$ and $I(Z; Y) > I(Z ; X) > 0$ and $I(Z ; X \mid Y) = 0$, then create $X \rightarrow Y$.]
5. Repeat steps 3–5 until no more arc directions can be assigned.

A variety of other algorithms are now available for fitting causal graph models even to very large multivariate data sets (Murphy, 2001; Tsamardinos et al., 2003; Wood et al., 2006; Silander and Myllymäki, 2006; Ellis and Wong, 2008).

We can summarize our proposed methodology for identifying potential causal exposure-response relations in large data sets, even if the relations are nonmonotonic, as follows. First, *preprocess the data* to remove any redundant variables and to eliminate any variables that occur after the response of interest or that are otherwise known not to be candidates for potential causal variables. (Redundant variables appear as the only nodes in each other's classification trees.) Next, *identify parents* of the response variable in the causal graph for that node. Finally, *fit a nonparametric model*, such as a classification tree, a nonparametric regression model, or simply the relevant conditional probability table (CPT) (possibly smoothed or approximated by simple regression functions), to the reduced data set consisting of the response variable – which is the dependent variable – and its parents. These steps can be implemented using commercially available classification tree and Bayesian network learning software products, such as KnowledgeSeeker™ and BayesiaLab™, respectively.

Current classification tree and Bayesian network-learning algorithms involve considerable sophistication and intensive computation. Many create multiple random subsets of training data and then use these subsamples to "vote" for the most informative predictors, thus reducing errors due to overtraining on a single data set and improving the robustness of model predictions. Another popular technique is to assess performance on multiple subsets of test data ("model cross-validation") to avoid overfitting models and underestimating prediction error rates. Although such algorithmic refinements increase the computational burden, they can significantly reduce the error rates in model predictions (Bauer and Kohavi, 1999). The techniques in this chapter and in Chapter 7 treat such computational statistics algorithms as black boxes; thus, further improvements in the algorithms for learning classification trees and causal graphs from data will not change how they are used in risk assessment.

Illustration for the Campylobacteriosis Case Control Data

The cross-tab information in Fig. 6.1 illustrates a single "split" (i.e., conditioning the dependent variable, *CASE STATUS*, on a single variable, *FAST FOOD PER WEEK*). But this is only one of many statistically significant splits, each having positive mutual information with the dependent variable. Figure 6.2 shows a more

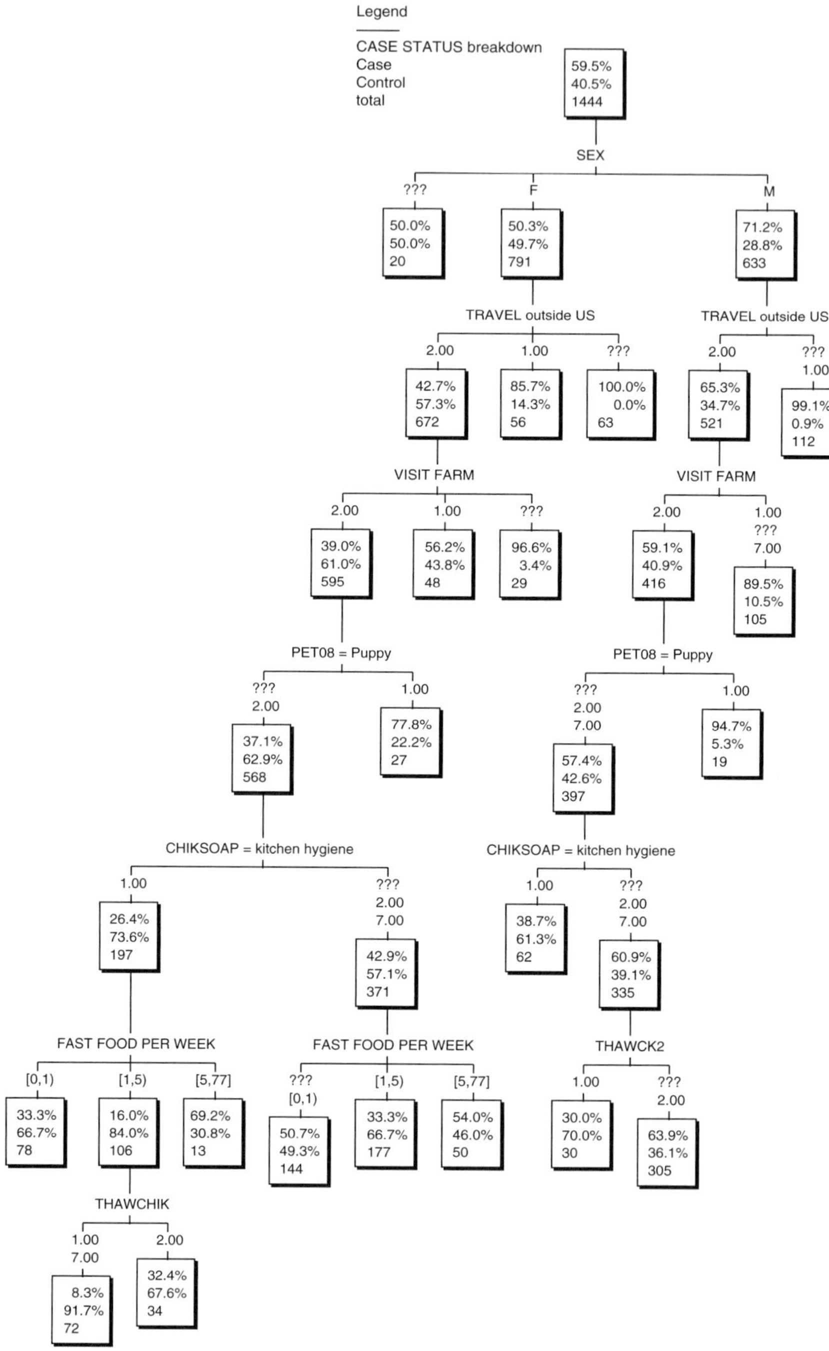

Fig. 6.2 A classification tree for campylobacteriosis risk

fully developed classification tree. All of the variables in this tree are parents of *CASE STATUS*, in that none can be eliminated by conditioning on other variables. (Potential children and descendants of *CASE STATUS*, mainly describing the duration and treatment of diarrhea, were pruned in the preprocessing step. Throughout the tree, variables are coded so that 1 = *Yes,* 2 = *No*, 7 and 77 = *don't know/no answer/refused to answer*. The KnowledgeSeeker^{TM} algorithm is computationally efficient, taking on the order of 10 seconds to develop each "split" in the tree when run on a laptop.)

In this tree, all potential confounding by other variables in the data set has automatically been eliminated, as discussed above. Thus, the statistically significant (but nonmonotonic) relation between *FAST FOOD PER WEEK* and risk (i.e., *CASE STATUS*) is potentially causal: It cannot be explained away by confounding with other variables in the data set. For example, consider the hypothesis that men, who intrinsically have greater susceptibility to campylobacteriosis than women, also eat at fast food restaurants more frequently, and that this explains the association between fast food dining and risk of campylobacteriosis. This hypothesis can be diagrammed as *FAST FOOD PER WEEK* $\leftarrow$ *SEX* $\rightarrow$ *CASE STATUS*. It is directly falsified by the classification tree in Fig. 6.2, since the *CASE STATUS* varies significantly with *FAST FOOD PER WEEK* even after conditioning on *SEX* = *F*, thus proving that *I*(*CASE STATUS* ; *FAST FOOD PER WEEK* | *SEX*) > 0. Other confounding-based explanations for the parents of *CASE STATUS* shown in Fig. 6.2 are similarly precluded by the data. (A few other variables, including drinking untreated water and having health insurance, were also identified as parents of *CASE STATUS* for small subpopulations but were pruned from the bottom of Fig. 6.2, leaving the tree shown. This was done to save space and because they affected only small fractions of the sample and did not appear in multiple parts of the tree, indicating that they had at most only very limited impacts.)

Also interesting is the set of variables that do *not* appear in the full classification tree for *CASE STATUS*. For example, it is well known that drinking raw milk is a risk factor for campylobacteriosis. Indeed, the tree-growing program lists it as a significant split, i.e., a variable having significant mutual information (and positive association, by any measure) with *CASE STATUS*. However, conditioning on *VISIT FARM* eliminates drinking raw milk as an additional parent of *CASE STATUS*: they belong to the same cluster of closely associated, partly redundant variables. Similarly, although *CHIKSOAP*, which records whether subjects reported using soap to wash after handling raw chicken in the kitchen, is a parent of *CASE STATUS*, buying, handling, thawing, and cooking raw chicken and eating chicken at home all belong to a cluster of tightly interrelated variables that are all associated with a reduced risk of being a case. (This cluster of variables is represented by *THAWCHICK* and *THAWCK2* in Fig. 6.2, referring to thawing chicken in any manner and thawing chicken in the refrigerator at home, respectively.) Thus, *CHIKSOAP* may be a marker for kitchen hygiene in general, rather than specifically for chicken-associated risk.

Figure 6.3 displays the *u*-shaped exposure-response relation identified in Fig. 6.2 for *FAST FOOD PER WEEK* and *CASE STATUS* in a more conventional (interaction

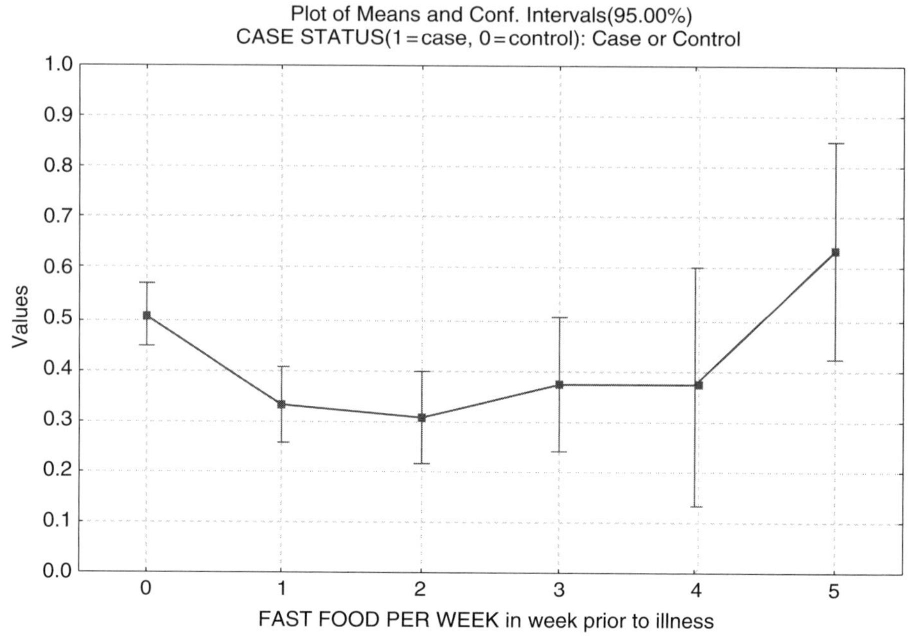

Fig. 6.3 A *u*-shaped exposure-response relation for campylobacteriosis in women

plot) format. *CASE STATUS* has been recoded in Fig. 6.3 so that 1 = case, 0 = control, as this is more usual than the 1 vs. 2 coding used in the original Centers for Disease Control data file. Table 6.1 summarizes the sizes of the different groups.

The reduction in risk between the group exposed to zero fast food meals per week and the group exposed to fast food once or twice per week is statistically significant ($p < 0.05$) by all standard tests. Although beyond the scope of the data, it is tempting to speculate that people with low exposures to fast food may have less acquired immunity to common pathogens such as *Campylobacter*, as previously noted for outbreaks associated with raw milk consumption (Blaser et al., 1987).

Table 6.1 Data for Fig. 6.3

Fast food meals in prior week	Fraction of exposed women that are campylobacteriosis cases	*N*
0	0.51	276
1	0.33	160
2	0.31	104
3	0.37	54
4	0.37	19
5	0.64	22
All groups	0.42	650

Conclusions

This chapter has proposed and illustrated information-theoretic methods for detecting potential causal nonlinear exposure-response relations in large data sets, by combining ideas from causal graph (or Bayesian network) modeling and classification tree analysis. Applied to a recent food safety case control data set, classification tree analysis successfully discovered a potentially causal (significantly informative, not confounded) *u*-shaped relation between the consumption of fast food by women and the resulting risk of campylobacteriosis. This sex-specific nonmonotonic relation has not previously been identified in analyses of this data using logistic regression modeling (e.g., Friedman et al., 2004).

In principle, classification trees can find arbitrarily shaped causal relations in other large data sets. The essential steps are

(a) *Identify informative variables* that help to predict the dependent variable (e.g., illness risk) of interest. This can be accomplished via classification tree analysis (even for nonmonotonic relations).
(b) *Eliminate variables (e.g., confounders, redundant variables, variables that follow the effect of interest in time)* whose mutual information with the dependent variable is fully explained away by the information contained in other variables, or that are inconsistent with the hypothesis of causality. This can be accomplished by conditional independence tests, e.g., using Bayesian network algorithms (including classification tree analysis of individual nodes in a Bayesian network).
(c) *Quantify the remaining relation* between the dependent variable and its parents using nonparametric methods (e.g., classification trees and conditional probability tables with nonparametric smoothing). The final relation, even if nonmonotonic, reveals the shape of potential causal relations between the dependent variable and a minimal set of predictors (its "parents" in a causal graph).

In practice, current classification tree algorithms and Bayesian network learning algorithms are practical even for data sets with thousands of records and variables, as run times are on the order of a few minutes on current laptop or desktop machines. Thus, these methods appear to be practical for identifying nonlinear relations even in large epidemiological data sets.

The methods discussed in this chapter show what information should be included in a risk model: information about the causal antecedents of the health effects (or other responses or consequences of interest) that the decision maker seeks to control by making decisions about controllable variables (e.g., preventable exposures). For example, if the total preventable campylobacteriosis risk depends primarily on fast food consumption by women, but is reduced by the preparation and consumption of home-cooked chicken, then risk management strategies intended to reduce campylobacteriosis rates should focus on reducing the risks from fast food consumption by women, rather than on reducing the risks from home-cooked chicken.

A risk model should then include gender, fast food consumption, and home-cooked chicken preparation and consumption as explanatory variables. Such information is crucial for avoiding the construction of causally irrelevant risk models such as the "$risk = K \times exposure$" model in the counterexamples in Chapter 5.

In summary, causal information provided by data-mining and causal modeling techniques can be invaluable in constructing and validating risk assessment models that show how risks will change when relevant causal drivers are changed – a primary goal of risk modeling. Chapter 7 focuses on how to use such data to test causal hypotheses and validate or replace causal assumptions in risk models.

Chapter 7
Overcoming Preconceptions and Confirmation Biases Using Data Mining

Data-mining methods such as classification tree analysis, conditional independence tests, and causal graphs can be used to discover possible causal relations in data sets, even if the relations are unknown a priori and involve nonlinearities and high-order interactions. Chapter 6 showed that information theory provided one possible common framework and set of principles for applying these methods to support causal inferences. This chapter examines how to apply these methods and related statistical techniques (such as Bayesian model averaging) to empirically test preexisting causal hypotheses, either supporting them by showing that they are consistent with data, or refuting them by showing that they are not. In the latter case, data-mining and modeling methods can also suggest improved causal hypotheses.

A powerful motivation for developing and using data-mining and causal modeling methods in risk assessment is that doing so can help to reveal unanticipated relations and identify conceptual blind spots in the initial understanding of risk. These methods encourage and support learning from data. They can help to overcome the common human tendency to impose one's a priori beliefs on the interpretation of data, discussed in Chapter 1. To recapitulate, most people – including scientists, risk analysts, and policy makers – make mistakes and exhibit powerful biases in reasoning about causation, if-then relations, and evidence. Common patterns include drawing decision-related conclusions and prematurely accepting specific causal hypotheses or explanations for observations, often based on narrative plausibility rather than on sound inference from factual evidence. There is also a strong and well-documented tendency to emphasize evidence that appears to confirm preexisting beliefs and causal judgments, while ignoring or discounting disconfirming evidence.

This chapter proposes that data-mining, statistical analysis, and causal modeling methods can help to overcome these tendencies. They can reveal unexpected truths about causal relations in data. To illustrate these methods, we consider a recently reported positive relation between a particular preventable exposure [of food animals to the streptogramin antibiotic virginiamycin (VM), used in poultry] and a particular health effect risk [resistance to VM-like (streptogramin) antibiotics in humans]. This reported association has been used to argue that poultry consumption causes increased resistance risks, that serious health impacts may result, and

L.A. Cox, Jr., *Risk Analysis of Complex and Uncertain Systems*,
International Series in Operations Research & Management Science 129,
DOI 10.1007/978-0-387-89014-2_7, © Springer Science+Business Media, LLC 2009

that, therefore, the use of VM in poultry should be restricted. However, the original study compared healthy vegetarians to hospitalized poultry consumers, raising the possibility that perhaps other, nonpoultry-related differences between healthy vegetarians and hospitalized patients might explain the observed differences in their resistance outcomes. Data mining and causal modeling can be used to explore this logical possibility.

Confirmation Bias in Causal Inferences

Recall from Chapters 1–3 that quantitative risk assessment (QRA) is largely about identifying and quantifying probabilistic *causal relations* between exposures and adverse consequences. Risk managers typically want to know how much of an effect (if any) reducing preventable exposures would have on reducing adverse human health consequences such as deaths or illnesses per year. QRA offers methods to help answer such questions.

Although causal relations are crucial to effective risk management, experimental psychology and neuropsychological studies demonstrate that, without using formal quantitative methods such as QRA, most people (including scientists) are prone to flawed and biased intuitive reasoning about causality and the relevance of evidence (Fugelsang and Dunbar, 2005). A common pattern, sometimes called "premature closure," is that individuals and groups tend to adopt prematurely causal hypotheses and conclusions to explain observations, based on inadequate information (Borrell-Carrio and Epstein, 2004). They then tend to seek confirming evidence and to ignore or underweight disconfirming evidence for the favored causal hypothesis – the phenomenon of *confirmation bias* (Fugelsang et al., 2004; Jonas et al., 2001). Experimental psychologists have also found that real-world reasoning about conditionals (if-then relations) and causality often differs from the prescriptions of formal logical analysis and modeling (Evans et al., 2007; Over et al., 2007; Oberauer, 2006).

If risk analysts fall into such "decision traps," they may end up publishing conclusions and advocating risk management actions that are not objectively justified and that then fail to achieve their stated objectives for reducing risks.

Example: The Wason Selection Task

A famous experiment investigates how people reason about conditionals and evidence, by showing subjects four cards with letters and numbers on them, such as A, 4, 7, D (Staller et al., 2000). Subjects are told that each card has a letter on one side and a number on the other. They are asked to identify the smallest subset of cards that must be turned over (revealing what is printed on their other sides) in order to decide whether the following hypothesis is correct:

H: "Any card that has a vowel on one side has an even number on its other side."

After some thought, most subjects correctly identify that it is necessary to turn over card A, in order to confirm whether the prediction from H is true (i.e., whether this card has an even number on its other side). But relatively few subjects spontaneously recognize that it is also necessary to turn over card 7, to verify that hypothesis H is not disconfirmed by the appearance of a vowel on its other side. (Many subjects also fail to recognize that turning over cards 4 or D would be irrelevant, because nothing that appears on their reverse sides can disconfirm hypothesis H.) Thus, even in this logically simple situation, it is difficult for most people to pinpoint the evidence needed to decide whether a simple stated hypothesis is true. (Performance improves when abstract logical problems are replaced by concrete problems involving social situations, such as detecting violations of an agreement to abide by an if-then rule. But risk analysis often deals with abstract, logical descriptions. For more on how performance on the Wason selection task varies across application domains, see www.psych.ucsb.edu/research/cep/socex/wason.htm#descriptive%20rule.)

The tendencies to seek confirming evidence, neglect disconfirming evidence, and overinterpret confirming evidence as support for a prior causal hypothesis are not confined to contrived abstract psychological experiments such as the Wason selection task. They also occur in important real-world decisions with significant financial or health consequences, as discussed in Chapter 1.

Example: Attributing Antibiotic Resistance to Specific Causes

A recently published study (Unicomb et al., 2006) noted that "The Australian government has prohibited the use of fluoroquinolones in food-producing animals" and "Among locally acquired infections, only 2% of isolates (range, 0%–8% in different states) were resistant to ciprofloxacin [a fluoroquinolone]." This is lower than the estimated corresponding average resistance rates for locally acquired infections in countries that have used fluoroquinolones for animal health (e.g., 6.4% in the United States) (Cox and Popken, 2006b). The authors interpret this difference causally: "The very low level of ciprofloxacin resistance in *C. jejuni* isolates likely reflects the success of Australia's policy of restricting use of fluoroquinolones in food-producing animals."

The presented evidence is certainly *consistent* with this causal hypothesis, much as turning over card A and finding an even number, or turning over card 4 and finding a vowel, would be consistent with hypothesis H in the Wason selection task. But more than such consistency is needed to support a causal hypothesis. For example, the same evidence may be consistent with different causal hypotheses. To support a unique causal interpretation of the data, other equally plausible or more plausible competing explanations should be eliminated (Grimes and Schulz, 2002). Formal mathematical models of causality (Greenland and Brumback, 2002) point out that if one variable [such as fluoroquinolone (FQ) use in animals] truly causes another (such as FQ resistance in human patients), then the association between them should not be *explained away* by any third variable. That is, conditioning on a third variable should not eliminate the association. Yet this particular study did not correct for the fact that, historically, "Australia prescribes proportionally less fluoroquinolones than other developed countries due to prescribing restrictions" (Baird, 1997). It did not address whether this low human use, rather than (or in addition to) low animal use, might explain the lower rate of FQ resistance in human patients in Australia. The suggested causal interpretation has therefore not been established by the data presented, even though it is consistent with it.

More generally, current widespread suspicion and concern that the use of animal antibiotics increases the frequency of antibiotic-resistant illnesses in humans provides a point of departure for many published interpretations of data and discussions of risk associated with particular "bug-drug" pairs. Yet authors seem not to be always mindful of the potential for confirmation bias and for invalid causal inferences in such settings, nor of the importance of investigating potential disconfirming evidence and plausible alternative explanations before drawing causal conclusions based on data that are consistent with alternative causal hypotheses (Shipley, 2000; Cox, 2001).

This chapter considers how current statistical models and methods can help to avoid potential confirmation biases (without introducing opposite preconceptions and biases) in interpreting animal antibiotic use and human health data causally. As a case study, we focus on a recent report by Kieke et al. (2006) that announced a positive relation between the use of virginiamycin (VM) (a member of the streptogramin class of antimicrobials) in poultry and resistance determinants or readily "inducible resistance" (discussed below) to VM-like (streptogramin) antibiotics in humans. The study concluded that "The results of the present investigation suggest that virginiamycin use in poultry contributes to human carriage of *E. faecium* that contains streptogramin resistance genes with readily inducible resistance." This appears to be a causal conclusion: that virginiamycin use in animals *contributes to* carriage of *E. faecium* with antibiotic resistance potential in humans. Based on this causal interpretation, an accompanying editorial (Frimodt-Moller and Hammerum, 2006) called for reduced use of virginiamycin use in food animals.

The following sections examine the evidence on which this suggested causal interpretation is based. They show that the data the authors analyzed do not justify this causal interpretation over others. Indeed, we shall see that conditional independence relations in the data (see Chapters 3 and 6) suggest a different causal explanation: that the "readily inducible resistance" defined and reported by Kieke et al. results from selection pressures associated with the hospital environment, rather than with food animals.

The existence of alternative plausible causal interpretations for the same data raises the central challenge addressed in this chapter: *To what extent can current statistical methods objectively determine which of several competing causal hypotheses is best supported by observed data, and how (if at all) can such data be used to identify and eliminate incorrect or unsupported explanations?* QRA methods typically seek to constrain as tightly as possible the set of causal interpretations that are consistent with or implied by available data, in part by using the data to identify and eliminate incorrect explanations where possible. Yet any claim to have done so using objective data analysis invites skepticism from those who favor the discarded causal theories. This gives particular value to methods of analysis that can produce valid conclusions using generally accepted statistical methods that can be independently replicated and verified by all parties.

The following sections seek to identify and illustrate such methods for VM and human streptogramin resistance. We first make some qualitative observations about the study design, and then we turn to quantitative risk modeling. Although we focus on the specific example of virginiamycin and streptogramin resistance, we believe that data analysis and modeling methods that avoid confirmation bias in causal attribution for antibiotic resistance are crucial for sound antimicrobial risk assessment and deserve wider development and application.

Study Design: Hospitalization Might Explain Observed Resistance Data

E. faecium is a commensal bacterium commonly found in human, bird, and animal intestines. It is not normally harmful and has been used as a probiotic dietary supplement for both humans and poultry. However, *E. faecium* can threaten seriously ill human patients, typically in intensive care units (ICU) of hospitals, via opportunistic infections. ICU patients with immune systems weakened by chemotherapy, organ transplants, AIDS, leukemia, or other conditions, or with surgical wounds or invasive medical devices, are at the greatest risk of opportunistic *E. faecium* infections. *E. faecium* bacteria have a high level of intrinsic and acquired resistance to some antibiotics, such as penicillin and vancomycin. Vancomycin-resistant *E. faecium* (VREF) have become prevalent in hospitals on several continents, and other drugs such as linezolid or the streptogramin combination Synercid® (quinupristin-dalfopristin) may be used to treat these VREF cases (Cox and Popken, 2006). This development has focused attention on the ongoing use of the streptogramin virginiamycin in food-producing animals.

Kieke et al. (2006) compared newly admitted (36 hours or less) hospitalized patients who ate poultry (and other meats) to healthy vegetarians in four communities. They concluded that human exposure to poultry was associated with the presence of quinupristin-dalfopristin (QD) resistance genes and experimentally inducible QD resistance in human fecal *E. faecium*. No actually QD-resistant *E. faecium* were found. QD is a streptogramin combination used in the human drug Synercid®, and resistance to the animal-use streptogramin drug virginiamycin confers cross-resistance to Synercid® in *E. faecium* (Donabedian et al., 2006). "Inducible QD resistance" is not (yet) a standard outcome measure and has no widely accepted definition (although it is well recognized that some resistance genes in some species of bacteria must be switched on by external stimuli, i.e., induced, in order to be expressed). Inducible QD resistance has not been shown to be a valid surrogate for clinically relevant QD resistance. The investigators generously provided us with their data (Belongia, 2007), enabling us to attempt to replicate and validate their results.

The study design, comparing hospitalized exposed cases to healthy controls, clearly creates a potential for uncontrolled confounding and noncausal statistical associations. As noted by Kieke et al., "[C]onfounding may have occurred, and other factors associated with vegetarian status may have contributed to the observed associations." Qualitatively, it seems plausible that frequently hospitalized patients may be exposed to nosocomial *E. faecium* and other hospital-associated conditions that healthy patients, including the vegetarians in this study, are not exposed to.

This study design suggests the following possible alternative explanation for the reported associations between resistance-related outcomes and exposure to poultry:

$$\textit{Resistance} \leftarrow \textit{Nosocomial exposures} \leftarrow \textit{Hospitalized patients} \rightarrow \textit{Poultry exposure}.$$

This causal graph highlights that the subjects who self-reported exposure to poultry meat (namely, the hospitalized patients) are also the ones exposed to nosocomial infections. Since nosocomial vancomycin-resistant *E. faecium* (VRE) infections are known to be associated with various types of resistance (Rice, 2006), it is unsurprising that poultry exposure may be *associated* with resistance in this study (because poultry eaters are hospitalized patients), even if poultry exposure does not necessarily *cause* increased resistance. Indeed, as noted by Kieke et al., "All PFGE patterns from humans and retail poultry were distinct, and no common clones were identified in both sources." This is consistent with the results of other recent studies. For example, Donabedian et al. (2006) concluded that "[Quinupristin-dalfopristin-resistant *E. faecium*] from humans did not have PFGE patterns similar to those from animal sources." Of course, the lack of a match might only indicate that the bacteria isolated from humans in these specific studies differed from the bacteria in these specific retail poultry samples. But, in general, no studies have demonstrated the transfer of quinupristin-dalfopristin resistant *E. faecium* from food animals to human patients; to date, studies that have searched for such a relation have not found it.

Kieke et al. (op cit.) reported that among participants without recent antibiotic use, "Carriage of *E. faecium* with *vatE* [a streptogramin resistance gene] was significantly associated with both touching raw poultry and higher poultry consumption in the combined hospital patient and vegetarian group." But this could simply reflect that hospitalized patients (many of whom had more than five physician visits in the previous year) have higher proportions of bacteria with resistance determinants than healthy subjects (i.e., vegetarians, in this study – who, of course, had little or no exposure to poultry). Interpreting the reported association as reflecting exposure to poultry rather than exposure to the hospital environment, or as a reason to "raise additional concerns regarding the continued use of virginiamycin in food animals" (Kieke et al., 2006), is unjustified if hospitalization, rather than transfer from poultry, explains the observed association between poultry exposure and carriage of *E. faecium* with *vatE*. Thus, a challenge for QRA is to determine which causal hypothesis is most consistent with the data.

Choice of Endpoints

As might be expected, the QD resistance prevalence is significantly higher for isolates from conventional retail poultry (56% prevalence) compared to antibiotic-free retail poultry (13%). Fortunately, resistance in poultry isolates was not observed to transfer to resistance in human isolates: "None of the human *E. faecium* isolates had constitutive resistance to quinupristin-dalfopristin" (Kieke et al., 2006). Antibiotic use on farms was also associated with significantly reduced *E. faecium* prevalence in retail poultry (48% for conventional retail poultry samples compared to 88% for antibiotic-free poultry samples).

A striking feature of the data is that *E. faecium* was isolated from 65% of vegetarians (65 out of 100) but from less than 19% of patients (the group that consumed and handled poultry) (105 out of 567). Moreover, the fraction of "susceptible" isolates is significantly *higher* for hospital patients than for vegetarians (24% vs. 12%). Thus, a randomly selected member of the poultry-exposed group (i.e., hospital patients) has probability $(105/567) \times (1 - 24\%) = 14\%$ of having reduced ("intermediate") QD susceptibility of *E. faecium*, compared to a much higher probability of $(65/100) \times (1 - 12\%) = 57\%$ for a randomly selected member of the nonpoultry-exposed group (i.e., vegetarians). Such calculations might be interpreted as suggesting that chicken eaters have *less* risk than vegetarians of carrying intermediate QD-susceptible *E. faecium*. These data refer to constitutive resistance and susceptibility (i.e., exhibited with no preexposure to virginiamycin). Also, most, but not quite all, hospital patients were self-reported consumers of chicken and other meats.

These calculations show how the same data set might be interpreted to support different conclusions about the human health effects of exposure to poultry, depending on the modeler's choice of outcome measures and comparisons to perform. Clearly, more objective methods of causal analysis and interpretation are desirable.

Quantitative Statistical Methods and Analysis

This section considers statistical models that can potentially help to resolve ambiguities and select among rival causal interpretations using relatively objective methods (especially, conditional independence tests).

Table 3.3 of Chapter 3 provides a checklist of potential threats to valid causal interpretations of the observed statistical association (left column) and some statistical methods developed to help avoid or eliminate these threats (right column). A "threat" to valid causal interpretation of an association is a potential noncausal explanation for it. Many of the statistical techniques in the right column have extensive technical literatures that have been previously surveyed, as well as implementation algorithms that are now widely available and included in standard statistical packages. In this chapter, therefore, we only briefly describe the main methods and software packages used to address selected threats relevant for this case study.

Applying this checklist to the case study identifies several potential statistical concerns. Table 7.1 lists these concerns and statistical methods to address them.

Table 7.1 Potential statistical problems and solutions

Potential problem	Solution approaches illustrated in this chapter
Model form selection	• Use nonparametric techniques (classification trees) • Use multiple model forms; test sensitivity of conclusions to model form
Variable selection (for exposure, responses, confounders, covariates)	• Bayesian model averaging (BMA) (Hoeting et al., 1999) • Classification tree analysis (see Chapter 6)
Coding of variables	• Use automatic quantization and partitioning techniques (e.g., classification trees) to set category boundaries • Do not discretize continuous variables
Missing data for explanatory variables	Use Bayesian multiple imputation
Selections of exposure and/or response metrics	Use multiple metrics; test sensitivity of conclusions to choice of metrics
Multiple testing/sample selection bias	Use all data, not just selected subsets

Sample Self-Selection Bias

Fewer than 40% of invited hospitalized subjects agreed to participate in the study. Study participants might differ systematically from those who did not agree to participate. We could not address this potential limitation, as the data had already been collected.

Bayesian Model Averaging (BMA) Avoids Variable Selection Bias

The conclusions in the case study (Kieke et al., 2006) are contingent on the validity of a regression model that selects a specific subset of predictors (including EAT-POULTRY; see Table 2 of the study) while excluding others (such as concurrent hospitalization status). It is now well known that, in such settings, *model selection bias* can exaggerate estimated effects and significance levels by ignoring *model uncertainty* about which variables to include as predictors, implicitly assuming that the selected model has the correct predictors (Hoeting et al., 1999). We examined this issue using Bayesian model averaging (BMA) software (Raftery et al., 2008) for the R statistical computing environment (R Development Core Team, 2008) to account for uncertainty about variable selection.

BMA assigns a posterior probability to each model in a *set* of plausible models (subsets of candidate variables), each with coefficients determined through standard regression procedures. It computes the posterior mean model coefficients for each variable and their standard errors. We used this information to compute posterior odds ratios and confidence intervals for each variable that account for model uncertainty. A key output of the algorithm, *probne0* (probability not equal to zero, shown in Tables 7.3, 7.4, and 7.5), gives the probability that each variable appears

Table 7.2 Counts of missing data in the Marshfield case study data set

Variable	Meaning	Missing data count
TouchBeef	Subject touched beef	3
TouchPork	Subject touched pork	4
BEEFTOT_MO	Times ate beef in prior month	4
PORKTOT_MO	Times ate pork in prior month	2
EATPOULTRY	Subject self-reported eating poultry	3
POULTRYTOT_MO	Times ate poultry in prior month	6
COOKOWN_MO	Times cooked own meal in prior month	3
TOUCHPOULTRY	Subject self-reported touching poultry	4
GRADE2	Highest grade or year of school	1
HOSP_GT1	Had >1 hospitalization in prior 12 mo	1
ICU	Was admitted to ICU in past 12 mo	1
LIVEBIRDEXP	Exposure to live turkey or chickens	2
Total		**34**

(i.e., has a coefficient significantly different from zero) in a randomly selected plausible model. We also computed *conditional odds ratios*, i.e., odds ratios conditioned on the variable appearing in a model.

Using Continuous Variables Avoids Dichotimization/Variable Coding Bias for Exposure

Kieke et al. dichotomized a continuous variable (POULTRYTOT_MO), describing the number of times per month that poultry is consumed, to obtain a binary indicator of the frequency of poultry consumption. Cases with poultry consumption above the median level were assigned a value of "high," while cases below the median level were assigned a value of "low." Vegetarians were a separate category.

Such a dichotomization of a continuous predictor can bias effects estimates (Streiner, 2002; Brenner and Loomis, 1994; Gustafson and Le, 2002; Royston et al., 2005; Ragland, 1992). At a minimum, alternative cut points should be used and the results presented (Brenner and Loomis, 1994). Royston et al. (2005) state that "dichotomization of continuous data is unnecessary for statistical analysis, and in particular, should not be applied to explanatory variables in regression models." We therefore reanalyzed the data keeping POULTYTOT_MO as a continuous variable.

Using Multiple Response Variable Definitions Avoids Results That Depend on Any Single Response Definition

The "median relative percentage of growth in the exposed group divided by that in the unexposed group" for cultured *E. faecium* under the experimental conditions of the case study, interpreted by Kieke et al. as a "measure of association for the inducible resistance models," has no known clinical relevance. We therefore considered several possible definitions of the response variable (see Appendix A) to determine whether the reported associations are robust to variations in definitions.

Table 7.3 BMA Output for full data set with the presence of *vatE* (VATE) as the response variable and *without* hospitalization status indicator included as a predictor (See text for discussion of the meanings of column headings)

Variable name	Probne0 = probability (%) that variable is a predictor	Posterior mean	Odds ratio	Lower 95% confidence limit	Upper 95% confidence limit	Conditional posterior mean
Intercept	100	–4.7348	NA	NA	NA	–4.7348
Hospital1	0					
Hospital2	0					
Hospital3	0					
Hospital4	0					
GENDER_E	0					
AGE	100	0.0486	1.0498	1.0253	1.0782	0.0486
Alcohol	0					
Dr_Visits2	0					
Dr_Visits3	0					
Dr_Visits4	0					
TouchBeef	0					
TouchPork	0					
CM_INDEX (a comorbidity index)	0					
ANYAB_BY29	57.2	0.9117	2.4886	1.5821	4.0140	1.5940
BEEFTOT_MO	27.5	0.0121	1.0122	1.0038	1.0217	0.0441
PORKTOT_MO	0					
EATPOULTRY	8.4	0.1776	1.1944	1.0760	1.3728	2.1266
POULTRYTOT_MO	0					
COOKOWN_MO	3.6	–0.0003	0.9997	0.9993	1.0001	–0.0083
TOUCHPOULTRY	0					
GRADE22	3.4	0.0445	1.0455	1.0176	1.0748	1.3158
GRADE23	0					
GRADE24	0					
WORK	0					
HOSP_GT1	0					
ICU	0					
LIVEBIRDEXP (indicates exposure to live birds)	0					
STATE_WI	0					
RACE2	0					

Table 7.4 BMA output for full data set with the presence of *vatE* (VATE) as the response variable and with hospitalization status indicator (HOSPITALIZED) included as a predictor

Variable name	Probne0 = probability (%) that variable is a predictor	Posterior mean	Odds ration	Lower 95% confidence limit	Upper 95% confidence limit	Conditional posterior mean
Intercept	100	–5.4665	NA	NA	NA	–5.4665
HOSPITALIZED	48.8	1.8606	6.43	2.33	68.93	3.8157
AGE	100	0.0376	1.04	1.01	1.07	0.0376
ANYAB_BY29	47.5	0.7000	2.01	1.37	3.02	1.4743
GRADE22	39.7	0.4442	1.56	1.12	2.19	1.1194

Nonparametric Methods and Multiple Alternative Regression Models Avoid Regression Model Form Selection/Misspecification Bias

The case study did not provide regression diagnostics or model validation results to indicate whether the reported associations are artifacts of regression model misspecification. We therefore reanalyzed the data using nonparametric (classification tree) methods, as well as some alternative parametric (regression) models, as follows.

- **Logistic regression analysis of *vatE* resistance gene data (VATE variable)** We modified the *bic.glm* (BMA for generalized linear modeling) function in R to use a specialized logistic regression algorithm, *logistf* (Heinz, 2008) that addresses bias and *separation*. Separation occurs in fitting a logistic regression model if the likelihood converges to a finite value while at least one parameter estimate diverges to (plus or minus) infinity. This can occur in small or sparse samples with highly predictive covariates. Without *logistf*, we found that variables such as EATPOULTRY did exhibit separation. EATPOULTRY has a value of 1 (or Yes, coded as Y in other analyses) for all but one patient in the subset of 45 *E. faecium*-colonized hospital patients without antibiotic use during the preceding month; for that patient, VATE = 0 (N or No). Thus, without correction, an inordinately large model weight is placed on EATPOULTRY as a predictor that separates that one VATE patient from others, but it also has an extremely large standard error.
- **Linear regression analysis of induced resistance values (REP_V_S_P)** With induced resistance (REP_V_S_P) as the response variable, we used BMA for multivariate linear regression to develop posterior mean regression coefficients for different variables, computed by the BMA *bicreg* R function for linear regression analysis, as inputs to a linear predictive model for the reponse REP_V_S_P. From these, we computed "adjusted ratios," similar to those in Table 4 of Kieke et al. (2006), using the methodology in Appendix A. We also provided the univariate "unadjusted ratios." (We extended the ratio concept to continuous nonindicator variables by using the conditions $\{> 0; = 0\}$ rather than $\{\text{Yes; No}\}$.)

Table 7.5 BMA output and resulting adjusted ratios with "induced resistance" (REP_V_S_P) as the response variable and with hospitalization status indicator included as a predictor

Variable name	Probne0 = probability that variable is a predictor (%)	Posterior mean	Conditional posterior mean	Unadjusted ratio	Adjusted ratio	Lower 95% confidence limit	Upper 95% confidence limit
Intercept	100	–3.8656	–3.8656	NA	NA	NA	NA
HOSPITALIZED	46.7	0.5427	1.1617	2.77	2.72	2.44	3.03
GENDER_E	0	0	0	1.42	1.69	1.45	1.97
AGE	28	0.0023	0.0083	NA	NA	NA	NA
Alcohol	0	0	0	0.82	0.78	0.65	0.94
Dr_Visits2	0	0	0	1.01	1.19	1.00	1.42
Dr_Visits3	0	0	0	1.51	1.60	1.35	1.88
Dr_Visits4	0	0	0	1.47	1.70	1.36	2.14
TouchBeef	100	–0.5865	–0.5865	1.23	1.57	1.37	1.80
TouchPork	12	0.0376	0.3140	1.52	1.50	1.28	1.75
CM_INDEX	6.6	–0.0032	–0.0476	2.61	2.49	2.20	2.81
ANYAB_BY29	76.4	0.3473	0.4545	3.69	2.36	2.08	2.68
BEEFTOT_MO	0	0	0	2.66	2.59	2.32	2.89
PORKTOT_MO	1.9	–0.0005	–0.0243	2.63	2.42	2.15	2.71
EATPOULTRY	54.4	0.5562	1.0218	2.77	2.70	2.42	3.01
POULTRYTOT_MO	0	0	0	2.57	2.69	2.41	3.00
COOKOWN_MO	0	0	0	0.19	0.42	0.36	0.50
TOUCHPOULTRY	1.4	0.0036	0.2485	1.23	1.60	1.39	1.84
GRADE22	19.4	0.1316	0.6765	1.80	1.76	1.50	2.07
GRADE23	14.9	0.1255	0.8403	0.96	0.86	0.71	1.04
GRADE24	14.9	0.1194	0.7994	0.76	0.52	0.45	0.59
WORK	0	0	0	0.63	0.63	0.54	0.74
HOSP_GT1	10.3	–0.0265	–0.2570	1.55	1.76	1.49	2.08
ICU	2.1	0.0068	0.3308	3.39	1.75	1.46	2.09
LIVEBIRDEXP	4	–0.0087	–0.2168	0.78	0.64	0.54	0.78
STATE_WI	0	0	0	1.20	1.38	1.17	1.63
RACE2	4.3	–0.0282	–0.6541	1.16	1.71	0.81	3.58

- **Classification tree analysis of VATE and REP_V_S_P**
  We used the nonparametric classification tree algorithm *rpart* (Recursive Partitioning and Regression Trees) in the R statistical software to recursively "split" response variables (VATE or REP_V_S_P) on the values of explanatory variables (or on ranges of values, for continuous and ordered-categorical variables) to "best" separate the conditional distributions of the response variables obtained by conditioning on the splits (using built-in criteria including mutual information between explanatory and response variables, and F-tests). Each leaf of a classification tree has a set of corresponding cases that match the description leading to that tip of the tree. These cases have an empirical joint frequency distribution of values for all variables. Classification trees are similar to regression models in that they have a single response variable and multiple explanatory variables. However, they can complement regression models by discovering nonlinear patterns, high-order interactions, and conditional independence relations in multivariate data (see Chapter 6). The *rpart* algorithm closely follows Breiman et al. (1984). To check our results with a different implementation, we also used the commercial KnowledgeSeeker™ classification tree software discussed in Chapter 6.

Bayesian Multiple Imputation Overcomes Biases from Missing Data

Table 7.2 summarizes missing data values in the case study data set (the "Marshfield" data set of Kieke et al., 2006; Belongia, 2007). A total of 14 records out of 170 (one record per subject) had at least one missing value, leaving 156 complete records. The set of 110 subjects who reported no prior use of antibiotics had six records with at least one missing value, leaving 104 complete records. The subset of 45 *hospitalized* patients with no prior use of antibiotics had five records with at least one missing value, leaving 40 complete records.

Many standard statistics packages have a default that simply deletes cases with missing data values. But this can introduce bias and reduce statistical power (Little and Rubin, 1987; Horton and Kleinman, 2007). Therefore, we used the *aregImpute* function of the Hmisc software package (Harrell, 2007) for the R open-source statistical language and environment. This function performs *multiple imputation* (using additive regression bootstrapping and predictive mean matching) to approximate drawing predicted values from a full Bayesian predictive distribution. We used the procedure on the full set of 170 records, and also for the two subsets of 110 with no reported prior antibiotic use and 45 hospitalized patients. [The SAS 9.1 software cited by Kieke et al. (2006) also includes multiple imputation functions, MIANALYZE and MI, but the article does not indicate whether they were used.]

Using All Data Avoids Multiple Testing/Multiple Comparisons/Subset Selection Bias

Kieke et al. (2006) analyzed a subset of subjects – those with no recent recorded antibiotic use. While this is common practice and common sense (since recent antibiotic consumption could select for antibiotic-resistant *E. faecium*), it does raise a statistical issue. The selection was made only *after* looking at the data and determining that, unlike other patients, "hospital patients without recent antibiotic use had an increased risk of carrying *E. faecium* isolates with *vatE* if they had touched raw poultry. . . . On the basis of these findings, results are reported for participants without recent antibiotic use." Deciding to report results on a subset of subjects because it has been found to support one's hypothesis clearly invalidates the use and interpretation of standard *p*-values and significance tests. It risks a strong form of confirmation bias. (Had the desired association held only in patients who had reported recent antibiotic use, the authors could have selected that subset instead.) To address this issue, we reanalyzed the data considering *all* subjects as well as the different subsets.

In summary, the statistical methods outlined here attempt to prevent confirmation bias from entering the analysis via any of the following routes:

- *Selecting variables* to support specific hypotheses, while excluding other variables (e.g., by including chicken-related variables but excluding hospital-related ones, when the two are correlated).
- *Selecting or constructing a specific response definition* (e.g., for "inducible resistance") that supports a specific hypothesis, while ignoring other response definitions (e.g., fraction of susceptible isolates) that do not support it.
- *Selecting a subset of subjects to* analyze (e.g., patients with no self-reported recent antibiotic use) that supports a specific hypothesis (e.g., that poultry exposure is associated with a QD resistance gene).
- *Selecting a single parametric model form* (e.g., logistic regression with some continuous variables such as POULTRYTOT_MO dichotomized), when different forms or nonparametric methods might give different conclusions.

We also attempted to avoid other potential biases (not necessarily resulting from choices made by the modeler, and hence not as subject to confirmation bias), by using multiple imputation for missing data and keeping POULTRYTOT_MO as a continuous variable.

Conditional Independence Tests Can Objectively Choose Among Rival Causal Models

The preceding measures may help to limit the potential influence of confirmation bias, but they cannot necessarily discriminate among rival causal models or hypotheses for explaining any associations that still persist in the absence of confirmation bias. Fortunately, as illustrated in the next section, classification tree analysis can also be used to test for *conditional independence* relations among variables, and these relations, in turn, can be used to test and discriminate among rival causal hypotheses.

Consider, for example, the following two alternative causal models:

$$\text{Model 1}: \textit{Nosocomial exposures} \leftarrow Hospitalized \leftarrow \textit{Poultry exposure} \rightarrow Resistance,$$
$$\text{Model 2}: Resistance \leftarrow \textit{Nosocomial exposures} \leftarrow Hospitalized \rightarrow \textit{Poultry exposure}.$$

In these models, the arrows indicate that each variable is statistically independent of its more remote ancestors, given the value of its parent (see Chapter 6). These alternative causal models make different, statistically testable, predictions about conditional independence relations among variables. Suppose that the variable *Resistance* is some measure or indicator of actual or potential QD resistance in *E. faecium*. (For example, it might be defined as a binary variable indicating the presence or absence of resistance genes, or as a continuous variable measuring the level of resistance in a specified test, etc.) Suppose that *Poultry exposure* is a variable summarizing self-reported exposure to poultry (e.g., as a binary indicator, a times-per-month measure of frequency, etc.). Finally, let *Hospitalized* be a binary indicator variable showing whether the respondent was hospitalized when the study questionnaire was administered, and let *Nosocomial exposures* be a variable indicating whether an isolate is of hospital origin.

According to both Model 1 and Model 2, any pair of these variables may be correlated. However, according to Model 1 (which may be loosely interpreted as implying that "Poultry exposure causes resistance"), but *not* according to Model 2 (loosely interpreted as "Hospitalization causes resistance"), *Resistance* should be conditionally independent of *Hospitalized*, given the value of *Poultry exposure*. Conversely, in Model 2, but not in Model 1, *Resistance* should be conditionally independent of *Poultry exposure*, given the value of *Hospitalized*. Thus, these models have quite different implications for conditional independence relations, and hence statistical methods for testing conditional independence relations can reveal which model (if either) is consistent with the data.

Although this example has considered using conditional independence tests to choose between two prespecified causal models, classification tree analysis can also be used to determine conditional independence relations in the absence of any prespecified hypothesis. Assuming that the data set is large and diverse enough to correctly reveal these relations, it is then possible to mathematically identify the possible causal graph models that are consistent with these empirically determined relations (see Chapters 3 and 6). This supports generating empirically driven causal theories, without any need to formulate an a priori hypothesis that can then be subject to confirmation bias.

Results of Quantitative Risk Assessment Modeling for vatE *Resistance Determinant*

Table 7.3 shows the main results of the BMA logistic regression analysis for the presence of the *vatE* resistance determinant (variable VATE), for the full set of 170 subjects with *E. faecium* isolates. The predictors that appear most often among the different plausible models are AGE, ANYAB_BY29 (an indicator of recent antibiotic use), and BEEFTOT_MO (the number of times beef is eaten per month). Less frequent are EATPOULTRY (self-reported poultry consumption; this verifies the association reported by Kieke et al.), COOKOWN_MO (the number of times cooked own meal per month, which has a mild protective effect), and GRADE22 (high school graduates). The most significant odds ratio is for the prior use of antibiotics (ANYAB_BY29). Of the others, all but COOKOWN_MO have significant odds ratios (confidence intervals do not include 1.0), but only slightly so. No individual hospital has a significant effect. (Hospital1-Hospital4 are binary indicator variables for the four hospitals that provided data for this study.)

As previously discussed, poultry consumption is strongly positively correlated with hospitalization: All but a few hospitalized cases ate poultry, while all nonhospitalized cases, being vegetarians, did not. We therefore created a single summary indicator variable, "HOSPITALIZED" with a value of 1 for hospitalized cases, and 0 otherwise. Table 7.4 shows the results of a BMA analysis when this hospitalization status variable is included as a predictor. (To save space, only the variables with nonzero inclusion probabilities are shown.) Now, COOKOWN_MO and EATPOULTRY (see Table 7.2 for variable definitions) drop out. Only HOSPITALIZED and ANYAB_BY29 have highly significant odds ratios. Thus, *when both exposure to poultry (EATPOULTRY) and hospitalization status (HOSPITALIZED) are allowed as candidate predictors, automated variable selection via BMA identifies HOSPITALIZED, but not EATPOULTRY, as a significant predictor of the presence of the vatE resistance determinant (VATE).* This is consistent with VATE being conditionally independent of EATPOULTRY, given HOSPITALIZED (i.e., with Model 2, but not Model 1, in the previous section).

To check this possibility without assuming any particular parametric regression model form (thus risking model misspecification biases), we also analyzed the full data by classification tree analysis. When HOSPITALIZED is excluded, then, similar to the BMA analysis, ANYAB_BY29, BEEFTOT_MO, and AGE are identified as significant predictors. Once prior antibiotic use (ANYAB29), AGE, and BEEFTOT_MO are accounted for (i.e., conditioned on), *no poultry-related variable appears as a predictor* of VATE. In other words, *resistance (VATE) is conditionally independent of poultry consumption, given AGE, BEEFTOT_MO, and ANYAB29.*

Importantly, the converse is not true: VATE is *not* conditionally independent of BEEFTOT_MO or ANYAB29, given poultry variables. For example, even after forming a tree by splitting first on EATPOULTRY and POULTRYTOT (the two poultry variables that are significantly associated with the response variable VATE), it is still the case that ANYAB29 (and BEEFTOT_MO_missing, meaning that a patient did not provide data about prior beef consumption) still enters the tree as an additional split. This asymmetry has strong implications for possible causal models. It implies that EATPOULTRY cannot be a direct parent of VATE in a causal graph, because its effects are fully "explained away" by the other variables with which it is correlated (BEEFTOT_MO, AGE, and ANYAB29).

When HOSPITALIZED is included in the data set, it becomes the most important predictor (first split). Figure 7.1 shows the resulting tree, generated by the commercial software KnowledgeSeeker^{TM}. This tree is read as follows. Each node contains three numbers. The bottommost number (e.g., 170 in the top node of the tree) shows the total number of cases described by that node. The middle number (e.g., 23.5%) is the percent of cases in the node that have VATE = Y (yes, the *vatE* streptogramin resistance gene was detected), and the top number (e.g., 76.5) is the percent of cases with VATE = N (no, it was not detected). These two percentages total to 100% at each node. The set of splits between the top node and any other node describe the cases at that node. For example, "HOSPITALIZED = N" is the description for the 65 cases with 0% having VATE = Y and 100% having VATE = N. "HOSPITALIZED = Y and BEEFTOT_MO = Y" is the description of the node with four cases, all of which have VATE = Y.

Classification trees are most often used to identify descriptions (i.e., conjunctions of variable values or ranges of values) that are highly predictive of or informative about the response variable (Breiman et al., 1984). However, they can also be used to reveal which variables (namely, those not in the tree) the response variable does *not* significantly depend on, given (i.e., after conditioning on) the subset of variables that *are* in the tree. This provides a statistical test for *conditional independence relations* in multivariate data sets – a staple of modern causal modeling (see Chapter 6). The response variable is conditionally independent (at least within the power of the classification tree algorithm to discover) of the variables not in the tree, conditioned on the ones that are in the tree. As discussed in Chapter 6, explanatory variables that cannot be forced out of a tree by conditioning on other variables (or, more generally, explanatory variables that the response variable cannot be made statistically independent of by conditioning on other variables) are candidates for having a potential direct causal relation with the response variable.

Figure 7.1 shows that *poultry consumption variables are not significant predictors of vatE risk after conditioning on hospitalization.* In other words, conditioning on hospitalization makes VATE conditionally independent of poultry variables. However, importantly, *the converse is not true.* If one splits (i.e., conditions) on EATPOULTRY first, then HOSPITALIZED (and BEEFTOT_MO) still enter the tree as significant predictors (Fig. 7.2). Thus, their effects are *not* explained away by correlation with EATPOULTRY. This asymmetry shows that poultry variables appeared to be significant predictors in the case study only because they acted as surrogates for hospitalization: Including HOSPITALIZED directly as a predictor eliminates the poultry consumption variables as significant predictors. This goes well beyond simply stating that hospitalization and poultry consumption are strongly associated or multicollinear with each other (so that either could act as a surrogate for the other, e.g., in multiple regression modeling with stepwise variable selection). It suggests that hospitalization is more fundamental than poultry consumption as a predictor of resistance, since including hospitalization as a predictor makes poultry variables redundant, but including poultry variables does not make hospitalization redundant.

Fig. 7.1 KnowledgeSeeker™ classification tree for full data set of 170 subjects with *E. faecium* isolates

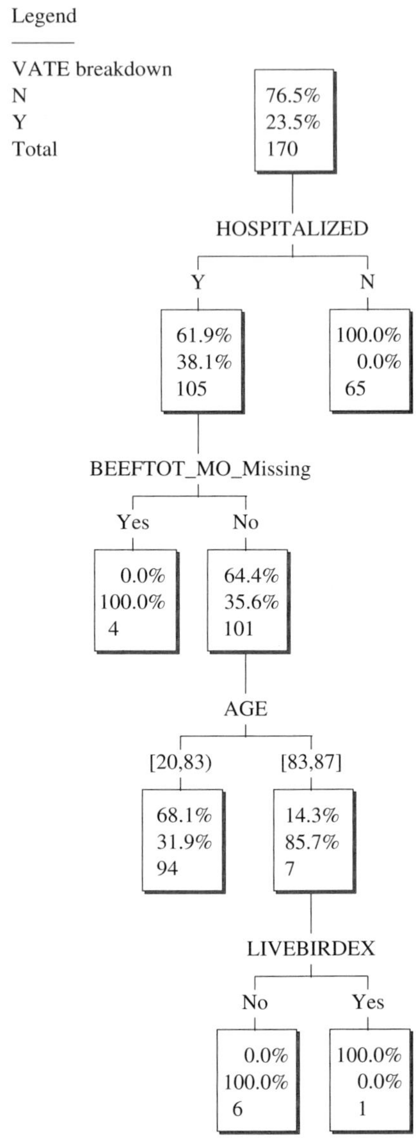

The pattern of conditional independence relations revealed by these classification tree analyses is consistent with the causal graph in Fig. 7.3. It is inconsistent with any causal graph (or causal hypothesis or interpretation) in which EATPOULTRY is a parent (potential direct cause) of VATE. ("NONVEG" is a mnemonic variable introduced here to stand for "nonvegetarian." By the study design, it is the same as HOSPITALIZED.)

Fig. 7.2 VATE is not conditionally independent of HOSPITALIZED, given EATPOULTRY

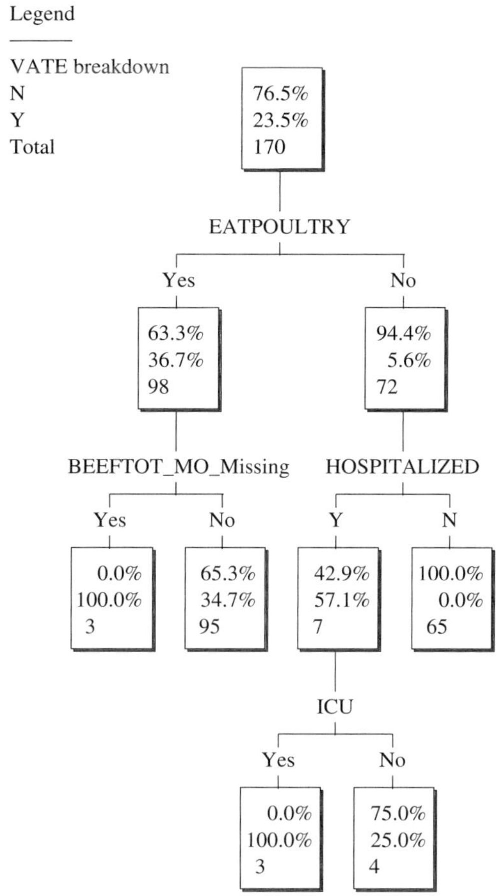

Results for Inducible Resistance

Table 7.5 shows the BMA linear regression results and ratios with induced resistance (REP_V_S_P) as the response variable. Again, HOSPITALIZED and EATPOULTRY enter the model with highly significant adjusted ratios. The sum of their inclusion probabilities is close to 100% in this table, implying that they are substitutes for each other. Classification tree analysis showed that *REP_V_S_P is also conditionally independent of all poultry variables after conditioning on nonpoultry variables.* (See Fig. 7.4, which is a tree for a continuous response variable. The

BEEFTOT_MO_missing ← VATE ← AGE
↑
EATPOULTRY ← NONVEG/HOSPITALIZED → ANYAB_BY29

Fig. 7.3 A causal graph model that is consistent with conditional independence relations revealed by classification tree analysis

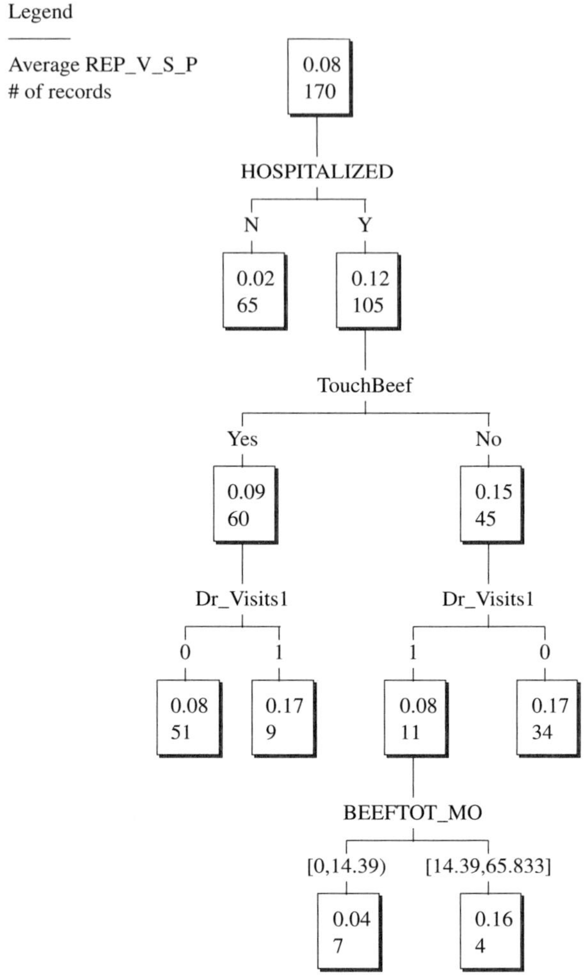

Fig. 7.4 KnowledgeSeeker™ classification tree for inducible resistance. (The upper number in each node is the conditional mean of REP_V_S_P)

upper number in each node is a conditional mean value for REP_V_S_P.) Again, however, HOSPITALIZED still enters as an explanatory variable even after conditioning on poultry variables. This empirical finding is inconsistent with any a priori causal hypothesis that poultry consumption increases the risk of inducible resistance.

Discussion and Implications for Previous Conclusions

The conditional independence relations identified in the case study data do *not* support a causal hypothesis that poultry consumption increases resistance risks. There was no significant relation between poultry exposure and VATE or inducible resistance after accounting for nonpoultry variables. Poultry-exposed subjects also had lower risks than other subjects for both *E. faecium* carriage and intermediate QD susceptibility.

The case study does not provide resistance results specifically for patients who might be harmed by QD resistance, i.e., *vanA* VREF patients (Cox and Popken, 2004b), so its clinical relevance is unclear. Its conclusion (Kieke et al., 2006) that "The results of the present study suggest that the FDA [draft risk assessment, FDA-CVM, 2004] model may underestimate the true risk of foodborne acquisition, because streptogramin resistance genes are commonly found in human fecal *E. faecium*" appears to be unwarranted by the data. The data do *not* show that foodborne resistance occurred, or that any nonzero resistance risk or clinical harm occurred. To the contrary, they support previous risk assessment modeling predictions that antibiotic-treated poultry might have *lower* levels of susceptible *E. faecium* than other poultry (Cox, 2005a), while finding no evidence of QD resistance in isolates from human patients.

Similarly, the case study data do not support the stated conclusion (Kieke et al., 2006) that "The presence of *vatE* and inducible streptogramin resistance in the endogenous fecal flora of newly hospitalized patients creates a genetic reservoir for the emergence of streptogramin-resistant, vancomycin-resistant *E. faecium* in the hospital environment." First, the study did not specifically examine streptogramin-resistant, vancomycin-resistant *E. faecium*, so it cannot support this conclusion. Second, many of the "newly hospitalized" patients had a record of many prior visits to physicians in the past year (and an unknown amount of prior exposure to hospital bacteria). The data do not suggest that these patients are reservoirs from which resistant bacteria enter the hospital environment. To the contrary, these patients may acquire resistant bacteria from the hospital environment, perhaps selected by the use of antibiotics in hospitals. Because the patients were newly admitted, resistance might not have been acquired yet on the current visit. Whether it might instead have resulted from previous exposures depends on details of hospitalization and exposure histories, transient colonization, and clearance that are not available in the current study. We do not seek to explain further why or how hospitalization fully accounts for resistance previously attributed to poultry in this case study; we simply emphasize that it does so (Figs. 7.1, 7.2, 7.3, and 7.4). Third, *vatE* has not been shown to be relevant to resistance in human patients. For example, it was found in *none* of 167 tested isolates of *E. faecalis* and *E. faecium* from bacteremia patients (Jones and Deshpande, 2004).

The speculation that future increased use of QD (i.e., Synercid™) combined with the presence of resistance genes "might facilitate the rapid emergence of streptogramin resistance" (Kieke et al., 2006) is likewise not supported by the study data, which showed zero QD resistance despite years of QD use in hospitals. It is also contradicted by a quantitative risk model for the emergence of resistance (Cox and Popken, 2004a).

Our analysis shows that many nonpoultry variables also have significant unexplained associations with induced resistance, including hospital location, age, beef consumption, pork consumption, higher education, and alcohol (the latter two with protective effects). In the 45-case subset of hospitalized patients, the only significant adjusted ratio occurred for the indicator variable for college graduates. Thus, "induced resistance" ratios may not be very meaningful (e.g., they do not indicate that poultry consumption is any more of a risk factor than educational status or other nonpoultry variables) and may reflect correlations to hospitalization status more than anything else.

In summary, although several strong causal conclusions and interpretations of the case study data have previously been presented that suggest increased human health risk from virginiamycin use in poultry (Kieke et al, 2006; Frimodt-Moller and Hammerum, 2006), statistical methods that seek to minimize confirmation bias and dependence on a priori causal hypotheses do not confirm those conclusions, but instead support an alternative causal interpretation in which hospitalized patients have increased resistance risks. Extending the Marshfield study to include *healthy* poultry eaters and *hospitalized* nonpoultry eaters (categories currently of size 0 and 4, respectively) would provide a better basis for the analysis of resistance element transfers (if any) from poultry to humans.

Summary and Conclusions

When adequate data are available, as in this case study, achieving the goal of objective causal inference may be helped by the techniques illustrated in this chapter, such as BMA, classification tree analysis, and conditional independence testing of causal assumptions and interpretations. The case study in this chapter illustrates that a risk assessor's choice of which predictive variables to include in a model can drive the risk attributed to specific variables, such as poultry consumption. It can make otherwise nonsignificant associations (e.g., between EATPOULTRY and VATE) appear significant, by excluding correlated variables (such as HOSPITALIZED) that explain them away.

To avoid such potential biases and dependencies on modeling choices, we have illustrated how techniques such as BMA, classification trees, and conditional independence tests can be used to bypass the human selection of variables, model forms, responses, subsets of subjects, and a priori causal hypotheses to test. Eliminating these human choices can reduce the opportunity for confirmation biases and increase the role and value of empirical data in revealing unexpected findings and causal relations.

In our analysis of the case study data, poultry consumption turned out to be a proxy for hospitalization. Rather than confirming prior causal interpretations that the study data "raise additional concerns regarding the continued use of virginiamycin in food animals" (Kieke et al., 2006), we found that the use of antibiotics is associated with a reduced prevalence of bacteria in retail meats. People who eat such meats have a lower prevalence of *E. faecium* and a higher proportion of QD-susceptible *E. faecium* than people who do not. The study data did not suggest that resistant bacteria or resistance determinants were transferred from poultry to human patients. The data are consistent with the causal hypothesis that hospitalization causes resistance (e.g., *vatE* presence) but are inconsistent with the alternative causal hypothesis that poultry consumption causes resistance.

Examining the data using conditional independence tests for potential causality revealed that poultry consumption acted as a surrogate for hospitalization in this study. After accounting for current hospitalization status, no evidence remains for a causal relationship between poultry consumption and increased streptogramin resistance. This example emphasizes both the importance and the practical possibility of analyzing and presenting quantitative risk information using data analysis techniques (such as Bayesian model averaging and conditional independence tests) that are as free as possible from potential selection, confirmation, and modeling biases.

The ability to use relatively objective and widely available modern statistical methods to test alternative causal models has the potential to revolutionize causal interpretations of risk assessment data. Perhaps the further development and use of such methods will eventually help to suppress confirmation bias and increase the value of empirical data in informing understanding of the causes of antibiotic resistance and other risks.

Appendix A: Computing Adjusted Ratios of Medians and Their Confidence Limits

Given the vector of posterior mean regression coefficients, $\hat{\beta}$ (from the BMA analysis), we calculate the vector of mean response values (for induced resistance REP_V_S_P) as

$$\text{estimated mean response: } \hat{y} = X\hat{\beta}, \tag{7.1}$$

$$\text{where } X = \left\{ \begin{matrix} 1 \; X_{11} \ldots X_{1p} \\ 1 \; X_{21} \ldots X_{2p} \\ : \; X_{31} \ldots X_{3p} \\ 1 \; X_{41} \ldots X_{np} \end{matrix} \right\}.$$

The X_{ij} are the values in the data matrix, after the adjustments described in the text, for each collection of n observations and p variables that were analyzed. As in Kieke et al. (2006), we have taken the natural logarithmic transform of $y =$ ln(REP_V_S_P). For a given variable, j, $j = 1, 2, 3, \ldots, p$, we can partition the response vector $\hat{y}$ into $\{\hat{y}_{pj}, \hat{y}_{0j}\}$, where $\hat{y}_{pj}$ is a vector of the responses, y_i, $i =$ 1,2,3,. . . , n, for X_{ij} positive (>0) and $\hat{y}_{0j}$ is a vector of the responses, y_i, where X_{ij} was equal to 0. This is a generalization of the criteria where X_{ij} is either "Yes" (1) or "No" (0), allowing the methodology to be applied to continuous variables as well. The "adjusted ratio" as defined in Kieke et al. and further generalized for all variables, $j = 1, 2, \ldots, p$, is then

$$\text{adjusted ratio}_j = \text{median}(\exp(\hat{y}_{pj}))/\text{median}(\exp(\hat{y}_{0j})). \tag{7.2}$$

To estimate varibility on the mean responses $\hat{y}_i$, we note that the standard formula for confidence intervals on the mean response is

$$\text{CI}(\hat{y}_i) = X_i.\hat{\beta} \pm t_{\alpha/2,n-k-1}{}^*\hat{\sigma}^*\sqrt{X_i.(X^T X)^{-1} X_i^T}, \tag{7.3}$$

where

- $X_i.$ is a row, *i*, of the data matrix, *X*,
- $t_{\alpha/2,n-k-1}$ is the *t*-statistic at confidence level (1–α) with *n*–*k*–1 degrees of freedom, where *n* is the number of observations in the data set and *k* is the number of variables with nonzero posterior mean coefficients,
- $\hat{\sigma}$ is the standard error of the regression estimate $= \sqrt{\frac{\sum(y_i - \hat{y}_i)^2}{n-k-1}}$, where y_i= ln(REP_V_S_P),
- $(X^T X)^{-1}$ is the variance-covariance matrix,
- $\hat{\sigma}^*\sqrt{X_i(X^T X)^{-1} X_i^T}$ is the standard deviation of the mean response for observation *i*.

To compute confidence limits on the adjusted ratios, we used simulation. Each iteration of the simulation generates a response vector, with each vector element, *i*, drawn from a *t*-distribution (t_{n-k-1}) with mean equal to the mean response, $\hat{y}_i$, and standard deviation equal to the standard deviation of the mean response for $\hat{y}_i$ as provided above. From each simulated response vector, we compute an adjusted ratio for each variable[1] as in Equation (7.2) above. We ran the simulation for 10,000 iterations to generate a large distribution for each variable's ratio. The lower confidence limit we report corresponds to the 0.025 quantile of the sample distribution, while the upper confidence limit corresponds to the 0.975 quantile of the sample distribution. The distributions appear to be approximately lognormal, but we have chosen to use the sample quantiles rather than quantiles of a fitted lognormal distribution, as this requires making fewer assumptions and we have a large sample to work with.

[1]Except for AGE, which has no 0 values.

Chapter 8
Estimating the Fraction of Disease Caused by One Component of a Complex Mixture: Bounds for Lung Cancer

Lung cancer illustrates many of the challenges of modeling cause and effect in very complex systems having only poorly understood causal mechanisms. When not enough is known to develop useful marginal and conditional probability distributions for uncertain quantities, it may be practical instead to develop *bounds* on uncertain quantities and causal relations. This chapter illustrates bounding for lung cancer risks. Other areas of quantitative modeling and operations research, from robust optimization to constraint logic programming, apply a similar insight: *It is often practical to use limited available data to develop bounds on the likely consequences of actions,* even if the data are not adequate to estimate informative, well-calibrated, probabilities for consequences. This chapter and Chapter 9 discuss risk bounds for uncertain complex systems. To illustrate how to develop bounds from data, we quantify bounds on preventable disease risks for two very different illnesses – lung cancer and penicillin-resistant bacterial infections, respectively.

Motivation: Estimating Fractions of Illnesses Preventable by Removing Specific Exposures

A current grand challenge for health risk analysis is how best to use "-omics" information – that is, mechanistic information and insights emerging from genomics, proteomics, transcriptomics, metabolomics, and so forth – to improve quantitative risk assessments (Aardema and MacGregor, 2002; Leighton, 2005). A better understanding of metabolic, cell signaling, and molecular-level pathways leading from exposures to biological effects does not necessarily translate easily to improved risk assessments and dose-response models. Nor does it always generate large data sets that allow data mining and empirical causal modeling (Chapters 6 and 7) to learn about causality and risks directly from data. Other approaches are needed to apply insights from the improved biological understanding of complex disease processes and systems to achieve improved health risk models. Analogous challenges arise for other complex systems: a better understanding of how a system works does not automatically imply a better ability to predict quantitatively how changing inputs will change probable outputs. Yet this is a key question for risk managers.

L.A. Cox, Jr., *Risk Analysis of Complex and Uncertain Systems*, International Series in Operations Research & Management Science 129,
DOI 10.1007/978-0-387-89014-2_8, © Springer Science+Business Media, LLC 2009

This chapter therefore considers how to adapt and apply probabilistic risk analysis (PRA) (Chapter 2) to molecular biological information about causal pathways, to obtain quantitative bounds on the fractions of health risks that can be prevented by removing specific exposures. We call these the *preventable fractions* of illness caused by those specific exposures. Motivating applications include estimating the fraction of occupational illnesses preventable by removing specific constituents of exposure (e.g., arsenic, cadmium, etc.) from a workplace and estimating the fraction of smoking-induced lung cancers that could be prevented by removing specific constituents or groups of constituents [e.g., polycyclic aromatic hydrocarbons (PAHs)] from the complex mixture of chemicals found in cigarette smoke. A special case of this latter problem – estimating the fraction of smoking-induced lung cancers that could be prevented by removing the contribution from a particularly well-studied PAH-related pathway – serves as a case study.

Why Not Use Population Attributable Fractions?

The population attributable fractions (PAFs) and attributable risks (ARs) discussed in most epidemiology textbooks are association-based. They do not necessarily address the *causal* question of how much illness would be prevented by removing particular exposures (although they are sometimes misinterpreted as doing so). The following two counterexamples illustrate the well-known limitations of this traditional epidemiological concept of attributable risk.

Example: Attribution of Risk to Consequences Instead of Causes

Let X, Y, Z, and $RISK$ be four binary 0-1 indicator variables, with the relations among them described by the following causal graph (Greenland and Brumback, 2002; Shipley, 2000):

$$Z \leftarrow Y \leftarrow X \rightarrow RISK.$$

The intended interpretation is that exposure X directly causes $RISK$ and also biomarker Y. Y, in turn, causes biomarker Z. For simplicity, "causes" is given a clear meaning in this example, specified by the following structural equations model:

$$\begin{aligned} E(RISK) &= 0.1X, \\ Y &= X, \\ Z &= Y. \end{aligned}$$

Changing any right-hand-side variable causes its corresponding left-hand-side variable to change to restore equality; thus, if exposure changes from $X = 1$ (exposed) to $X = 0$ (unexposed), then $E(RISK) = \Pr(RISK = 1)$ will change from 0.1 to 0 (and Y and Z will also change from 1 to 0). In this sense, the exogenous variable X directly causes *RISK*. On the other hand, exogenously changing Z (or Y) will have no such effect on *RISK*: they are consequences, rather than causes, of X, and are not causes of *RISK*. Yet attributable risk calculations necessarily assign X, Y, and Z identical attributable risks (of 100% each), since they have identical values. Intuitively, what one wants is *not* a method that assigns X, Y, and Z identical attributable risks, but rather one that assigns the *preventable fractions* as follows: 100% for X and 0% for Y and Z, reflecting their asymmetric roles in causing *RISK*.

Example: Positive Attributable Risk is Compatible with Negative Causation

The following table shows hypothetical risks to people exposed to four logically possible combinations of two binary factors, A and B. Each factor has only two levels, present and absent.

	Factor B absent	Factor B present
Factor A absent	0	0.2
Factor A present	0.1	0.7

In this table, the attributable risk for factor A is positive, since A is positively associated with risk at each level of B. A traditional calculation based on the difference in risks with and without a factor present would attribute $(0.7 - 0.1)/0.7 = 86\%$ of the risk for someone exposed to both A and B to factor B; and $(0.7 - 0.2)/0.7 = 71\%$ to factor A. (However, removing both factors would not eliminate $86\% + 71\% = 157\%$ of the total risk.) That A has a positive "attributable risk" and a positive association with risk at each level of B (and overall) does *not* necessarily imply that removing A would reduce risk; thus, these fractions do not answer the practical question of how much (or what fraction of) risk would be prevented by removing each exposure. Indeed, risk might increase if A were removed.

For example, suppose that people who are not exposed to A are more likely to become exposed to B; then reducing exposure to A might increase exposure to B, and hence risk. As a less trivial example, suppose that factor A is "takes baby aspirin each day," that factor B is "has dangerous levels of C-reactive protein (CRP)," and that the risk in question is the risk of myocardial infarction (MI). While it may be true that only the subset of people with both factors present ($A = B = 1$) are at maximum risk of a heart attack (since people diagnosed as being at greatest risk may start taking baby aspirin), this does *not* imply that taking baby aspirin each day increases the risk of MI. Indeed, it may *reduce* the risk (the causal effect is protective), e.g., from 0.9 to 0.7, yet still be *associated* with increased risk (attributable risk is positive), simply because only maximum-risk people self-select to take aspirin. In this example, "being at maximum risk" is a hidden or latent variable, not included in the table, that both explains the simultaneous presence of factors A and B and also causes the high risk of heart attack. The 0.9 for maximum-risk people who do not take aspirin will never be observed if all maximum-risk people do, in fact, take aspirin; thus, the table does not include the information needed to identify the causal impact of taking aspirin. (Hidden variables need not be as blatant as the intentional self-selection of exposures to reduce risks. A genetic polymorphism such as a CYP450 alteration that makes people both more likely to consume a product and more likely to suffer an ill effect, whether or not they consume the product, can induce a positive association and attributable risk for exposure to the product, even if consumption of the product leaves risk unchanged or reduces it.) Thus, attributable-risk calculations do not, in general, answer the question that risk managers care most about: *How would risks change if specific exposures were removed?* A different approach, more grounded in the understanding of causal mechanisms, is needed to answer this question.

Theory: Paths, Event Probabilities, Bounds on Causation

One of the most useful informal ideas in health risk analysis is that of a *causal path*, i.e., a sequence of events, each of which is enabled (and perhaps hastened) when its predecessors have been completed, and the conjunction of which suffices to create an undesired health effect. This concept is related both to the *sufficient causes* framework in epidemiology (Rothman and Greenland, 2005; Hoffman et al., 2005) and also to *path sets* in systems reliability theory and fault tree analysis (Chang et al., 2004; Wreathall and Nemeth, 2004), as well as to *Horn clauses* in expert systems for reasoning about which events will occur (Martins and Mendes, 2001). In each framework, a *conjunction of events* (the "path set") suffices to cause a particular "top event" of concern (e.g., a heart attack or a lung tumor – the top event is denoted generically by the 0-1 indicator variable *RISK* in the preceding examples).

Any fault tree (or more general Boolean logical system) for determining whether the *RISK* top event occurs, based on the occurrences of its predecessors, can be expressed in a standard "disjunctive normal form" (Brafman, 2004), i.e., as a set of alternative path sets, any of which suffices to cause the top event. The probability of the top event, *E*(*RISK*), can then be calculated (or closely approximated) from the probabilities of events in its path sets via well-developed combinatorial reliability algorithms; and the change in this probability [i.e., in *E*(*RISK*)] when a particular event (e.g., indicating a specific exposure) is prevented from happening can be used to quantify various measures of the "importance" of that event in causing the top event (e.g., Chang et al., 2004). Moving from these concepts to a measure of the reduction in *E*(*RISK*) (i.e., a reduction in the expected fraction of cases in which *RISK* will equal 1 instead of 0) when a particular event or path set is blocked or removed might seem to be a small step and a useful contribution to defining and quantifying preventable fractions.

However, using this seemingly simple intuitive idea in health risk analysis confronts formidable practical and conceptual challenges. A complete path set (i.e., a set of events whose conjunction genuinely suffices to cause the top event) may not be known; the set of all such path sets is also often unknown; crucial events may not be observable with available data; the "occurrence" of key events (such as adduct formation or cell proliferation) is often a matter of degree, rather than a simple binary event; events may be transient (e.g., damage to a cell may be repaired); and the exact meaning of "cause" is notoriously hard to define satisfactorily. Despite these difficulties, useful progress can be made – and many philosophical pitfalls about the meaning of causation can be avoided – by using observed *co-occurrence frequencies* of specific biological events and conditions to estimate *upper bounds* on the increase in probability of a top event created by a specific component of exposure.

The main idea is to apply logical inequality constraints, such as the following, to individual-level molecular epidemiological data to obtain solid upper bounds on incremental probabilities of causation:

$$\Pr(X \textit{ caused } Y) \leq \Pr(X \textit{ preceded } Y) \leq \Pr(X \text{ and } Y \textit{ both occurred}). \quad (8.1)$$

Here, X and Y represent events or conditions that can be observed in individuals. (These inequality rules are written in the past tense because we envision applying them to a data set recording what has happened in each of many individuals. However, tense is really irrelevant insofar as the same rules apply to past, present, and future samples from the data set.) Pr(X and Y both occurred) refers to the joint probability that both X and Y occur in a randomly sampled individual.

It is tempting to give Pr(X caused Y) a specific interpretation, e.g., as the (counterfactual) probability that, in a randomly selected individual, both X and Y have occurred, but Y would not have occurred had X not occurred (i.e., "but for" X, in legal parlance). However, a key point is that *it is not necessary to specify a unique definition of "caused" to apply these inequalities,* since, for example, the probability that X caused Y (in any individual) cannot be greater than the probability that X and Y both occurred in that individual, for *any* of various possible definitions of "caused." In practical applications, it is often possible to estimate the right-hand-side term, Pr(X and Y both occurred), for a randomly selected individual in a population from data on the *frequency of co-occurrences* of X and Y in individuals in the study population. When multiple events form a causal chain or pathway, the preceding inequalities can be extended in obvious ways such as the following:

$$\Pr((X \text{ causes } Y) \,\&\, (Y \text{causes } Z) \text{ in an individual})$$
$$\leq \Pr(X \,\&\, Y \,\&\, Z \text{ all occur in that individual}).$$

The traditional definition of "attributable fraction" as the relative difference between the occurrence probabilities of a top event when exposure is present vs. when it is absent has the following counterpart as an inequality for causal statements:

$$\Pr(X \text{ caused } Y|Y \text{ occurred}) \leq [\Pr(Y|X) - \Pr(Y|\text{not } X)]/\Pr(Y|X) \tag{8.2}$$

[assuming that the right-hand side is nonnegative, i.e., that $\Pr(Y \mid X) > \Pr(Y \mid \text{not } X)$]. The right-hand side of (8.2) is a common textbook definition of the *attributable fraction* of Y due to X. It attempts to capture the intuition that the probability that X is the cause of Y is the probability that Y occurs when X is present, but that Y would not have "happened anyway," without (or "but for") the presence of X. [This interpretation works well and can be made rigorous in important special cases, such as when X is one of several *competing risks*, representing mutually exclusive sources, any one of which can cause Y by itself via a single-hit mechanism (Cox, 1985).] However, as illustrated by the baby aspirin counterexample in the section "Why Not Use Population Attributable Fractions?," the difference of conditional probabilities in the numerator on the right side of (8.2) does not provide an adequate general definition of the incremental risk of Y caused by X, since it does not allow for the possibility of hidden causes and confounders that create an association between X and Y even if X does not itself cause Y (and may even prevent or inhibit the occurrence of Y). Hence, we treat (8.2) not as a logical definition but as a potentially useful plausible upper bound for Pr(X causes $Y \mid Y$ occurs) in some applications. It is most applicable and useful when the difference $[\Pr(Y \mid X) - \Pr(Y \mid \text{not } X)]$ is, in fact, due entirely or primarily to the causal impact of X on the occurrence probability of Y. Such situations may be easier to identify for specific biological causal mechanisms than for organism-level responses in a population.

A Bayesian Motivation for the Attributable Fraction Formula

Here is an optional Bayesian rationale for the attributable fraction formula (8.2) using a classical competing-risks framework (i.e., assuming that mutually exclusive causes compete to be the first to make a detectable effect Y occur). Nothing else depends on this interpretation and discussion, but it may be useful in motivating (8.2) for some readers.

Bayes' rule suggests that

$$\begin{aligned}\Pr(X \text{ caused } Y|Y \text{ occurred}) = {} & \Pr(Y \text{occurred } |X \text{ caused } Y) \\ & * \Pr(X \text{ caused } Y)/\Pr(Y \text{ occurred}),\end{aligned}$$

where Pr(Y occurred) = Pr(X caused Y to occur) + Pr(*not-X* caused Y to occur).

Now clearly, Pr(Y occurred | X caused Y to occur) = 1. Substituting this into the above simplifies it to

$$\begin{aligned}\Pr(X \text{ caused } Y|Y \text{ occurred}) &= \Pr(X \text{ caused } Y)/\Pr(Y \text{ occurred}) \\ &= \Pr(X \text{ caused} Y)/[\Pr(X \text{ caused } Y \text{ to occur}) \\ &\quad + \Pr(not\text{-}X \text{ caused } Y \text{ to occur})].\end{aligned}$$

We adopt the following conditional probability notation:

- $\Pr(Y \mid X) = \Pr(Y$ is caused by something, given that X is present).
- $\Pr(Y \mid not\text{-}X) = \Pr(Y$ is caused by something else, given that X is not present) $= \Pr(not\text{-}X$ caused $Y)$. (Note that this step uses the competing-risk assumption of mutually exclusive, collectively exhaustive causes.)

Then the total probability that Y occurs when X is present $= \Pr(Y \mid X) = \Pr(X$ caused Y to occur) + $\Pr(not\text{-}X$ caused Y to occur) $= \Pr(X$ caused $Y) + \Pr(Y \mid not\text{-}X)$. Solving gives

$$\Pr(X \text{ caused } Y) = \Pr(Y|X) - \Pr(Y|not\text{-}X).$$

Substituting this into the numerator of the earlier equation

$$\begin{aligned}\Pr(X \text{ caused } Y|Y \text{ occurred}) &= \Pr(X \text{ caused } Y)/\Pr(Y \text{ occurred}) \\ &= \Pr(X \text{ caused } Y)/[\Pr(X \text{ caused } Y \text{ to occur}) \\ &\quad + \Pr(not\text{-}X \text{ caused } Y \text{ to occur})]\end{aligned}$$

yields

$$\begin{aligned}\Pr(X \text{ caused } Y|Y \text{ occurred}) = {} & [\Pr(Y|X) - \Pr(Y|\text{not-}X)]/[\Pr(X \text{ caused } Y \text{ to occur}) \\ & + \Pr(not\text{-}X \text{ caused } Y \text{ to occur})].\end{aligned}$$

The denominator here is just [$\Pr(X$ caused Y to occur) + $\Pr(not\text{-}X$ caused Y to occur)] $= \Pr(Y \mid X)$ (from above). Making this substitution gives us the final result:

$$\Pr(X \text{ caused } Y|Y \text{ occurred}) = [\Pr(Y|X) - \Pr(Y|\text{not } X)]/\Pr(Y|X).$$

This is just the right side of inequality (8.2).

This derivation yields an equality, because it assumes that the difference [$\Pr(Y \mid X) - \Pr(Y \mid \text{not } X)$] is entirely due to the difference in intensities (i.e., the expected occurrences per unit time) of Y when X is present compared to when X is absent. In other words, the contribution of X is assumed to be to hasten the occurrence of Y – a causal effect. [Intuitively, imagine that N different sources shoot invisible "bullets" or "hits" at some target. The first bullet to strike the target causes damage Y, which can be detected. Source X generates hits with intensity $h(X)$, measured in average hits per unit time generated by source X. The total intensity of hits arriving at the target is the sum of the intensities from all N sources. Under classical competing-risks assumptions, given that the target is hit, i.e., that Y occurs, the probability that X caused it is just the expected fraction of hits from X = (expected hits from all sources including X – expected hits from all sources other than X)/(total expected hits from all sources including X). This provides another, perhaps more physically intuitive, motivation for the [$\Pr(Y \mid X) - \Pr(Y \mid \text{not } X)$]/$\Pr(Y \mid X)$ formula.] However, in reality, some of the positive association between the presence of *X* and the occurrence rate of *Y* may come from confounding, rather than from a true causal effect (extra hits per unit time) contributed by *X*. This motivates replacing the equality with an inequality, as in (8.2).

The above derivation assumes that all possible causes are known. But if this is not true, then the fraction of hits contributed by source *X* may be overstated (because of omitted hits from unknown competing sources that should have been included in the denominator, diluting *X*'s contribution). This also motivates use of the inequality.

The following section examines how bounds (8.1) and (8.2) might be applied to molecular epidemiological data to quantify the maximum plausible contribution to the causation of lung cancer made by one particularly well-studied causal pathway. The problem of calculating top event probabilities in networks (or fault trees) of multiple, overlapping path sets from probabilities of the individual events has already been well addressed in the systems reliability literature (Chang et al., 2004). Hence, we focus on the simplest case of a single pathway, and on the conceptual and practical challenges of estimating quantitative bounds on its contribution to the occurrence of lung cancer in smokers, based on co-occurrence frequencies of relevant events in molecular epidemiological data.

The Smoking-PAH-BPDE-p53-Lung Cancer Causal Pathway

Figure 8.1 outlines one of the best-studied specific causal mechanisms that have been proposed for how cigarette smoking increases human lung cancer risk (Denissenko et al., 1996; Rojas et al., 2004). In brief, enzymes in the cytochrome P450 superfamily (especially CYP1A1) oxidize PAHs in cigarette smoke to form molecules of highly reactive metabolites such as benzo[a]pyrene diol epoxide (BPDE). These reactive molecules bind to DNA at specific "hot-spot" codons of the p53 tumor suppressor gene in smokers' lungs, forming BPDE-DNA adducts that induce characteristic G:C → T:A transversions and inactivate or impair the functioning of p53. The loss of p53 function plays multiple well-documented roles in allowing cells to escape from normal cell cycle controls (Olivier et al., 2004). By disrupting cell cycle checkpoints (leading to increased chromosomal instability and aneuploidy) and conferring resistance to apoptosis, the inactivation of p53 contributes to the increased malignant transformation of tumor precursor cells. Figure 8.1 summarizes this mechanistic account and shows some of the related genetic polymorphisms that may contribute to variability in individual risks from smoking.

This mechanistic hypothesis is so well established that benzo[a]pyrene has sometimes been described as "the bullet of the 'smoking gun'...[which] has been known for a long time to be involved in the causation of lung cancer" (Alexandrov et al., 2002). The hypothesis makes considerable intuitive sense in light of evidence on p53 and cancer. More than half of all human cancers exhibit mutations in the p53 tumor suppressor gene, and more than three quarters of these p53 mutations involve a missense substitution of a single incorrect amino acid (Glazko et al., 2004). In

Smoking
⇓
Benzo(a)pyrene (B(a)P) and other carcinogenic PAHs inhaled in cigarette smoke
⇓ CYP1A1, GSTM1 genotypes (Alexandrov et al., 2002)
benzo[a]pyrene diol epoxide (BPDE) and reactive oxygen species (ROS) formed in lung
⇓ polymorphisms affecting DNA repair capacity
BPDE-DNA adducts formed in lung cells
⇓ p53 codon 72 polymorphism
G to T transversions in p53 "hot spot" codons in lung cells ⇒ mutation "signature"
⇓
p53 inactivated in lung cells
⇓
Increased risk of lung cancer

Fig. 8.1 A proposed mechanism of lung cancer caused by PAHs from smoking

lung cancers (but not in cancers of the skin, colon, and other organs), the most characteristic p53 mutation is a G → T transversion, and this "signature" alteration is strongly and specifically associated with cigarette smoking (Alexandrov et al., 2002; Denissenko et al., 1996; Rojas et al., 2004). Moreover, the spectrum of mutations induced by BPDE in bronchial epithelial cells is similar to the spectrum of mutations observed in p53 genes from human lung tumors (Rojas et al., 2004). In addition, cultured peripheral lymphocytes from lung cancer patients exposed to BPDE accumulate more BPDE-DNA adducts than corresponding cells from control subjects, perhaps due to less effective DNA repair processes (Li et al., 2001a). [However, similar findings of elevated BPDE-DNA adduct rates among peripheral lymphocytes from head and neck squamous cell carcinoma patients (Li et al., 2001b) and in lymphoblasts from breast cancer patients (Motykiewicz et al., 2002) suggest that elevated adduct levels do not necessarily reflect an early step in causing lung cancer, but rather may themselves be caused by underlying conditions such as decreased DNA repair rates that also contribute to carcinogenesis.]

The hypothesis that smoking-induced p53 mutations cause smoking-induced lung cancers is strengthened by the specificity of these mutations in the lung cancer types that are most clearly associated with smoking. As summarized by Campling and el-Dairey (2003),

> The frequency of p53 alterations in lung cancer is highest in those subtypes of bronchial carcinomas that are most consistently associated with smoking, especially SCLC [small cell lung carcinomas] and squamous cell carcinomas. The frequency is lower in adenocarcinomas, in which the association with smoking, although present, is not as strong. The frequency of p53 abnormalities is higher in patients with greater cumulative tobacco exposure. Tobacco-specific carcinogens, in particular BPDE, cause a unique spectrum of p53 mutations, quite distinct from those found in cancers that are not associated with smoking. This characteristic genetic "signature" may persist even decades following smoking cessation.

Quantitatively, Rodin and Rodin (2002) reported "that the frequency of p53 G → T transversions in lung cancer of smokers is about three times higher than their frequency in lung cancer of nonsmokers and in most other smoke-unrelated cancers." [They also state that, "In contrast, the frequency of C $\rightarrow$ A transversions, the DNA-strand mirror counterpart of G $\rightarrow$ T transversions, appears to be similar in virtually all human cancers. Along with other data, this strand bias leads us to suggest that smoking may inhibit repair of G $\rightarrow$ T primary lesions on the non-transcribed strand." They conjecture that "a direct mutagenic action is not the only smoke-associated cause of the prevalence of this class of p53 mutations in lung cancer" (Rodin and Rodin, 2004). Others contest this view, arguing that, instead, "the abundance and sequence specificity of G $\rightarrow$ T transversions in lung tumors is best explained by a direct mutagenic action of PAH compounds present in cigarette smoke" (Pfeifer and Hainaut, 2003). We will attempt to estimate plausible upper bounds on the size of the contribution of the pathway in Fig. 8.1 to smoker lung cancer risk, independent of the contested details (e.g., direct mutation vs. inhibited repair) of the mechanisms involved.]

Applying the Theory: Quantifying the Contribution of the Smoking-PAH-BPDE-p53 Pathway to Lung Cancer Risk

A Simple Theoretical Calculation Using Causal Fractions

Intuitively, one might seek to estimate the fraction of smoker lung cancers caused by the mechanism in Fig. 8.1 by multiplying together the following three components: (1) the fraction of smokers' lung cancers that are caused by damage of p53, as shown near the bottom of Fig. 8.1 (since, without such p53 damage, the mechanism in Fig. 8.1 cannot apply); (2) the fraction of these lung cancers in which the inactivation of p53 is caused specifically by BPDE-DNA adducts at p53 codons; (3) the fraction of these BPDE-DNA adducts that, in turn, are caused by inhalation of PAHs in cigarette smoke. This can be expressed as follows:

$$\text{Preventable fraction}_{\text{PAH-BPDE-P53}} = f_{\text{p53}} \times f_{\text{BPDE-p53}} \times f_{\text{smoking-BPDE}}, \tag{8.3}$$

where

- Preventable $\text{fraction}_{\text{PAH-BPDE-P53}}$ = fraction of smoker lung cancers *caused by* the postulated mechanism in Fig. 8.1, i.e., inhaled PAHs causing BPDE-DNA adducts on p53 codons of target lung cells, leading to p53 inactivation,
- f_{p53} = fraction of smoker lung cancers *caused by* p53 damage,
- $f_{\text{BPDE-p53}}$ = fraction of this p53 damage (in lung cancers caused by p53 inactivation) *caused by* BPDE-DNA adducts at p53 codons,
- $f_{\text{smoking-BPDE}}$ = fraction of these BPDE-DNA adducts (in lung cancers caused by p53 damage from BPDE-DNA adducts) *caused by* PAHs from cigarette smoke.

Formal justification for this intuitively appealing product formula comes from the theory of causal graphs discussed in Chapters 3, 6, and 7 (Shipley, 2000; Greenland

and Brumback, 2002). The joint probability of any conjunction of events, say, A, B, C, ..., can be factored as Pr(A & B & C & ...) = Pr(A)Pr(B | A)Pr(C | A & B).... The sequence of factors on the right side of (8.3) is just such a chain of conditional probabilities (upon identifying the fraction of cases in which a condition holds as the probability that it holds for a randomly selected case). Each factor is conditioned on all those that precede it. The product (8.3) represents the joint probability that (a) a lung cancer is caused by p53 inactivation and/or other p53 damage associated with the G $\rightarrow$ T transversions, *and* (b) this p53 damage is caused by BPDE-DNA adducts, *and* (c) these adducts are caused by PAHs from smoking. These three conditions are jointly necessary for PAHs to cause lung cancer via the causal mechanism in Fig. 8.1. They may not be sufficient: there may be other necessary conditions that are not represented in Equation (8.3). If so, then this formula tends to *overestimate* the PAH-preventable fraction (since multiplying by further fractions between 0 and 1, representing the conditional probabilities of other necessary conditions, can only further decrease the product). But this is acceptable for establishing a plausible upper bound on the PAH-preventable fraction. Conversely, if PAHs cause lung cancer by paths other than the one shown in Fig. 8.1, then Equation (8.3) could *underestimate* the PAH-preventable fraction. Thus, Equation (8.3) should be interpreted specifically as giving the PAH-preventable fraction for the hypothesized causal mechanism in Fig. 8.1.

Application of the seemingly simple concept in (8.3) is frustrated by the fact that none of the three component quantities to be multiplied is known; moreover, all involve the term "caused by" (italicized), which has not been given any operational definition. It may be difficult or impossible to obtain these quantities from available data, which measure observable effects rather than causes. To overcome these obstacles, we apply the upper-bounding approach explained in the earlier section "Theory: Paths, Event Probabilities, Bounds on Causation." Table 8.1 summarizes the key steps and results.

The logic and data for this calculation are as follows.

Step 1: Replace Causal Fractions with Fractions Based on Occurrence Rates

First, each "caused by" condition in the first column is replaced by a "with" (or "among") condition in the second column, thereby replacing unobservable quantities that refer to causation with quantities that, in principle, can be estimated from data on observable occurrence frequencies of molecular events. Specifically, "f_{p53} = fraction of smoker lung cancers *caused by* p53 damage (indicated by G $\rightarrow$ T transversions)" is replaced by "F_{p53} = fraction of smoker lung cancers *with* p53 G $\rightarrow$ T transversions." Inequality (8.1) justifies this replacement: smoking-induced p53 damage cannot cause lung cancer unless such damage actually occurs. (The occurrence rate of *all* damage associated with the "signature" p53 G $\rightarrow$ T transversions in Fig. 8.1 is an upper-bound surrogate for the occurrence rate of the

Table 8.1 Preventable fraction calculation for the pathway in Fig. 8.1

Component/meaning	Upper-bound surrogate	Estimate and source
f_{p53} = fraction of smoker lung cancers *caused by* p53 damage (indicated by G → T transversions)	F_{p53} = fraction of smoker lung cancers *with* p53 G → T transversions	0.30 for smokers; 0.12 for nonsmokers (Pfeifer et al., 2002)
$f_{BPDE\text{-}p53}$ = fraction of this p53 damage *caused by* BPDE-DNA adducts	$F_{BPDE\text{-}p53}$ = fraction of smoker lung cancers *with* p53 G → T transversions *and* BPDE-DNA adducts	0.4 = (0.12 BPDE-DNA fraction in smokers, Alexandrov et al., 2002)/(0.3 p53 G → T fraction, Pfeifer et al., 2002)
$f_{smoking\text{-}BPDE}$ = fraction of these BPDE-DNA adducts *caused by* PAHs from cigarette smoke	$F_{smoking\text{-}BPDE}$ = smoking-attributable fraction of BPDE-DNA adducts *among* smokers compared to nonsmokers	0.60 = (2.5 – 1)/2.5; 2.5 = estimated ratio of BPDE-DNA adducts in smokers vs. nonsmokers (Lodovici et al., 1998)
Preventable $\text{fraction}_{PAH\text{-}BPDE\text{-}P53} = f_{p53} \times f_{BPDE\text{-}p53} \times f_{smoking\text{-}BPDE}$ = fraction of smoker lung cancers *caused by* inhaled PAHs causing BPDE-DNA adducts at p53 codons, causing p53 inactivation	Preventable $\text{fraction}_{PAH\text{-}BPDE\text{-}P53} \le F_{p53} \times F_{BPDE\text{-}p53} \times F_{smoking\text{-}BPDE}$	$\le 0.30 \times 0.40 \times 0.60 = 0.07$

specific damage that causes lung cancer in Fig. 8.1; thus, G → T transversions may only be markers for other types of smoking-induced p53 damage and/or may themselves increase cancer risk.)

Similarly, "$f_{BPDE\text{-}p53}$ = fraction of this p53 damage *caused by* BPDE-DNA adducts" is replaced by the upper-bound surrogate "$F_{BPDE\text{-}p53}$ = fraction of smoker lung cancers *with* p53 G → T transversions *and* BPDE-DNA adducts." That is, the frequency of *causation* of p53 damage by BPDE-DNA adducts in a population of smokers cannot exceed the frequency of the *co-occurrence* of p53 damage and BPDE-DNA adducts. [In practice, both changes are long-lasting, making observing them practicable. Lodovici et al. (1998) report BPDE-DNA adducts lasting for decades; and Campling and el-Dairey (2003) report a similar long persistence of mutational spectra in p53 due to BPDE-DNA adducts.] Consistent with using upper bounding to deal with unknowns, we require only that BPDE-DNA adducts be present in the lung, but not that they be present specifically on hot spots of the p53 gene (as postulated in Fig. 8.1), since the probability of the latter event can be no greater than the probability of the former, and the probability of the former is easier to estimate from available data.

The attributable fraction inequality in Equation (8.2) motivates replacing "$f_{smoking\text{-}BPDE}$ = fraction of these BPDE-DNA adducts (i.e., of the BPDE-DNA adducts causing p53 damage leading to lung cancer, among smokers with such damage) *caused by* PAHs from cigarette smoke" with "$F_{\text{smoking-BPDE}}$ = smoking-attributable fraction of BPDE-DNA adducts among smokers compared to nonsmokers." This formula treats the *entire* excess risk of BPDE-DNA adducts observed among smokers compared to nonsmokers as being caused by PAHs from smoking (rather than by something else, such as hidden confounders or other associated exposures) in order to obtain an upper-bound surrogate for the causal contribution from the pathway in Fig. 8.1. (Both the smokers and the nonsmokers being compared in this attributable fraction calculation should ideally be lung cancer patients with p53 damage caused by BPDE-DNA adducts.)

The attributable fraction formula on the right side of inequality (8.2) is theoretically exactly correct [and (8.2) becomes an equality] in a "competing-risks" model (Wohlfahrt et al., 1999). In this framework: (a) BPDE-DNA adducts arise from multiple sources of PAH exposures (e.g., smoking, background air pollution, ETS, cooking, diesel exhaust, etc.), each generating BPDE-DNA adducts according to a Poisson arrival process. (b) One of these adducts is ultimately considered (in principle) to be "the" cause of the p53 damage that eventually leads to a smoker's lung cancer. Interpreting X as the event that "The BPDE-DNA adduct leading to lung cancer came from smoking" and Y as the event "A BPDE-DNA adduct leading to lung cancer came from some source," the right side of (8.2) simply gives the *fraction* of adducts that come from smoking rather than other sources. This is also the *probability* that the crucial cancer-causing adduct came from smoking rather than other sources, assuming exchangeable molecules and that smokers and nonsmokers have identical exposures except for smoking. (This may be conservative, i.e., overestimate the smoking contribution, if smoking is associated with other increased exposures.) Of course, there is no assumption that the formation of this critical adduct is the only event required to cause lung cancer, but it is considered to be one necessary event among several (possibly incompletely known) events that are jointly sufficient to cause lung cancer.

Outside the framework of the competing-risks model, Formula (8.2) can also potentially be justified (although not necessarily interpreted as a probability) by normative axioms for assigning fractional "shares" in causation to different sources. Even for complex nonlinear dose-response mechanisms with interacting sources, for which competing-risks assumptions fail, it often makes sense that the fraction of BPDE-DNA adducts that should be considered to be caused by smoking (according to this axiomatic framework) is just the fraction of BPDE molecules that come from smoking instead of from other sources (Cox, 1985).

A more direct and pragmatic justification for using this fraction in preventable risk calculations is that if all smoking-related BPDE-DNA adducts were removed or prevented, the rate of such adducts in smoker lungs presumably would not fall below the background rate in nonsmokers. Hence, only the difference between the BPDE-DNA adduct rate in smokers compared to that in nonsmokers is relevant for quantifying the contribution of smoking to excess lung cancer risks mediated by these adducts. [We view this traditional justification for the attributable fraction formula (8.2) not as a logical or mathematical necessity, but rather as a common-sense assertion about how the relevant biological processes of adduct formation is thought to work. While in theory it might misestimate preventable fractions, as in the baby aspirin example, this may be less likely for specific pathways and events, such as BPDE-DNA adduct formation at p53 codons, than for overall exposure-effect relations in population-level epidemiological data. We therefore use inequality (8.2) to estimate $F_{\text{smoking-BPDE}}$.]

Finally, the fraction of smoker lung cancers caused by the sequence of events in Fig. 8.1 (i.e., that could be prevented by preventing events on this pathway, e.g., by removing exposure to PAHs from cigarette smoke) is estimated using the product of the upper-bound surrogates just described:

$$\text{Preventable fraction}_{\text{PAH-BPDE-P53}} \leq F_{\text{p53}} \times F_{\text{BPDE-p53}} \times F_{\text{smoking-BPDE}}. \quad (8.4)$$

Step 2: Quantify Occurrence Rates Using Molecular-Level Data

The next task is to use available molecular epidemiological data to quantify the three factors F_{p53}, $F_{\text{BPDE-p53}}$, and $F_{\text{smoking-BPDE}}$.

For F_{p53}, Pfeifer et al. (2002) report that "The prevalence of G to T transversions is 30% in smokers' lung cancer but only 12% in lung cancers of nonsmokers. ... Recent studies have indicated that there is a strong coincidence of G to T transversion hotspots in lung cancers and sites of preferential formation of PAH adducts along the p53 gene." Similarly, Rodin and Rodin (2002) state,

> The high frequency of G $\rightarrow$ T transversions in the p53 gene is a distinctive feature of lung cancer patients with a smoking history and is commonly believed to reflect the direct mutagenic signature of polycyclic aromatic hydrocarbon (PAH) adducts along the gene. Using the April 2000 update of the p53 mutation database of the International Agency for Research on Cancer together with the primary literature, we confirm that the frequency of p53 G $\rightarrow$ T transversions in lung cancer of smokers is about three times higher than their frequency in lung cancer of nonsmokers and in most other smoke-unrelated cancers.

Adopting the more precise absolute numbers of Pfeifer et al. (2002) for purposes of illustration, we estimate $\boldsymbol{F_{\text{p53}} = 0.30}$ for smoker lung cancers, as compared to only 0.12 for nonsmoker lung cancers.

$F_{\text{smoking-BPDE}}$ can be estimated from measured values of BPDE-DNA adduct rates in smokers compared to nonsmokers. Lodovici et al. (1998) report average total BPDE-DNA adduct levels of 4.46 per 10^8 bases in smokers, compared to 4.04 in ex-smokers and 1.76 in nonsmokers. The ratio of BPDE-DNA adduct levels in smokers and nonsmokers in this study is thus 4.46/1.76 = 2.5, completely consistent with the ratio of p53 damage (G $\rightarrow$ T transversions) in lung cancers for smokers compared to nonsmokers from Pfeifer et al. (2002), which was also (0.30/0.12) = 2.5. [A ratio of 2 could have been estimated by using B(a)P measurements as a rough surrogate for BPDE-DNA adducts. Goldman et al. (2001) state that the "Concentration of DNA adducts is considered the biologically effective dose of the parent PAH. Because accumulation of parent PAH compounds in the lung tissue precedes formation of DNA adducts, it is intuitive that the lipophilic parent PAHs would be a good estimate of dose in the target organ. . . . Smoking increased the concentration of five PAHs including benzo(a)pyrene, which increased approximately 2-fold." However, the availability of data directly on BPDE-DNA adduct rates makes it unnecessary to use precursor PAH levels as surrogates for adduct levels. Nonetheless, it is reassuring that these different approaches give roughly consonant answers.]

Using 2.5 as the estimated ratio of BPDE-DNA adducts in smokers compared to nonsmokers and applying the attributable fraction formula (8.2) yields the following estimated fraction of BPDE-DNA adducts preventable by removing smoking exposure: $\boldsymbol{F}_{\textbf{smoking-BPDE}} = (2.5 - 1)/2.5 = \mathbf{0.6}$. Similarly, applying (8.2) to the p53 G → T transversion rates in lung cancers of smokers compared to nonsmokers would yield the following estimated fraction of p53 G → T transversions preventable by removing smoking exposure: $(0.30 - 0.12)/0.3 = 0.6$. Again, although such consonance does not prove the validity of the calculations, it is reassuring that there is no obvious inconsistency that would invalidate the causal chain assumptions in Fig. 8.1, i.e., that PAH levels translate to BPDE-DNA adduct rates, which in turn cause p53 damage indicated by p53 G → T transversions. We appeal to this consonance to support the use of 2.5 as the estimated ratio of relevant (lung-cancer causing) BPDE-p53 adducts in smokers compared to nonsmokers, which implies $\boldsymbol{F}_{\textbf{smoking-BPDE}} = (2.5 - 1)/2.5 = \mathbf{0.6}$.

It only remains to quantify $F_{\text{BPDE-p53}}$, the fraction of smoker lung cancers with both p53 G → T transversions *and* BPDE-DNA adducts (since these are the cases in which BPDE-DNA adducts might have *caused* the p53 damage indicated by the p53 G → T transversions). We could find no data on this specific co-occurrence frequency, but we can use upper bounding to bridge this data gap.

Table 8.1 of Alexandrov et al. (2002) summarizes data from multiple studies and countries on the fraction of all smokers with BPDE-DNA adducts in their lungs. For the United States, this fraction is about $31/256 = 0.12$ (pooling across multiple studies), perhaps reflecting the roles of genetic polymorphisms (such as CYP1A1 and GSTM1 genotypes) in determining individual susceptibility to the creation of such adducts (ibid.). Although the fraction of these cases that also have p53 G → T transversions is unknown, a logical upper-bound estimate can be obtained by assuming that this unknown fraction is 1, i.e., that *all* smokers with BPDE-DNA adducts also have p53 G → T transversions. [This extreme assumption is not ruled out by the available data, as (0.12 for smokers with BPDE-DNA adducts) < (0.30 for smokers with lung cancers and p53 G → T transversions). Empirical justification for the implicit assumption that the 12% BPDE-DNA adduct rate from the data surveyed by Alexandrov et al. (2002) for *all* smokers can be applied to smokers *with lung cancers* comes from limited available data. Gyorffy et al. (2004) report PAH-DNA adduct levels in the range of 2.6–6.2 adducts per 10^8 nucleotides among smokers (as well as nonsmokers) *with lung cancer,* quite consistent with the mean BPDE-DNA adduct level of 4.46 per 10^8 bases reported for *all* smokers by Lodovici et al. (1998). Therefore, we tentatively assume that BPDE-DNA adduct rates are similar among all smokers and smokers with lung cancer. Because this assumption is uncertain, in sensitivity analyses, we will consider the trivial (uninformed by data) logical upper bound of $F_{\text{BPDE-p53}} \leq 1$.]

These assumptions imply an estimate of $\boldsymbol{F_{\text{BPDE-p53}} \leq 0.12/0.30 = 0.40}$, based on the worst-case assumption that *all* of the estimated 12% of smokers with lung cancer and BPDE-DNA adducts also belong to the 30% of smokers with lung cancer who have p53 G $\rightarrow$ T transversions. (When data become available specifically on the co-occurrence frequency of these conditions, this upper bound may be reduced. Rather than $F_{\text{BPDE-p53}} \leq 0.40$, an estimate of $F_{\text{BPDE-p53}} = 0.12 \times 0.30 = 0.036$ would result if it were assumed that having BPDE-DNA adducts and having p53 G $\rightarrow$ T transversions in lung cancers are statistically independent events, rather than that the former implies the latter. The true but unknown value for $F_{\text{BPDE-p53}}$ may lie between the values for these two extreme cases, i.e., between 0.036 and 0.4.)

Step 3: Combine Upper-Bound Surrogate Fractions for Events in a Path Set

The final step in calculating preventable fractions combines the values of the upper-bound surrogates for the causal events that jointly suffice to produce the health effect of interest – here, lung cancer. For the single-path model in Fig. 8.1, as shown in (8.4), the calculation consists of simply multiplying the estimated values of F_{p53}, $F_{\text{BPDE-p53}}$, and $F_{\text{smoking-BPDE}}$. The result is

$$\begin{aligned}\text{Preventable fraction}_{\text{PAH-BPDE-P53}} &\leq F_{\text{p53}} \times F_{\text{BPDE-p53}} \leq F_{\text{smoking-BPDE}} \\ &= 0.3 \times 0.4 \times 0.6 = 0.07.\end{aligned}$$

More generally, one would multiply event probabilities – estimated from occurrence fractions – along causal chains or path sets, sum such results over disjoint path sets for the same top event, and apply inclusion-exclusion approximations or other combinatorial methods if path sets are not disjoint; see, e.g., Chang et al. (2004). Since the model in Fig. 8.1 has only one path set, only multiplication is required in this case.

An intuitive interpretation of this calculation is as follows. Only 30% of smokers who develop lung cancers have p53 G $\rightarrow$ T transversions, and 40% of these 30% (i.e., 40% $\times$ 30% = 12%) would occur even without smoking, as evidenced by the 12% rate of p53 G $\rightarrow$ T transversions among nonsmoker lung cancer patients (Pfeifer et al., 2002). Therefore, only the remaining 60% $\times$ 30% = 18% of smoker lung cancers with p53 G $\rightarrow$ T transversions (the specific group identified as relevant in Fig. 8.1) could have been caused by BPDE-DNA adducts from smoking-related PAHs. But at most 40% (= 0.12/0.30) of these cases (i.e., smokers with p53 G $\rightarrow$ T transversions and lung cancer) are expected to have BPDE-DNA adducts [upon combining data of Alexandrov et al. (2002) showing a 12% BPDE-DNA adduct prevalence rate with data of Pfeifer et al. (2002) showing a 30% prevalence rate of p53 G $\rightarrow$ T transversions among smokers with lung cancer], even assuming that *every* case with BPDE-DNA adducts becomes one of the cases with resulting p53 G $\rightarrow$ T transversions leading to lung cancer. Therefore, at most 40% $\times$ 18% =

7% of smokers might experience the conjunction of events or conditions (i.e., lung cancer, p53 G $\rightarrow$ T transversions, and BPDE-DNA adducts from smoking) shown in Fig. 8.1. Even assuming that, in *all* of these cases, this conjunction is caused by smoking (and would be prevented if exposure to PAHs in smoking were removed), the preventable fraction cannot exceed the total fraction of smokers affected, i.e., about 7%. When data permit, the empirical conditional probability (or fraction) of p53 G $\rightarrow$ T transversions among smokers with lung cancer and BPDE-DNA adducts should replace this logical upper bound of 100%, leading to a refined estimate.

Uncertainties and Sensitivities

The conclusion that the preventable fraction for the smoking-PAH-BPDE-p53-lung cancer pathway in Fig. 8.1 is probably not more than about 7% could potentially be refined by replacing the point estimates of F_{p53}, $F_{\text{BPDE-p53}}$, and $F_{\text{smoking-BPDE}}$ with probability distributions and applying Monte Carlo uncertainty analysis. However, the main logic is sufficiently straightforward – multiplying several fractions between 0 and 1 together, even when their exact values are uncertain and not all relevant fractions have been included, quickly yields relatively small numbers – so that it is probably more illuminating to simply discuss the major uncertainties and sensitivities.

The largest uncertainty is about the value of $F_{\text{BPDE-p53}}$, the fraction of smokers with both BPDE-DNA adducts and p53 damage (used as an upper-bound surrogate for the fraction of smokers in whom BPDE-DNA adducts cause p53 damage leading to lung cancer). Although we estimated $F_{\text{BPDE-p53}}$ as 0.40, this reflects an extreme assumption that *all* smokers with BPDE-DNA adducts and lung cancer have p53 damage (with the G $\rightarrow$ T transversions characteristic of smoking-induced damage), even though only about 30% of all smokers with lung cancer have such damage (Pfeifer et al., 2002). Thus, the 7% estimate for the preventable fraction might well be too high, possibly by a factor of $1/0.30 = 3.3$. On the other hand, even replacing the estimated bound $F_{\text{BPDE-p53}} \leq 0.40$ based on the data and assumptions documented in the preceding section with the noninformative bound $F_{\text{BPDE-p53}} \leq 1$ would increase the preventable fraction estimate to only $0.3 \times 0.6 = 0.18$, still representing a minority of all smoker lung cancers.

Using upper statistical confidence limits in place of point estimates would slightly increase the 7% estimate, but this effect is relatively small. For example, in the data of Alexandrov et al. (2002) on the fraction of all smokers with BPDE-DNA adducts in the United States, using a 95% binomial upper confidence limit would only increase the estimate from $31/256 = 0.12$ to 0.16. Probably a more important limitation is that each of the estimated values for the factors F_{p53}, $F_{\text{BPDE-p53}}$, and $F_{\text{smoking-BPDE}}$ is based on only a limited number of studies, as documented in the preceding section and Table 8.1. Further studies might provide different estimates for these factors. Ideally, comparison, reconciliation (if needed), and meta-analysis of many individual studies could increase confidence in the robustness (or, conversely, reveal the sensitivity to selected studies) of the results presented here. Until such additional studies become available, the results summarized in Table 8.1 should be viewed as illustrating preventable fraction calculations for the path in Fig. 8.1 using the limited data available now, but with the understanding that future studies may lead to changes in these estimates.

Finally, we have assumed throughout that the mechanism in Fig. 8.1 is correct, and we have focused entirely on quantifying the potential contribution (preventable fraction) of smoker lung cancers from this pathway. In reality, PAHs might operate through other (perhaps as-yet unknown) pathways. This uncertainty implies that the estimate of a 7% preventable fraction only means that *either* the causal pathway in Fig. 8.1 is not the only one (or even the main one) by which PAHs in cigarette smoke cause an increased risk of lung cancer (i.e., other pathways remain to be discovered), *or* that the contribution of PAHs to smoker lung cancer risk is fairly modest (e.g., not more than about 7%). In either case, it appears that the mechanism in Fig. 8.1 may not be "the bullet in the smoking gun" that it has sometimes been described as being.

Discussion

Previous recent biochemical research that did not rely on quantitative risk assessment (QRA) has strongly suggested that "The high formation of BPDE-N(2)-dG adducts in bronchial epithelial cells and investigations showing that the profile of mutations induced by BPDE in these cells is similar to that seen in the p53 gene isolated from human lung tumors implicates benzo[a]pyrene as [an] important carcinogen in tobacco-induced lung cancer in human beings" (Rojas et al., 2004). Preventable fraction calculations of the type illustrated in this chapter can help to estimate how *quantitatively important* such a pathway is in explaining (and, if blocked, in preventing) smokers' lung cancer risk. The calculations summarized in Table 8.1 suggest that removing or blocking this causal pathway would accomplish at most only a modest (probably less than 7%) reduction in smokers' risk of lung cancer. This strengthens previous suggestions that mechanisms and constituents of tobacco smoke other than the B[a]P pathway in Fig. 8.1 are important in determining lung cancer risk (Pott and Heinrich, 1990) and that mechanisms other than genotoxic effects of PAH metabolites on DNA (marked by BPDE-DNA adducts), especially proliferation and progression, probably dominate lung cancer causation by smoking (Hazelton et al., 2005).

To the extent that our quantitative findings contrast with conventional wisdom that PAHs and BPDE adducts on p53 play a dominant causal role in smoking-caused lung cancer, it is worth asking why the difference arises and whether concordant evidence supports it. A potentially useful lesson is the distinction between the *strength of evidence* for a causal mechanism and the *strength of the effect* of that mechanism on cancer risk. The evidence supporting the pathway in Fig. 8.1 is strong, but its quantitative impact on risk appears to be quite limited, implying that more work remains to be done to identify constituents of cigarette smoke whose removal would greatly reduce lung cancer risk.

Data for p53 mutations in breast cancer provide concordant evidence of a clear but limited role for p53 mutations. For example, Conway et al. (2002) report p53 mutation levels of 36.5% in the breast tumors of smokers as compared to 23.6% in nonsmokers. This ratio of 1.6 is comparable to the ratio of about 2.0 for PAH-DNA adducts in the lungs of smokers compared to nonsmokers. [After adjustment for age, race, menopausal status, clinical stage, tumor size, and family history of breast cancer, current smokers were significantly more likely to harbor any p53 mutation (OR = 2.11, 95% CI, 1.17–3.78), while the odds ratio for a G:C-T:A transversion was reported to be 10.53 (95% CI, 1.77–62.55).] Yet the association of breast cancer with smoking is at most very weak (IARC, 2004), suggesting that p53 mutations associated with smoking are not a major cause of breast cancer, despite their clear association.

The main result of our quantitative assessment applied to PAHs is that the preventable fraction of lung cancers among smokers for the specific mechanism in Fig. 8.1 is probably not larger than about 7%. Since, overall, smoking may increase the risk of lung cancer by a factor of 10 or more (e.g., Stellman et al., 2001), this relatively modest estimated contribution suggests that other biological mechanisms and causal pathways probably play more important roles (and hence must be considered in the rational design of significantly less risky cigarettes or other tobacco products). Tobacco industry interest in p53 mutations need not be seen as "efforts to challenge the science linking smoking and lung cancer" or as seeking "to cast doubt on the link between smoking and p53 mutations" (Bitton et al., 2005). Rather, if the p53-mediated effects of PAHs are indeed not the "bullet of the smoking gun" (Alexandrov et al., 2002) that many researchers have hoped, then perhaps the methods of causal effect bounding illustrated here can help to identify other fruitful candidates for tobacco smoke constituents (or sets of constituents) that are linked to smoking-induced lung cancer risk by stronger effects on risk as well as by strong scientific evidence of a causal relation.

Conclusions

Current epidemiology textbooks often interpret population attributable fractions based on 2 × 2 tables or logistic regression models of exposure-response associations as *preventable fractions,* i.e., as fractions of illnesses in a population that would be prevented if exposure were removed. In general, as suggested by the counterexamples and case studies in Chapters 5–7, this causal interpretation is not correct. Not only is statistical association not necessarily an indication of causation, but in general it does *not* identify how much risk would be prevented by removing specific constituents of complex exposures. This chapter has therefore illustrated how to calculate useful bounds on preventable fractions, having valid causal interpretations, from the partial molecular epidemiological and biological information that is often available in practice. Combining probabilistic risk assessment (PRA) concepts with inequality constraints for the relations between event probabilities and causation (such as that the probability that exposure *X causes* response *Y* cannot exceed the probability that exposure *X precedes* response *Y*, or the probability that both *X* and *Y occur*) yields quantitative bounds for the potential contribution to causation from specific causal pathways.

We illustrated estimating an upper bound on the contribution to lung cancer risk made by a specific, much-discussed causal pathway that links smoking to PAH (specifically, BPDE-DNA) adducts at hot-spot codons at p53 in lung cells. The result is a surprisingly small preventable fraction (of perhaps 7% or less) for this pathway, suggesting that it will be important to consider other mechanisms and non-PAH constituents of tobacco smoke in designing less risky tobacco-based products. Upper bounding provides a relatively straightforward way to use molecular epidemiological information on the *occurrence rates* of biological events (estimated from data) to estimate the upper bounds on contributions to *causation* (i.e., preventable fractions in the population) that are of chief interest for risk management and product design decisions.

Chapter 9
Bounding Resistance Risks for Penicillin

Chapter 8 calculated a plausible upper bound for the fraction of disease preventable by blocking a specific causal pathway in a complex, uncertain biological system, using smoking-induced lung cancer as an example. This example required considering relevant biological knowledge and biomarker data in some detail. But plausible upper bounds on preventable risks can also be developed using much less detailed knowledge, and data that are relatively easy to obtain and understand, for many other systems. This chapter develops bounding calculations for preventable fractions in a very different complex uncertain system: the highly uncertain set of pathways and mechanisms (if any) leading from the use of penicillin in food animals to resistance to penicillin drugs in human patients. Unlike the campylobacteriosis and virginiamycin case studies in Chapters 6 and 7, respectively, no individual-level exposure and response data are provided for this penicillin risk assessment. Instead, the challenge in this chapter is to estimate how much human health harm might be prevented by removing a source of exposure – one that is not known with great confidence to harm human health at all – using only readily available, high-level data, such as the total number of illnesses per year and statistics on strains of bacteria found in humans and animals. Chapter 5 has already discussed some of the pitfalls of using high-level data carelessly (e.g., by fitting a reduced-form model such as "$risk = K \times exposure$" and then misinterpreting it as a predictively valid causal relation). This chapter develops methods for drawing more useful inferences from such data, accepting that only rough bounds, rather than precise predictions, can be supported in the absence of more detailed data.

Background, Hazard Identification and Scope: Reducing Ampicillin-Resistant *E. faecium* (AREF) Infections in ICU Patients

Penicillin-based drugs are approved for use in food animals in the United States to treat, control, and prevent diseases and, to a lesser extent, to improve growth rates (FDA-CVM, 2007; Sechen, 2006; AHI, 2006). Concerns that penicillin use might increase the risk of antibiotic resistance in human enterococcal infections from non-human sources, thus leading to increased morbidity and mortality (WHO, 2005),

L.A. Cox, Jr., *Risk Analysis of Complex and Uncertain Systems*, International Series in Operations Research & Management Science 129,
DOI 10.1007/978-0-387-89014-2_9,

have made approved feed usages of penicillins in food animals a controversial topic for several decades in the United States (IOM, 1989; FDA, 2000, 2003). Penicillin drugs are considered "critically important" for animal health (FAO/WHO/OIE, 2008), implying that withdrawing approvals would likely have negative consequences for animal health. The following sections develop a plausible upper bound on the potential for the continued use of penicillin drugs in food animals to harm human health by increasing the number of antibiotic-resistant enterococcal infections in human patients. After summarizing the relevant background for the hazard of greatest concern – infection of intensive care unit (ICU) patients with ampicillin-resistant *E. faecium* (AREF) bacteria – the following sections focus on quantifying the fraction of such resistant infections that might be prevented by discontinuing the use of penicillin drugs in food animals.

Enterococci are commensal gram-positive bacteria found in the intestinal flora of most healthy birds and mammals, including people. In humans, enterococci typically comprise not more than 1% of the microflora of adults (FDA-CVM, 2004) and are normally harmless. The *Enterococcus* genus has 17 species, but most human clinical isolates are either *E. faecalis* (74–90%) or *E. faecium* (5–16%) (Varman et al., 2006). *E. faecalis* infections are responsible for most clinical enterococcal infections, but penicillin (or ampicillin) resistance is rare in *E. faecalis* isolates from food animals and almost nonexistent in *E. faecalis* isolates from retail meats (Hayes et al., 2003, 2004; McGowan et al., 2006; NARMS, 2005). Furthermore, ampicillin remains highly effective against clinical *E. faecalis* (Jones et al., 2004). Therefore, our risk assessment focuses on the identified hazard of penicillin/ampicillin resistance among human *E. faecium* infections.

Risk to human health arises because some strains of enterococci may become opportunistic pathogens, potentially resistant to multiple drugs, that infect patients who are already seriously ill (typically in ICUs) with immune systems weakened by organ transplants, chemotherapy, AIDS, or other causes. Indeed, enterococcal infection is the second most common hospital-acquired infection in the United States (Varman et al., 2006). These infections can prolong illness and increase patient mortality. Vancomycin-resistant enterococci (VRE) are of particular concern because of their virulence and resistance to even some recently developed antibiotics. Vancomycin-resistant *E. faecium* (VREF) can cause serious and often fatal disease in vulnerable populations, such as liver transplant patients and patients with hematologic malignancies (Rice, 2001).

Many enterococcal infections, including VRE, resolve without antimicrobial treatment (Varman et al., 2006; Rice, 2001). In severe cases for which antimicrobial treatment is provided, penicillin and ampicillin are often the leading choices. Other drugs that are also effective against human enterococcal infections include gentamicin, vancomycin, quinupristin-dalfopristin (Synercid®), linezolid (Zyvox®), tigecycline (Tygacil®), and nitrofurantion. These can be used in patients with allergy or high-level resistance to penicillin and ampicillin. Many ampicillin-resistant cases can also be treated successfully with high doses of ampicillin, either alone or in combination with drugs such as gentamicin or streptomycin (Varman et al., 2006; Murray, 2000).

Most *E. faecium* infections in ICU patients in the United States are now resistant to vancomycin (Edmond et al., 1999; Jones et al., 2004). Patients with vancomycin-resistant *E. faecium* (VREF) have worse outcomes than those with vancomycin-susceptible strains – longer hospital stays and higher mortality (Webb et al., 2001). As noted by Rice (2001), *virtually all VREF cases are also ampicillin-resistant:* "More than 95% of VRE recovered in the United States are *E. faecium*; virtually all are resistant to high levels of ampicillin." Hence, our risk assessment treats VREF as being (at least approximately) a subset of ampicillin-resistant *E. faecium* (AREF). One suggested explanation is "that the close association of the vancomycin and ampicillin resistance phenotypes, at least in VanB-type VRE, is explainable by their inclusion within large, transferable genetic elements" (Hanrahan et al., 2000, p. 1350). However, widespread ampicillin resistance appeared in 1982 (Fortun et al., 2002), while vancomycin resistance appeared in the early 1990s in *E. faecium* (Murray, 2000).

Since most VREF are AREF (although many AREF are not VREF), and assuming that changes in animal *penicillin* use would not significantly affect *vancomycin* resistance (consistent with historical data), we focus on human (ICU patient) infections with *vancomycin-susceptible* strains of *ampicillin-resistant E. faecium.* Presumably, this is the subpopulation that might experience decreased ampicillin resistance if discontinuing animal penicillin drugs were to replace some AREF cases with ampicillin-susceptible cases. For patients with VREF, we assume that AREF would persist (due to the observed co-occurrence of AREF in VREF strains), so that no benefit from reduced AREF would be achieved for these patients.

The following sections seek to quantify potentially preventable AREF cases and the human health benefits that might be created if these AREF cases were prevented (made ampicillin-susceptible) by discontinuing penicillin drug uses in food animals. The quantitative risk assessment draws on recent advances in sequencing technology that enable more precise strain groupings and epidemiological analyses than have previously been possible.

Methods and Data: Upper Bounds for Preventable Mortalities

Recognizing that a farm-to-fork ("forward chaining") model is not practical for AREF, due to data and knowledge gaps in release, exposure, and dose-response relations, we instead start with more readily available human data on ICU case loads and resistance rates, similar to the approach in Cox and Popken, 2004b. We then work backward to estimate a plausible upper bound on the annual number of human patient mortalities that might be prevented by discontinuing penicillin use in food animals.

For the purposes of conservative (i.e., upper-bound) risk assessment, we define a *potentially preventable mortality* to occur whenever the following conditions hold: (1) An ICU patient dies, following (2) an *E. faecium* infection that (3) is resistant to ampicillin (AREF) (and hence might have benefited had ampicillin resistance been prevented). The infection was (4) vancomycin-susceptible (and hence might have also been ampicillin-susceptible, had it not been for penicillin use in food animals), (5) not known to have been contracted from the hospital environment (and hence might have been prevented by actions external to the hospital, such as elimination of AREFs from food animals), (6) could have come from food animals (i.e., has a genotype or resistance determinants of the types found in food animals). (7) The patient tolerated penicillin (i.e., was not allergic, and hence might have benefited from ampicillin, had it not been for resistance). We propose that the conjunction of these seven conditions should be interpreted as *necessary*for a mortality to have been caused (with nonnegligible probability) by resistance due to the use of penicillin in food animals, even though it is not *sufficient* (e.g., the infecting strain might have had some other origin than food animals, or the patient might have died anyway, even if the infection had been ampicillin-susceptible). Accordingly, the following sections estimate a plausible upper bound on annual preventable mortalities from AREF infections based on the following product of factors:

preventable AREF mortalities per year
$\leq$ (total number of ICU infections per year)
$\times$ (fraction caused by E. faecium)
$\times$ (fraction of ICU E. faecium infections that are AREF and exogenous, i.e., not known to be of nosocomial origin)
$\times$ (fraction of these exogenous AREF cases that are vancomycin-susceptible)
$\times$ (fraction of vancomycin-susceptible exogenous AREF cases that might have come from food animals)
$\times$ (fraction of these cases that are penicillin-tolerant)
$\times$ (excess mortality rate for AREF cases compared to ASEF cases).

That is, we first quantify the expected annual number of AREF cases in the United States that might benefit from ampicillin treatment if food animal uses of penicillin were halted (i.e., cases that are penicillin-tolerant and vancomycin-susceptible and that might have been caused by resistance determinants from food animals). Then we multiply this number by the excess mortality rate for resistant as opposed to susceptible cases.

Estimated Number of ICU Infections per Year

Enterococcal infections are generally limited to already hospitalized individuals. *E. faecium* infections are frequently associated with nosocomial bloodstream infections occurring within intensive care units (ICUs). A recent FDA risk assessment for virginiamycin (FDA-CVM, 2004) provided the following two approximate estimates:

- N = annual number of ICU infections = 104,372.5 based on bloodstream infections,
- N = 315,000 based on septicemia cases.

The study does not weight these alternatives. To be conservative (i.e., to maximize estimated risks), we will use the larger estimate, N = 315,000 cases/year. [Patients with severely complicated urinary tract infections (UTIs) are also sometimes treated with intravenous antibiotics, including combinations of gentamicin and ampicillin, but ceftriaxone may be substituted for ampicillin if needed, and oral antibiotics (e.g., trimethoprim, cephalosporins, nitrofurantoin, or ciprofloxacin) are used in the vast majority of cases (http://adam.about.com/reports/000036_7.htm). We therefore do not include UTI cases in this assessment.]

Fraction of ICU Infections Caused by E. faecium

The proportion of ICU infections that are caused by *E. faecium* can be estimated with the help of the following two fractions from the same FDA-CVM study:

- $P_{\text{ent}} = 0.10$ = fraction of ICU infections caused by *Enterococcus*. [Wisplinghoff et al. (2004) provide an estimate of 0.09. To be conservative, we use the higher estimate of 0.10.]
- $P_{\text{EF,ent}} = 0.25$ = fraction of enterococcal infections caused by *E. faecium*.

The product of these two factors, $f_{\text{EF}} = P_{\text{ent}}{}^{*}P_{\text{EF,ent}} = 0.025$, is the estimated fraction of ICU infections caused by *E. faecium*. An approximate value for the expected annual rate of *E. faecium* infections can then be obtained via the equation

$$\begin{aligned} &\text{expected annual number of } \textit{E. faecium} \text{ infections} = N^{*}f_{\text{EF}} = N^{*}P^{*}_{\text{ent}}P_{\text{EF,ent}} \\ &\quad \le 315,000^{*}0.025 = 7,875\ \textit{E. faecium} \text{ infections/year.} \end{aligned}$$

Fraction of ICU E. faecium *Infections That Are Ampicillin-Resistant and Exogenous (Nonnosocomial)*

Most ampicillin-resistant *E. faecium* infections are contracted nosocomially. Indeed, it is possible that few or none originate in food animals. For example, as stated by Kuhn et al. (2005), in Europe "it seems that animal-associated VRE probably reflect the former use of avoparcin in animal production, whereas VRE in human-associated samples may be a result of antibiotic use in hospitals." Since nosocomial transmission is a hospital-specific problem that can often be eliminated by rigorous control measures, we restrict our risk assessment to exogenous (nonnosocomial) cases that are potentially attributable to food animals. (If this restriction is dropped in sensitivity analysis, the effect is simply to divide the estimated annual impacts by the nonnosocomial fraction of cases, which increases them approximately sixfold.)

Cox and Popken (2004) use data from several studies in the 1990s to estimate an approximate mean value of 0.17 for the fraction of exogenous cases in the United States. More recent investigations of the molecular epidemiology of drug-resistant *E. faecium* infections suggest that, if anything, this proportion may have decreased since the 1990s as a particular hospital-adapted clone of *E. faecium* called CC17 has spread widely in hospitals in the United States and elsewhere (Leavis et al., 2006; Top et al., 2007). Approximately 88% of *E. faecium* isolates from hospital outbreaks ($n = 32$) belong to Complex-17, compared to 59% of all clinical isolates ($n = 162$), 23% of isolates from hospital surveillance in Australia, Europe, and North and South America ($n = 64$), 5% of community isolates ($n = 57$), and 1% of isolates from animal surveillance ($n = 96$, including bison, calves, cats and dogs, ostriches, poultry, pigs, and rodents in Africa and Europe) (Leavis et al., 2006). In the United States, too, Complex-17 and a closely associated clade of hospital-associated strains dominate the epidemiology of AREF (Leavis et al., 2007; Deshpande et al., 2007). To be conservative, we continue to assume that the fraction of exogenous cases is **0.17**, although acknowledging that this fraction may be declining significantly as these hospital-associated strains accounts for a higher proportion of all *E. faecium* infections.

The fraction of exogenous cases that are ampicillin-resistant can be estimated from data in Table 1 of Willems et al. (2005). Pooling all non-outbreak, non-Complex-17 cases (where "non-outbreak" cases are the sum of clinical and hospital surveillance isolates) gives a total of 20 ampicillin-resistant cases out of 107 total cases, for a resistance fraction of 20/107 = 0.187 ampicillin-resistant cases per non-outbreak case. In summary, the estimated expected annual number of ampicillin-resistant, exogenously caused (i.e., nonnosocomial) *E. faecium* infections in the United States is no more than

$$(7,875\ E.\,faecium\ \text{infections/year})^*(0.17\ \text{nonnosocomial fraction})^*(0.187\ \text{ampicillin-resistant fraction}) = 7875^*0.17^*0.187 = 250\ \text{exogenous ampicillin-resistant}\ E.\ faecium\ \text{(AREF) infections per year.}$$

Fraction of Vancomycin-Susceptible Cases

Assuming that almost all vancomycin-resistant strains of *E. faecium* in the United States are also ampicillin-resistant (but not vice versa) (Rice, 2001), the relatively recent data of Jones et al. (2004) show that, in the United States, about 14% of *E. faecium* strains are ampicillin-resistant *and* vancomycin-susceptible. Specifically, 90.3% of *E. faecium* isolates were resistant to ampicillin and 76.3% of *E. faecium* isolates were resistant to vancomycin. The difference is 0.903 – 0.763 = 0.14. This is an estimate of the fraction of *E. faecium* isolates that are AREF but not VREF – in other words, the vancomycin-susceptible AREF of interest for our risk assessment.

Thus, 0.14/0.903 = **0.155** is the estimated fraction of AREF that are vancomycin-susceptible. Using this point estimate yields

expected exogenous ampicillin-resistant and vancomycin-susceptible cases per year
$\leq 250^{*}0.155 = 38.75$ vancomycin-susceptible AREF infections/year.

This should be considered an upper bound. For example, research by Suppola et al. (1999) suggests that Van A and Van B *E. faecium* incorporate into an *endemic* vancomycin-susceptible AREF strain.

Fraction of Exogenous Cases Potentially from Food Animals

As reviewed above, genetic similarities between ampicillin-resistant strains found in non-outbreak *E. faecium* infections among hospitalized patients (most of which carry the *esp* virulence gene and other distinctive genes) and strains found in food animals (most of which do not) is weak (Kuhn et al., 2005; Leavis et al., 2006, 2007). No clear empirical attribution of hospital cases to food animals can be made based on these data.

Figure 9.1 summarizes data that suggest a possible upper-bound quantitative estimate for the contribution of strains of *E. faecium* found in food animals to strains found in non-outbreak (non-Complex-17) human patient isolates. The figure represents inferred patterns of evolutionary descent among multiple strains of *E. faecium*. Each number represents a sequence type (ST). Lines connect sequence types that differ in only one of seven "housekeeping genes." The relative sizes of the circles represent the relative prevalences of the sequence types. In the relatively few cases where human patient and food animal (pig and poultry) clusters overlap, the strains falling in the overlap might have come from a common environmental source (e.g., soil or water) or might be due to a "reverse causation" flow from humans to pigs via surface water, flies, pets, or other paths (Guardabassi and Dalsgaard, 2004; Macovei and Zurek, 2006; Rodrigues et al., 2002).

The Multi Locus Sequence Typing website (www.mlst.net – curator: Rob Willems) provides a database of 490 *E. faecium* samples – a subset of those used to generate Fig. 1 of Leavis et al. (2006). The data indicate 87 unique sequence types among "Clin_Isol" and "Hosp_Surv" (human, non-outbreak, noncommunity) clusters. (We do not consider the "Hosp_Outbreaks" category, since these are assumed to fall into the nosocomially transmitted group. We also do not consider the "Human_comm" category, as these are noninfectious strains found in healthy individuals.) Two of these 87 sequence types (26 and 32) are shared with poultry and four (5, 6, 18, 133) are shared with pigs. If we assume, conservatively, that *all*shared types represent transmission from food animals to human patients (rather than from people to animals, or to both from common environmental sources such as soil, water, flies, or birds), then an estimate of the fraction of strains in human patients that might originate in food animals is 6/87 = 0.069. Using this point estimate reduces the above estimate to

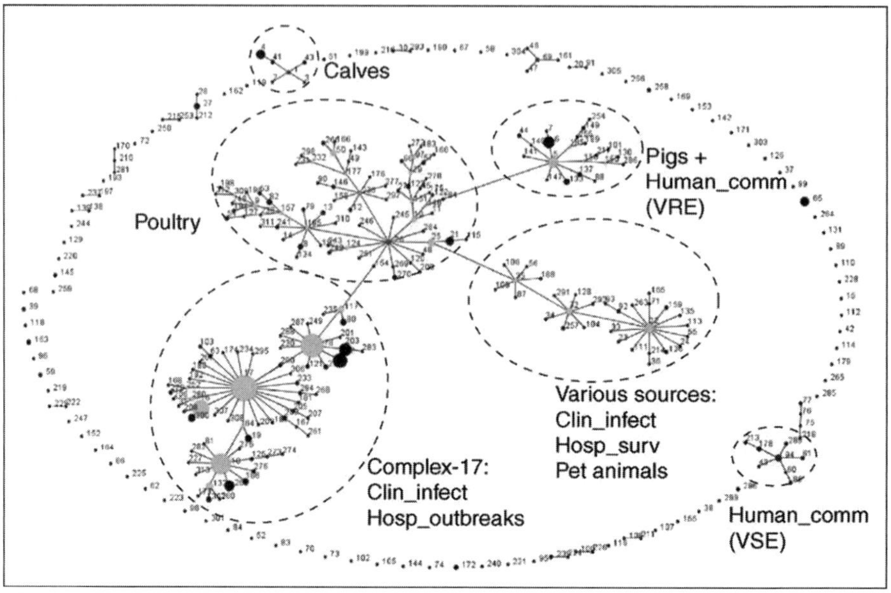

Fig. 9.1 Isolates from human hospital patients belong to a disjoint cluster from isolates in pigs and healthy human VRE and are almost disjoint from poultry isolates
A population snapshot of 855 *E. faecium* isolates on the basis of MLST allelic profiles using the eBURST algorithm (Leavis et al., 2006). This snapshot shows all clonal complexes, singletons, and patterns of evolutionary descent. The relative size of the circles indicates their prevalence in the MLST database (http://www.mlst.net/). Numbers correspond to the sequence types (STs), and lines connect single locus variants: STs that differ in only one of the seven housekeeping genes. CC17, the major subpopulation representing hospital outbreaks and clinical infections, is indicated, as well as the source of other major subgroups. Annotations: Clin_infect, isolates from clinical sites (mainly blood) from hospitalized patients; Human_comm, feces isolates from human volunteers not connected to hospitals; Hosp_outbreak, isolates from hospital outbreaks; Hosp_surv, feces isolates from hospitalized patients without an enterococcal infection and not associated with an enterococcal hospital outbreak; VSE, vancomycin-susceptible enterococci. *Source:* Figure reproduced with permission from Leavis et al. (2006)

expected exogenous vancomycin-susceptible and ampicillin-resistant cases per year from food animals $\leq$ (38.75 exogenous vancomycin-susceptible and ampicillin-resistant cases per year) $\times$ (fraction of not more than 6/87 from food animals) $\leq$ 2.67 exogenous vancomycin-susceptible and ampicillin-resistant infections per year from food animals.

Penicillin Allergies

Hospitalized patients who are allergic to penicillin cannot have their enterococcal infections treated with penicillin or ampicillin. Since such patients are not harmed by penicillin resistance, we need to exclude them from risk calculations. A large U.S. study of hospitalized patients requiring antimicrobials found that 15.6% reported an allergy to penicillin (Lee et al., 2000). (This exceeds the average in the general population, which is expected.) The remaining $(1 - 0.156) = 0.844$ of patients corresponds to

expected exogenous vancomycin-susceptible ampicillin-resistant cases
per year from food animals, in penicillin-tolerant patients $\leq 0.844 \times 2.67 = 2.25$.

Excess Mortalities

The next step is to calculate the increase in human health harm – especially, increased mortality – among the 2.25 expected cases per year calculated in the previous steps. Fortun et al. (2002) report no statistically significant differences in outcomes between ampicillin-resistant cases and ampicillin-susceptible cases. They state that

> There were no significant differences in the outcome of patients with ampicillin-resistant and - susceptible strains. We did not find significant differences in mortality between the two groups. Overall mortality in patients with bacteremia caused by ampicillin-resistant and - susceptible *E. faecium* was 34% and 21%, respectively (OR: 2.1; 95% CI: 0.47–9.95). Mortality attributed to bacteremia was 21% and 15%, respectively (OR: 1.5; 95% CI: 0.27–8.85) (p. 4).

To obtain a nonzero risk estimate despite the reported absence of statistically significant differences in mortality, we make the conservative assumption that the numerical difference in bacteremia-attributed mortality rates between patients with ampicillin-resistant and - susceptible strains reflects a true causal effect (i.e., that resistance does cause a $21\% - 15\% = 6\%$ increase in absolute mortality risk, per patient per infection). In other words, we treat the statistically nonsignificant difference as a true difference caused by ampicillin resistance (but acknowledge that this is not the original authors' interpretation and that the true difference could be as small as zero). This assumption provides a possible basis for calculating a nonzero human health risk from ampicillin resistance.

With this assumption, the expected annual excess mortality risk caused by ampicillin resistance becomes

expected excess mortalities per year (for the entire U.S. population) caused by
exogenous vancomycin-susceptible and ampicillin-resistant *E. faecium*
infections, assumed to originate from food animals, in penicillin-tolerant patients
$\leq 2.25 \times 0.06 = 0.135$ excess mortalities/year.

In reality, annual mortality risks from AREF are likely to be much smaller than this, as patient-cultured isolates would typically be screened for resistance prior to treatment (standard procedure indicated for such infections) and then patients with AREF would be treated with other drugs such as gentamicin, vancomycin, quinupristin-dalfopristin, linezolid, daptomycin, tigecycline, or nitrofurantoin. In addition, the above mortality estimate does not address the quality-adjusted life-years (QALYs) lost due to potentially preventable resistance. The patients at risk are severely ill (usually, immunocompromised) patients such as leukemia, transplant, and AIDS patients. Thus, the hypothesized increased mortality risk (per infection with ampicillin-resistant vancomycin-susceptible *E. faecium*) represents fewer quality-adjusted life-years (QALYs) lost than would be the case for otherwise healthy patients. We have therefore not attempted to quantify QALY impacts.

Results Summary, Sensitivity, and Uncertainty Analysis

Table 9.1 summarizes key parameter estimates, calculations, assumptions, and resulting risk estimates from this study. When presenting point estimates, it is customary to also present interval estimates to inform decision makers about the plausible range of estimated values. In the present analysis, however, the key uncertainties have little to do with statistical sampling error, and they are not adequately characterized by confidence limits. Rather, they arise from uncertainty about the validity and conservatism of the assumptions in Table 9.1.

Qualitatively, the main uncertainty is about whether a nonzero risk to human health exists from animal use of penicillin drugs. We have assumed that there is, but there is no clear empirical proof that the risk is nonzero. To bridge this knowledge gap, Table 9.1 incorporates several conservative qualitative assumptions that jointly imply that the risk is nonzero. Other quantitative parameter values presented, and their implied risk estimate of $\leq$0.135 excess mortalities/year, are intended to be realistic, data-driven values (rather than extreme upper bounds or 95% upper confidence limits) *contingent* on these conservative qualitative assumptions.

The most important conservative elements in Table 9.1 are the following qualitative assumptions:

- *The transfer of ampicillin resistance from food animal bacteria to bacteria infecting human patients occurs.* The assumption that ampicillin-resistant strains and/or determinants are transferred from strains in food animals to human ICU patients is fundamental to the assessment in Table 9.1. Such a transfer has never been shown to occur, but may be possible, based on the similarities described in Fig. 9.1.
- *Withdrawing animal drug use would immediately and completely prevent the problem.* Table 9.1 assumes that halting penicillin use in food animals would immediately eliminate all ampicillin resistance from the cases in Table 9.1. This is a deliberately extreme assumption. In reality, halting use might have little or no impact on the already very low levels of ampicillin resistance.

Table 9.1 Summary of risk calculation

Factor	More conservative value	Less conservative value	Source
N = ICU infections/year	$N = 315{,}000$	$N = 104{,}372.5$	FDA-CVM (2004)
P_{ent} = fraction of ICU infections caused by *Enterococcus*	0.10	0.09 (Wisplinghoffet al., 2004)	FDA-CVM (2004)
$P_{\text{EF,ent}}$ = fraction of enterococcal infections caused by *E. faecium.*	0.25		FDA-CVM (2004)
Fraction of enterococcal infections caused by *E. faecium* that are exogenous (nonnosocomial)	≤0.17		Cox and Popken (2004) (May be smaller now due to spread of CC-17)
Fraction of exogenous cases that are ampicillin-resistant	0.187		Willems et al. (2005)
Fraction of exogenous ampicillin-resistant cases that are vancomycin-susceptible	0.155		Jones et al. (2004)
Fraction of exogenous ampicillin-resistant vancomycin-susceptible cases possibly from food animals	0–0.069 (0.069 assumed)		Data of Leavis et al. (2006)
Fraction of exogenous ampicillin-resistant cases with penicillin-tolerant host	0.844		Lee et al. (2000)
Fraction of these cases that would become ampicillin-susceptible if penicillin use in food animals were terminated	0.00–1.00. (1 is assumed.)		Conservative assumption
Increase in mortality risk per case, due to ampicillin resistance	0.00–0.06. (0.06 is assumed.)		Fortun et al. (2002), conservative assumption
RISK = ≤0.18 potential excess mortalities/year	315,000*0.10*0.25*0.17*0.187*0.155*0.069*0.844*0.06 = 0.135	104,372.5*0.09*0.25*0.17*0.187*0.155*0.069*0.844*0.06 ≈ 0.04 mortalities/year	Product of preceding factors

- *Resistance increases patient mortality.* The assumption that ampicillin resistance causes an increase in the mortality rates of the patients in Table 9.1 is made even though, in reality, no statistically significant difference in mortality rates has been found between resistant and nonresistant cases (Fortun et al., 2002).

With these assumptions, the calculations in Table 9.1 predict that excess mortalities per year in the entire U.S. population could be as high as 0.135, or about one excess mortality expected once every seven to eight years on average, if current conditions persist. This risk is concentrated among ICU patients already at a high risk of such infections. With less conservative assumptions, the estimated risk falls to about 0.04 excess mortalities per year, i.e., about one excess mortality every 25 years in the United States under current conditions. The multiplicative calculation in Table 9.1 makes sensitivity analysis of these results to changes in the values of specific factors especially straightforward: The final risk estimate is directly proportional to each factor listed.

The more conservative risk estimate of 0.135 excess mortalities per year equates to an average individual risk rate in the most at-risk group (ICU patients) of approximately $0.135/315{,}000 = 4.3 \times 10^{-7}$ excess mortalities per ICU patient. For the U.S. population as a whole, this corresponds to an average individual risk of approximately $0.135/300\text{E}6 = 4.5 \times 10^{-10}$ excess fatalities per person-year, or a lifetime risk of about $80 \times (6 \times 10^{-10}) = 3.6 \times 10^{-8}$ excess risk of mortality per lifetime (for an assumed 80-year lifetime). This is well below the risk level of 1×10^{-6} (1 per million lifetimes) sometimes cited as a threshold for concern for carcinogens in the environment. If the less conservative risk estimate of 0.04 excess mortalities per year is used, these individual and population risks are reduced by a factor of 0.04/0.135, or more than threefold. If one or more of the key qualitative assumptions listed above are violated, then the true risk could be as low as zero.

The most important uncertainty in this analysis is discrete: Is the preventable fraction of risk positive or is it zero? Such uncertainty is not well characterized by a confidence interval. Nonetheless, it may be useful to consider a rough upper bound on how large the true risk might be. Markov's inequality for nonnegative random variables (see Table 2.1) gives a crude estimate if we assume that the risk estimates in Table 9.1 represent expected values. In this case, a (possibly extreme) upper bound on the true but unknown risk is that it has at most a 5% probability of exceeding the point estimates (0.135 or 0.04 excess mortalities per year) by more than 20-fold. To the extent that these point estimates are biased upward by the assumptions listed in Table 9.1, upper bounds based on Markov's inequality will be even more conservative.

Summary and Conclusions

This chapter has considered the possibility that penicillin-based drugs approved for use in food animals in the United States might increase the incidence of ampicillin-resistant *Enterococcus faecium* (AREF) of animal origin in human infections, leading to increased hospitalization and mortality due to a reduced response to ampicillin or penicillin. Upper bounding was used to assess the risks from the continued use of penicillin-based drugs in food animals in the United States, using several assumptions to overcome current scientific uncertainties and data gaps. Multiplying the total at-risk population of intensive care unit patients by a series of estimated factors suggests that not more than 0.037 excess mortalities per year (under conservative assumptions) to 0.18 excess mortalities per year (under very conservative assumptions) might be prevented in the whole U.S. population if the current use of penicillin-based drugs in food animals were discontinued and if this successfully reduced the prevalence of antibiotic-resistant *E. faecium* infections among ICU patients. These calculations suggest that the current penicillin usage in food animals in the United States presents very low (possibly zero) human health risks. Such information may be useful to risk managers and policy makers, even though no fully probabilistic model was developed, since adequate data to create and validate such a model are not available.

Removing penicillin drugs from the animal drug market has long been proposed as a risk management option by those for whom uncertain human health risks loom larger, in the absence of quantification, than animal health benefits. But significant baseline resistance among antibiotic-free animals (Patton et al., 2006), along with small potential human health benefits even if removal immediately eliminated all preventable cases of resistance in human patients (as in Table 9.1), suggests that even complete product removal would not detectably improve human health. Increasing the surveillance of food animal-associated enterococci and tracking their penicillin/ampicillin resistances (NARMS, 2005), as well as complying with judicious use guidelines for practitioners and producers (AVMA, 2008), may suffice to protect human health against the current small risks without compromising the health of food animals. Although concerns about penicillin use in food animals and the potential transfer of resistance to humans have been debated for decades, current knowledge and data, as analyzed in Table 9.1, suggest that ongoing penicillin use in food animals in the United States creates at most only relatively minor risks to human health. Quantitatively, these risks appear unlikely to exceed one potentially preventable mortality in the U.S. population roughly every 7–25 years. The true value could be smaller; it might be zero if one or more of the various key conservative qualitative assumptions discussed earlier are incorrect.

Knowing that an uncertain but preventable risk is either zero or relatively small (say, on the order of one or fewer occurrence every few billion person-years, as in this example) may provide a decision maker with enough information to make a clear decision about what to do today – and to make the same decision as would be made if complete information about risk were available. In this case, the additional information in a full probability distribution for consequences contains no incremental value for improving decisions. The simpler bounding risk assessment, with its partial characterization of risk, suffices as well as would a more elaborate, fully probabilistic, risk model to support high-quality decision making. That the risk might be zero or might be positive (but small) does not inhibit an optimal decision from being made today, if the same decision (e.g., to allow prudent uses of penicillin drugs in food animals) is optimal in either case. Only if future information might reveal which side of a decision boundary the uncertain risk falls on (e.g., only if it might reveal a large enough risk to humans to justify reducing animal uses) might it be worth acquiring additional information before making a decision. If all contingencies lead to the same best decision today, then the value of information from discovering which is correct is zero, for purposes of improving that decision.

Chapter 10 explores further how to use multiple alternative assumption sets, when the correct one is unknown, to provide enough information to support current decision needs, even though better information may become available later. Chapter 14 discusses the value of information (VOI) calculations further.

Chapter 10
Confronting Uncertain Causal Mechanisms – Portfolios of Possibilities

This chapter returns to smoking and lung cancer risk as a fruitful example of a complex system with many uncertainties (and, as discussed in Chapter 11, nonlinearities) in its input-output (dose-response) relations. These uncertainties, complexities, and nonlinearities raise important challenges for quantitative risk assessment (QRA) modeling. The challenge confronted in this chapter is how to estimate the potential effects on lung cancer of removing a specific constituent, cadmium (Cd), from cigarette smoke, given the very incomplete scientific information available now about its possible modes of carcinogenic action. Not enough is known about how cadmium affects lung cancer to allow useful bounds on risk to be established using biomarkers, as in Chapter 8. A different strategy is needed for QRA.

The QRA in this chapter is based on a *set of alternative effects estimates*, derived from (a) a rough mathematical model describing how different cell populations in the lung might be affected by cadmium in cigarette smoke; and (b) alternative sets of assumptions about how Cd in cigarette smoke might affect the model parameters, thus increasing lung cancer risk. Rather than estimating unmeasured model parameters by traditional statistical curve fitting (adjusting the parameter values to match model input-output predictions to epidemiological or animal experimental tumor data), this chapter instead proposes rough estimates of parameter values based on their biological interpretations and on data about lung cancer risks associated with corresponding genetic polymorphisms. This knowledge-rich QRA modeling can succeed if sufficient knowledge of potential causal mechanisms is available, even if data points are lacking to support traditional statistical parameter estimation.

The resulting knowledge-driven parameter estimates are admittedly uncertain and approximate. But they suggest a potentially useful *portfolio approach* to estimating the impacts of removing Cd that gives robust conclusions. It views Cd as creating a portfolio of uncertain health impacts that can be expressed as biologically independent relative risk factors having clear mechanistic interpretations. The factors reflect the possible effects on biologically interpretable parameters such as normal lung stem cell proliferation rate; DNA repair inhibition in normal stem cells affected by initiating events; the proliferation, promotion, and progression of initiated (i.e., premalignant) cells; and death or sparing of initiated and malignant cells as they are further transformed to become fully tumorigenic. (The following

L.A. Cox, Jr., *Risk Analysis of Complex and Uncertain Systems*,
International Series in Operations Research & Management Science 129,
DOI 10.1007/978-0-387-89014-2_10,

sections explain these parameters more fully.) Given such a portfolio of possible causal mechanisms, and the resulting effects of Cd on lung cancer risk, it is easy to calculate the probabilities that removing Cd will have total effects of different sizes, starting from estimated probabilities that each individual mechanism holds. Even if the probabilities of the individual causal mechanisms are very uncertain (as seems realistic), approximate bounds for them, provided using expert judgment, allow corresponding bounds to be calculated for the probability that removing Cd would prevent at least a specified fraction of smoking-induced lung cancers, taking into account the contributions from the whole portfolio of mechanisms. This type of judgment-based approximate probabilistic analysis can provide a useful basis for deciding whether to investigate further the possible health benefits of reducing Cd. At the same time, the use of bounds on subjective probabilities indicates the substantial uncertainty about possible effects, given the current incomplete scientific knowledge about causal mechanisms.

Background: Cadmium and Smoking Risk

In 1994, Hertz-Picciotto and Hu presented two regulatory-type risk assessments of human lung cancer mortality risks from cadmium (Cd) in cigarette smoke. Based on linear no-threshold statistical risk models applied to rat data and epidemiological data, they concluded that, depending on factors such as the fraction of sidestream smoke inhaled:

- "The epidemiologic data predicted that 1 to 18 lung cancer deaths per 10,000 smokers may be attributable to inhaled cadmium in cigarette smoke, or approximately 0.2% to 1.6% of smoking-induced lung cancer deaths. Upper 95% bounds on these figures are 7 to 95 lung cancer deaths or 1.6% to 8.8% of smoking-related deaths."
- "The rodent data predicted that 80 to 416 lung cancer deaths per 10,000 smokers (95% upper bounds: 136–707) or 13% to 47% (23–81%) of smoking-induced lung cancer mortality may be attributable to cadmium in cigarette smoke."
- "Linear extrapolation from human data appears to provide plausible estimates of risk at low doses. Considering the large number of carcinogens present in cigarette smoke, the extrapolation from rodents appears to overestimate human risks."

These numbers were developed from traditional statistical risk models, which assume that risk increases smoothly with cumulative exposure. More recent epidemiological data suggest that this may not be an accurate assumption for cadmium (Jarup et al., 1998; Sorahan and Esmen, 2004), so the fractions of lung cancer cases due to Cd estimated by Hertz-Picciotto and Hu may not be accurate either. Nonetheless, they raise a key question for predictive risk assessment: *To what extent would reducing Cd in cigarette smoke reduce the risks of lung cancer from cigarette smoking?* The answer depends on the exposure levels and accumulation of Cd in target tissues as well as on the correct (but currently very uncertain) dose-response relation for Cd in the context of other substances to which a smoker is exposed.

This chapter reconsiders the causal relation between Cd in cigarettes and the risk of lung cancer in smokers in light of advances since 1994 in the understanding of potential biological mechanisms by which Cd may cause lung cancer. Multiple *in vitro* and *ex vivo* studies have suggested possible mechanisms, yet scientific understanding of the mechanisms of Cd-induced lung carcinogenesis remains very incomplete. At the same time, new and updated epidemiological studies have raised questions about some earlier reports associating Cd exposures with lung cancer, especially because of the apparent lack of a strong, consistent pattern of increasing risk with increasing cumulative exposure (e.g., Sorahan and Esmen, 2004). We therefore reexamine the possible relation between Cd in cigarette smoke and human lung cancer risk in light of current scientific knowledge and data, without making the traditional regulatory risk modeling assumptions used by Hertz-Picciotto and Hu and without assuming that lung cancer risk association is necessarily a smooth, monotonically increasing function of cumulative Cd exposure. Current scientific uncertainties make it impossible to give a confident deterministic prediction of how much difference removing Cd from cigarettes would make to lung cancer risk. Any useful (and any honest) answer must reflect the fact that the predicted risk depends greatly on what assumptions are made... and the correct set of assumptions is unknown. Moreover, in contrast to the situation with penicillin drugs in Chapter 9, different assumption sets do *not* all lead to predicted risks small enough (or large enough) to imply the same decision. We therefore propose and illustrate methods for assessing the extent to which removing Cd from tobacco should be expected to reduce lung cancer risk to smokers, given the incomplete but useful scientific information now available.

Previous Cadmium-Lung Cancer Risk Studies

Cadmium Compounds are Rat Lung Carcinogens

Sufficiently high and prolonged exposures to aerosols of cadmium (Cd) compounds including cadmium chloride ($CdCl_2$); cadmium oxide (CdO), particles of which are readily solubilized in the lung, though not in water, and appear to be relatively toxic compared to other Cd compounds (Glaser et al., 1986; Grose et al., 1987); cadmium sulfide (CdS), which is cleared by alveolar macrophages and may be only about 10% as potent as CdO in causing inflammatory responses (Oberdorster, 1992); and cadmium sulfate are known to increase the frequency of pulmonary adenocarcinomas in rats (Takenaka et al., 1983; Oberdorster, 1992; Heinreich, 1992; Waalkes, 2003). Similar increases in lung cancer risk are not observed in hamsters or mice, even at doses that induce lung cell inflammation and cell proliferation, and even though mice retain about twice as great a Cd dose per gram of lung as rats. The high susceptibility of rats may in part reflect the lower basal levels and induced expression in the lungs of the protective protein metallothionein (MT) (Oberdorster et al., 1994; McKenna et al., 1998).

Epidemiological Data are Inconclusive

Whether cadmium exposures increase the risk of lung cancer in humans remains uncertain. Cd has been classified as a human carcinogen (IARC, 1993), based largely on statistical modeling of epidemiological data (e.g., Stayner et al., 1992). However, available epidemiological data are somewhat ambiguous, with conclusions about an association between Cd exposure and lung cancer risk varying greatly depending on the data sets, statistical models, and exposure reconstruction assumptions used (see Table 10.1). Contemporaneous exposures to other chemicals (e.g., nickel and arsenic compounds) have confounded a clear causal interpretation of past studies (Sorahan and Lancashire, 1997; Sorahan and Esmen, 2004), while the absence of a clear relation between cumulative exposure and the risk of lung cancer in most data sets (e.g., Jarup et al., 1998; Sorahan and Esmen, 2004) poses a puzzle for traditional models of carcinogenesis if cadmium is indeed a human carcinogen.

Table 10.1 summarizes the conclusions from several important cadmium epidemiology studies, illustrating the conflicting interpretations that have been suggested. The influential studies by Stayner and co-authors that reported clear dose-response relations did so only for *estimated* exposures and *model-predicted* risks, using statistical models that assumed a smooth, typically monotonic, relation between them. As shown by Stayner et al., the raw data suggest different, non-monotonic exposure-response patterns (e.g., *u*-shaped for Hispanics and *n*-shaped for non-Hispanics). Thus, while suggestive, the results from the statistical risk models are not necessarily valid: the reported dose-response relations may be artifacts of questionable data aggregation and modeling assumptions, rather than true empirical relations.

Conversely, the absence of a consistent exposure-response relation noted by other investigators need not imply that Cd is not a human lung carcinogen. Rather, Cd may induce lung cancer by nontraditional (e.g., nongenotoxic) mechanisms, with cancer-increasing effects that are independent of cumulative exposure over the ranges found in occupational studies. The following sections explain how such nontraditional mechanisms and exposure-response relations may operate for cadmium.

Pharmacokinetic Data Show That Smoking Increases Cadmium Levels in the Human Lung

Despite the mixed and inconclusive epidemiological evidence in Table 10.1, several lines of biological evidence suggest that Cd in cigarette smoke may be a human lung

Table 10.1 Selected findings from cadmium lung cancer epidemiology (emphases added)

"**For lung cancer, adjusted hazard ratio was 1.70 (1.13–2.57, p = 0.011) for a doubling of 24-h urinary cadmium excretion,** 4.17 (1.21–14.4, $p = 0.024$) for residence in the high-exposure area versus the low-exposure area, and 1.57 (1.11–2.24, $p = 0.012$) for a doubling of cadmium concentration in soil."	Nawrot et al. (2006)
"Estimated **cumulative cadmium exposures were not related to risks of lung cancer** or risks of chronic obstructive pulmonary diseases, even when exposure histories were lagged first by 10, then by 20 years. CONCLUSIONS: The study findings do not support the hypotheses that cadmium compounds are human lung carcinogens."	Sorahan and Esmen (2004)
"Overall, considering the results of the most recent studies does not suggest that the effect of cadmium on lung cancer increases with improvement of the study design but points to a lower relative risk in the groups exposed to cadmium in the absence of arsenic and nickel."	Verougstraete et al. (2003)
"There was an **increased overall risk for lung cancer, but no exposure-response relation**between cumulative exposure to cadmium or nickel and risk of lung cancer."	Jarup et al. (1998)
"A **statistically significant dose-response relationship** was evident in nearly all of the regression models evaluated. Based on our analyses, the lifetime excess lung cancer risk at the current Occupational Safety and Health Administration standard for cadmium fumes of 100 micrograms/m^3 is approximately 50–111 lung cancer deaths per 1,000 workers exposed to cadmium for 45 years."	Stayner et al. (1992)
"The **increasing risk** of lung cancer associated with increasing duration of employment **could not be accounted for by cadmium** and did not appear to be restricted to any particular process or department."	Ades and Kazantzis (1988)
"This new analysis suggests that long term, high level exposure to cadmium is associated with an **increased risk of cancer**. The role of concomitant exposure to nickel needs further study."	Elinder et al. (1985)

carcinogen. For pharmacokinetics, autopsy data show that human smokers accumulate about an eightfold increase in the lung concentrations of Cd compared to nonsmokers, with the elevated Cd levels returning to nonsmoker levels only slowly, e.g., in about 22 years after the cessation of smoking (Paakko et al., 1989). (For comparison, in mice and rats, the increase in lung Cd in smoke-exposed animals is about five- to sixfold; see Gairola and Wagner, 1991.) Cd levels have also been found to be significantly higher in lung cancer tumor tissue than in lung tissue from other lung surgery patients (Kollmeier et al., 1990) and autopsied smelter workers dying with lung cancer had median lung Cd levels about fivefold greater than the levels in workers dying from other cancers, although the causal interpretation of such associations is not self-evident (Gerhardsson et al., 1986). While humans and other primates have basal metallothionein (MT) levels about an order of magnitude higher than in rats, which could reduce the risk from Cd, they also have much longer retention times for Cd in the lungs (by a factor of 10 in experiments reported by Oberdorster, 1992), and recent evidence suggests that smoking may saturate MT, at least in alveolar macrophages, leaving the remaining Cd dose available to interact with cells and perhaps cause carcinogenic damage (Grasseschi et al., 2003).

Considerable *in vitro* human and animal cell studies and some *in vivo* animal data, reviewed below, suggest that many biological processes involved in carcinogenesis may be affected by increases in Cd, especially in the context of other smoking-induced effects on lung cell populations. Thus, the substantial (approximately eightfold) excesses of Cd that accumulate in the lungs of smokers under real exposure conditions provide reason to further investigate how and whether Cd from cigarette smoke might increase lung cancer rates.

Biological Mechanisms of Cadmium Lung Carcinogenesis

A Transition Model Simplifies the Description of Cadmium-Induced Lung Carcinogenesis

Several different pharmacodynamic and biological mechanisms have been proposed recently, each with at least some experimental support, to explain how Cd exposures may increase lung cancer rates in rats, and possibly in humans. To help organize available data on Cd-induced lung carcinogenesis, Fig. 10.1presents a conceptual model of carcinogenesis. The parameters of the model, explained next, can be affected by Cd exposures, and the resulting changes in their values summarize how Cd affects cancer risk.

This framework is based on previous two-stage clonal expansion (TSCE) models of lung carcinogenesis (e.g., Hazelton et al., 2005; Moolgavkar et al., 1993; Luebeck et al., 1999). It makes the following modifications to accommodate Cd-specific data:

- *A "preinitiated" compartment is inserted* between the normal stem cell and initiated stem cell compartments to allow convenient modeling of exposure-related effects on DNA repair (e.g., Potts et al., 2003). DNA damage (e.g., due to oxidative damage or covalent binding of reactive metabolite molecules to DNA) may cause a normal stem cell to undergo a transition to the preinitiated compartment. This transition rate is denoted as *a*. Its units are the expected transitions per normal stem cell at risk per unit time. Like other parameters in the model, *a* may be time-varying and may depend on exposure, levels of detoxifying enzymes and protective proteins, and other factors.

$$\begin{array}{ccccccccc} & r, a & & f & & c & & h & \\ \text{Normal} & \Leftrightarrow & \text{Pre-initiated} & \Rightarrow & \text{Initiated} & \Rightarrow & \text{Malignant} & \Rightarrow & \text{Tumor} \\ \Downarrow(D-B) & & & & \Downarrow(d-b) & & \Downarrow(e-g) & & \end{array}$$

Fig. 10.1 A conceptual model of carcinogenesis

- *Damaged cells in the preinitiated compartment are repaired* at an average rate of r repairs per damaged cell per unit time. However, if a damaged cell divides before being repaired, then the DNA damage is "locked in" and the daughter cell and its progeny belong to the initiated compartment. The probability that somatically heritable damage is locked in before it is repaired is $p = f/(f + r)$, where f is the effective rate of mitogenesis (viewed as a transition that competes with repair for a preinitiated cell). While Fig. 10.1 shows f and r as two separate parameters, only the ratio p is needed to calculate the fraction of preinitiated cells that become initiated.
- *Separate "malignant" and "tumor" compartments*. A "malignant" cell is defined here as one that is immortal (unless killed by cytotoxic damage), but not invasive, as for carcinoma *in situ*. By contrast, a fully developed "tumor" cell exhibits additional tumorigenic properties such as invasiveness and continued autonomous growth.
- *Exposure-dependent latency times:* The hazard rate or hazard function h for the random time for a malignant cell to become fully tumorigenic may depend on exposures. Similarly, the transition rate c for converting or transforming initiated to malignant cells may be exposure-dependent. This enables the modeling of situations in which Cd exposures hasten or inhibit the transformation of cells to fully tumorigenic cells. [Cd compounds can cause either of these opposite effects *in vitro*, depending on relatively small changes in exposure concentration and duration (e.g., Abshire et al., 1996; Takiguchi et al., 2003).]
- *Exposure may affect the birth and death rates of the normal stem cell compartment as well as of the initiated cell compartment.* The net death rate per cell per unit time in the normal stem cell compartment is denoted by $(D - B)$, while the net death rate in the initiated cell compartment is denoted by $(d - b)$, where D and d denote the absolute death (or differentiation) rates and B and b denote the absolute birth rates of cells in these two compartments, respectively. These parameters are needed to represent experimental evidence that Cd exposures can affect the proliferation rates (Bajpai et al., 1999; Hart et al., 1989; Palmer et al., 1975) and apoptosis rates (Lag et al., 2002; Shih et al., 2003; Shin et al., 2004) of normal cells as well as of initiated cells (Fang et al., 2002; Hart et al., 2001). Rather than modeling normal stem cell kinetics in detail, it may be simpler and sufficiently accurate to recognize that this compartment undergoes homeostatic self-regulation (Cox, 1992) and that the net effect of chronic exposure to Cd on normal stem cell kinetics may be a simple percentage change in the number of normal stem cells at risk of initiating damage.
- *Malignant cells may undergo exposure-dependent cytotoxic death* and/or exposure-related selective survival and growth compared to other cells. The net death rate per malignant cell per unit time in the normal stem cell compartment is denoted by $(e - g)$, i.e., the difference of (possibly dose-dependent) "exit" and "growth" rates per malignant cell per unit time. This feature is included to facilitate the representation of experimental evidence that Cd exposure can inhibit lung tumor growth and metastasis (Waalkes and Diwan, 1999), perhaps in part by cell-specific cytotoxicity, e.g., due to a deficiency of MT in malignant cells (Waalkes et al., 1993).

- The effects of exposures on growth rates and transition rate parameters are *not* assumed to follow any specific parametric form. Instead, the percentage change in each parameter, considered as a function of Cd dose (or, in the context of smoking, considered as a function of smoking intensity and duration for a given Cd content in the tobacco), is treated as a quantity to be estimated empirically, e.g., from animal or human smoking data and/or from relevant *in vitro* data and extrapolation assumptions. This flexibility is required to represent experimental evidence from *in vitro* cell lines that relatively slight (e.g., 2- to 10-fold) changes in Cd concentration can reverse the directions of Cd impacts on such parameters as

  (a) *Proliferation rates,* from stimulatory to inhibitory (Beyersmann and Hechtenberg, 1997).
  (b) *Tumorigenic progression and transformation rates,* from stimulating malignant progression to inhibiting it (Abshire et al., 1996).
  (c) *Production of glutathione* (GSH) in a lung cell line, from stimulation to depletion (Gaubin et al., 2000), as has also been reported for lung epithelial lining fluid GSH levels of chronic vs. acutely exposed cigarette smokers *in vivo* (Rahman and MacNee, 1999).
  (d) Similarly, extending the duration of Cd exposure from 1 week to 10 weeks reverses the direction of its impacts of *DNA methylation* (from hypomethylation to hypermethylation) and *MT expression* in a rat liver cell line (Takiguchi et al., 2003).

Such reversals emphasize the need for flexibility in modeling the directions and magnitudes of Cd exposure effects on model parameters, depending on the intensity and duration of exposure. None of them is consistent with a linear relation between Cd exposure and changes in transition rates.

In summary, the modifications to the standard TSCE model introduced here are made to allow the representation of relevant data on Cd-induced lung carcinogenesis and cell effects.

The quantitative variables N, P, I, M, and T will be used to refer to the numbers of normal, preinitiated, initiated, malignant, and tumorigenic cells, respectively, in Fig. 10.1. These are dynamic variables, i.e., they may change over time, at rates that reflect the other model quantities (transition rates and birth-death rates), which in turn may be affected by Cd exposures and other exposures.

Cadmium Can Affect Lung Carcinogenesis via Multiple Mechanisms

In the framework of Fig. 10.1, a carcinogen acts by increasing one or more of the following seven quantities:

- N = *number of susceptible normal stem cells* available for an initiating transformation. Carcinogens may increase their birth rate, B (either directly, by upregulating growth signals or their receptors, or indirectly, by cytotoxic damage to more mature cells that stimulates the compensating proliferation of stem cells), or decrease the death rate, D.
- a = *preinitiation rate* (N to P transition rate) per susceptible normal stem cell per unit time. A classical initiator might increase a by increasing concentrations of reactive oxygen species (ROS) or metabolites available to bind to and damage DNA, either by increasing their production or by slowing their detoxification and removal.
- p = *initiation probability* that initiation events become locked in by mitosis before they can be repaired. Exposure to an initiator carcinogen increases p if it increases the effective rate of mitogenesis, f, or decreases the effective repair rate, r, for preinitiated cells.
- I = *number of initiated stem cells* available for malignant transformation. A promoter carcinogen can reduce the net death rate ($d - b$) (if $d > b$; or increase the net birth rate, $b - d$, if $b > d$), leading to the increased survival and proliferation ("promotion") of initiated clones.
- M = *number of malignant stem cells* available for further tumorigenic transformation by decreasing their net death rate ($e - f$). This might be done by increasing their proliferation rate, by inactivating tumor suppressor genes such as p53, or by cytotoxic selection of malignant cells in competition with normal stem cells.
- *Effective malignant transformation (or "progression") rates* c and h, i.e., the rates of transformation from I to M (per initiated cell per unit time) and from M to T, respectively. For example, this might occur by inactivating tumor suppressor genes, or by cytotoxic selection of more transformed cells in competition with less transformed cells.

Table 10.2 summarizes such effects for Cd. In this table, a symbol such as "$N\uparrow$" is interpreted to mean "N increases," while "$N \rightarrow I$" is interpreted as "the flux of cells from N to I (i.e., number of cells per unit time making this transition) increases." Each of the generic mechanisms of carcinogenesis just described has a more detailed counterpart for Cd in Table 10.2. The following paragraphs summarize several aspects of Cd biology that may be relevant for understanding its possible roles in smoking-related lung cancer.

Smoking and Cd Exposures Stimulate Reactive Oxygen Species (ROS) Production

Cd^{2+} itself is only weakly genotoxic. Unlike some other heavy metal carcinogens (CuII, CrVI, NiII, CoII), Cd is not a good redox transition metal (does not accept or donate electrons) under physiological conditions, and hence does little direct genotoxic damage (Waisberg et al., 2003).

Table 10.2 Survey of cadmium's biological effects potentially relevant for lung cancer

Effect (Symbol/Meaning)	Biological mechanisms and evidence for cadmium
$N\uparrow$ *Proliferation* of normal stem cells: Increase *N*.	• Smoking and Cd exposure each increase the proliferation of lung cells and lung cellularity. "Hallmarks of Cd adaptation include hyperplasia and hypertrophy of type II alveolar epithelial stem cells" (Hart et al., 2001).
Note: Although Cd stimulates lung cell proliferation, this also occurs in mice, which do not develop Cd-induced lung cancer (McKenna et al., 1998). Thus, the relevance of such proliferation per se to lung cancer risk is unclear.	• *In vitro*, "About 1 μM Cd stimulates DNA synthesis and cell proliferation in various cell lines, whereas more elevated concentrations are inhibitory" (Beyersmann and Hechtenberg, 1997). Beyersmann (2002) discusses evidence that Cd stimulates cell proliferation *by activating proto-oncogenes* and/or by interfering with genes downregulating cell growth. • *In vivo*, Cd increased the total number of cells in rat lungs, inducing an inflammatory response and interstitial fibroblastic proliferation (Bajpai et al., 1999). The pretreatment of male Lewis rats with Cd aerosols increased the number of alveolar epithelial type II cells (Hart et al., 1989). "Exposure to $CdCl_2$ aerosol evokes a wave of cellular proliferation in rat lung" (Palmer et al., 1975). "Acute exposure to aerosolized 0.005 M $CdCl_2$... can lead to a fivefold increase in rat lung cellularity at 4 days" (Asvadi and Hayes, 1978).
$N \rightarrow P$ *Initiation:* Increase flux of initiated cells, *Nap.*	• Indirect genotoxic damage ($a\uparrow$) may occur through Cd-induced ROS production, e.g., in mitochondria (Wang et al., 2004).
$P \rightarrow I$ *Initiation:* Increase flux of initiated cells, *Nap.*	• *Indirect genotoxic damage* ($r\downarrow$, $p\uparrow$) occurs through the inhibition of DNA repair, e.g., by the downregulation of OGG1. • Mitogenic stimuli ($f\uparrow$) by the modulation of gene expression and signal transduction (Waisberg et al., 2003).
$I\uparrow$ *Promotion:* Proliferation of initiated cells, due to decrease in $(d - b)$. *Note:* Cd induction of proto-oncogenes may be ROS-mediated (Joseph et al., 2001).	• In Balb/3T3 A31 cells *in vitro*: "Cd may activate oncogenes such as c-myc, mdm2, and cellular tumor antigen p53" [$b\uparrow$ and/or $g\uparrow$ in Fig. 10.1].

Table 10.2 (continued)

Effect (Symbol/Meaning)	Biological mechanisms and evidence for cadmium
	• Inhibit tumor suppressor genes such as wild-type p53 and p27 [$d\downarrow$ and/or $e\downarrow$ in Fig. 10.1], and thus "accelerate the proliferation of initiated cells" (Fang et al., 2002). "Apoptotic cell death, following an oxidant challenge, is significantly attenuated in Cd-adapted cells" (Hart et al., 2001) [$d\downarrow$].
$I \rightarrow M$ *Progression* of initiated to malignant cells.	• Cd-induced cell transformation may in part result from gene amplification of c-myc and c-jun. Some Cd-transformed cells may possess neoplastic potential resulting from genomic instability (Spruill et al., 2002) [$c\uparrow$].
$M\uparrow$ *Growth* of malignant clone.	• "Suppressed apoptosis could leave pre-neoplastic or neoplastic cells alive, favor their clonal expansion, and ultimately promote tumor development (Hart et al., 2001) [$e\downarrow$].
$M \rightarrow T$ *Transformation* of malignant clone.	• Cd can transform cell lines to malignant types *in vitro* (Abshire et al., 1996; Takiguchi et al., 2003) [$h\uparrow$].
Note: These mechanisms are still hypothetical.	• "Cd-induced disruption of E-cadherin dependent cell–cell junctions may represent a pivotal step in both the initiation of cancer by cadmium and in the tumor promoting actions of cadmium" (Pearson and Prozialeck, 2001).

However, it can indirectly induce oxidative damage in cells and stimulate the production of reactive oxygen species (ROS) in mouse peritoneal macrophages *in vivo* (Ramirez and Gimenez, 2003) and in liver, brain, and heart cell mitochondria (perhaps by the transfer of an electron to O_2 to form superoxide) (Wang et al., 2004; Pourahmad et al., 2003). Cd adaptation in cultured alveolar epithelial cells reduces oxidant-induced DNA damage (but also DNA repair capability), but such oxidative damage occurs in nonadapted cells exposed to Cd (Potts et al., 2001).

Under some conditions, Cd depletes glutathione (GSH) and protein-bound sulfhydryl groups, thus increasing the net production of ROS such as superoxide ion, hydroxyl radicals, and H_2O_2 that can damage DNA, alter gene expression, and activate protein kinase C (PKC) cell signaling (Stohs et al., 2001). However, Cd exposure increases GSH in rat lung cells (Shukla et al., 2000). In humans, chronic smokers typically have increased GSH levels (Rahman and MacNee, 1999). Enzymes with substrates in the GSH metabolic pathway have been associated with susceptibility to lung cancer risk among smokers (Yang et al., 2004; Jourenkova-Mironova et al., 1998). For example, it has been reported that the risk of lung cancer increases steeply with pack-years of smoking among GSTT1-null smokers, but not among other smokers (Hou et al., 2001). Such polymorphism data are still being clarified and disputed, however.

ROS production and oxidative damage have been considered important in the toxicity of both mainstream and sidestream cigarette smoke (St. Clair et al., 1994) as well as of cadmium *in vitro* (Yang et al., 1997). It is not yet clear just how strong a role such damage actually plays in lung carcinogenesis, nor how much incremental difference Cd makes to ROS production (and to related processes, such as GSH stimulation or depletion) in the overall context of cigarette smoking. Oxidants in cigarette smoke (probably including oxidants generated by the reactive metabolites hydroquinone and catechol) are associated with tumor promoter activity. They can activate PKC signaling and contribute to increased tumor cell invasion and metastasis of cancer cells to mouse lungs (Gopalakrishna et al., 1994). In general, both Cd exposure and smoking may induce proliferation in lung cells, at least in part through oxidative damage.

Cadmium Inhibits DNA Repair and Is a Co-Carcinogen for PAHs

A common form of ROS-induced oxidative damage to DNA is the highly mutagenic 7,8-dihydro-8-oxoguanine adduct, which is commonly repaired by the base excise repair enzyme 8-oxoguanine glycosylase I (OGG1) before the damage is locked in by cell division. The loss of heterozygosity of the OGG1 gene is frequently observed in lung tumors (Chevillard et al., 1998) and other cancers (Shinmura and Yokota, 2001), and genetic polymorphisms in the OGG1 gene have been associated with an increased risk of lung cancer in several studies (see Table 10.3). In OGG1 knockout mice, lung tumors develop spontaneously after about 18 months, at a rate five times higher than in wild-type mice (unless another gene, mth1, whose corresponding

Table 10.3 Survey of selected results on genetic polymorphisms and lung cancer risks

Polymorphism	Presumed role	Lung cancer relative risks (1 = no effect)	References
• CYP1A1-Ile462Val	Phase I metabolic activation of carcinogens	3 (1.51–5.91) for lung cancer, 5 (2.03–11.6) for adenocarcinoma	Hung et al. (2003)
		1 for adenocarcinoma	Sunaga et al. (2002)
• CYP2A13 variant allele		0.41 (0.23–0.71) for adenocarcinoma	Wang H. et al. (2003)
• CYP2A6*4 variant		0.23 (0.08–0.67) for lung cancer, no scc or sc were observed	Ariyoshi et al. (2002)
NAD(P)H quinone oxidoreductase (NQO1) NQO1-Pro/Pro vs. Ser/Ser	Activate carcinogens	2.49 (1.17–5.32) for adenocarcinomas in smokers only	Lin et al. (2003)
		2.15 for adenocarcinoma	Sunaga et al. (2002)
Manganese superoxide dismutase (MnSOD)	Decrease ROS	1	Lin et al. (2003)
Glutathione-S-transferase P1 (GSTP1)	Detoxify carcinogens	1.63 SCC (0.96–2.74)	Lin et al. (2003)
		1 for lung cancer	Schneider et al. (2004)
		1 for lung adenocarcinoma	Wang et al. (2003)
GSTT1-null vs. -positive	Detoxify carcinogens	1.69 (1.12–2.56) for lung cancer	Chan-Yeung et al. (2004)
		1.61	Sunaga et al. (2002)
		2.4 (1.31–4.41) for lung cancer; 5 for scc	Sorensen et al. (2004)
		1 for lung cancer	Salagovic et al. (1998)
		1 for lung adenocarcinoma	Wang et al. (2003)
GSTM1-null	Detoxify carcinogens	1.7 for lung cancer (1.0–2.7)	Salagovic et al. (1998)
		0.32 (0.12–0.82) for scc	Lewis et al. (2002)
GSTM1 null & GSTP1	Detoxify carcinogens	2.4 for lung adenocarcinoma (1.1–5.1)	Wang et al. (2003)
		6.9 for sc (1.6–30.2)	Stucker et al. (2002)

Table 10.3 (continued)

Polymorphism	Presumed role	Lung cancer relative risks (1 = no effect)	References
GSTM1-null genotype & CYP2E1 wild type		3.0 for lung adenocarcinoma	Wang et al. (2003)
OGG1-Ser326Cys 2 hOGG1 326Cys alleles OGG1 variants	DNA repair	1 for adenocarcinoma 2.1 (1.2–3.7) possible for lung cancer 1.9 (1.2–2.9) for non sc lung cancers	Sunaga et al. (2002) Le Marchand (2002) Park et al. (2004)
p53, codon 72	Tumor suppressor	2.47 (1.34–4.54) for one nucleotide polymorphic allele (Pro); 3.88 (1.16–13.39) for ProPro 1.45 (1.01–2.06)	Irarrazabal et al. (2003) Fan et al. (2000)
TP53 ProPro genotype		16.4 (1.77–151.7) for heavy smokers	Hiraki et al. (2003)

Key: scc = squamous cell carcinoma; sc = small cell carcinoma; 95% confidence intervals in parentheses.

enzyme is involved in hydrolyzing 8-oxo-dGTP, is also knocked out) (Sakumi et al., 2003). Preliminary epidemiological evidence suggests that OGG1 may also modulate lung cancer risks from PAH-rich coal emissions in humans *in vivo* among nonsmoking but highly exposed women in China (Lan et al., 2004).

Cd exposure inhibits OGG1-mediated DNA repair in several experimental systems. A single exposure of adult male Lewis rats to Cd aerosol has been shown to cause a time- and dose-dependent downregulation in the pulmonary levels of OGG1 *in vivo* (Potts et al., 2001). In cultured alveolar epithelial cells, Cd inhibits the repair of H_2O_2-induced oxidative DNA damage (Potts et al., 2001), in part by downregulating OGG1 expression (Potts et al., 2003) and inhibiting the base excision repair enzymes OGG1 and endonuclease III (Hart et al., 2001).

In vitro, noncytotoxic levels of Cd enhance the genotoxicity of direct-acting mutagens (Beyersmann and Hechtenberg, 1997), including benzo[a]pyrene [B(a)P] and perhaps other polycyclic aromatic hydrocarbons (PAHs) found in cigarette smoke, by inhibiting the repair of relevant B(a)P-DNA adducts (decreasing r in Fig. 10.1). Cd may also alter p53 and p21 growth regulation and hasten the mitosis of damaged cells, in effect, increasing f as well as decreasing r in Fig. 10.1 (Mukherjee et al., 2004). Cd inhibits the repair of benzo[a]pyrene diol epoxide (BPDE)-DNA adducts, which inactivate the p53 tumor suppressor gene and are the PAH adducts most clearly associated with an increased lung cancer risk (Rojas et al., 2004; Li et al., 2001b). Specific *G* to *T* transversion mutations at "hot spots" of the p53 gene occur commonly in smoking PAH-induced lung cancers, but relatively rarely in other cancers (Hernandez-Boussard and Hainaut, 1998; Alexandrov et al., 2002); they are specifically associated with BPDE damage (Lewis and Parry, 2004). BPDE-DNA adduct and PAH levels are elevated two- to threefold in the lungs of smokers compared to nonsmokers (Lodovici et al., 1998). Thus, by slowing the repair of BPDE-DNA adducts (Mukherjee et al., 2004) or other DNA damage, Cd may contribute to the carcinogenicity of cigarette smoke. The quantitative impact of this mechanism has not been determined, however.

This co-carcinogenicity of Cd may help to explain why epidemiological studies do not always find a clear relation between cumulative Cd exposure and the risk of lung cancer. If the risk of lung cancer is driven primarily by contemporaneous exposures to other direct-acting carcinogens, then measuring Cd exposure levels will not provide adequate information from which to predict the increase in lung cancer risk that it enables. In addition, Cd may have some tumor-inhibiting effects (Waalkes and Diwan, 1999) as well as carcinogenic and co-carcinogenic effects, and these effects may not be related in any simple way to cumulative exposure.

Quantifying Potential Cadmium Effects on Lung Cancer Risk

Table 10.2 provides qualitative information on how Cd might increase lung cancer risks. Such information is useful for the hazard identification component of risk assessment. This section undertakes a more speculative quantitative calculation of how large these risks might be, using the organizing framework in Fig. 10.1 and evidence from genetic polymorphism studies on the approximate magnitudes of different biological effects in increasing lung cancer risks.

Polymorphism Evidence on Lung Cancer Risks from Different Mechanisms

Table 10.3 summarizes results from several studies that examine the relative risks of lung cancer among people with various types of genetic polymorphisms. While there are still inconsistencies among studies and some results and interpretations are disputed, it appears likely that genotypes with (a) a relatively high *metabolic activation* of carcinogens [e.g., via wild-type cytochrome (CYP) and/or production of ROS], (b) a reduced capacity to sequester or *detoxify* them (e.g., due to GSTT1 null polymorphism, perhaps in combination with other glutathione-inhibiting polymorphisms), (c) a diminished *DNA repair* capacity (e.g., from OGG1 variants), and/or (d) reduced or inactivated *tumor suppression* (e.g., p53) defenses, have an increased risk of lung cancer in general and of smoking-associated lung cancer in particular. To the extent that exposures to carcinogens reproduce or mimic these effects, the relative risks associated with various combinations of polymorphisms (the third column in Table 10.3) may provide a useful rough indication of the risk increases to be expected from exposures.

Other genetic polymorphisms not shown in Table 10.3, such as myeloperoxidase (MPO) gene mutation, microsomal epoxide hydrolase (EH) polymorphism, and polymorphisms in repair genes other than OGG1, have also been associated with significant increases in lung cancer risk. The intent of Table 10.3 is not to list all such polymorphisms, but to provide initial information (and ranges of disagreement) about the approximate sizes of effects associated with changes in the key processes of carcinogen activation, removal, and damage control (DNA repair or tumor suppression) that are plausibly also important in Cd-induced lung carcinogenesis. The main conclusion is that most individual effect (relative risk) estimates are between about 1.5 and 4, although some fall outside this range.

Frequencies of the genetic polymorphisms in Table 10.3 may differ for smokers and nonsmokers. For example, people with intact CYP2A6 alleles that make them especially susceptible to CYP-activated carcinogens may also be disproportionately likely to smoke [Fujieda et al., 2004; Ariyoshi et al., 2002, for male Japanese. Loriot et al. (2001) present contrary analysis and findings for a French population.] Thus, the carcinogenic effects of smoking, including any Cd-mediated ones, may take place in hosts who already have one or more predisposing factors from Table 10.3 favoring the development of lung cancer.

Quasi-Steady-State Analysis

For purposes of quantitative risk assessment, it is usual to analyze TSCE models and their generalizations by focusing on the probability distribution of the *first passage times* of cells through the network of compartments representing the model. Appendix A offers a simplified *steady-state analysis* of cell population changes induced by smoking, coupled with a probabilistic analysis of the remaining, relatively slow, changes, which are modeled as one-time shifts in corresponding parameter values from their previous (nonsmoking) levels to new (smoking or smoking-without-Cd) levels. This *quasi-steady-state analysis* is intended as a potentially useful heuristic for estimating the potential contribution of Cd to a chronic smoker's risk of lung cancer. It assumes that smoking-induced changes in model parameter values take place in a relatively small (negligible) fraction of the time that smoking lasts, and then remain in effect at their new levels for the duration of the smoker's life. Thus, dynamics and transients arising from interruption or cessation of chronic smoking are not modeled.

When a person begins chronic smoking, several changes occur in lung cell populations, including a roughly threefold initial increase in alveolar cellularity by some measures (Mancini et al., 1993); changes in metabolizing enzymes (CYP, GST/GSH, MPO, EH, etc.); an eventual increase in PAH-induced genotoxic damage (e.g., BPDE-DNA adduct formation) of perhaps two- to fourfold in lung cells (Lodovici et al., 1998); increases in p53 inactivation (at least in a substantial proportion of smokers); and alterations in DNA repair, apoptosis, and proliferation rates, and adaptation of surviving cells. The sequence and timing of these changes and their implications for cancer risk may depend on an individual's genotype, age, and co-exposure history, as well as on smoking intensity – and perhaps, for some of the changes, on the levels of Cd in the lung. Cd levels in the lung gradually increase as smoking continues, eventually reaching levels nearly an order of magnitude greater than in nonsmokers (Paakko et al., 1989). As Cd levels in the lung increase, they may cause ("switch on") Cd concentration-dependent effects in susceptible lung cells.

Table 10.4 outlines the main effects of smoking, interpreted in terms of the model parameters in Fig. 10.1, and suggests rough estimates of the plausible sizes of these effects using a relative risk (RR) framework, discussed further in Appendix A. The rightmost column uses the same relative risk framework to describe possible reductions in the smoking-induced risks of lung cancer from eliminating Cd from cigarette smoke. The relative risk numbers in Table 10.4 are to be interpreted as follows. The third column gives estimated increases due to smoking for the Fig. 10.1 model parameters N, a, p (and its components r and f), c, $1/(c + d - b)$, and $1/(h + e - g)$. Increases are expressed via relative risk factors of the form $RR = (1 + x)$. The product of these factors is the total relative risk due to smoking. The fourth column in Table 10.4 gives estimates of the factors by which these factors might be reduced if Cd were removed from cigarette smoke. Multiplying them gives the total reduction ($RR < 1$) in lung cancer risk for a smoker of non-Cd cigarettes compared to a smoker of current cigarettes. For example, based on the estimated values (1, 1, 0.75, 0.9, 0.8, 0.8) for fractional reductions in $[N, a, p, c, 1/(c + d - b), 1/(h + e - g)]$, respectively, if Cd were removed, the total estimated reduction in smoker's risk would be $RR = 0.75 * 0.9 * 0.8 * 0.8 = 0.43$, i.e., the production of tumorigenic cells would be predicted to fall by more than half. (Changes in h are not included in Table 10.4, because changes in this process can be thought of as primarily affecting the latency period, rather than the total production of tumor cells, unless the process

Table 10.4 Initial estimates of smoking and Cd relative risk factors, by mechanism

Change in parameter	Interpretations and mechanisms	Suggested effect of chronic smoking on lung cancer relative risk (1 = no effect)	Suggested effect if Cd is removed (1 = no effect)
$N\uparrow$	Proliferative stimulus	1–3? (Asvadi and Hayes, 1978; Mancini et al., 1993)	RR = 1? (Cd swamped by other smoke effects)
$a\uparrow$	Preinitiation rate, e.g., from ROS production	2 (based on BPDE adducts, Lodovici et al., 1998)	RR = 1? (See Ramirez and Gimenez, 2003)
$p\uparrow$	p = repair failure probability = $f/(f + r)$	Net effect for p: RR = 2?	RR = 0.75? (Assumes that removing Cd would reduce total flux p^*a^*N by 25%, from 2x to 1.5x nonsmoking level)
$r\downarrow$	DNA repair inhibition	• 2, estimated from OGG1 polymorphism (Park et al., 2004; Le Marchand 2002) and OGG variability (Paz-Elizur, 2003) data	<1? (Waisberg et al.,2003)
$f\uparrow$	Mitogenic stimulus	• >1? (Waisberg et al., 2003)	
$c\uparrow$	Initiated cells progress	RR = 1.5 for all smoking? (Hazelton et al., 2001)	RR = 0.9 (Spruill et al., 2002)
$b\uparrow$	Initiated cells proliferate	>1? (Hazelton et al., 2001)	(Hart et al., 2001; Fang et al., 2002; Joseph et al., 2001)
$d\downarrow$	Initiated cells survive		
$1/(c + d - b)$	*Net effect: I*↑	Net effect: RR = 1.5?	RR = 0.8
$g\uparrow$	Malignant clone expands	Net effect: 1.5–4 based on analogy to effects of p53 polymorphism (e.g., Irarrazabal et al., 2003)	
$e\downarrow$	Malignant cells survive		RR = 0.8? (Hart et al., 2001; Joseph et al., 2001 for ROS-induced proto-oncogenes)
$1/(h + e - g)$	*Net effect: M*↑		
$h\uparrow$	Malignant cells transform to aggressive tumor cells	Reduces time to tumor (Minna et al., 2002)	(Abshire et al., 1996; Takiguchi et al., 2003)
$h\downarrow$	Tumor cells killed by Cd	Increases time to tumor	(Waalkes and Diwan, 1999)

is right-censored by the host's death.) The possibility that Cd may kill tumor cells (Waalkes and Diwan, 1999) is noted, but the emphasis of the model is on preventing the production of tumor cells, rather than on attempting to use Cd to kill them.

The effects estimates in the two rightmost columns of Table 10.4 are intended as informed guesses that are perhaps more likely to underestimate than to overestimate the health benefits of removing Cd (since, unless reducing Cd will create a significant reduction in health risks, it may be prudent to invest resources elsewhere). Given this conservative bias, the potential effects on smoker risk of removing Cd from tobacco products were estimated as follows:

- *N:* RR = 1 for Cd effects on *N* (i.e., no effect on number of stem cells at risk).
- *a:* Similarly, the value of RR = 1 for *a* reflects an assumption that increases in initiation rates (e.g., due to oxidative DNA damage caused by Cd-induced ROS) from cadmium are negligible. *In vitro* data (e.g., Ramirez and Gimenez, 2003) give some reason to question this assumption, as indicated by the question mark after it in Table 10.4. (Removing Cd might reduce *a*.)
- *p:* The *in vitro* evidence that Cd suppresses DNA repair is strong. This may account for at least a twofold increase in the lung cancer RR among smokers, if the effect is comparable to that from an OGG1 polymorphism that also inhibits DNA repair. If Cd accounts for 25% of this total effect, then reducing Cd might decrease *p* to 75% of its preremoval value (RR = 0.75) due to improved DNA repair. In addition, the mitogenic effects of Cd on damaged cells may increase *f* and hence increase *p* (Waisberg et al.,2003), but because of the uncertainties in extrapolating from *in vitro* to *in vivo* settings, Table 10.4 takes no credit for this possibility, treating *f* as unchanged by Cd.
- N^*a^*p: Combining the preceding three parameters, we assume that removing Cd could decrease the production of initiated cells per unit time by 25%. Although this effect is shown as being mediated by a 25% decrease in *p*, the product form makes it unnecessary to commit to this specific hypothesis (i.e., the decrease in N^*a^*p is all that matters, not the specific decreases in each of its three components).
- $c/(c + d - b)$: Under certain conditions (Appendix A), including a positive denominator, this ratio can be interpreted as the expected number of malignant cells eventually produced per initiated cell entering *I*. It may be interpreted roughly as the probability that an initiated cell (or the clone descending from it) survives and eventually becomes malignant. It has an estimated reduction from removing Cd of 0.9*0.8 = 0.72. This reflects considerable *in vitro* evidence that Cd both increases *c* and *b* and reduces *d* (and hence $d - b$). The traditional TSCE model applied to epidemiological data also suggests that Cd exposure may increase the proliferation and transformation of initiated cells (Stayner et al., 1992). Thus, several lines of evidence suggest that Cd increases this parameter, so that removing Cd may reduce it.
- $1/(h + e - g)$: The value of 0.8 assumed for the reduction in this ratio if Cd is removed reflects the fact that a smaller fraction of malignant cells is expected to become fully tumorigenic in the absence of Cd. However, this is based primar-

ily on *in vitro* experimental data, and the magnitude of the effect *in vivo* in the context of smoking is very uncertain at present.

In summary, the effects estimates in Table 10.4 are based on approximations and judgments made in light of data that are, in many cases, informative but far from conclusive. The results suggest, but do not prove, that there may be a significant reduction in lung cancer risk by removing Cd from cigarette smoke.

A Portfolio Approach to Estimating the Preventable Fraction of Risk for Cd

To further assess the uncertain fraction of smoker lung cancer risk that might be prevented by removing Cd, while allowing for the possibility that some or all of the postulated effects in Table 10.4 (especially those based on *in vitro* evidence and animal data) may turn out not to hold for human smokers *in vivo* under real conditions, it is useful to view the multiple effects as a *portfolio* of potential causal impacts. The components of the portfolio are individually uncertain. Yet they may collectively represent an opportunity to achieve a significant reduction in risk by removing or reducing Cd in cigarette smoke. Using a portfolio perspective, simple calculations can be used to bound the uncertain health impact of removing Cd.

To illustrate, suppose that removing Cd is judged to have at least a 10% subjective probability of achieving at least a 10% reduction in each of the following five factors [N, a, p, $c/(c + d - b)$, $h/(h + e - g)$] independently (since it affects each of them, if at all, by different biological mechanisms, as shown in Table 10.4). Given no other assumptions, the probability that at least one of these five factors will in fact turn out to be affected by Cd removal is

$$\text{Pr(at least 10\% risk reduction)} = 1 - \text{Pr(no effect)} = 1 - (0.9)^5 = 0.41.$$

Thus, even if we assume that there is a 90% probability that each separate factor will turn out to be completely unaffected by removing Cd, there is still a better than 40% probability that the risk to smokers caused by the portfolio of Cd effects will be significantly reduced (combined RR ≤ 0.9) by removing Cd.

More sophisticated probabilistic modeling (e.g., addressing the nine individual parameters N, a, b, c, d, e, f, g, and r) could generate more detailed probabilistic descriptions of the likelihoods of effects of different sizes. Even without such calculations, however, a key point is clear: Because Cd potentially acts through many separate parameters to increase cancer risk, it is plausible that removing Cd from cigarette smoke could have a significant beneficial impact on reducing risk. This conclusion holds even though there is a great deal of uncertainty about the validity of each proposed mechanism considered in isolation. (This conclusion could fail if other constituents of cigarette smoke acted through and saturated the same biological mechanisms as Cd, but this does not appear to be the case based on the mechanisms in Table 10.4.)

Discussion and Conclusions

This chapter has focused on quantifying the uncertain effects on lung cancer risk of removing cadmium from cigarette smoke, given the incomplete scientific information currently available about its mode(s) of action. Building on earlier stochastic models of carcinogenesis, the QRA model proposed here represents smoking effects on steady-state normal stem cell numbers; the inhibition of DNA repair, initiation, and competition between mitosis and DNA repair in determining an initiated cell's fate; the proliferation, promotion, and progression of initiated cells; and the death or sparing of initiated and malignant cells as they are further transformed to become fully tumorigenic.

Rather than estimating unmeasured model parameters by fitting model-predicted curves to epidemiological or animal tumor data, this chapter has relied on initial rough estimates of parameters based on their biological interpretations and on the literature describing the biological mechanisms and effects among humans and animals exposed to Cd compounds and to cigarette smoke. Genetic polymorphism data were used to suggest plausible ranges of quantitative health impacts for particular biological mechanisms. The resulting parameter estimates are admittedly uncertain and approximate. Yet they suggest that viewing Cd as creating a portfolio of uncertain health impacts, expressed as biologically independent relative risk factors having clear mechanistic interpretations, can provide useful estimates of the likely impacts of removing Cd. Because Cd can act through many distinct biological mechanisms, it appears plausible (subjective probability greater than 40%) that removing Cd from cigarette smoke would reduce smoker risks of lung cancer by at least 10%, although it is possible (i.e., consistent with what is known with confidence) that the true effect could be much larger or smaller (or zero). Conservative estimates and assumptions made in this calculation suggest that the true impact could be greater for some smokers.

This conclusion – that a substantial reduction in risk if Cd is removed is plausible but not certain – is robust to several scientific uncertainties about cadmium and smoking effects. It generally is consistent with earlier conclusions from the regulatory risk assessments of Hertz-Picciotto and Hu (1994), although using very different reasoning, data, and modeling methods. The conclusion that there is a subjective probability of over 40% that removing Cd from cigarette smoke would reduce smoker risks of lung cancer by at least 10% indicates both the high uncertainty about the probable consequences and the fact that a significant reduction in risk is plausible in light of current knowledge. This conclusion amounts to little more than a quantitative expression of "Maybe!" as an answer to the question of whether removing Cd would reduce the risk substantially (e.g., by 10% or more). It is a relatively weak conclusion, compared to the quantitative bounds on risk provided by biomarker data and causal modeling in Chapters 8 and 9. But it may be the strongest answer warranted by the current very incomplete knowledge, and it may provide enough information to help decide whether it is worthwhile to investigate further the potential human health benefits of reducing cadmium in cigarette smoke.

Appendix A: Relative Risk Framework

Figure 10.1 corresponds to the following system of mass-balance ordinary differential equations (using deterministic equations for the means of the underlying stochastic variables and focusing on expected values and rates of increase in model quantities I, M, and T, rather than on first passage times):

- $N =$ constant (with a value that may depend on smoking),
- $dP/dt = a^*N - (r + f)P(t)$,
- $dI/dt = f^*P(t) - (c + d - b)I(t)$,
- $dM/dt = c^*I(t) - (h + e - g)M(t)$,
- $dT/dt = h^*M(t)$.

The first of these equations sets the size of the normal lung stem cell compartment to a constant, N, that is assumed to be homeostatically maintained and that may depend on smoking behavior. Thus, if transients are ignored, normal stem cell numbers are assumed to remain steady at some level, N, during smoking (reflecting assumed homeostasis). While there is good evidence that smoking increases lung cellularity, it is not clear exactly how this affects the numbers of lung stem cells at risk of carcinogenic transformations. Hence, in Table 10.4, this entry ranges from 1 (no smoking-induced amplification) to 3 [the value of an early increase in cellularity noted by Mancini et al. (1993), although not specifically for stem cells].

The remaining equations express the identity that the rate of growth in each compartment at each moment is the difference between the total inflow from all sources (typically, cell births in the compartment plus new immigrations from the preceding compartment) and the outflows to all destinations (typically cell death, differentiation, or transition to a subsequent compartment). Arguably, the rate parameter h is excessively simplified, as the acquisition of fully tumorigenic properties may require over 30 events (both clonal genetic alterations and epigenetic lesions such as promoter hypermethylations) that can occur in different ways and orders and that can lead to different specific histological types of lung cancer (Minna et al., 2002; Wistuba et al., 2001). The simplistic final equation, $dT/dt = h^*M(t)$, emphasizes that the formation of such fully tumorigenic cells, however complex, is driven and limited by the number of early-stage malignant cells from M entering the process.

Rather than solving the above dynamic system for time-varying values of $P(t)$, $I(t)$, $M(t)$, and $T(t)$, we consider steady-state solutions for P, I, and M by setting their time derivatives equal to zero and solving the resulting algebraic system, yielding

- $P = a^*N/(r + f)$,
- $I = [f^*P(t)/(c + d - b)] = (p^*a^*N)/(c + d - b)$ [using the definition $p = f/(f + r)$],
- $M = [c^*I(t)/(h + e - g)] = (c^*p^*a^*N)/[(c + d - b)^*(h + e - g)]$
- $= (p^*a^*N)/\{[1 + (d - b)/c]^*[(h + e - g)]\}$.

These solutions are physically meaningful if the initiated and early malignant cell populations do not spontaneously grow without bound (as fully tumorigenic cells would do) but instead tend to become extinct, i.e., if $c + d > b$ and $h + e > g$. (This differs from the assumptions of most TSCE models, in which the net birth rates are assumed to remain positive. For lung cancer, premalignant clonal expansion and *in situ* carcinoma both appear to have self-limited growth.) However, the preceding formulas provide key components of the transient solutions even if the net birth rates are positive (Hazelton et al., 2005). If the net birth rates are negative, so that $c + d > b$ and $h + e > g$, then the formulas are well motivated; otherwise, they provide only a heuristic guide to the contributions of different mechanisms to cancer risks, and a time-varying analysis is required for more accurate results. Subject to these caveats, the expected steady-state production of new tumorigenic cells per unit time can be written as a product:

$$\begin{aligned}&\text{production rate of tumorigenic cells} = dT/dt = h^{*}M(t)\\&= (p^{*}a^{*}N)^{*}[c/(c+d-b)]^{*}[h/(h+e-g)]\\&= (p^{*}a^{*}N)/\{[1+(d-b)/c]^{*}[1+(e-g)/h]\}\end{aligned}$$

An increase in any of the components p, a, N, $[c/(c + d - b)]$, or $[h/(h + e - g)]$ by a factor of $(1 + x)$ will multiply the production rate of malignant cells by the same factor, motivating the use of relative risk factors in Table 10.4. Interpretively, $(p^{*}a^{*}N)$ is just the flux of new initiated cells created per unit time, while $[c/(c + d - b)]^{*}[h/(h + e - g)]$ reflects the expected fraction of tumor cells eventually exiting the promotion-progression pipeline for each initiated cell entering it.

Chapter 11
Determining What Can Be Predicted: Identifiability

One of the best developed ways to predict how changing inputs to a complex system will change its probable outputs is to *simulate* the behavior of the system. Modern simulation modeling software environments (such as MATLAB/SIMULINK®, or STELLA/ITHINK® for continuous simulation, and SIMUL8® for discrete-event simulation) make the mechanics of simulation model building and use relatively straightforward. Stochastic simulation risk models have been developed for business, engineering, biological, social, and economic systems. (Agent-based simulation models have also been developed for complex social and economic systems, but this chapter focuses on continuous simulation.)

Before a system can be simulated, it must be described. Systems are typically described in simulation models as compositions of subsystems or components, linked by input-output flows and/or information signals, with known input-output specifications. (These may include conditional probabilities of outputs given inputs, in stochastic models.) Even if the parameters describing the input-output relations for model components or subsystems are not known, however, it may be possible to identify them from input-output data for the system as a whole. Whether this can be done – whether model parameters can be uniquely determined, even in principle, from sufficiently large and diverse samples of observed system input-output data – is the issue of *identifiability*.

This chapter examines identifiability for a stochastic dynamic model of lung carcinogenesis. The major goal is to estimate exposure-dependent transition rates for cells among different compartments of the model, together with net exposure-dependent proliferation rates for premalignant cells, using a combination of (a) biological knowledge and assumptions about the structure of the system (its compartments and transitions among them) and (b) aggregate epidemiological data on observed causal relations between smoking histories and increased risks of lung cancer. The main technical challenge is how to use a combination of knowledge and data to constrain possible parameter values, and how to answer the mathematical question of whether the constraints determine the parameter values uniquely. It turns out that the answer is "almost": it is not possible to identify the model parameters uniquely using only high-level epidemiological data, but it *is* possible to narrow down the possibilities to a few discrete possibilities (only two alternatives, in the

L.A. Cox, Jr., *Risk Analysis of Complex and Uncertain Systems*, International Series in Operations Research & Management Science 129, DOI 10.1007/978-0-387-89014-2_11,

lung cancer model presented here). Even a small amount of additional, lower-level biological information then suffices to figure out which possibility is correct.

Identifiability

The problem of the identifiability of complex disease models from data is fundamental in biologically based risk assessment, disease modeling, and many other areas of applied risk assessment. Identifiability deals with whether model parameters and corresponding input-output relations and predictions can be uniquely determined from observations under ideal conditions of arbitrarily many, arbitrarily accurate, observations. If the answer is no, then no amount of statistical ingenuity can uniquely determine the values of model parameters from real data.

Example 1: A Simple Example of Nonidentifiability

Suppose that the administered dose, X, affects the internal dose, Y, and that Y in turn affects the mutation rate Z in some population of cells, via the following structural equations:

$$Y = a^* X,$$

$$Z = b^* Y.$$

Here, a and b are parameters to be estimated from data. If only the input X and the output Z are measured, then it is impossible to uniquely identify the separate values of a and b from the observed pairs of (X, Z) values. Only the *reduced parameter* given by their product, $c = a^*b$, can be uniquely identified. (Indeed, $c = a^*b$ can be estimated as the regression coefficient in the least-squares regression of Z values against X values with zero intercept.) However, this reduced parameter suffices to predict Z values from X values via the reduced-form equation

$$Z = c^* X.$$

Thus, even though a and b cannot be uniquely identified from (X, Z) data, the input-output relation between X and Z is uniquely identifiable from such data. This is all that is needed for many practical purposes.

Example 2: Unique Identifiability in a Two-Stage Clonal Expansion Model

In the following example, cells independently and randomly make transitions from a "normal" state ("compartment"), N, into an intermediate compartment, I, and thence

$$\begin{array}{ccccc} & & \Downarrow b & & \\ N & \Rightarrow & I & \Rightarrow & M \\ & \mu_1 & & \mu_2 & \\ & & \Downarrow d & & \end{array}$$

Fig. 11.1 A two-stage clonal expansion model (TSCE) for carcinogenesis. μ_1 is the transition rate (per normal cell per unit time) from N to I; μ_2 is the transition rate (per intermediate cell per unit time) from I to M; b is the rate (per intermediate cell per unit time) at which new cells in I are born; and d is the rate (per intermediate cell per unit time) at which cells in I die or make transitions to other states that do not further participate in the process

to a second ("malignant") compartment, M. They may also undergo a net proliferation in compartment I at some rate $g > 0$. Perhaps surprisingly, observing the time course $M(t)$ alone at only a few distinct times is sufficient to uniquely identify several parameters in this model as well as to uniquely predict the future of $M(t)$.

Figure 11.1 depicts the compartments and transitions among them.

A corresponding system of ordinary differential equations (ODEs) for the expected sizes of compartments N, I, and M over time is

$$\begin{aligned} dN(t)/dt &= -\mu_1 N(t), \\ dI(t)/dt &= \mu_1 N(t) + (b - d - \mu_2)I(t) = \mu_1 N(t) + (g - \mu_2)I(t), \\ dM(t) &= \mu_2 I(t), \end{aligned}$$

where

- $N(t)$ = expected number of normal stem cells at time t,
- $I(t)$ = expected number of intermediate cells at time t,
- $M(t)$ = expected number of malignant cells at time t.

The reduced parameter $B = (b - d) - \mu_2 = g - \mu_2$ (where $g = b - d$ is the net birth rate in compartment I) is the net growth rate per intermediate cell per unit time, i.e., the birth rate minus the rate of losses due to death/differentiation and malignant conversion.

Note on the use of unconditional vs. conditional expected values. The use of unconditional expected values, rather than conditional expected values given no tumor cells formed to date, has become fairly common for modeling purposes (e.g., Little, 1995), even though it may be important to use "exact" (conditional) rather than "approximate" (unconditional) formulas in data analysis and statistical inference when fitting stochastic models of carcinogenesis to data (e.g., Hazelton et al., 2005). Once model parameter values have been estimated from data (or exogenously specified in theoretical models where parameter values are treated as known quantities), the resulting model can be used to predict unconditional expected sizes of cell populations over time. Doing so allows direct comparison to some previous models based on epidemiological data that quantify the preceding model parameters (e.g., Schollnberger et al., 2006). The use of the expected number of tumor cells $M(t)$ as an indicator of risk is discussed further in the following section. The main results of this chapter have also been verified via stochastic simulation modeling (using Crystal Ball™) with the age-specific hazard function for time until the first malignant cell, rather than expected number of malignant cells, as the risk variable.

The preceding system of ODEs with initial conditions $(N(0), I(0), M(0)) = (N_0, 0, 0)$ has the time-varying solution (provided B, $-\mu_1$, and 0 have distinct values)

$$N(t) = N_0 e^{-\mu 1t},$$

$$I(t) = N_0\mu_1[e^{Bt}] - e^{-\mu 1t}]/(B + \mu_1),$$

$$M(t) = N_0\mu_1\mu_2[e^{-\mu 1t}/((B + \mu_1)\mu_1) + e^{Bt}/((B + \mu_1)B) - 1/(B\mu_1)].$$

(Exact solutions exist for the "degenerate" cases when two or more of B, $-\mu_1$, and 0 coincide. They can be obtained by taking limits of the generic solution as the value of one variable approaches the other.) Treating the final equation as a nonlinear regression model for $M(t)$ with t as the independent variable, the parameters μ_1 and B can be uniquely estimated from data consisting of (t, $M(t)$) pairs (since μ_1 and B are distinct rate constants in a sum of exponentials plus a constant, and terms in a sum of exponentials are linearly independent). The reduced parameter $N_0\mu_2$ is also uniquely identifiable from (t, $M(t)$) data, although N_0 and μ_2 are not separately identifiable [since only their product affects $M(t)$]. From only three observed values of $M(t)$ at different times, known with sufficient precision, one can in principle uniquely identify the vector (μ_1, B, $N_0\mu_2$), and hence uniquely predict the future time course of $M(t)$ [as well as the time courses of the "hidden" variables $N(t)$ and $I(t)$, apart from the scaling constant N_0]. Similarly, if μ_1, B, and $N_0\mu_2$ are linear functions of an exposure variable x (with distinct slope coefficients), then the six coefficients (slopes and intercepts) relating these reduced parameters to x are also uniquely identifiable from data consisting of x values and (t, $M(t)$) pairs [since (μ_1, B, $N_0\mu_2$) is uniquely identifiable from (t, $M(t)$) data for each value of x].

In practice, confident estimation of the reduced parameters from a few data points may be complicated by numerical problems (ill conditioning of the regression equations) and limited precision of measurements, but mathematical identifiability does not preclude the possibility of estimating the entire time course of $M(t)$ from as few as three data points at nonzero times.

A substantial technical literature has addressed the topic of identifiability in both linear systems (Audoly et al., 1998) and nonlinear systems (Audoly et al., 2001), and a complete understanding of when unique identifiability is possible has been developed for several important classes of compartmental models (ibid.) However, in these models, the available data consist of input-output experiments in which the experimenter adds input quantities (e.g., radioactive tracers) to some compartments and observes the resulting time courses of outputs (e.g., radioactive counts) in one or more compartments. By contrast, this chapter examines identifiability when inputs consist of exposures that affect the pharmacodynamic *transition rates* of cells among compartments and/or cell proliferation rates within compartments. Observed outputs consist only of malignant cells or tumors at the end of a chain of compartments, $M(t)$. There is no opportunity to directly inject exogenous inputs to any compartment; rather, one can only observe (or perhaps change) exposure conditions and collect observations on $M(t)$.

For this setting, we introduce a new identifiability result showing that effects of exposure on transition rates and proliferation rates can be identified as belonging to a *discrete set of alternative possible combinations of values*, based on the types of $(x, t, M(t))$ data discussed in the preceding example. That is, the parameter vector cannot be uniquely identified, but it can be identified as belonging to a discrete set of possible parameter vectors, each of which is fully consistent with all preintervention data. Unlike the innocuous cases of nonunique identifiability in Examples 1 and 2, where the nonuniqueness does not carry over to reduced parameters or to model predictions, these different discrete possibilities can predict significantly different future values for $M(t)$ following an intervention that changes the effects of exposure on one or more transition rates. They thus create discrete uncertainty about how such an intervention (perhaps corresponding to a change in the mixture of carcinogenic exposures caused by a hazardous product, workplace, or environment) will affect exposure-related cancer risks. Such discrete uncertainties (i.e., cases in which cancer risk models can predict that one of a small finite set of quite different possible risks will result, but with no ability to predict which one) are directly relevant to the uncertainty characterization step of quantitative risk assessments for interventions that affect pharmacodynamic dose-response relations.

This result implies that no amount of additional experimental or epidemiological data on age-specific tumor rates for different exposure groups can reduce the discrete uncertainty about how interventions will affect cancer risk. Rather, biological data that reveal which and how much specific transition rates are affected by the intervention are required to uniquely predict which member of the set of alternative possible predictions that are consistent with past data will actually occur.

These results hold for a broad class of *multistage clonal expansion (MSCE)* models that generalize both the two-stage clonal expansion (TSCE) model in Fig. 11.1 and also the traditional linearized multistage (LMS) family of cancer dose-response models (viewed as stochastic transition processes in which cells make random transitions from each stage to the next, as in Armitage and Doll, 1954; Moolgavkar, 1978; Cox, 1995; Little, 1995; Little et al., 2002; and Ritter et al., 2003; see Fig. 11.2). This class of models is described next.

$$\begin{array}{c} \Downarrow g_0 \ \Downarrow g_1 \ \ \Downarrow g_2 \ldots \ \Downarrow g_n \\ N \Rightarrow I_1 \Rightarrow I_2 \ldots \Rightarrow I_n \Rightarrow M, \\ \mu_1 \quad \mu_2 \ldots \quad \mu_n \quad \mu_{n+1} \end{array}$$

Fig. 11.2 General multistage clonal expansion model with n intermediate compartments

Multistage Clonal Expansion (MSCE) Models of Carcinogenesis

Figure 11.2 diagrams the structure of the models of carcinogenesis addressed in this chapter. In this figure, n intermediate compartments intervene between the normal and malignant cell populations. The case $n = 1$ is the TSCE model. (The normal compartment N may also undergo net growth while an individual matures, but it is often assumed to remain approximately constant in adulthood.)

μ_i is the transition rate (per cell per unit time) from the predecessor of compartment I_i into compartment I_i. g_i is the net birth (proliferation) rate for compartment I_i. Proliferation may take place in one or more of these intermediate compartments. If all $g_i = 0$ and all μ_i are linear functions of dose, then Fig. 11.2 reduces to a traditional linearized multistage (LMS) model, meaning a sequence of stages in which the transition rates of cells from each stage to the next are assumed to consist of a background (zero-exposure) rate plus a dose-dependent increase that is proportional to the biologically effective dose at that stage (Armitage and Doll, 1954; Moolgavkar, 1978; Cox, 1995; Little, 1995; Little et al., 2002; Ritter et al., 2003). We will refer to the general class of compartmental models with the structure in Fig. 11.2 as *multistage clonal expansion (MSCE) models* since they combine multiple stages with the possibility of clonal expansion at more than one stage.

A system of ODEs corresponding to Fig. 11.2 is

$$
\begin{aligned}
dN(t)/dt &= \mu_1 N(t),\\
dI_1(t)/dt &= \mu_1 N(t) + (g_1 - \mu_2)I_1(t).\\
dI_k(t)/dt &= \mu_k I_{k-1}(t) + (g_k - \mu_{k+1})I_k(t) \text{ for } k = 2, 3, \ldots, n,\\
dM(t) &= \mu_{n+1} I_n(t).
\end{aligned}
$$

This system describes the dynamic evolution of the expected sizes of the cell populations in the successive compartments, $N(t)$, $I_k(t)$ for each intermediate stage k, and $M(t)$. Only the last compartment, M, gives rise to observable tumors and tumor-related deaths. The use of expected values is motivated as follows. First, fully stochastic simulation (using random jump processes modeled as nonhomogeneous Poisson processes to replace the deterministic changes in expected values) confirms that the above system of ODEs correctly describes the evolution of the mean values of compartment sizes in many realistic situations. Second, since all state variables [$N(t)$, $I_k(t)$, $M(t)$] are nonnegative, Markov's inequality holds. This implies that the random number of malignant cells formed by any time is unlikely to be very much larger than the expected value $M(t)$, in the sense that, when the expected value $M(t)$ is small (e.g., less than 0.10), the probability that one or more malignant cells has been formed is also small (not more than 0.10). [Conversely, if $M(t)$ is much larger than 1, then the probability that no malignant cells have yet been formed will be close to zero. However, this is less relevant to lung cancer.]

The TSCE model has been notably successful in describing a wide variety of epidemiological and experimental data sets (e.g., Hazelton et al., 2005; Schollnberger et al., 2006). Allowing for the possibility of proliferation at more than one stage may therefore be somewhat controversial, in that it suggests a needless complication of the TSCE framework. Yet recent molecular biological data (Park et al., 1999; Pan et al., 2005; Panov, 2005; Wistuba et al., 2000; Wistuba and Gazdar, 2006) strongly suggest that multiple biologically distinct phases of premalignant clonal expansion occur in many lung cancers. Lung cancer biologists have identified an initial "patch" phase in which premalignant gene-level changes, such as the loss of heterozygosity at chromosome 3p (LOH3p) and the hypermethylation of the promoter regions of some tumor suppressor genes, occur and expand clonally at multiple locations within the lung, forming small local patches of altered cells (Park et al., 1999; Wistuba et al., 2000; Wistuba and Gazdar, 2006). The formation of patches is followed by an additional premalignant "field cancerization" phase in which the clonal expansion of cell populations with several additional gene-level alterations occurs (Pan et al., 2005; Panov, 2005; Wistuba and Gazdar, 2006). The clonally expanding field may ultimately come to be widely distributed throughout one or both lungs (e.g., Franklin et al., 1997). [For squamous cell carcinomas, additional premalignant, local proliferation may also occur as carcinoma *in situ*, but this does not necessarily occur for other types of lung cancer, such as adenocarcinomas (Wistuba and Gazdar, 2006).] Such "field cancerization" now appears to be more than a speculative theory, at least for lung cancers, as distinct phases of progressive molecular changes followed by clonal expansion have been observed using molecular biological techniques (Panov, 2005; Wistuba and Gazdar, 2006). Thus, a mathematical model (MSCE) that allows clonal expansion to take place at more than one stage appears to be potentially useful, at least for describing some lung cancers.

Following Schollnberger et al. (2006) and traditional linearized multistage modeling practice, we assume that the biologically effective dose, $s(t)$, affects the model's pharmacodynamic parameters through linear equations:

$$\mu_k(t) = b_k + q_k{}^* s(t),$$

(pharmacodynamics model)

$$g_k(t) = g_{k0} + g_{k1}{}^* s(t).$$

Here, b_k and g_{k0} are the "background" (zero-exposure) values for the transition and growth rates μ_k and g_k, respectively, while q_k and g_{k1} are potency or slope factors for the increases in these rates in the presence of exposure that creates the internal dose $s(t)$. The internal dose $s(t)$ is defined to make the above linear pharmacodynamics equations hold. Nonlinear pharmacokinetic effects such as the saturation or induction of enzymes are accounted for using an internal dose equation or model of the form $s(t) = f[x(t)]$, where f can be a nonlinear function of the administered dose $x(t)$.

The mathematical structure of the family of models investigated in this chapter may be summarized as follows. Define the state vector

$$\mathbf{z}(t) = [N(t), I_1(t), \ldots, I_n(t), M(t)]'$$

and take the initial conditions to be $\mathbf{z}(0) = [1, 0, \ldots, 0]'$ ($'$ denotes transpose). In other words, initially all cells start in the "normal" compartment and the number of cells is scaled so that $N(0) = 1$. [In many models, $N(t)$ is assumed to grow linearly from age 0 to age 20. In this case, we will scale the population size so that $N = 1$ at age 20 and redefine t as years after age 20. This is appropriate if transitions out of compartment N prior to age 20 are small enough to be neglected.] The evolution of $\mathbf{z}(t)$ over later times as a function of doses is described by the following model equations:

$$d\mathbf{z}(t)/dt = \mathbf{A}(t)\mathbf{z}(t) \text{ with initial condition } \mathbf{z}(0) = [1, 0, \ldots, 0]'.$$

The $(n + 2)$ by $(n + 2)$ transition matrix $\mathbf{A}(t)$ is the sum of background and dose-dependent terms:

$$A(t) = \mathbf{B} + \mathbf{Q}s(t),$$

where $\mathbf{B}$ and $\mathbf{Q}$ are the following constant-parameter matrices:

$$\mathbf{B} = \begin{matrix} -b_1 & 0 & 0 & \ldots & 0 \\ b_1 & (g_{10} - b_2) & 0 & \ldots & 0 \\ 0 & b_2 & (g_{20-b_3}) & \ldots & 0 \\ 0 & 0 & b_3 & \ldots & 0 \\ & \ldots & & & \end{matrix}$$

$$\mathbf{Q} = \begin{matrix} -q_1 & 0 & 0 & \ldots & 0 \\ q_1 & (g_{11} - q_2) & 0 & \ldots & 0 \\ 0 & q_2 & (g_{21-q_3}) & \ldots & 0 \\ 0 & 0 & q_3 & \ldots & 0 \\ & \ldots & & & \end{matrix}$$

When exposure generates a constant internal dose rate s between times t_0 and t_1, then $\mathbf{z}(t)$ for any time t between t_0 and t_1 is given by the exact solution

$$\mathbf{Z}(t) = e^{(\mathbf{B}+\mathbf{Q}s)(t-t_0)}\,\mathbf{z}(t_0) \text{ for } t_0 \leq t \leq t_1.$$

In this equation, $e^{(\mathbf{B}+\mathbf{Q}s)(t-t_0)}$ is the matrix exponential function

$$e^{\mathbf{A}} = \exp(\mathbf{A}) = \mathbf{I} + \mathbf{A} + \mathbf{A}^2/2! + \cdots + \mathbf{A}^k/k! + \ldots$$

with $\mathbf{A} = (\mathbf{B} + \mathbf{Q}s)(t - t_0)$. [$\mathbf{I}$ denotes the unit $(n + 2)$ by $(n + 2)$ matrix.] A predictive model for $M(t)$, the expected number of tumor cells formed by time t when exposure is held constant at s and the initial state is $[1, 0, 0, \ldots, 0]'$, is found by extracting the last component of $\mathbf{z}(t)$:

$$M(t) = [0,\ 0,\ \ldots,\ 1]e^{(\mathbf{B}+\mathbf{Q}_S)(t-t_0)}[1,\ 0,\ \ldots,\ 0]',$$

which is the lower left entry in the matrix $e^{(\mathbf{B}+\mathbf{Q}_S)(t-t_0)}$. This is only a symbolic solution, however, since $e^{(\mathbf{B}+\mathbf{Q}_S)(t-t_0)}$ remains to be evaluated. The following result provides an explicit solution to be used for an analysis of identifiability.

Theorem 1. *The solution to the system of ordinary differential equations*
$dz(t)/dt = \mathbf{A}z(t)$
with initial condition $z(0) = [1, 0, \ldots, 0]'$ and a constant transition rate matrix A defined as above is
$z_k(t) = w_k[exp(a_{11}t)/w_{1k} + \ldots + exp(a_{kk}t)/w_{kk}]$,
where w_k is the product of the $(k-1)$ off-diagonal elements $a_{j+1,j}$ for $j = 1, 2, \ldots, (k-1)$ and w_{kk} is the product of the $(k-1)$ differences $(a_{jj} - a_{ii})$ for all $i \neq j$ and $i \leq k$, assuming that all of these differences are nonzero:
$w_k = a_{21}\, a_{32} \ldots a_{k,k\text{-}1}$ for $k > 1$, and $w_1 = 1$,
$w_{jk} = (a_{jj} - a_{11})(a_{jj} - a_{22}) \ldots (a_{jj} - a_{kk})$ for $k > 1$, and $w_{11} = 1$;
the product for w_{jk} excludes the factor $(a_{jj} - a_{jj})$.

Proof. See Appendix A.

[Explicit formulas for any set of initial conditions can be obtained by taking linear combinations of solutions for initial conditions of the form $\mathbf{z}(0) = [0, \ldots, 0, 1, 0, \ldots, 0]'$ (the 1 is in the *i*th place), since ignoring the first *i*–1 rows and columns of **A** allows Theorem 1 to be applied to the remaining subsystem.]

Theorem 1 implies that in an MSCE model with an $r \times r$ transition matrix **A**, the population of the last compartment, denoted by $M(t)$, is given by

$$M(t) = w[\exp(a_{11}t)/w_1 + \cdots + \exp(a_{rr}t)/w_r],$$

where a_{kk} is the *k*th diagonal element of **A**, *w* is the product of the $(r-1)$ off-diagonal elements $a_{k,k-1}$ for $k = 2, 3, \ldots, r$–1 and w_k is the product of the $r-1$ differences $a_{kk} - a_{jj}$ for all $j \neq k$. (For small to moderate *r*, such as $r = 5$, this solution can also be obtained via a symbolic mathematics program, e.g., using the *Mathematica* commands "A = {{a, 0, 0, 0, 0}, {a1, b, 0, 0, 0}, {0, b1, c, 0, 0}, {0, 0, c1, d, 0}, {0, 0, 0, d1, e}}; Simplify[MatrixExp[A t][[5,1]]].")

This expression for $M(t)$ has a high degree of symmetry in the parameters. To describe this symmetry, we will now adopt the more explicit notation $M_{\mathbf{A}}(t)$ to refer to the value of $M(t)$ implied by transition matrix **A**. If σ is any permutation of the indices $(1, 2, \ldots, r)$ and $(b_2, b_3, \ldots, b_r)$ is any *r*–1 tuple for which $b_2 b_3 \ldots b_r = w$, then the matrix $\mathbf{A}^{\sigma}$ obtained by permuting the diagonal elements of **A** according to σ and replacing its off-diagonal elements $(a_{21}, a_{32}, \ldots, a_{k,k-1})$ by $(b_2, b_3, \ldots, b_r)$ gives the same input-output relation as **A**.

Corollary. $M_A(t) = M_{A\sigma}(t)$ *for all t.*

Proof. In the notation of Theorem 1, w_k is the same for both matrices, by construction. The effect of σ on $w_{jk} = (a_{jj} - a_{11})(a_{jj} - a_{22}) \ldots (a_{jj} - a_{kk})$ is simultaneously to change the order of the factors and to replace each a_{jj} by a_{ii} where $i = \sigma(j)$. The result is w_{ik}. Therefore, σ changes $\exp(a_{jj}t)/w_{jk}$ into $\exp(a_{ii}t)/w_{ik}$, implying that it merely alters the order of the summation $\exp(a_{11}t)/w_{1k} + \ldots + \exp(a_{kk}t)/w_{kk}$, thereby leaving its value invariant. This exhibits $M_{\mathbf{A}}(t)$ and $M_{\mathbf{A}\sigma}(t)$ each as the product of w_k and the same sum, so they are equal. QED.

If flows are conserved (e.g., for transitions among compartments without proliferation), then additional constraints relate the diagonal elements of **A** to its off-diagonal ones, restricting the symmetries of the corollary to a subgroup that preserves them. For example, in a four-compartment model ($n = 2$) with conservation of flows (so that $a_{11} + a_{21} = a_{22} + a_{32} = 0$ and $a_{44} = 0$) and all diagonal elements distinct, the only allowed permutations of $(a_{11,} a_{22}, a_{33}, a_{44})$ must fix $a_{44} = 0$ and so may be considered to be permutations of the first three elements. Thus, $M_{\mathbf{A}}(t)$ determines the coefficients of **A** only up to arbitrary permutations of its first three diagonal elements, implying that there are six distinct combinations of parameter values (one for each permutation) that are all perfectly consistent with observed $(t, M(t))$ data.

Nonunique Identifiability of Multistage Models from Input-Output Data

The corollary implies, in the special case that $\mathbf{A} = \mathbf{Q}^*s$ (so that background rates are negligible compared to rates during exposure) or if $\mathbf{BQ} = \mathbf{QB}$ (implying that ratios of exposure-related rates are the same as ratios of background rates), that *the elements of* **A** *cannot be uniquely identified from any possible constant-dose input-output pairs,* i.e., from observations of the form $(x, t, M(t))$ specifying how the expected number of malignant cells, $M(t)$, varies with time or age, t, for different exposure levels, x. [The condition $\mathbf{BQ} = \mathbf{QB}$ includes the case where exposure is modeled as $\mathbf{A} = \mathbf{Q}(\beta + s)$ with β a constant.] Instead, all that can be identified from such data is an unordered set of $n + 2$ diagonal elements a_{jj}, $j = 1, 2, \ldots, n + 2$, and the product w of the off-diagonal elements.

Example 3: Counting 5 × 5 Matrices with Sign Restrictions

Consider a model represented by the 5 × 5 transition matrix

$$A = \begin{matrix} -a & 0 & 0 & 0 & 0 \\ a_1 & -b & 0 & 0 & 0 \\ 0 & b_1 & -c & 0 & 0 \\ 0 & 0 & c_1 & d & 0 \\ 0 & 0 & 0 & d_1 & e \end{matrix}$$

having $n = 3$ intermediate compartments, proliferation taking place only in the penultimate compartment ($d > 0$), and positive flow rates between each earlier stage and its successor (i.e., $a > 0$, $b > 0$, and $c > 0$). The flow balance constraints $-a + a_1 = -b + b_1 = -c + c_1 = e = 0$ imply that e cannot be interchanged with any of $\{a, b, c, d\}$ (since e is zero and they are not), leaving only permutations of $\{-a, -b, -c, d\}$ involved in the symmetries. Since d is positive in this model (due to proliferation) and a, b, c are not, we can distinguish d from the other parameters a priori. The permutations must therefore preserve the partition $(\{a, b, c\}, \{d\}, \{e\})$, yielding $3!^*1!^*1! = 6$ equivalent possibilities altogether. In general, an intervention that affects a subset of the transitions by reducing their dose-dependent components may generate a finite set of alternative predictions, all of which are fully consistent with past data.

These mathematical results for special cases are perhaps of limited practical interest when the special conditions used to prove them (e.g., negligible background rates or constant exposures) do not hold. But simulations suggest that this main result – of a small number of significantly different predictions being equally consistent with all past epidemiological data – also holds to a close approximation even for more realistic nonnegligible background rates and time-varying exposures. More than one model may fit all possible epidemiological data approximately equally well, yet the different models may make significantly different predictions for the effects of interventions that change how exposure affects stage-specific transition rates. We leave as an open problem the further mathematical characterization of this phenomenon, and instead illustrate it with the following simulation example.

Example 4: Two Equally Likely Effects of Reducing a Transition Rate

Setting: Suppose that the age-specific risk for exposed individuals prior to an exposure-reducing intervention is described by the following model. There are four sequential stages (which may, if desired, be interpreted for lung cancers as N = normal, P = patch, F = field, and M = malignant), with net proliferation occurring in the third (F) stage. $N(t)$ grows linearly (at a normalized rate of 5 per year) from 0 at age 0 to a normalized value of 100 at age 20 (ignoring transitions to P). The background (zero-exposure) transition rates among the successive pairs of stages are $b_{NP} = 0.00006$, $b_{PF} = 0.05$, and $b_{FM} = 0.00008$, and the background proliferation rate in stage F is $b_F = 0.08$. (For this small example, we abandon previous notations in favor of a more mnemonic notation in which b_{NP}, for example, denotes the background transition rate from the first stage, N, to the second stage, P.) Exposure increases each of these rates. A maximal (response-saturating) exposure increases the background level b_k to an increased level $(1 + R_k)^*b_k = b_k + R_k{}^*b_k = b_k + \Delta b_k$, where $\Delta b_k = R_k{}^*b_k$ is the size of the increase due to exposure. The estimated values of these R_k parameters are as follows: $R_{NP} = 2.0$, $R_{PF} = 0.2$, $R_{FM} = 2.19$, and $R_F = 0.19$ for a maximally exposed individual. For an unexposed individual, of course, the R_k values are all 0. [These numerical parameter values were selected to approximately reproduce the exposure-$M(t)$ input-output relations of the RIVM-3 fit of the TSCE model of Schollnberger et al. (2006) describing smoking-associated lung cancer for someone smoking three or more packs per day; see Cox and Huber (2007) for details.] Appendix B lists the model equations for this example. (The equations include a 0–1 variable called "Switch?" that has the effect of interchanging μ_{NP} and μ_{PF} when it has the value 1.)

In summary, the conceptual model is determined by a 4-tuple of parameters:

$$[\mu_{NP}, \mu_{FM} - \mu_F, \mu_{FM}] = [(1 + R_{NP})^* b_{NP}, (1 + R_{PF})^* b_{PF}, (1 + R_{FM})^* b_{FM} - (1 + R_F)^* b_F, (1 + R_{FM})^* b_{FM}].$$

In the notation of Example 3, with $n = 2$ instead of 3, we can write this 4-tuple $[\mu_{NP}, \mu_{PF}, \mu_{FM} - \mu_F, \mu_{FM}]$ as $[a, b, c, c_1]$. Their numerical values are $[a, b, c, c_1]$ = [0.00006, 0.05, –0.07992, 0.00008] for an unexposed (never smoked) individual, with $R_{NP} = R_{PF} = R_{FM} = R_F = 0$ and $[a, b, c, c_1]$ = [3.0*0.00006, 1.2*0.05, 3.19*0.00008 – 1.9*0.08, 3.19*0.00008] = [0.00018, 0.06, –0.1517448, 0.0002552] for a maximally exposed individual.

Problem *Predict the effect on* M(t) *of an intervention that eliminates the effect of exposure on the first transition rate* μ_{NP} *[in effect reducing (1 +* R_{NP}*) from 3 to 1 for a maximally exposed individual. For example, this might be the effect of removing from a complex mixture exposure-specific constituent(s) that increase* μ_{NP} *from* b_{NP} *to (1 +* R_{NP}*)**b_{NP}*.]*.

Solution For constant lifetime exposure, the corollary allows arbitrary permutations of $\{a, b, c\}$, provided $a_1 b_1 c_1 = abc_1 = abc(c_1/c)$ remains constant. Thus, whenever a permutation replaces c with a, for example, then c_1 has to be replaced by ac_1/c. This implies that six possible choices of $[a, b, c, c_1]$ give identical (perfect) fits to any exposure/$M(t)$ input-output histories generated by the model. Prior to the intervention, these six possible values are as follows:

- $[a, b, c, c_1]$ = [0.00018, 0.06, –0.1517448, 0.0002552],
- $[a, c, b, b^* c_1/c]$ = [0.00018, –0.1517448, 0.06, –0.0001009],
- $[b, a, c, c_1]$ = [0.06, 0.00018, –0.1517448, 0.0002552],
- $[b, c, a, a^* c_1/c]$ = [0.06, –0.1517448, 0.00018, –0.00000030272],
- $[c, a, b, b^* c_1/c]$ = [–0.1517448, 0.00018, 0.06, –0.0001009],
- $[c, b, a, a^* c_1/c]$ = [–0.1517448, 0.06, 0.00018, –0.00000030272].

Since a, b, and c_1 correspond to transition rates ($a = \mu_{NP}$, $b = \mu_{PF}$, $c_1 = \mu_{FM}$), they must be nonnegative. This eliminates four of the six possibilities, leaving only these two:

$$[a, b, c, c_1] = [0.00018,\ 0.06,\ -0.1517448,\ 0.0002552],$$
$$[b, a, c, c_1] = [0.06,\ 0.00018,\ -0.1517448,\ 0.0002552].$$

Following the intervention, the value of the transition rate μ_{NP} is assumed to be its background (zero-exposure) level. This is $b_{NP} = 0.00006$ if $\mu_{NP} = 0.00018$ before intervention and is $b_{NP} = 0.05$ if $\mu_{NP} = 0.06$ before intervention. Thus, two possible sets of parameter values following the intervention are

- $[a, b, c, c_1] = [0.00006, 0.06, -0.1517448, 0.0002552]$ if $a = 0.00018$ before intervention, or
- $[a, b, c, c_1] = [0.05, 0.00018, -0.1517448, 0.0002552]$ if $a = 0.06$ before intervention.

Which of these two possibilities is the correct one makes a dramatic difference in the risk reduction achieved by the intervention. For the first set of parameter values, $M(t)$ is reduced by approximately two thirds at older ages, but it is reduced by only about one sixth at older ages if the second set of values is correct.

Simulation shows similar conclusions even if exposure is time-varying. For example, the top curve in Fig. 11.3 gives the age-specific $M(t)$ values *before* the intervention is implemented if $a = 0.00018$ and $b = 0.06$, while the bottom curve in Fig. 11.3 shows $M(t)$ if these two parameter values are interchanged, $b = 0.00018$ and $a = 0.06$. In this figure, exposure starts at age 20, ends at age 60, and is large enough so that the internal dose is 25% of its maximum possible (saturated) value. Despite the nonnegligible background risk and time-varying exposure history, the two predicted curves are close enough to each other so that it is plausible that epidemiological data cannot distinguish between them. [This is a fortiori the case in a fully stochastic analysis, where there is substantial variability of the stochastic jump processes for cumulative malignant cells around its mean value, $M(t)$.]

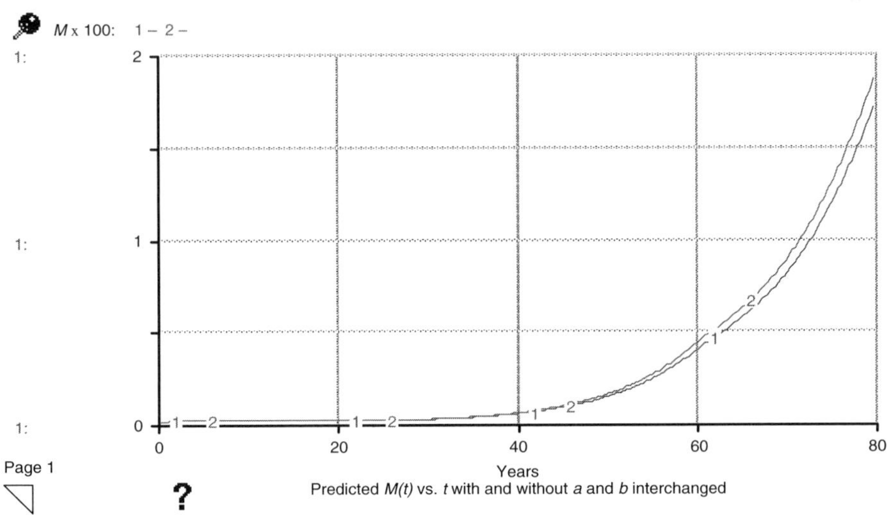

Fig. 11.3 Predicted $M(t)$ curves for the two permutations of parameters (a, b) are closely similar, even for time-varying exposures. [The vertical axis shows $M(t) \times 100$.])

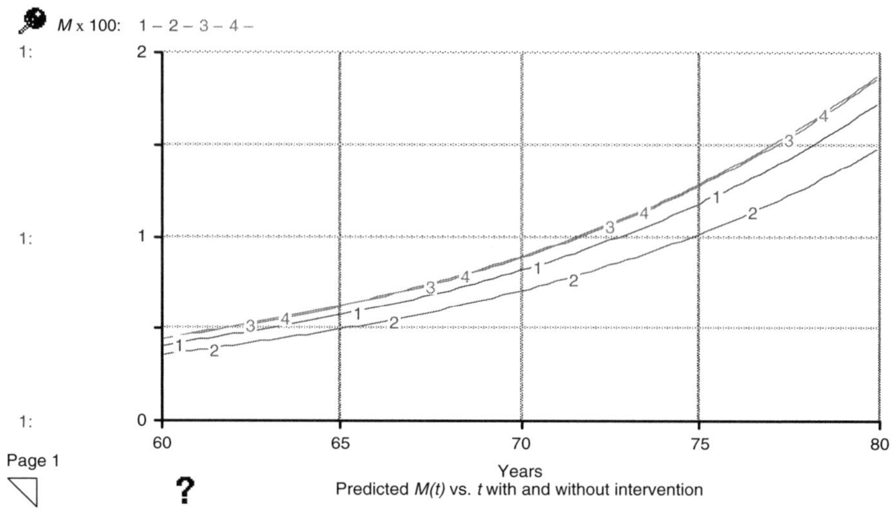

Fig. 11.4 Predicted effects of a risk-reducing intervention differ significantly for the two permutations of parameters (a, b) in Fig. 11.3, even though these two permutations predict very similar exposure-$M(t)$ input-output relations in the absence of intervention [The vertical axis shows $M(t) \times 100$. The x-axis runs from 60 to 80 years to better display the effects.])

On the other hand, Fig. 11.4 shows that very different predictions arise from the two different parameter permutations in Fig. 11.3, $(a, b) = (0.00018, 0.06)$ and $(a, b) = (0.06, 0.00018)$, even though they give very similar input-output predictions in the absence of interventions that change the dose-response relation. The four curves in Fig. 11.4 show the $M(t)$ curves (time courses) predicted by these two models with and without an intervention that reduces parameter $a = \mu_{NP}$ to its background level, b_{NP}. Curves 1 and 2 show the pre- and postintervention $M(t)$ curves, respectively, for an intervention that reduces μ_{NP} from $(1 + R_{NP})^* b_{NP} = 0.00018$ to $b_{NP} = 0.00006$. The resulting reduction in $M(t)$ is clear. By contrast, curves 3 and 4 show the pre- and postintervention $M(t)$ curves, respectively, for an intervention that reduces μ_{NP} from $(1 + R_{NP})^* b_{NP} = 0.06$ to $b_{NP} = 0.05$. Curves 3 and 4 almost coincide: For this pair of parameter values, the intervention creates almost no reduction in $M(t)$.

In summary, this example shows that the ability to use the MSCE model to predict the effect on $M(t)$ of an intervention that reduces R_{NP} from 2 to 0 is limited: The effect can be either negligible or significant, depending on which permutation, $(a, b) = (0.00018, 0.06)$ or $(a, b) = (0.06, 0.00018)$, is correct.

Prior to implementing the intervention, it is impossible to deduce from study of the risk curves M(t) for different exposure histories which of these two alternative effects will be caused by intervention, as both are approximately equally consistent with all preintervention data. To predict which will occur, it is necessary to supplement the observational data on exposure-response history pairs with biological measurements that indicate whether $a = 0.00018$ or $a = 0.06$. Since these two rates differ by a factor of more than 300-fold, even relatively imprecise measurements may suffice to discriminate between them.

Discussion and Conclusions

This chapter has introduced a new identifiability result for multistage clonal expansion (MSCE) models of carcinogenesis, showing that such models can lead to *discrete sets of alternative predictions* for the effects of interventions that change how the dose affects the response at one or a few stages of the carcinogenic process. These alternative predictions, which are all exactly or approximately equally consistent with all preintervention data, arise from discrete symmetries of the dynamic equations describing the models. The symmetries are exact in certain special cases (e.g., for constant-lifetime exposures), but are only approximate for time-varying exposures. However, the conclusion remains that different MSCE models related to each other via groups of symmetries of the model parameters can be approximately equally consistent with all past epidemiological data yet yield significantly different predictions for the effects of interventions on risk, as indicated by $M(t)$. This suggests the potential importance of biological data that can help to discriminate among the symmetric mathematical possibilities that fit the epidemiological data equally well. Even imprecise biological measurements or comparisons of rates can potentially identify the correct combination of parameter values if they reveal the approximate relative magnitudes of the transition rates for different stages.

The mathematical results presented in this chapter are limited to special cases, as stated in Theorem 1. However, simulation suggests that similar results hold under less restrictive conditions. A worthwhile direction for further research is better mathematical characterization of the extent to which alternative MSCE models that fit epidemiological data approximately equally well, and that make identical (or almost identical) predictions for effects of changes in exposure histories, can predict very different effects for changes that affect pharmacodynamics. To facilitate additional investigation, Appendix B gives the ITHINK™ simulation language script used for Example 4.

The following comments may help to clarify the interpretation of the results of this chapter and to place them in the context of previously known results.

1. The main results (Theorem 1, its corollary, and Example 4) concern *discrete symmetries* (exact or approximate) and the resulting *discrete indeterminacy* of predictions for the effects of interventions that change dose-response relations. In other words, they show that uncertainty about postintervention risks is sometimes best characterized by a few discrete alternatives rather than by, for example, a mean and a confidence interval.
2. These results are thus quite different from more familiar results involving continuous aspects of identifiability. In particular, they are *not* simply extensions of the types of nonunique identifiability results illustrated in Examples 1 and 2

[e.g., where only a product of factors is uniquely identified but not the separate factors; or, more generally, where it can be determined that vectors of "reduced parameters" belong to a smooth manifold (Audoly et al., 2001) but they cannot be further uniquely identified from observed data].

3. All of the discrete symmetric models in Theorem 1 make identical predictions (under the conditions in Theorem 1) or approximately identical predictions (more generally, as in Fig. 11.3) for effects of interventions that only affect exposure levels. (They also make the same predictions for interventions that only affect pharmacokinetics, i.e., the biologically effective doses produced by exposures.) However, they make different predictions for interventions that change the pharmacodynamic actions of exposures (e.g., by removing from a complex mixture one or more constituents that act on specific stages of an MSCE model). In other words, the discrete indeterminacy results discussed in this chapter apply *specifically to interventions that change the dose-response relation* by changing pharmacodynamic parameters (the R_k). They do *not* apply to interventions that only affect exposures. For such interventions, unique predictions typically *are* possible, despite the discrete indeterminacy of the models, since the different symmetric models then make the same predictions.
4. The results in this chapter do not contradict previous results for the TSCE model (e.g., Arora et al., 1993). For the TSCE model with $n = 1$ intermediate stage, the number of discrete symmetries is $1!1!1! = 1$. Thus, the set of alternative symmetric models reduces to a single one in this case. The reduced parameters needed to make predictions can also typically be uniquely identified from data, as in Example 2. As soon as one or more additional stages are added, however, the discrete indeterminacy that is the subject of this chapter becomes possible, as illustrated in Example 4.
5. The class of MSCE models discussed in this chapter does not assume or imply linearity of dose-response relations. First, the internal doses $s(t)$ can be nonlinearly related to administered doses $x(t)$ through a relation $s(t) = f[x(t)]$. Second, even though we follow the literature in assuming that biologically effective doses affect pharmacodynamic parameters linearly through the equations

$$\mu_k(t) = b_k + q_k{}^* s(t) \qquad \text{(pharmacodynamics model)},$$
$$g_k(t) = g_{k0} + g_{k1}{}^* s(t),$$

this does not imply that the dose-response relation for the expected number of malignant cells formed by a fixed age, such as $t = 80$ years, is linear. [Indeed, for the model in Example 4, the values of $M(80)$ for $s = 0, 0.2, 0.4, 0.6, 0.8$, and 1 are 0.77, 1.46, 2.70, 4.86, 8.65, and 15.24, respectively. Thus, doubling s from 0.4 to 0.8 more than doubles $M(80)$, from 2.70 to 8.65; and less than doubling s from 0.6 to 1 more than triples $M(80)$, from 4.86 to 15.24.]
6. The results are not artifacts of incomplete (deterministic) analysis; i.e., fully stochastic simulation modeling also supports Theorem 1 and its corollary. Additional stochastic simulation using Crystal Ball™ also confirms in many cases that the conclusions of Example 4 for $M(t)$ also hold for survival and hazard func-

tions for the random time until the first malignant cell is formed. Thus, although this chapter has focused on the mean value $M(t)$ of the random number of malignant cells formed by age t, which can be analyzed by deterministic ODEs, it appears a fully stochastic analysis would not qualitatively change the results. Fully stochastic (discrete-event) simulation modeling confirms that the solutions for $M(t)$ provided by the ODE models studied in this chapter accurately describe mean compartment sizes in the corresponding stochastic transition models for many biologically realistic values of the transition rates.

7. Analytically, the expected value $M(t)$ is closely related to the random time at which the first malignant cell is formed, as follows. Let $X(t)$ denote the random number of malignant cells formed by age t. Since $X(t)$ is nonnegative, Markov's inequality holds: $\Pr[X(t) \geq k] \leq E[X(t)]/k = M(t)/k$. Letting $k = 1$ for the first malignant cell yields Pr(at least one malignant cell is formed by time t) $\leq M(t)$.
8. The results of this chapter do not contradict previous analyses of exact and approximate solutions to linear multistage models (Cox, 1995; Ritter et al., 2003). Previous analyses that conclude that exchanging the order of stages does not affect the predicted risk for a given exposure are fully consistent with the current results. It is only when interventions affect pharmacodynamics at one or a few stages (thus breaking the symmetry) that the discrete indeterminacy results in Theorem 1 and its corollary arise.

The analysis and conclusions of Theorem 1 can be extended to other multistage compartmental models. These may include disease models in which patients progress through a linear sequence or chain of stages, as well as similar "catenary" models used in classical pharmacokinetics, if only concentrations in the last stage (compartment) are observed. The main conclusion, that several alternative sets of parameter values may be consistent with past epidemiological observations, yet predict quite different risks for interventions that affect the pharmacodynamic parameters of dose-response relations, implies that curve fitting and parameter estimation, while undeniably useful, do not always suffice to make unique predictions for risks in new situations. Some more detailed causal knowledge, such as knowing which constituents affect which transition rates, may be essential for predicting how changing the composition of exposure will change exposure-related risks.

Appendix A: Proof of Theorem 1

The following succinct proof is due to Professor William Huber (see the acknowledgments at the start of the book).

For $k > 1$, the w's satisfy recursive relationships

(*) $w_k = a_{k,k-1} w_{k-1}$ and

(**) $1/w_{jk-1} + a_{kk}/w_{jk} = a_{jj}/w_{jk}$ for all j between 1 and k–1.

These are immediate from the definitions; (**) is just a rearrangement of the relationship $(a_{jj} - a_{kk})w_{jk\text{-}1} = w_{jk}$. For $k = 1$, the theorem is trivial. For $k > 1$, we calculate

$$\begin{aligned}(\mathbf{Az}(t))_k &= a_{k,k-1}z_{k-1}(t) + a_{kk}z_k(t)\\ &= a_{k,k-1}w_{k-1}[\exp(a_{11}t)/w_{1\,k-1} + \cdots + \exp(a_{k-1,k-1}t)/w_{k-1,k-1}]\\ &\quad + a_{kk}w_k[\exp(a_{11}t)/w_{1\,k} + \cdots + \exp(a_{kk}t)/w_{kk}].\end{aligned}$$

Apply (*) to the first term and distribute a_{kk} over the second term to obtain

$$\begin{aligned}&w_k[\exp(a_{11}t)/w_{1\,k-1} + \cdots + \exp(a_{k-1,k-1}t)/w_{k-1,k-1}] + w_k[a_{kk}\exp(a_{11}t)/w_{1\,k}\\ &\quad + \cdots + a_{kk}\exp(a_{kk}t)/w_{kk}].\end{aligned}$$

Finally, upon factoring out w_k, collecting the coefficients for each exponential, and applying (**) to the first k–1 of them, the derivative of $z_k(t)$ becomes recognizable:

$$\begin{aligned}&w_k[(1/w_{1\,k-1} + a_{kk}/w_{1\,k})\exp(a_{11}t) + \cdots + (1/w_{k-1,k-1} + a_{kk}/w_{kk})\exp(a_{k-1,k-1}t)\\ &\quad + (a_{kk}/w_{1\,k})\exp(a_{kk}t)] = w_k[(a_{11}/w_{1\,k})\exp(a_{11}t) + \cdots + (a_{kk}/w_{kk})\exp(a_{kk}t)]\\ &\quad = dz_k(t)/dt.\end{aligned}$$

This demonstrates that $\mathbf{z}(t)$ satisfies the system of equations. It remains to show that it also satisfies the initial condition. For $k = 1$, the value is $z_1(0) = 1^*[\exp(a_{11}{}^*0)] = 1$, as desired. For larger values of k, we need to show that

$$\begin{aligned}0 = z_k(0) &= \exp(a_{11}{}^*0)/w_{1\,k} + \cdots + \exp(a_{kk}{}^*0)/w_{kk}\\ &= 1/w_{1\,k} + \cdots + 1/w_{kk}.\end{aligned}$$

The partial fraction expansion of $1/w_{1k} + \ldots + 1/w_{kk}$ is a sum whose terms are in the form $u_{ikj}/(a_{ii} - a_{jj})$. By inspection, we obtain

$$u_{ikj} = 1/\Pi(a_{ii} - a_{ll}) - 1/\Pi(a_{jj} - a_{ll}),$$

with the products extending over all l between 1 and k but skipping i and j. The products are subtracted, not added, because $(a_{ii} - a_{jj})$ appears in w_{ik} while $(a_{jj} - a_{ii}) = -(a_{ii} - a_{jj})$ appears in w_{jk} with the opposite sign. Evidently, the limiting value of u_{ikj} as a_{ii} and a_{jj} become equal is zero, because the two products approach a common finite value. Thus, $(a_{ii} - a_{jj})$ is a factor of u_{ikj}, implying $1/w_{1k} + \ldots + 1/w_{kk}$ has a "removable singularity" on the set $a_{ii} = a_{jj}$. Since i and j were arbitrary, we conclude that $z_k(0)$ has no singularities at all and therefore is really a polynomial. The proof is finished by observing that $z_k(0)$ approaches zero whenever any a_{ii} becomes arbitrarily large, which for a polynomial can occur only when it is identically zero. QED.

Appendix B: Listing of ITHINK™ Model Equations for the Example in Figure 11.3

Compartment M
M(t) = M(t – dt) + (flow_FM) * dt
INIT M = 0
flow_FM = F*(bFM + delta_FM)

Compartment F
F(t) = F(t – dt) + (flow_PF + proliferation_F – flow_FM) * dt
INIT F =
INFLOWS to Compartment F:
flow_PF = P*switch?*(bNP + deltaNP) + P*(1 – switch?)*(bPF + deltaPF)
proliferation_F = F*(bF + delta_bF)

Compartment P
P(t) = P(t – dt) + (flow_NP – flow_PF) * dt
INIT P = 0
INFLOWS to Compartment P:
flow_NP = N*(1 – switch?)*(bNP + deltaNP) + N*switch?*(bPF + deltaPF)
OUTFLOWS:
flow_PF = P*switch?*(bNP + deltaNP) + P*(1 – switch?)*(bPF + deltaPF)

Compartment N
INFLOWS to Compartment N:
N(t) = N(t – dt) + (growth – flow_NP) * dt
INIT N =
growth = if (TIME < 20) then 100/20 else 0
OUTFLOWS:
flow_NP = N*(1 – switch?)*(bNP + deltaNP) + N*switch?*(bPF + deltaPF)

FORMULAS AND PARAMETERS

bF = 0.08 {0.08}
bFM = 0.00008
bNP = 0.00006
bPF = 0.05
deltaNP = bNP*RNP*exposed?
deltaPF = bPF*RPF*exposed?
delta_bF = bF*RF*exposed?
delta_FM = bFM*RFM*exposed?
end_time = 60
exposed? = if ((TIME >= start_time) and (TIME <= end_time)) then exposure_factor else 0
exposure_factor = 1 {fraction of saturation exposure intternal dose}
M_x_100 = M*100
RF = 0.9

RFM = 2.19
RNP = 2 {2}
RPF = 0.2 {0.2}
start_time = 20
switch? = 0

Part IV
Applications and Extensions

Chapter 12
Predicting the Effects of Changes: Could Removing Arsenic from Tobacco Smoke Significantly Reduce Smoker Risks of Lung Cancer?

The remainder of the book applies principles from earlier chapters to several challenging quantitative risk assessment (QRA) problems for complex, uncertain, and nonlinear systems. This chapter returns to the problem of predicting how removing a specific constituent (arsenic) from a complex mixture (cigarette smoke) would affect lung cancer risks. This goes beyond the bounding and portfolio QRAs in Chapters 8 and 10 by applying the systems dynamics model in Chapter 11 to obtain explicit quantitative results. Rather than only estimating bounds for the probable changes in consequences, this chapter predicts specific quantitative reductions in risk, *contingent* on specified assumptions about causal mechanisms. Quantitative sensitivity analysis shows how predicted risk reductions (under stated assumptions) and preventable fractions of risk change as key assumptions are changed.

Biologically Based Risk Assessment Modeling

Biologically based risk assessment (BBRA) models apply systems dynamics models [represented by systems of ordinary differential equations (ODEs) and algebraic formulas, as in Chapter 11] to predict the responses of biological systems to doses of various agents. Such models can potentially enable risk assessors to predict biological responses to exposures, using pharmacokinetics and pharmacodynamics data, even before they obtain relevant epidemiological data. To deliver on this potential, however, it is necessary to build a model linking exposures to responses. This is challenging if detailed causal mechanisms are only partly understood.

This chapter shows how to use available (very incomplete) molecular biological data to estimate dose-response relations and to predict the fractions of cancers that would be prevented if a particular source or constituent of current exposure were removed or reduced. As discussed in Chapter 8, such *preventable fractions* cannot be quantified correctly in general using traditional epidemiological methods, such as population attributable fractions, but only by using relevant causal biological knowledge and data. This chapter focuses on how to use data on the frequencies of molecular and gene-level changes in exposed vs. unexposed subjects, together with knowledge of which changes occur approximately where in the carcinogenic

L.A. Cox, Jr., *Risk Analysis of Complex and Uncertain Systems*, International Series in Operations Research & Management Science 129, DOI 10.1007/978-0-387-89014-2_12,

process, to infer the approximate effects of exposure on the underlying rates at which these events occur. For concreteness, we illustrate the proposed method by predicting the fraction of smoker lung cancers that could be prevented by removing another constituent, arsenic, from tobacco smoke.

Although available data do not allow a unique, definitive answer, quantitative risk modeling with available information can help to answer pragmatic questions such as

1. *What is the plausible range of risk reductions* in lung cancer risk (consistent with available knowledge) caused by removing arsenic from cigarette smoke?
2. *How sensitive is the predicted reduction in risk* to incomplete removal of arsenic?
3. *What are the most important scientific uncertainties,* whose resolution would most improve the ability to quantify risk reductions from removing arsenic?

The QRA in this chapter can address such questions constructively, despite the many scientific uncertainties that remain.

The following sections show how to construct a plausible estimate of potential risk reductions, using available knowledge and data, and using specified assumptions to bridge remaining knowledge gaps. This allows currently available toxicological data and knowledge to help determine which constituents and modes of action contribute most to exposure-related cancer risk, while recognizing that improved models, mechanistic knowledge, and data may lead to better models and narrower uncertainty intervals for answers in the future.

Arsenic as a Potential Human Lung Carcinogen

Many epidemiological and *in vitro* studies have suggested that arsenic exposure may increase the risk of lung cancer in humans. The available evidence is not conclusive, however. Table 12.1 illustrates some contrasting conclusions from different epidemiological studies, including reports of sublinear, linear, and superlinear relations between exposure to arsenic compounds and lung cancer risk.

The dependence of lung cancer risk on arsenic exposure and cigarette smoking has also been investigated and speculated about for nearly 50 years (Holland et al., 1959). A hypothesis that changes in the levels of arsenic in cigarettes might help to explain the corresponding historical changes in lung cancer rates has been advanced several times (e.g., Buechley, 1963) but is difficult to appraise without direct experimental investigation. Recent studies have shown significant synergy between exposure to arsenic in drinking water and the risk of lung cancer in smokers. For example, for an epidemiological study in Taiwan, Chen et al. (2004) stated that "The etiologic fraction of lung cancer attributable to the joint exposure of ingested arsenic and cigarette smoking ranged from 32% to 55%. The synergy indices ranged from 1.62 to 2.52, indicating a synergistic effect of ingested arsenic and cigarette smoking on lung cancer." Hazelton et al. (2001) also reported that "Tobacco and arsenic dominate the attributable risk for lung cancer" in a population of tin miners, with

Table 12.1 Ambiguous epidemiological evidence on arsenic and lung cancer

Exposure-response relation (emphases added)	Study
"A **positive** dose-response relationship was found between cumulative arsenic exposure [estimated intensity but not duration] and lung cancer mortality with an overall SMR of 372... ." (1989)	Jarup (1989) Swedish smelter workers. **OR = 1.07** [1.02–1.1] among smokers (Lundstrom, 2006)
"There is compelling evidence in support of **positive associations** between arsenic and risk of both lung and bladder cancers, and between cadmium and lung cancer risk."	Navarro-Silvera and Rohan (2007)
"[R]espiratory cancer mortality increased **linearly** with increasing cumulative exposure group, ranging from two to nine times expected."	Lee-Feldstein (1986), for Montana smelter workers exposed to arsenic trioxide
"[S]tudies with quantitative data are consistent with a **supralinear** dose-response relationship" (or with confounding by smoking, or with mismodeled exposure errors).	Hertz-Picciotto and Smith (1993), for Chinese miners
"The additional follow up confirms the earlier finding that at low doses the increments in death rates for respiratory cancer for a given increment in dose are greater than at high doses."	Enterline et al. (1995) for Tacoma, Washington, copper smelters
Sublinear or Hormetic (U-shaped): "Recent epidemiological studies have shown that the relative risk for cancer among populations exposed to $\leq$ 60 ppb As in their drinking water is often **lower** than the risk for the unexposed control population." (Snow et al., 2005)	Snow et al. (2005) Schoen et al. (2004) (no clear effect in the United States for drinking water) Lamm et al. (2006) (Taiwanese villages)

substantial joint effects. Likewise, experimental evidence in male Syrian golden hamsters shows that aerosolized arsenic and cigarette smoke interact synergistically in depleting glutathione and producing oxidative DNA damage in the lungs (Hays et al., 2006).

"That alteration of DNA methylation by arsenic offers a plausible, unified hypothesis for the carcinogenic mechanism of action of arsenic" was proposed in the 1990s (Mass and Wang, 1997). Supporting epidemiological evidence includes a study by Chanda et al. (2006) that found that "Significant DNA hypermethylation of [the] promoter region of p53 gene was observed in DNA of arsenic-exposed people compared to control subjects. This hypermethylation showed a dose-response relationship. ... Significant hypermethylation of gene p16 was also observed in cases of arsenicosis exposed to high level of arsenic."

In mice, too, the hypermethylation of specific tumor suppressor genes has been linked to arsenic-associated lung cancer risk. Cui et al. (2006) reported that

> Arsenic was accumulated dose dependently in the lung tissues of iAs(V)-exposed [A/J] mice. ... [T]he rate of poorly differentiated lung adenocarcinoma was much higher in iAs(V)-exposed mice than in the control. Methylation rates appeared to be higher in a dose-related tendency in lung tumors from iAs(V)-exposed mice compared to the control.

> Lower or loss of p16INK4a and RASSF1A expression was found in lung tumors from iAs(V)-exposed mice, compared to that in nontumor lung tissues from both control and iAs(V)-exposed mice, and this reduced or lost expression was in accordance with hypermethylation of the genes. In conclusion, iAs(V) exposure increased lung tumor incidence and multiplicity in A/J mice. Epigenetic changes of tumor suppressor genes such as p16INK4a and RASSF1A are involved in the iAs(V)-induced lung carcinogenesis.

Comparing gene methylation patterns specifically in human smoker lung cancer patients to those in nonsmoker lung cancer patients suggests that some, but not all, of the mouse findings are relevant to smoking-induced lung cancer in humans. Specifically, methylation of the promoter region of RASSF1a does *not* appear to differ significantly between the tumors (primarily non-small cell lung cancers, NSCLCs) of smokers and nonsmokers (e.g., Liu et al., 2007). But hypermethylation of the promoter region of p16INK4a (more briefly described as "p16 methylation") *does* appear to be elevated (both in frequency and in levels) in the specific types of lung tumors that are most strongly associated with smoking: adenocarcinomas (ADCs) with *K-ras* mutations.

Recent research has revealed two competing oncogenic pathways for human lung adenocarcinomas: a *K-ras* mediated pathway that is associated with smoking and an epidermal growth factor receptor (EGFR)-mediated pathway that is not (Wistuba and Gazdar, 2006). Pack-years smoked and duration of smoking are highly significantly associated with p16INK4a methylation in NSLC tumors ($p = 0.007$ and $p = 0.0009$, respectively), and such methylation is negatively associated with time since smoking cessation ($p = 0.03$, all p-values for Wilcoxon rank sum tests) (Kim et al., 2001). In turn, p16INK4a methylation significantly increases ADCs in the *K-ras* mediated (smoking-associated) pathway (the odds ratio for a *K-ras* mutation among adenocarcinomas with methylated p16INK4a is 4.93, 95% CI, 1.54–15.7) at the cost of ADCs in the EGFR pathway (the EGFR mutation odds ratio among adenocarcinomas with methylated p16INK4a = 0.07, 95% CI, 0.02–0.33) (Toyooka et al., 2006). Thus, p16INK4a methylation significantly increases the probability of the *K-ras* mediated pathway and significantly decreases the probability of the competing EGFR pathway among human smokers.

The p16INK4a protein has been identified as a tumor suppressor [inhibiting cyclin-dependent kinases CDK4 and CDK6 from initiating phosphorylation of the retinoblastoma (Rb) tumor suppressor protein] that can cause damaged cells to stop dividing, arresting them in stage G1 of the mitotic cycle and thus removing them from the pool of proliferating cells at risk of further carcinogenic transformation (Ohtani et al., 2004). Such *oncogene-induced senescence*, probably mediated by the interactions of inflammatory cytokines (especially IL-6 and IL-8) as well as p16INK4a protein, is an important mechanism that normally protects cells against cancer (Kuilman et al., 2008). It is tempting to speculate that gene silencing of p16 by hypermethylation slows or prevents initiated (premalignant) cells from entering senescence, thereby increasing the risk of further carcinogenic transformations, e.g., due to *K-ras*.

In light of this evidence, the remainder of this chapter focuses on the following question: *Assuming that arsenic in cigarette smoke increases lung cancer risk specifically by hypermethylating the promoter region of p16INK4a, how would reducing or eliminating arsenic in tobacco smoke change lung cancer risks from smoking?* The larger question addressed via this example is, how can such specific molecular-biological knowledge about causal mechanisms of carcinogenesis be used to inform and improve quantitative risk assessment?

It might at first seem that applying such mechanistic information to risk assessment requires unrealistic amounts of other information before its implications for risk can be determined. For example, if cigarette smoke contains other constituents that play the same functional role as arsenic (namely, hypermethylating the promoter region of p16INK4a, in this example) and that fully saturate this mechanism, then removing arsenic would have no impact on risk. However, epidemiological and experimental data indicate that inhalation exposure to arsenic increases lung cancer rates among smokers (Hays et al., 2006; Jones et al., 2007). If this is correct, then smoking alone does not saturate arsenic-induced lung carcinogenesis, and the challenge of figuring out how much reduction in lung cancer risk might be caused by reducing arsenic in cigarette smoke remains. To this, we now turn.

Data, Methods, and Models

A Multistage Clonal Expansion (MSCE) Framework for Lung Field Cancerization

The following discussion defines the fraction of smoker lung cancers "caused by" arsenic in cigarette smoke as the fraction of smoker lung cancers that would be prevented if arsenic (As) were removed from cigarette smoke (see Chapter 8). This fraction will be estimated with the help of a multistage clonal expansion (MSCE) model for lung carcinogenesis (see Chapter 11), diagrammed in Fig. 12.1.

This framework represents lung cancer as resulting from an accumulation of gene-level damage. (See Table 12.2 for a summary of several smoking-associated gene-level events, both genetic and epigenetic.) Accumulating damage causes three main transitions at the tissue level, as follows.

Normal-to-patch transition: The first ($N \Rightarrow P$) transition in Fig. 12.1 assumes that, like other epithelial tissues, normal lung tissue is organized into many relatively small compartments of cells, each of which independently may undergo transformations of its cells (Michor et al., 2003; Panov, 2005). Lung cancer biologists have identified an initial "patch" phase in which early premalignant gene-level changes [such as loss of heterozygosity at chromosome 3p (LOH 3p) or hyperme-

$$\Downarrow \mu_F$$

$$N \Rightarrow P \Rightarrow F \Rightarrow M$$

$$\mu_{NP} \quad \mu_{PF} \quad \mu_{FM}$$

Fig. 12.1 MSCE model for lung carcinogenesis

Table 12.2 Suggested structure of some smoking-related gene-level changes in lung cancer

Change (location in process)	SCLC	NSCLC	Smoking effects	References
Module 1 (Hyperplasia, patches, $N \Rightarrow P$)				
Telomere shortening			Shortens peripheral blood lymphocyte telomeres	Morla et al. (2006)
Telomerase deregulation			Telomerase activity strongly correlated with pack-years	Yim et al. (2007)
RASSF1A (3p21.3) small deletions			3p21.3 deletions are significantly higher in smokers than in nonsmokers	Zienolddiny et al. (2001)
RASSF1A (3p21.3) promoter hypermethylation	0.9–1	0.30–0.40	Associated with age at start of smoking (Kim et al., 2004)	Meuwissen and Berns (2005)
RASSF1A (3p21.3) LOH			3p21 LOH is higher in smokers than in nonsmokers	Hirao et al. (2001)
Neoangiogenesis			Nicotine can induce *in vitro* Smoking induces *in vivo*	Dasgupta et al. (2006) Hiroshima et al. (2002)
Module 2 (Dysplasia, field expansion, $P \Rightarrow F$)				
FHIT expression decreased (3p14.2); loss of FHIT protein		0.87 SCC 0.57 ADC	*0.75 for NSCLCs of smokers vs. 0.39 for nonsmokers*	Sozzi et al. (1998)
• FHIT mutation	0.8	0.4		
• FHIT deletions			3p14.2 deletions are more frequent among smokers	Zienolddiny et al. (2001)
• FHIT promoter hypermethylation			*For SCC, 0.45 in smokers vs. 0.13 in nonsmokers*	Kim (2004)
RARβ (3p24) reduced expression	0.4			
• RARβ promoter hypermethylation (Zochbauer-Muller et al., 2001)	0.7	0.4	*For NSCLC: 0.31 in smokers vs. 0.10 in nonsmokers*	Tomizawa et al. (2004)

Table 12.2 (continued)

Change (location in process)	SCLC	NSCLC	Smoking effects	References
Larger 3p deletions				
p16 (9p21) expression abnormal; disruption of p16INK4a-Cyclin D1-CDK4-RB pathway		0.46		Spanakis et al. (1999)
• p16INK4a methylation • p16INK4a inactivated by promoter hypermethylation = 0.25 for NSCLC (Zochbauer-Muller, 2001)		0.41 SCC 0.22 ADC	Increases with pack-years, decreases with time since quitting; *0.32 in nonsmokers vs. 0.63 in smokers*	Kim (2001); Yanagawa et al. (2003); Liu et al. (2006)
• p16INK4a mutated • p16 allelic loss	< 0.05	0.2–0.5	*4/14 in nonsmoker NSCLCs; 23/33 in smokers*	Wistuba and Gazdar (2006) Sanchez-Cespedes et al. (2001)
• 9p21 deletions			9p21 deletions are more frequent among smokers	Zienolddiny et al. (2001)
• RB (13q14) expression compromised	> 0.90	0.15–0.30	pRB and p16 expression are negatively correlated	Pan et al. (2005); Geradts et al. (1999)
p53 inactivation (17p13) *Note:* Missense mutations (mainly G → T transversions) at p53 "hot spots" (especially, codon 157) are associated with smoking (Wistuba and Gazdar, 2006)			"TP53 mutations were detected in *47.5% never, 55.6% former, and 77.4% current smokers"* "G:C–>T:A mutations in TP53 are more frequently observed in smoking-associated lung cancers"	Le Calvez et al. (2005) Mechanic et al. (2005)
Module 3 (Conversion, Proliferation, Progression; Carcinoma *in situ*)				
• TP53 mutations	< 0.01	0.15–0.2 (up to 0.5 in ADCs)	Associated with smoking	Tam et al. (2006); Wistuba and Gazdar (2006)
• RAS mutations (smoking-associated G → T transversions at codon 12 of *K-ras* are the most usual RAS mutations in ADCs)			*2/27 in nonsmokers; 8/27 in smokers*	Slebos et al. (1991)

thylation of the promoter regions of some tumor suppressor genes] lead to altered subpopulations of cells that expand clonally within their own small compartments. These changes can form many small local "patches" of altered cells, widely distributed in the lung (Park et al., 1999; Pan et al., 2005; Panov, 2005; Schollnberger et al., 2006; Wistuba et al., 2000; Wistuba and Gazdar, 2006). Thus, Fig. 12.1 posits that clonal compartments of cells within the lung epithelium make transitions from being occupied by normal (unaltered) cells (*N*) to being occupied by altered cells, forming clonal or subclonal foci called *patches* (*P*). Biologically, the patch cells typically have genetic and/or epigenetic alterations such as LOH 3p deletion or promoter hypermethylation of the RASSF1A gene at 3p21.3 (Table 12.2). For modeling purposes, the defining characteristic of patch cells is that they can outcompete and replace normal cells within a local clonal compartment, but they do not spread beyond the compartment boundaries. The transition of compartments from "normal" (*N*) to "patch" (*P*) states is depicted graphically as $N \Rightarrow P$.

Such transitions can be modeled in greater detail as Moran stochastic processes with explicit transformation and repair rates (Michor et al., 2003; Nowak et al., 2006). However, mathematical analysis (ibid.) shows that the outcome of such a stochastic process is that altered cells soon come to occupy either the whole compartment or none of it, depending on which of these two absorbing boundaries is reached first following the creation of the first altered cell. In this chapter, therefore, the details of the competing transformation and repair processes are suppressed and the whole compartment is viewed as making an irreversible transition from *N* to *P* when its normal cells are replaced by patch cells. (Transients in which a compartment is temporarily occupied by some normal and some patch cells are ignored, since we treat the compartment transitions from normal to patch as being approximately instantaneous on the time scale of years to decades most relevant for carcinogenesis.)

Patch-to-field transition: The formation of patches is followed by a further premalignant "field cancerization" phase in which additional gene-level preneoplastic changes lead to a clonal expanding "field" of proliferating cells that can spread across compartment or patch boundaries, outcompeting and replacing normal cells (and perhaps also patch cells) in adjacent compartments or patches (Pan et al., 2005; Panov, 2005; Wistuba and Gazdar, 2006). The expanding population of field cells may eventually come to occupy a large area in one or both lungs (e.g., Franklin et al., 1997; Sikkink et al., 2003). Field growth is not cancerous, however: There is no invasion of tissue, and expansion occurs by outcompeting other cells in replacing cells that die or terminally differentiate.

Such "field cancerization" now appears to be more than a speculative theory, at least for lung cancers, as distinct phases of progressive molecular changes followed by clonal expansion have been observed using molecular biological techniques (Panov, 2005; Wistuba and Gazdar, 2006). Preneoplastic changes that produce a field could include decreased expression of the fragile histidine triad (FHIT) gene at 3p14.2 and of the retinoic acid receptor beta (RARβ) gene at 3p24; disruption of the p16INK4a-Cyclin D1-CDK4-RB control pathway by promoter hypermethylation, deletion, loss of heterozygosity (LOH), or mutation of p16INK4a (at 9p21) or by compromised expression of the retinoblastoma tumor suppressor gene (TSG) RB at 13q14; and inactivation of p53 at 17p13 (Table 12.2). Figure 12.1 depicts the transition of compartments from "patch" (P) to "field" (F) states graphically as $P \Rightarrow F$.

Field-to-malignant transition: The final major transition in Fig. 12.1 is from field cells to malignant cells. Malignant cells include small cell lung cancer cells (SCLC) and non-small cell lung cancer (NSCLC) cells, which, in turn, include the adenocarcinoma (ADC) and squamous cell carcinoma (SCC) subtypes. The transition from field to malignant cells may be accompanied by the activation of oncogenes (e.g., caused by *K-ras* mutations at codon 12 in ADCs) and the inactivation of additional TSGs (e.g., due to TP53 mutations). Figure 12.1 denotes this transition as $F \Rightarrow M$.

A Mathematical Model of Field Carcinogenesis

In the conceptual framework of Fig. 12.1, exposures to carcinogens increase cancer risk by stochastically increasing the random numbers of tumor cells formed by any given age, by hastening one or more of the three transition rates from N to P (denoted by μ_{NP}), from P to F (denoted by μ_{PF}), or from F to M (denoted by μ_{FM}) and/or by increasing the net growth rate (denoted by μ_F) of the field cell population. Other specific effects, such as disruptions of telomerase regulation and telomere length, DNA repair, proliferation rates, or apoptosis rates, are subsumed into the transitions among these four compartments for modeling purposes. The parameter vector $\mu = [\mu_{PF}, \mu_{FM}, \mu_F, \mu_{FM}]$ thus determines the expected number of lung tumor cells formed at any particular age [and, indeed, the entire stochastic process for the state vector (N, P, F, M) over time], with exposure affecting lung cancer risk by affecting one or more of these parameters.

This conceptual framework can be developed into a full quantitative model for the dose-time-response relationship between chemical exposure and lung cancer risk, as in Chapter 11. For the reader's convenience, we recapitulate the main modeling steps here. The flow balance equations corresponding to Fig. 12.1 are as follows:

$$
\begin{aligned}
dN/dt &= -\mu_{NP}N(t),\\
dP/dt &= \mu_{NP}N(t) - \mu_{PF}P(t),\\
dF/dt &= \mu_{PF}P(t) + (\mu_F - \mu_{FM})F(t),\\
dM/dt &= \mu_{FM}F(t),
\end{aligned}
$$

where $N(t)$,$P(t)$, and $F(t)$ are the expected sizes of the normal, patch, and field states at time t, respectively. This system of equations can be solved for the expected number of tumor cells formed by age t (Cox and Huber, 2007). For a constant average daily smoking exposure starting at age 20 with approximately 100% of cells and compartments in state N at age $t = 20$ years, the solution for the expected cumulative number of tumor cells by age $20 + t$ years, or the dose-time-response model, is

$$M(t) = (d/c)(1 + \alpha e^{-at} + \beta e^{-bt} + \gamma e^{-ct}).$$

For ease of readability, the substitutions $a = \mu_{NP}$, $b = \mu_{PF}$, $c = \mu_{FM} - \mu_F$, and $d = \mu_{FM}$ have been made and the following reduced parameters have been used:

$$\alpha = bc/(a-b)(c-a);\ \beta = ac/(a-b)(b-c);\ \gamma = ab/(c-a)(b-c).$$

[Recall from Chapter 11 that Markov's inequality implies that the probability that the actual random number of malignant cells formed is much greater than this expected value is small when $M(t)$ is small, e.g., for $M(t) < 0.1$). Fully stochastic (discrete-event) simulation confirms that this expected value analysis is accurate (ibid.).] The symmetries in this solution imply that if [a, b, c, d] is any set of values for [μ_{NP}, μ_{PF}, μ_{FM}–μ_F, μ_{FM}] that fits available data, then so is [b, a, c, d]. [As discussed in Chapter 11, based on Cox and Huber, 2007, six different sets of parameter values, related to each other as members of a symmetry group, provide identical predictions of $M(t)$ for any smoking history $s(t)$, but four of them can be eliminated when $\mu_{FM} - \mu_F < 0$, as implying unrealistic negative values for some transition rates.]

Following Schollnberger et al., the dose dependence of transition rates is incorporated into the model via the linear formulas

$$\mu_k(t) = b_k + q_k s(t) \text{ for } k = NP,\ PF,\ F, \text{ and } FM.$$

Here, $s(t)$ (discussed next) reflects the biologically effective dose of carcinogens acting on transition rates at time t, b_k is the background (zero-exposure) rate for μ_k, and q_k is a potency or slope factor quantifying the increase in μ_k per unit increase in $s(t)$.

The internal dose variable $s(t)$ is defined to make the relations between the internal doses and the transition rates linear. To reproduce the input-output relations of the model presented by Schollnberger et al. (2006) for smoking and lung cancer, we estimate $s(t)$ as the following nonlinear function of $x(t)$, the smoking intensity at time$t$:

$$s(t) = (1 - e^{-p^* x(t)})/p = \text{biologically effective dose at time } t \text{ from } x(t) \text{ cigarettes/day},$$

where $p = 0.1$ is a *saturation parameter*. $s(t)$ may be interpreted as the "effective number" of cigarettes per day at time t, measured in terms of the resulting internal dose that affects the transition rates. For very low levels of exposure, $s(t) \approx x(t)$, but the function is concave and saturates at about two packs per day. For example, when $x(t) = 1$ cigarette/day, $s(t) = 0.95$, but when $x(t) = 2$ cigarettes/day, $s(t) = 1.81$; when $x(t) = 10$ cigarettes/day, $s(t) = 6.32$; and when $x(t) = 30$ cigarettes/day, $s(t) = 9.5$. For arbitrarily large values of $x(t)$, the internal dose approaches an upper-bound asymptote of $s(t) = 10$. [The estimated value $p = 0.1$ and the form of the saturation function were selected to achieve this asymptote, based on input-output relations for the model of Schollnberger et al. (2006); see Appendix A.]

Model parameter values were estimated to provide a good visual fit (minimizing the maximum absolute deviation) to the input-output predictions of the epidemiologically based TSCE model of Schollnberger et al., for a broad range of smoking histories, including 0, 10, 15, 30, and 60 cigarettes per day smoked from ages 20, 30, or 40 to ages 60, 70, 75, or 80, with all simulations run through age 80. Simulations numerically solve the above system of ODEs to calculate $M(t) = (d/c)(1+ \alpha e^{-at}+ \beta e^{-bt}+\gamma e^{-ct})$ for different ages t. Equations of the best-fit TSCE model for males and females combined (fit RIVM3 in Table 6 of Schollnberger et al.) were used to simulate time courses of $M(t)$ for different smoking histories $s(t)$. Model parameter values were selected to closely reproduce these courses. This yielded the following fully specified model (with all parameter values quantified):

$$\begin{aligned}[\mu_{NP}(t),\ \mu_{PF}(t),\ \mu_{F}(t),\ \mu_{FM}(t)] &= [b_{NP},\ b_{PF},\ b_{F},\ b_{FM}] \\ &\quad + [q_{NP},\ q_{PF},\ q_{F}, q_{FM}]s(t) \\ &= [0.05,\ 0.00006,\ 0.08,\ 0.00008] \\ &\quad + [0.001,\ 0.000012,\ 0.0072,\ 0.0000176]s(t),\end{aligned}$$

where $s(t) = (1 - e^{-0.1*x(t)})/0.1$.

The initial condition for the model is $[N, P, F, M] = [100, 0, 0, 0]$ at time $t = 20$ years, i.e., the size of the normal compartment, N, is normalized to 100 at age 20. Mathematical analysis (Cox and Huber, 2007) reveals that exchanging the values of $\mu_{NP}(t)$ and $\mu_{PF}(t)$ leaves predicted input-output relations nearly unchanged, so that an a priori equally valid set of parameter estimates (yielding identical predictions for all constant-exposure scenarios) is

$$\begin{aligned}[\mu_{NP}(t),\ \mu_{PF}(t),\ \mu_{F}(t),\ \mu_{FM}(t)] &= [b_{NP},\ b_{PF},\ b_{F},\ b_{FM}] \\ &\quad + [q_{NP},\ q_{PF},\ q_{F}, q_{FM}]s(t) \\ &= [0.00006,\ 0.05,\ 0.08,\ 0.00008] \\ &\quad + [0.000012,\ 0.001,\ 0.0072,\ 0.0000176]s(t).\end{aligned}$$

However, we shall see below how available data on prevalence of hypermethylation of the promoter region of p16INK4a can be used to identify the first set of parameter estimates, but not the second, as being consistent with biomarker data.

Modeling the Effects on Lung Cancer Risk of Reductions in Carcinogenic Constituents

Exposure to cigarette smoke significantly increases the frequencies of several gene-level changes found in lung tumors, as shown by the italicized numbers in the "Smoking Effects" column of Table 12.2. If reducing arsenic in tobacco smoke slows the rates of some of these events, then it might also slow the rate – and hence reduce the lifetime risk – of developing some lung cancers. The preceding MSCE model provides the following way to quantify this potential risk reduction in lung cancer risk from reducing one or more constituents of cigarette smoke:

Step 1: Identify known or assumed causal mechanisms or mode(s) of action, i.e., where in the carcinogenic process in Fig. 12.1 the constituents act. Table 12.2 postulates that p16INK4a hypermethylation, the assumed mechanism for arsenic-induced lung cancer, increases the transition rate μ_{PF} (it is in module 2 of Table 12.2). This corresponds to an increased rate of entry of premalignant P cells into the clonal expansion ("field") state F. This is generally consistent with the biological role of p16INK4a in cell cycle regulation (as part of the p16INK4a-Cyclin D1-CDK4-RB control axis), and with *in vitro* evidence that arsenic promotes the proliferation of human lung adenocarcinoma cell lines with mutated p53 (Liao et al., 2007).

Step 2: Estimate the change in each affected transition rate that would be caused by reducing the constituent(s) in cigarette smoke. The effect of removing arsenic can be modeled as a reduction in the transition rate μ_{PF} from a preremoval level of $b_{PF} + q_{PF}\ s(t)$ to a postremoval level of $b_{PF} + gq_{PF}\ s(t)$, where g is a number between 0 and 1, interpreted as the fraction of the smoking effect on increasing μ_{PF} that remains after arsenic is removed. The maximum possible effect size is obtained by setting $g = 0$, implying that removing arsenic would restore μ_{PF} to its background level.

Step 3: Estimate the reduction in risk caused by the estimated reduction in transition rates. A computer simulation model (listed in Appendix B) that numerically solves the system of ODEs for the MSCE model and calculates $M(t)$ for t between 20 and 80 years was used to quantify the reductions in age-specific risks if different transition rates are reduced by various amounts. Table 12.3, prepared using this simulation model, shows how reductions in the four transition rates μ_k for $k = NP$, PF, F, and FM map to corresponding reductions in risk, as measured by a fractional reduction in $M(80)$, the expected number of malignant cells formed by age 80. Note that reducing μ_{PF} or μ_{FM} reduces risk proportionally, i.e., by the same factor. Reducing μ_{NP} produces less-than-proportional reductions in risk. For example, a 20% reduction in μ_{NP} (i.e., multiplying it by a factor of 0.80) creates only a 14% reduction in risk (risk reduction factor = 0.86). Reductions in μ_F can produce much more-than-proportional reductions in risk. For example, a 10% reduction in μ_F (reduction by a factor of 0.9) creates a 40% reduction in risk (risk reduction factor = 0.60). Most relevantly for arsenic, reductions in μ_{PF} produce proportional reductions in risk.

The emphasized entries in Table 12.3 show the approximate maximum possible reductions in risk that can be achieved by reducing each transition rate to its background (zero-exposure) level. For example, with b_{NP}= 0.05 and $q_{NP} = 0.001$, the value of μ_{NP} at $x = 60$ cigarettes per day (for which $s = 10$) is $\mu_{NP} = b_{NP} + q_{NP} * 10 = 0.05 + 0.01 = 0.06$. At 0 cigarettes per day, this value falls to $\mu_{NP} =$

Table 12.3 Risk reductions achieved by given reductions in transition rates

If transition rate is multiplied by the factor in this column...	... Then the lung cancer risk *M* (80) is multiplied by the fraction in these columns (depending on which of these four transition rates is reduced by the factor shown in the left column)			
	μ_{NP}	μ_{PF}	μ_F	μ_{FM}
1 (no reduction)	1.00	1.00	1.00	1.00
0.90	0.93	0.90	0.60	0.90
0.80	***0.86***	0.80	0.36	0.80
0.70	0.77	0.70	0.23	0.70
0.60	0.69	0.60	0.15	0.60
0.50	0.59	0.50	***0.10***	0.50
0.40	0.49	0.40	0.07	0.40
0.30	0.39	***0.30***	0.05	***0.30***
0.20	0.27	0.20	0.04	0.20
0.10	0.14	0.10	0.03	0.10
0.00	0.00	0.00	0.02	0.00

Note: The base case risk is 0.15 at age 80, from 60 cigarettes/day from age 20 to 60.

$b_{NP} = 0.05$, which is $5/6 = 0.83$ of its maximum value; thus, the entry in the row for 0.8 (the nearest value to 0.83 in the table) is emphasized in the column for μ_{NP}. Because the background term accounts for a relatively small proportion of the maximum (saturated) value of μ_{PF}, reducing this parameter (e.g., by removing arsenic) can potentially cause a relatively large reduction in risk, by a factor of up to 1/3.

Although this chapter focuses on a single risk-reducing mechanism (the decreased hypermethylation of p16INK4a, assumed to reduce μ_{PF}), it is worth noting that *risk-reducing effects from reductions in multiple transition rates simultaneously combine approximately multiplicatively* for a wide range of parameter values. For example, reducing each of μ_{PF} and μ_{FM} by 10% (i.e., by a factor of 0.9 each) reduces risk by a factor of 0.81. Reducing each of μ_{PF} and μ_F by 10% reduces risk by 0.9*0.6 = 0.54, the product of their separate effects.

In principle, steps 1–3 and Table 12.3 allow the prediction of risk reductions obtained by reducing specific constituents in cigarette smoke. (Interventions that affect multiple transition rates simultaneously can also be modeled, using the simulation model in Appendix B.) However, in practice, it is necessary to find data to quantify, or at least bound, the change in transition rates caused by the proposed reductions in constituents, such as arsenic, in our example. Biomarker data can provide the needed information, as illustrated next.

Linking Biomarker Data to Model Transition Parameters

NSCLCs constitute about 80% of lung cancers (Panov, 2005). Jarmalaite et al. (2003) reported that methylation of p16 occurred only in NSCLCs, with a significantly higher rate in former smokers as compared to current smokers ($p = 0.035$). Liu et al. (2006) reported that, in a study of NSCLCs, hypermethylation of the promoter region of p16INK4a was found in 51 of 81 lung tumors from smokers (63%). This was nearly twice the rate found in lung tumors from never-smokers (13 of 41 lung tumors, 32%).

These data show that substantial methylation occurs even among nonsmoker tumors, but that smoking has a highly significant ($P = 0.001$) association with increased methylation. Indeed, the increase from 32 to 63% is great enough so that the hypothesis that eliminating smoking would reduce μ_{PF} by at most a factor of 5/6 (see the discussion in step 2 above) cannot be maintained, at least if methylation of p16INK4a is assumed to account for a substantial proportion of smoking-associated increases in μ_{PF}. On the other hand, it *is* compatible with a maximum reduction in μ_{PF} by a factor of 1/3. Hence, we tentatively assume that the correct model is as follows:

MSCE Model 1

$$\begin{aligned}
&[\mu_{NP}(t),\ \mu_{PF}(t),\ \mu_F(t),\ \mu_{FM}(t)]\\
&\quad = [b_{NP},\ b_{PF},\ b_F,\ b_{FM}]\\
&\qquad +[q_{NP},\ q_{PF},\ q_F,\ q_{FM}]s(t)\\
&\quad = [0.05,\ 0.00006,\ 0.08,\ 0.00008]\\
&\qquad +[0.001,\ 0.000012,\ 0.0072,\ 0.0000176]s(t),\\
&s(t) = (1 - e^{-0.1^* x(t)})/0.1
\end{aligned}$$

To complete the model specification and match the input-output relations of the model of Schollnberger et al. (2006), we also assume that the normal compartment grows linearly between ages 0 and 20 years, to reflect growth from child to adult (Cox and Huber, 2007), with carcinogenic transitions assumed to be negligible, $\mu_{NP}(t) \approx 0$, for ages 0–10. Appendices A and B list the full sets of model equations for the TSCE model and the MSCE model of field carcinogenesis, respectively. Table 12.3 was prepared using this MSCE model, implemented in the ITHINK^{TM} 7.0 modeling environment.

Results

The data of Liu et al. (2006) can be used to help quantify the effect of arsenic exposure on increasing μ_{PF}, with the help of the following simplifying assumptions:

- Attribute *all* of the reported increase in p16INK4a methylation prevalence in tumors of smokers compared to nonsmokers to arsenic in cigarette smoke (a deliberate worst-case assumption).
- Assume that arsenic increases the rate of p16INK4a methylation by enough to explain the entire increase in observed methylation prevalence from 32 to 63%.
- Obviously, p16INK4a methylation is not always necessary for causation of NSCLC tumors, since many NSCLC tumors do not have this transformation. Therefore, we adopt a simple *competing-risks* model in which either (a) p16INK4a methylation or (b) some other cause(s) (unspecified, but that

accomplishes a similar role, making p16INK4a methylation unnecessary) contributes to the formation of NSCLCs.

- Let B denote the background rate at which p16INK4a methylation takes place in the absence of smoking, and let A denote the rate of the alternative competing cause(s). Then the probability that p16INK4a methylation contributes to a randomly selected NSCLC in a never-smoker (i.e., that p16INK4a methylation occurs before the competing risk in a cell line that eventually becomes NSCLC) is, from competing-risks theory, $B/(A + B)$. Equating this to 0.32 based on the data of Liu et al. yields

$$B/(A + B) = 0.32 \rightarrow B = 0.32A + 0.32B \rightarrow 0.68B = 0.32A \rightarrow A/B = 0.68/0.32 = 2.125.$$

- If exposure to arsenic in cigarette smoke increases the background p16INK4a methylation rate B by a factor of K that, in turn, increases the methylation probability for NSCLCs from $B/(A + B) = 0.32$ to $KB/(A + KB) = 0.63$, in keeping with the data of Liu et al., then the value of K can be determined from the data, as follows:

$$KB/(A + KB) = 0.63 \rightarrow KB(1 - 0.63) = 0.63A \rightarrow K = 0.63(A/B)/0.37 = (0.63{*}2.125)/0.37 = 3.62$$

- Based on this competing-risks model, assume that arsenic in cigarette smoke (at the levels experienced by the smokers in the study of Liu et al., 2006) increases the p16INK4a methylation rate in NSCLCs by a factor of $K = 3.62$. Further assuming that it leaves the competing-risks rate A unchanged and that one or the other of these two competing pathways accounts for the transition rate μ_{PF}, the net impact on μ_{PF} is to increase it from a background rate of $A + B = b_{PF} = 0.00006$ to a smoking-exposed level of $A + KB = A + 3.62B = b_{PF} + q_{PF}\, s = 0.00006 + 0.000012s$, where s reflects the smoking level. (This calculation assumes that the smoking model parameters apply specifically to NSCLCs. Since NSCLCs constitute a large majority – about 80% – of all lung cancers, this assumption is likely to be at least roughly correct.)
- Now we have enough data to solve for the remaining unknowns (subject to the assumptions and approximations already discussed). Combining $A/B = 2.125$ with the model constraint $A + B = b_{PF} = 0.00006$ leads to the solution $A + B = 2.125B + B = 3.125B = 0.00006$, so $B = 0.00006/3.125 = 1.92\text{E-}5$ and $A = 0.00006 - B = 0.00006 - 0.0000192 = 4\text{E-}5$. The smoking-exposed level of $\mu_{PF} = A + KB = A + 3.62\,B = 4\text{E-}5 + 3.62{*}1.92\text{E-}5 = 0.00011$ corresponds to a smoking level such that $b_{PF} + q_{PF}\, s = 0.00006 + 0.000012\, s = 0.00011$, i.e., $s = (0.00011 - 0.00006)/0.000012 \approx 4$. Inverting the formula $s = (1 - e^{-0.1*x})/0.1$ to obtain $x = -10 \ln(1 - 0.1s)$ and substituting $s = 4$ gives a corresponding smoking level of $x \approx 5$ cigarettes per day – a number that, reassuringly, is of the right order of magnitude to be plausible.

In summary, these calculations suggest that MSCE Model 1, with q_{PF} interpreted as reflecting the potency of arsenic in increasing $\mu_{PF}(t)$ via hypermethylation of the promoter region of p16INK4a, is at least roughly consistent with the empirical data of Liu et al. (2006) on the prevalence of such hypermethylation in NSCLC tumors of smokers and nonsmokers.

If this interpretation is not wildly wrong – that is, *if* MSCE Model 1 provides an approximately correct description of lung carcinogenesis, and *if* arsenic in cigarette smoke is indeed responsible for most of the increase in p16INK4a promoter region hypermethylation among smokers compared to nonsmokers, and *if* this increase in p16 methylation, in turn, causes most of the estimated increase in μ_{PF} among smokers – *then* the proportionality between the reduction in μ_{PF} and the reduction in risk in Table 12.3 implies that removing arsenic from cigarette smoke could have a large impact on lung cancer risk. It could reduce by up to 2/3 the expected number of tumor cells formed by age 80 for someone who smokes 60 cigarettes per day from age 20 to 60.

Limitations of Modeling Assumptions and Calculations

The preceding modeling and calculations are admittedly rough. They ignore interindividual heterogeneity in exposures and in parameter values; skip uncertainty, variability, and sensitivity analyses; incorporate the extreme assumption that arsenic in cigarette smoke accounts for all of the estimated increase in $\mu_{PF}(t)$ among smokers via increased methylation of p16INK4a; and gloss over important distinctions among lung tumors types, on the grounds that most lung tumors are NSCLCs. Nonetheless, that the model is at least roughly consistent with available data from the single study of Liu et al. (2006) suggests that it may provide a useful starting point estimate of the potential impact on lung cancer risk of removing arsenic. The upper-bounding assumption that arsenic in cigarette smoke accounts for the entire increase in $\mu_{PF}(t)$ among smokers compared to nonsmokers can perhaps be replaced by a refined estimate of the true fraction of this increase that is caused by arsenic when additional data become available on smoking levels, arsenic exposure levels, and corresponding p16 methylation levels. Until then, sensitivity analyses will be used to explore how the provisional conclusion that removing arsenic could remove a large proportion of malignant cells (up to two thirds of them, for the smoking scenario in Table 12.3) changes if less extreme assumptions are made.

Sensitivities, Uncertainties, Implications, and Conclusions

Based on the preceding model and calculations, what can a risk assessor tell a decision maker who wants to know whether removing arsenic from cigarettes might significantly reduce lung cancer risks to smokers? Clearly, a key scientific uncertainty is the assumption that arsenic affects lung cancer risk by increasing p16 methylation, and that this drives increases in μ_{PF}, the rate at which patch cells enter the clonal expansion phase, F. Although the calculations in the previous section suggest that this account is roughly consistent with the data of Liu et al. (2006), MSCE Model 1 is far from being well validated. On the other hand, despite these important uncertainties and knowledge gaps, we can now address some of the pragmatic questions that decision makers must struggle with. For example:

1. The question "*What is the plausible range of risk reductions in lung cancer risk from removing arsenic from cigarette smoke?*" can be answered as follows: Up to two thirds of lung cancer malignant cells (or at least of NSCLC malignant cells) might be prevented, under the assumptions of MSCE Model 1, for the smoking scenario in Table 12.3. (The simulation model in Appendix B can be used to prepare analogous tables for other smoking scenarios.) This upper bound may later prove to be too high (point 2 below), as it is contingent on a worst-case assumption that arsenic accounts for the full smoking-related increase in μ_{PF}. However, that such a high estimate is at least roughly consistent with current mechanistic knowledge and epidemiological data (as captured in the simulation model) and biomarker data (Liu et al.,2006) may justify further investigation of this relationship (point 3 below).
2. A quantitative answer to "*How sensitive is the reduction in risk to less-than-full removal of arsenic?*" is as follows. If removing arsenic leaves intact a fraction $g > 0$ of the smoking-related increase in rate μ_{PF}, then the finding of proportionality between a reduction in μ_{PF} and a resulting reduction in risk (see Table 12.3) implies that the revised, smaller upper bound will become $1 - [(b_{PF} + 10gq_{PF})/(b_{PF} + 10q_{PF})] = 1 - (0.00006 + 0.00012\ g)/0.00018$ for a smoking scenario with $s(t) = 10$, e.g., with $x = 60$ cigarettes/day. For example, if removing arsenic reduces the smoking-related increase in μ_{PF} by only 50% instead of by 100% ($g = 0.5$ instead of 0), then the new upper bound for risk reduction will become 1/3 instead of 2/3.
3. Similarly, the finding of a linear relation between the reductions in μ_{PF} (t) and reductions in risk (see Table 12.2) permits a quantitative answer to the question "*How sensitive are estimated risk reductions to the presence of other p16 methylating agents in cigarette smoke?*" If we relax the worst-case assumption that arsenic accounts for *all* of the estimated increase in $\mu_{PF}(t)$ among smokers (via increased methylation of p16INK4a), and assume instead that arsenic accounts for only a fraction f of the increase [with other agents in cigarette smoke, such as tobacco-specific nitrosamines, accounting for the remaining $(1 - f)$ of the increase], then the new upper bound for the predicted reduction in risk from removing arsenic would be $(2/3)f$ instead of 2/3. This simple sensitivity analysis also provides a straightforward way to incorporate expert judgments into the analysis to help inform investment decisions about the possible benefits of removing arsenic. For example, if an expert believes that other (nonarsenic) constituents of cigarette smoke account for at least a fraction $\geq (1 - f)$ of the increase in methylation of p16INK4a, then this can be used to tighten the upper-bound estimate on the risk reduction from removing arsenic, from 2/3 to $\leq (2/3)f$.

4. The question "*What are the most important scientific uncertainties, that, if resolved, would most improve the ability to quantify risk reductions from removing arsenic from cigarette smoke?*" can be at least partly addressed by suggesting several lines of experimental investigation, as follows. First, perform experiments to validate (or falsify and replace) the hypothesis that removing arsenic would reduce p16INK4a promoter region hypermethylation to its background level. Second, perform experiments to estimate the impact of p16INK4a hypermethylation on μ_{PF}. For example, what fraction of field cells (identified experimentally as in Wistuba and Gazdar, 2006) have both hypermethylated p16INK4a and patch markers such as LOH 3p (Table 12.2)? What fraction of premalignant lung epithelial cells with hypermethylated p16INK4a are also involved in clonal expansion in the lung? Experimental answers to these questions can potentially help to validate or refine the modeling assumptions and the resulting plausible range for preventable lung cancers caused by arsenic estimated in previous sections.

In summary, the model discussed in this chapter provides a constructive way to (a) use incomplete data to constrain the plausible range of risk reductions from removing specific constituents from a complex mixture (cigarette smoke); and (b) quantify the sensitivity of results to changes in assumptions. Even if the biological assumptions about arsenic in this chapter are eventually replaced with better ones, the ability to quantify risk reductions from interventions that act through different model parameters (Table 12.3) may still prove useful for various constituents. Although the MSCE model invites refinements (and resolution of current key scientific uncertainties) to obtain improved bounds, it also shows how available data can support interim conclusions for decision makers trying to assess whether removing particular carcinogenic constituents, such as arsenic, could create a significant reduction in human health risks. Based on the model in this chapter, the current (interim) answer for arsenic is: Yes, with reservations. Considerable work remains to be done to determine the proper extent, if any, of the reservations.

Appendix A: Listing for TSCE Model of Smoking and Lung Cancer

Intermediate_cells(t) = Intermediate_cells(t – dt) + (initiation + promotion – conversion) * dt
INIT Intermediate_cells = 0

INFLOWS:
initiation = mu1 s*Normal_cells
promotion = Intermediate_cells*es
OUTFLOWS:
conversion = mu2 s*Intermediate_cells
Malignant_cells_TSCE(t) = Malignant_cells_TSCE(t – dt) + (conversion) * dt
INIT Malignant_cells_TSCE = 0

INFLOWS:
conversion = mu2 s*Intermediate_cells
Normal_cells(t) = Normal_cells(t – dt) + (development – initiation) * dt
INIT Normal_cells = 0

INFLOWS:
development = if (TIME < 20) then (1E7/20) else 0
OUTFLOWS:
initiation = mu1 s*Normal_cells
e0 = 6.5E-2 {Schollberg, 2006, joint fit for males and females}
e1 = m1 {Schollberg, 2006, joint fit for males and females}
e2 = 1.19 {Schollberg, 2006, joint fit for males and females}
es = e0*(1 + fse1e2)
fse1e2 = e2*(1 – exp(–(e1/e2)*s))
fsm1m2 = m2*(1 – exp(–(m1/m2)*s))
m1 = 0.15 {Schollberg, 2006, joint fit for males and females}
m2 = 1.83 {Schollberg, 2006, joint fit for males and females}
Malignant_cells_TSCE_x_100 = Malignant_cells_TSCE*100
mu0 = 1.87E-7 {Schollberg, 2006, joint fit for males and females}
mu01 = mu0
mu02 = mu0
mu1 s = mu01
mu2 s = mu02*(1 + fsm1m2)

Appendix B: Listing for MSCE Lung Cancer Model with Field Carcinogenesis

F(t) = F(t – dt) + (fPF_new + net_births – fFM_new) * dt
INIT F = 0

INFLOWS:
fPF_new = P*muPF
net_births = muF*F
OUTFLOWS:
fFM_new = F*muFM
M(t) = M(t – dt) + (fFM_new) * dt
INIT M = 0

INFLOWS:
fFM_new = F*muFM
N(t) = N(t – dt) + (development_2 – fNP_new) * dt
INIT N = 0

INFLOWS:
development_2 = if (TIME < 20) then (100/20) else 0

```
OUTFLOWS:
fNP_new = N*muNP
P(t) = P(t – dt) + (fNP_new – fPF_new) * dt
INIT P = 0

INFLOWS:
fNP_new = N*muNP
OUTFLOWS:
fPF_new = P*muPF
bF = 0.08
bFM = 0.00008
bNP = 0.05
bPF = 0.00006
effective_internal_dose_s = (1 – exp(–0.1*s))/0.1
muF = (bF + qF*effective_internal_dose_s)
muFM = (bFM + qFM*effective_internal_dose_s)
muNP = if (TIME < 11) then 0 else (bNP + qNP*effective_internal_dose_s)
muPF = (bPF + qPF*effective_internal_dose_s)
qF = 0.0072
qFM = 0.0000176
qNP = 0.001
qPF = 0.000012
s = if ((TIME >= start_age) and (TIME < stop_age)) then x else 0
start_age = 20
stop_age = 60
x = 60
```

Chapter 13
Simplifying Complex Dynamic Networks: A Model of Protease Imbalance and COPD Dynamic Dose-Response

The risk models in previous chapters have emphasized causal processes described by directed acyclic graphs (Bayesian network models) and by multistage carcinogenic processes in which exposure hastens the net transition rates and increases the net proliferation rates of affected cells. More is required to predict the probable consequences of interventions in feedback-control systems, where changing the levels of controllable inputs can affect not only the specific subprocesses targeted by the interventions, but also the equilibrium levels of other variables throughout the entire interconnected system.

A dynamic system's response to attempted control actions can be notoriously counterintuitive. Students of systems dynamics in business and management science learn this through simulation games (such as the notorious "MIT beer game," which models instability in supply chains). Students of traffic network engineering learn it through Braess's paradox, in which adding capacity to a network to relieve congestion shifts the system to a new, more congested, equilibrium! In health risk assessment, disease risk may respond unexpectedly to exposures that alter feedback loops that normally maintain organs, tissues, and organisms in homeostatic equilibrium.

Simulation can predict the response of a feedback-control system to changes in inputs if an accurate simulation model of the system is available. But if only the overall structure of a system is known – its main subsystems and which ones affect others, but not the quantitative details of their input-output behaviors – then it may be impossible to develop a simulation model using standard simulation software packages. Accurate predictive simulation modeling is not possible for highly uncertain systems if it requires more detailed quantitative information than is available about the behaviors of system components and subsystems. In addition, simulation often produces more detailed answers than the decision maker cares about, including simulation of transient responses, as well as of changes in equilibrium. In many applications, it is desirable to produce less detailed outputs from less detailed inputs. Mathematical economics has developed a way to predict the probable consequences of changing inputs when many details of the modeled systems are unavailable. This is the method of *comparative statics*, which is inspired by Le Chatelier's principle in thermodynamics (Samuelson, 1947). It uses conditions that must hold in equilibrium to compare the equilibria that hold before and after an exogenous change or intervention.

L.A. Cox, Jr., *Risk Analysis of Complex and Uncertain Systems*, International Series in Operations Research & Management Science 129,
DOI 10.1007/978-0-387-89014-2_13,

This chapter develops a comparative statics analysis of how cigarette smoking increases the risk of chronic obstructive lung disease (COPD), a disease characterized by a dynamic imbalance among the levels of enzymes called *proteases*, which digest proteins (including lung tissue), and the levels of *antiproteases*, which normally hold proteases in check. In normal equilibrium, these two sets of enzymes are in dynamic balance. In COPD, the proteases dominate, and lung tissue is gradually destroyed. Comparative statics provides a useful mathematical tool for understanding the shift from a normal, homeostatic, nonsmoking, dynamic equilibrium to an abnormal, postsmoking, dynamic equilibrium in which the normal balance of proteases and antiproteases is permanently disrupted. This chapter also illustrates an important technique for reducing large models with many unknown parameters, represented by networks of interacting homeostatic feedback-control processes, to much smaller models with only a few "reduced parameters" that must be estimated from data. Such reductions in model complexity make possible the analysis of equilibria for large dynamic systems.

Background on COPD

Chronic obstructive pulmonary disease (COPD) is a smoking-associated degenerative lung disease – treatable but not reversible – that culminates in potentially lethal clinical conditions such as chronic bronchitis and emphysema. Smoking produces gradually accelerating loss of lung function in COPD patients, as measured by forced expiratory volume in 1 second (FEV1) and other measures. Quitting smoking slows the rate of further deterioration in lung function compared to its trajectory in continuing smokers, but does not reverse past losses or prevent continued deterioration. With few exceptions – most notably, a sensitive subpopulation with a genetic deficiency in the production of the antiprotease alpha-1-antitrypsin (Abboud and Vimalanathan, 2008) – strong genetic risk factors for COPD are not yet well understood. Why only some smokers (fewer than 20%) progress to severe destruction and clinical COPD (Sharafkhaneh et al., 2008), even though essentially all smokers suffer lung injury and inflammation, and why COPD and associated airway inflammation and lung tissue destruction continue, even after smoking ceases, have long been important puzzles in COPD research (e.g., Hogg, 2006).

This chapter develops and applies methods of dose-response modeling for COPD, based on a model of how long-term exposure to cigarette smoke (CS) disrupts normal homeostasis in a particular way in the network of protease (protein-digesting) and antiprotease (protease-inhibiting) enzymes mediating interactions among lung cells and the extracellular matrix (ECM) in lung tissue. Once homeostasis has been disrupted in this way, the behavior of the network shifts permanently to a new regime in which proteases dominate antiproteases, leading to the ongoing net destruction of lung tissue and degradation of the ECM. Although the qualitative hypothesis of a "protease-antiprotease imbalance" has been prominent in COPD research for decades (Abboud and Vimalanathan, 2008), this is, to our knowledge, the first mathematical model showing how this imbalance can become self-sustaining in the absence of continued exposure. The model has interesting implications for the dynamic relation between exposure and risk of COPD.

A Flow Process Network Model of Protease-Antiprotease Imbalance in COPD

Figure 13.1 presents the main features of the model. The logic is as follows. Smoke-activated alveolar macrophages (*M*) release several molecular signals (such as IL-8, LTB_4, and GRO-α). Figure 13.1 does not show these explicitly, but their effects are indicated (using the graphical notation of the ITHINK^{TM} modeling environment used to draw Fig. 13.1) by thin "information arrows" directed from compartments (state variables, represented by boxes) to flows (thick arrows pointing into or out of compartment boxes). Thus, the stocks in some compartments can affect inflow rates to, and outflow rates from, other compartments.

Some signals from *M* (notably, IL-8) stimulate an increase in lung neutrophils (*N*), largely through chemoattraction (Tudhope et al., 2008). Other signals upregulate the production of the protease macrophage elastase, MMP-12 (*P* in Fig. 13.1). (MMP-12 is produced not only by alveolar macrophages, but also by other lung cells, such as dendritic cells.) MMP-12, in turn, degrades lung tissue, especially elastic fibers, forming elastin fragments (*F*) (Maeno et al., 2007). Elastin fragments potently attract monocytes, which then mature into additional macrophages.

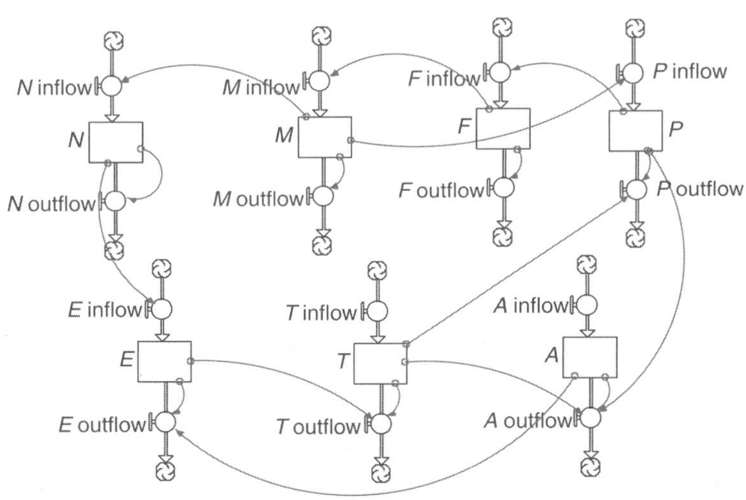

Fig. 13.1 A dynamic model of protease-antiprotease imbalance. $P =$ macrophage elastase (matrix metalloproteinase-12), $F =$ elastin fragments, $M =$ macrophages, $N =$ neutrophils, $E =$ neutrophil elastase, $T =$ tissue inhibitor of metalloproteinase-1, $A =$ alpha-1-antitrypsin. The model is implemented in the ITHINK^{TM} continuous simulation modeling environment

Thus, the following *positive feedback loop* is created: $P\uparrow \rightarrow F\uparrow \rightarrow M\uparrow \rightarrow P\uparrow$ (Sharafkhaneh et al., 2008; Maeno et al., 2007). (Here, up arrows represent upregulation; horizontal arrows indicate causation.) This loop is shown in the upper right of Fig. 13.3.

Increased MMP-12 (*P*) also increases the removal rate of the antiprotease α-1 antitrypsin (*A*), reducing its equilibrium level (Houghton et al., 2006). This, in turn, reduces the removal rate and increases the level of the neutrophil-derived protease, neutrophil elastase (*E*). MMP-induced degradation of the extracellular matrix (ECM) also liberates collagen-derived peptides that attract yet more neutrophils (*N*) into the lung (van Houwelingen et al., 2008). With more neutrophils and more neutrophil-elastase (*E*) per neutrophil, levels of *E* rise; this, in turn, increases the removal rate and reduces the level of the antiprotease TIMP-1 (tissue inhibitor of metalloproteinase-1) (*T*) (Houghton et al., 2006; Shapiro et al., 2003). The decrease in TIMP-1 (*T*) reduces the removal rate and increases the equilibrium level of macrophage elastase (*P*) (Shapiro et al., 2003). The bottom half of Fig. 13.1 shows this second positive feedback loop:

$$P\uparrow \rightarrow A\downarrow \rightarrow E\uparrow \rightarrow T\downarrow \rightarrow P\uparrow .$$

A third positive feedback loop through *P* is created by the fact that neutrophils produce neutrophil elastase: $P\uparrow \rightarrow F\uparrow \rightarrow M\uparrow \rightarrow N\uparrow \rightarrow E\uparrow \rightarrow T\downarrow \rightarrow P\uparrow$.

Thus, once *P* starts to increase, several positive feedback loops tend to amplify the increase [as well as the levels of macrophages (*M*), neutrophils (*N*), neutrophil elastase (*E*), and elastin fragments (*F*), all of which are positively involved with *P* in the feedback loops] until a new equilibrium is reached.

Yet despite these positive feedback loops, most cigarette smokers do not develop COPD (Sharafkhaneh et al., 2008). Analyzing the equations corresponding to the process flow network in Fig. 13.1 suggests a possible explanation in terms of the existence of two distinct, locally stable, equilibria, as detailed in the next section.

The model in Fig. 13.1 is admittedly a very high-level, simplified, and aggregate description of selected processes of importance in COPD. Many of the links in this diagram could in principle be recursively expanded into entire pathways or subnetworks, down to the molecular level. Figure 13.1 deliberately ignores all lower-level details and many short-term transients. It represents the roles of crucial cell populations, such as CD4+ and CD8+ T lymphocytes (Borchers et al., 2007; Maeno et al., 2007) and dendritic cells, only implicitly, by their effects on the quantities in Fig. 13.1. Other consequences of activated neutrophils and macrophages are not described, including the increased apoptosis of alveolar cells (due to the release of cytokines such as IL-18, which upregulates the production of caspases and cathepsins that increase apoptosis of epithelial cells) and changes in macrophage phenotypes that reduce their ability to recognize, respond appropriately to, and remove apoptotic cells. Although apoptosis of alveolar cells without a compensating increase in production, leading to a net destruction of alveolar tissue, may be critical in causing emphysema (Demedts et al., 2006b), Fig. 13.1 only considers the production of elastin fragments, and not other correlated events such as apoptosis-driven tissue destruction. Thus, Fig. 13.1 represents only a skeleton or core of a description of protease-antiprotease imbalance in COPD-related emphysema.

Despite these limitations, the model of protease-antiprotease imbalance in Fig. 13.1 has considerable support from experimental data. Table 13.1 summarizes the results of recent experimental investigations of COPD etiology – or, more accurately, smoking-associated emphysema-like changes in mice. [There are no fully satisfactory animal models for human COPD (Churg and Wright, 2007). Animals, unlike humans, do not develop a degenerative version of the disease that persists even after smoke exposure ceases (Wright et al., 2008). Why this is so has been a mystery in COPD research; a possible explanation is suggested below.]

Table 13.1 Evidence for causal links in Fig. 13.1, from experiments in mice

Link	Comments on more detailed pathways and evidence for links	References
$P\uparrow \rightarrow F\uparrow$	"[T]he CD8+ T cell product, IFN-gamma-inducible protein-10, induces production of macrophage elastase (matrix metalloproteinase 12) that degrades elastin, both causing lung destruction directly and generating elastin fragments that serve as monocyte chemokines augmenting macrophage-mediated lung destruction."	Maeno et al. (2007)
$F\uparrow \rightarrow M\uparrow$	"This EF [elastin fragments, *F*] chemotactic signal ultimately leads to enhanced lung tissue destruction by placing more macrophages, and thus more MMP-12, within close proximity of the airspace."	Houghton et al. (2006)
	"Elastin fragment antagonism in this model abrogated both macrophage accumulation and airspace enlargement."	Shapiro et al. (2003)
$M\uparrow \rightarrow P\uparrow$	"Moreover, there was less macrophage elastase activity secondary to decreased macrophage accumulation in neutrophil elastase-deficient mice."	Shapiro et al. (2003)
$M\uparrow \rightarrow N\uparrow$	Activated alveolar macrophages (*M*) release cytokines TNF-α and GM-CSF that stimulate production of IL-8, a potent chemoattractant for recruiting neutrophils (*N*) to the lung, via a p38 MAPK pathway. (TNF-α and GM-CSF also prime the neutrophils.)	Tudhope et al. (2008)
$N\uparrow \rightarrow E\uparrow$	Production of neutrophil elastase (*E*) by lung neutrophils is regulated in part by SLPI via an NFκB-dependent pathway that is compromised in Nrf-2 knockout mice.	Iizuka (2005)
$E\uparrow \rightarrow T\downarrow$	"[Neutrophil elastase, *E*] is capable of degrading the tissue inhibitors of metalloproteinases (TIMPs), the major inhibitors of MMP-12.	Houghton et al. (2006)
$T\downarrow \rightarrow P\uparrow$	Thus it appears that these proteinases are able to indirectly augment each other's function by degrading one another's inhibitors."	
	"Each elastase inactivated the endogenous inhibitor of the other, with neutrophil elastase degrading tissue inhibitor of metalloproteinase-1, and macrophage elastase degrading α1-antitrypsin."	Shapiro et al. (2003)
$P\uparrow \rightarrow A\downarrow$ $A\downarrow \rightarrow E\uparrow$	"Furthermore, MMP-12 degrades α1-antitrypsin, whose major function is to inhibit NE."	Houghton et al. (2006)

The overall causal structure in Fig. 13.1 shows MMP-12 (P) as the unique node common to all three positive feedback loops. Consistent with this postulated linchpin role, mice deficient in MMP-12 (unlike wild-type mice) do not develop emphysema or an increase in lung macrophages when exposed to cigarette smoke (CS) (Hautamaki et al., 1997; Houghton et al., 2006). Likewise, mice with a CD8+ T-cell deficiency that inhibits normal increases in MMP-12 during CS exposure are protected from macrophage and neutrophil accumulation and emphysema (Maeno et al., 2007), as are senescence-accelerated mice with inhibited MMP-12 production (Mori et al., 2008). MMP-12 increases appear to be both necessary and perhaps sufficient (Hautamaki et al., 1997) for long-term CS exposures to induce emphysema in mice.

In humans, too, MMP-12 is significantly elevated in the lungs of COPD patients compared to healthy smokers, and even more so compared to levels in never-smokers (Molet et al., 2005; Babusyte et al., 2007). Both tobacco smoke and wood smoke increase MMP-12 levels in human subjects *in vivo* (Montaño et al., 2004). Other MMPs, such as MMP-2, MMP-8, and MMP-9, may also be important in human COPD (Vernooy et al., 2004); thus, for humans, P in Fig. 13.1 may be a surrogate for several strongly correlated MMPs, of which MMP-12 is the most clearly identified.

In summary, while the model in Fig. 13.1 is admittedly simplified, and much more detailed and complete models of COPD pathogenesis should and doubtless will be developed, Fig. 13.1 does capture key causal relations that appear to be empirically important in mice and that are consistent with human data. The following sections analyze the implications of this model for smoking-COPD dose-response relations.

Mathematical Analysis of the Protease-Antiprotease Network

Table 13.2 summarizes a system of ordinary differential equations (ODEs) corresponding to Fig. 13.1. These model equations incorporate some simplifying assumptions and approximations, such as that the inflow (production) of new elastin fragments (F) is approximately proportional to P (MMP-12), that the fractional degradation rate of elastin fragments per unit time is approximately constant, and that neutrophil elastase is produced at a rate proportional to the number of neutrophils (N) and is removed at a fractional rate proportional to alpha-1-antitrypsin (A). The complete dynamic model has seven variables (P, F, M, N, A, E, T), seven equations, and 18 parameters (a_{FP}, d_F, b_M, a_{MF}, d_M, b_N, a_{NM}, d_N, b_A, e_{AP}, a_{EN}, e_{EA}, b_T, e_{TE}, b_P, a_{PM}, d_P, e_{PT}). The notation is that b_j is the unstimulated inflow ("birth" of new material) into compartment j; a_{ji} is the additional inflow into compartment j per unit time stimulated by each unit of material in predecessor compartment i (thus, a_{ji} is the potency of i in stimulating production of j, assuming for simplicity an approximately directly proportional relation, e.g., because all levels are well below saturation); d_j is the unstimulated fractional rate of clearance or outflow ("death")

Table 13.2 Model equations and simplifications

Link	Model equation	Equilibrium simplification	Justification
$P\uparrow \rightarrow F\uparrow$	$dF/dt = a_{FP}P - d_F F$	$F = P$	See text.
$F\uparrow \rightarrow M\uparrow$	$dM/dt = b_M + a_{MF}F - d_M M$	$M = w_M P + (1 - w_M) = w_M(P - 1) + 1$, where $w_M = a_{MF}/(a_{MF} + b_M)$	Use $F = P$ & initial equilibrium condition $b_M + a_{MF} = d_M$.
$M\uparrow \rightarrow N\uparrow$	$dN/dt = b_N + a_{NM}M - d_N N$	$N = w_N w_M(P - 1) + 1$, where $w_N = a_{NM}/(a_{NM} + b_N)$	Similar to formula for M.
$P\uparrow \rightarrow A\downarrow$	$dA/dt = b_A - e_{AP}AP$	$A = 1/P$	For $A = P = 1$ and $dA/dt = 0$, $b_A = e_{AP}$, so $(1 - AP) = 0$.
$N\uparrow \rightarrow E\uparrow$ $A\downarrow \rightarrow E\uparrow$	$dE/dt = a_{EN}N - e_{EA}AE$	$E = NP = [w_N w_M(P - 1) + 1]P$	In initial equilibrium, $N = AE$. Then, use $A = 1/P$.
$E\uparrow\rightarrow T\downarrow$	$dT/dt = b_T - e_{TE}TE$	$T = 1/E = 1/[w_N w_M(P - 1) + 1]P$	Similar to formula for A.
$M\uparrow \rightarrow P\uparrow$ $T\downarrow \rightarrow P\uparrow$	$dP/dt = b_P + a_{PM}M - d_P P - e_{PT}PT$	$b_P + a_{PM}M = d_P P + e_{PT}PT$	$dP/dt = 0$ in equilibrium.

of material from compartment j (per unit of material in j per unit time); and e_{ji} is the potency of i in stimulating additional removal of j.

The "Equilibrium Simplification" column of Table 13.2 shows how these equations simplify in equilibrium, so that the equilibrium values of all six variables other than P are determined from the value of P and only two reduced parameters, w_M and w_N. To accomplish this reduction from 18 to two unknown parameters, we first normalize all state variables (i.e., compartment levels), without loss of generality, to make their levels 1 in the initial (unexposed) equilibrium. Then, setting time derivatives to zero in equilibrium, it follows that $dF/dt = a_{FP}P - d_F F$, for example, implies the initial equilibrium condition: $0 = a_{FP}P - d_F F = a_{FP} - d_F$ (using the initial condition normalization, $F = P = 1$). Hence, $a_{FP} = d_F$. If P now changes away from its initial equilibrium value of $P = 1$ to some new value, then the new equilibrium condition $a_{FP}P - d_F F = 0$ (together with the parameter value constraint $a_{FP} = d_F$) implies $a_{FP}(P - F) = 0$, or $F = P$. Other equilibrium simplifications in Table 13.2 are derived similarly, as sketched in the "Justification" column.

If exposure to cigarette smoke (CS) causes an increase in P (Maeno et al., 2007), then the formulas in the "Equilibrium Simplification" column of Table 13.2 show how all other quantities will adjust to the new level of P. The equilibrium inflows and outflows for P can also be studied as functions of P:

$$\textit{Inflow to } P = (b_P + a_{PM} - a_{PM}w_M) + a_{PM}w_M P, \tag{13.1a}$$

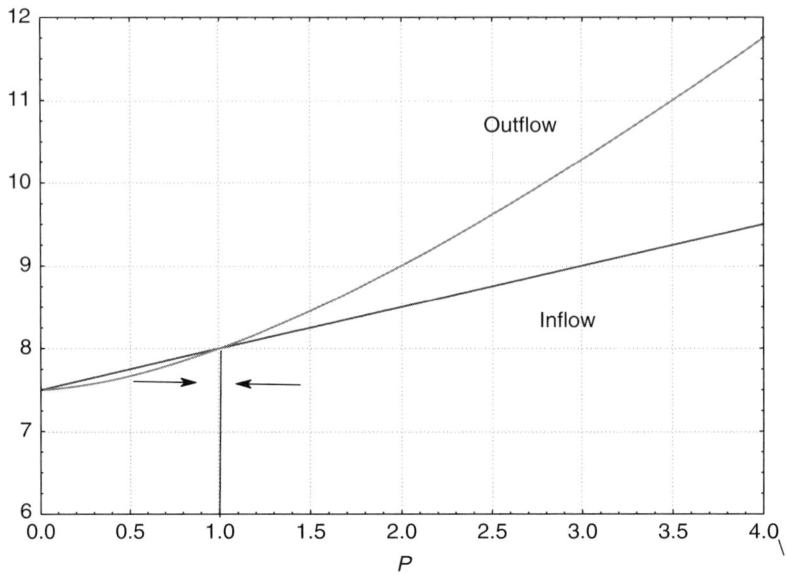

Fig. 13.2 Unique stable "healthy" equilibrium at $P^* = 1$. (Parameter values: $b_P + a_{PM} = 8$, $a_{PM}w_M = 0.5$, $d_P = 2$, $w_N w_M = 0.2$.)

$$\textit{Outflow} \text{ from } P = d_P P + e_{PT}/[w_N w_M(P-1)+1], \tag{13.1b}$$

where $w_M = a_{MF}/(a_{MF} + b_M)$ and $w_N = a_{NM}/(a_{NM} + b_N)$.

Inflow is a straight-line function of P with y-intercept $(b_P + a_{PM} - a_{PM}w_M)$ and slope $a_{PM}w_M$. *Outflow* asymptotically approaches a straight line with slope d_P, for large values of P. [Its y-intercept is $e_{PT}/(1 - w_N w_M)$, where $e_{PT} = b_P + a_{PM} - d_P$ in the initial normalized equilibrium with all variables equal to 1. Note that if b_P is exogenously increased by an exposure that increases inflow, however, e_{PT} need not change.] Equating (13.1a) and (13.1b) gives a quadratic equation. The two curves intersect at $P = 1$, and perhaps at one other equilibrium point, P^*.

An equilibrium point P^* is locally stable if *Inflow* intersects *Outflow* from above at P^*, i.e., if *Inflow* > *Outflow* for $P < P^*$ and *Inflow* < *Outflow* for $P > P^*$, since then P will increase when it is less than P^* and will decrease when it is greater than P^*. If *Inflow* intersects *Outflow* from below, then the equilibrium where they intersect is unstable: Any small change in P will be amplified. Since it is not biologically realistic for P to increase without bound, we assume a maximum possible inflow rate, $\textit{Inflow}_{\max}$, determined by rate-limiting steps in the production rate of P. In other words, $\textit{Inflow}_{\max}$ is the saturation level for inflow of P.

As an example, Fig. 13.2 plots *Inflow* and *Outflow* as functions of P for the following parameter values: $b_P + a_{PM} = 8$, $a_{PM}w_M = 0.5$, $d_P = 2$, $w_N w_M = 0.2$. [These parameter values were selected simply to provide a concrete illustration. As discussed below, there are essentially two types of possible equilibrium behaviors of the system, illustrated in Figs. 13.2 and 13.3, respectively, due to the fact that

Equation (13.2) below is quadratic. The precise values of the parameters do not matter in illustrating these possibilities.] In this case, $P^* = 1$ is the unique stable equilibrium. To the left of 1, *Inflow* exceeds *Outflow*; to the right, *Outflow* exceeds *Inflow*. Thus, any exogenous one-shot change in the level of P (e.g., due to a transient increase or decrease in b_p) will be undone as the system operates to restore the homeostatic equilibrium.

If exposure to cigarette smoke shifts the *Inflow* line upward and/or the *Outflow* curve downward during exposure (e.g., by increasing b_P), then their intersection point – the unique stable equilibrium – will shift rightward, to higher levels of P. Other variables [elastin fragments (F), macrophages (M), neutrophils (N), and neutrophil elastase (E)] will also shift to new, higher levels during exposure, according to the formulas in Table 13.2. [CS may also act directly to increase some of these elements, as well as acting indirectly via the feedback loops in Fig. 13.1 when P increases (Churg et al., 2002).] Once exposure stops, if the curves return to their original positions (e.g., if b_P falls back to its original level), then the presmoking equilibrium will be restored.

In Fig. 13.2, *Inflow* only slightly exceeds *Outflow* for P values between 0 and 1. The gap between *Inflow* and *Outflow* can be increased at all points other than $P = 1$, where they intersect, by decreasing the slope of the *Inflow* line ($a_{PM}w_M$). Geometrically, decreasing $a_{PM}w_M$ while holding other parameters fixed pivots the *Inflow* line clockwise (flatter slope) about the fixed point ($P = 1$, *Inflow* $= b_P + a_{PM}$).

Conversely, increasing the slope of the *Inflow* line ($a_{PM}w_M$) rotates it counterclockwise. Increasing the slope enough to make it steeper than that of the outflow curve has a dramatic qualitative effect on the behavior of the system, as illustrated in Fig. 13.3. Here, $a_{PM}w_M$ has been increased from 0.5 to 1.5. (The *Outflow* curve is the same as in Fig. 13.2 but is now shown on an expanded scale.) Now, $P = 1$ is an *unstable equilibrium.* Any increase of P above 1 is amplified by the positive feedback loops in Fig. 13.1, leading to further increases until a new stable equilibrium is reached at $P = 8$. Any smoking-related upward shift in the *Inflow* line (or downward shift in the *Outflow* curve) will shift this new stable equilibrium still further rightward (and will bring $P = 1$ into its basin of attraction, so that P will spontaneously increase above 1).

Even if these vertical shifts are undone when smoking ceases, *the system will not return to its presmoking equilibrium* at $P = 1$ while the increase in slope $a_{PM}w_M$ persists. Instead, it will settle at the new stable equilibrium level ($P = 8$ in Fig. 13.3) following the cessation of exposure.

Figure 13.4 plots essentially the same information in a different way. It shows the *net inflow* into P, defined as the difference between *Inflow* and *Outflow*, as a function of the level of P:

$$\begin{aligned}
NetInflow &= (Inflow - Outflow) = dP/dt \\
&= (b_P + a_{PM} - a_{PM}w_M) + a_{PM}w_M P - d_P P - e_{PT}/[w_N w_M(P-1)+1] \\
&= b_P + a_{PM}(1 - w_M) + (a_{PM}w_M - d_P)P - e_{PT}/[w_N w_M(P-1)+1].
\end{aligned}$$

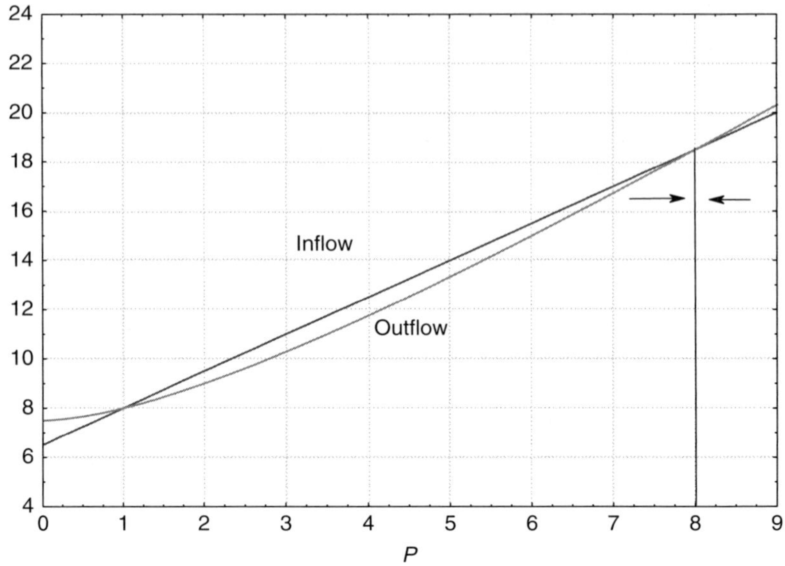

Fig. 13.3 Unstable equilibrium at $P = 1$, stable "COPD" equilibrium at $P = 8$. (Parameter values: $b_P + a_{PM} = 8$, $a_{PM}w_M = 1.5$, $d_P = 2$, $w_N w_M = 0.2$.)

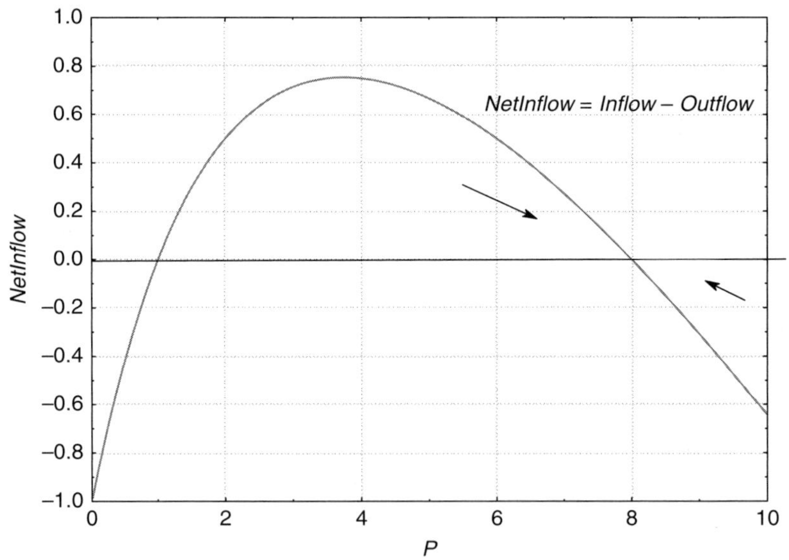

Fig. 13.4 *Net inflow* for the model in Fig. 13.3 is zero at equilibrium points

In this diagram, the equilibrium points occur where *NetInflow* = 0. An equilibrium is stable if the *NetInflow* curve cuts the x-axis (*NetInflow* = 0) from above.

If the slope of the *Inflow* line is increased further (so that $a_{PM}w_M > d_P$), then it will lie above the *Outflow* curve at all points to the right of $P = 1$. In this case, similar to Fig. 13.4, any initial increase in P will be amplified until a new, saturated, stable equilibrium is reached with *Inflow* = *Outflow* = *Inflow*$_{\max}$. (Any reduction in P below 1 would also be amplified until $P = 0$, but this does not occur if exposures increase P.)

Some Possible Implications for Experimental and Clinical COPD

Figs 13.2, 13.3, and 13.4 illustrate two different possibilities for how the protease-antiprotease network in Fig. 13.1 can respond to exogenously caused changes in the level of P. Figure 13.2 is the case of a *stable unique equilibrium system*, in which the *NetInflow* curve (not shown) cuts the x-axis exactly once, from above, with a negative slope. An exogenous increase in the *Inflow* line (e.g., in b_P) shifts this equilibrium rightward, but returning b_P to its original level will restore P to its original equilibrium. This is qualitatively similar to experimental emphysema observed in animal models (Churg and Wright, 2007).

By contrast, Fig. 13.3 illustrates the case of *two equilibria*, an unstable one to the left and a stable one to the right. In this system, an exogenous increase in b_P (shifting the *NetInflow* curve up) would shift the left equilibrium further leftward and the right one further rightward. A system initially at $P = 1$ will spontaneously move rightward to the new ("COPD") equilibrium, and stay there. Restoring b_P to its initial level will *not* restore such a system to its initial (left) equilibrium. P and other variables that depend positively on it (F, M, N, E) will remain permanently elevated (although at lower levels when smoking cessation brings b_P back down than if smoking continues). This pattern is qualitatively similar to that of smoking-induced COPD observed in human patients.

What can turn a system of the first (stable) type into one of the second type? It suffices to increase the slope of the *Inflow* line, e.g., by increasing the parameter a_{PM}, representing the average production of MMP-12 per alveolar macrophage. Thus, it is tempting to conjecture that lasting smoking-induced changes in human macrophage phenotypes (Löfdahl et al., 2006) explain the induction of irreversible COPD. For this explanation to be biologically plausible, however, COPD patients would have to have steeper *Inflow* lines than healthy smokers. *This is a testable implication,* as discussed in the next section.

To recapitulate our hypothesis for how smoking causes COPD, if smoking gradually and permanently tilts the *Inflow* line in Fig. 13.2 upward until it becomes steeper than the outflow curve at $P = 1$, it will destabilize the normal ("healthy") equilibrium and induce a new, stable ("COPD") equilibrium to the right of it, as in Figs. 13.3 and 13.4. (The location of the new equilibrium value of P will be at the smaller of the intersection of the *Outflow* curve with the tilted *Inflow* line or with the horizontal $Inflow_{\max}$ line, not shown in Figs. 13.3 and 13.4.) If smoking not only tilts the *Inflow* line upward but also reversibly increases its height [e.g., due to an increase in b_P during smoking, as CS causes CD8+ T-cell-mediated increases in MMP-12 production (Maeno et al., 2007)], then, after smoking ceases, the levels of P and other quantities that depend positively on it would decline to the new stable equilibrium level, but not to the initial equilibrium level.

Is the Model Consistent with Available Human Data?

Directly identifying the theoretical *Inflow* and *Outflow* curves from clinical or epidemiological data may be difficult or impossible, because smoking affects many model input parameters simultaneously (e.g., Churg et al., 2002). However, the hypothesis that smoking tilts the *Inflow* line permanently upward can nonetheless be assessed from limited human data. Recall that the slope of the *Inflow* line is proportional to parameter a_{PM}, representing the *average production of MMP-12 per alveolar macrophage*. Thus, we can ask whether, in clinical reality, MMP-12 production is permanently elevated in COPD patients compared to non-COPD smokers (and, a fortiori, compared to never-smokers).

Table 13.3 summarizes data from two recent studies on MMP-12 levels measured in bronchoalveolar lavage (BAL) and induced sputum (IS) of present, former, and never-smokers. These data show a roughly 1.5- to twofold increase in the measured MMP-12 levels in healthy smokers compared to never-smokers; and a further roughly three- to fourfold increase in the measured MMP-12 levels among COPD smokers compared to healthy smokers. Much of this increase persists among COPD ex-smokers. On the other hand, macrophage counts are not significantly elevated in IS from COPD patients compared to IS from nonsmokers (Rufino et al., 2007). This pattern is therefore very consistent with the hypothesis that *MMP-12 per macrophage is increased among COPD patients,* corresponding to an upward tilt in the *Inflow* line. [Demedts et al. (2006a) also reported MMP-12 enzymatic activity 30 times higher in patients with COPD than in controls.])

Table 13.3 MMP-12 levels are elevated in COPD compared to non-COPD smokers

Population	MMP-12 protein levels in induced sputum (IS)	Number of BAL MMP-12 + macrophages
COPD patients (IS) COPD smokers (BAL)	17.5 ng/mg	1.6×10^6/ml
Ex-smokers (IS) COPD ex-smokers (BAL)	6.1 ng/mg	0.9×10^6/ml
Healthy smokers	6.7 ng/mg	0.4×10^6/ml
Never-smokers	4.2 ng/mg	0.2×10^6/ml
	Data: Demedts et al. (2006a)	*Data:* Babusyte et al. (2007)

Other studies have confirmed that COPD patients produce more MMP-12 than controls (Molet et al., 2005) and that macrophage phenotypes differ in COPD patients compared to healthy smokers (and in healthy smokers compared to non-smokers), but do not appear to change following smoking cessation (Domagała-Kulawik, 2003, 2006). Molet et al. (2005) found that

> The number of MMP-12 expressing macrophages together with the staining intensity was higher in BAL samples from COPD patients than in control subjects. Similar results were noted in bronchial biopsies with higher MMP-12 expression in COPD subjects than in controls. Enhanced MMP-12 level was also observed in BAL fluids from patient with COPD in comparison to control subjects. [T]his study demonstrated that COPD patients produce greater quantities of MMP-12 than controls, which may be a critical step in the pathogenesis of COPD and emphysema.

CS exposure causes other lasting changes in macrophage phenotypes among COPD patients. For example, Li et al. (2006) found significantly increased NF-κB activity induced by TNF-α in alveolar macrophages from patients with COPD. Neutrophil elastase (E) cleaves specific (phosphatidylserine) receptors on macrophages, reducing their ability to recognize and remove apoptotic cells and to resolve inflammation (Demedts et al., 2006b).

In summary, it seems plausible that CS exposure in humans might lastingly alter macrophage phenotypes, increasing the average MMP-12 produced per activated alveolar macrophage (M) and thus steepening the *Inflow* line. For patients with sufficiently responsive macrophages (e.g., due to normal phenotypic variation in the population), the *Inflow* line may become steeper than the *Outflow* curve in the vicinity of $P = 1$ in Fig. 13.2, leading to a shift from normal to COPD equilibrium.

Whether these changes account for a CS-induced shift from a stable, normal, healthy equilibrium to a new stable COPD equilibrium remains to be empirically verified. Some additional testable predictions that might help to confirm or refute the proposed conceptual model are as follows:

- *Heterogeneity:* Smokers who will develop COPD exhibit larger increases in MMP-12 per macrophage following CS exposure than smokers who do not develop COPD. (This prediction and the following ones are contingent on the hypothesis that the specific mechanism by which the *Inflow* line becomes steeper than the *Outflow* curve at $P = 1$ in Fig. 13.2, for COPD-prone smokers, is a shift in alveolar macrophage phenotypes toward ones that produce more MMP-12-per-macrophage.)
- *Reversibility:* Smokers who quit before the MMP-12 per macrophage becomes too large (or, more generally, before the *Inflow* line becomes steeper than the *Outflow* curve at $P = 1$ in Fig. 13.2) will experience largely *reversible* symptoms, with P and other model variables returning to approximately their initial levels.
- *Species differences:* Mice and other experimental animals (e.g., rats, guinea pigs) that develop reversible inflammation and emphysema-like symptoms only in response to continuing CS exposure have *Inflow* lines that never become steeper than the *Outflow* curve, i.e., MMP-12 per macrophage never becomes large enough to trigger irreversible COPD. If a genetically engineered strain were

developed with a higher MMP-12 per macrophage, irreversible COPD could be induced in these species.
- *Dose-response:* If COPD occurs (and is irreversible) once the cumulative shift in the distribution of alveolar macrophage phenotypes makes the *Inflow* line steeper than the *Outflow* curve at $P = 1$ in Fig. 13.2, then a COPD-prone smoker will develop COPD when *cumulative* smoke exposure passes a threshold.

If this proposed model is substantially correct, then, in the COPD equilibrium, elevated levels of MMP-12 (P) are predicted to persist even after smoking ceases, although at lower levels than during continued smoking. (The COPD equilibrium is shifted even further right during continued smoking if smoking shifts the *NetInflow* curve upward. If the *NetInflow* curve then shifts back downward when smoking ceases, then P levels decline from the levels reached during smoking to the new stable equilibrium corresponding to the right x-intercept of the *NetInflow* curve.) This prediction is consistent with the empirical data in Table 13.3. Other variables that depend on P are predicted to follow similar time courses. These include alveolar macrophages (M) and their products, such as IL-8, IL-18, caspases, and TNF-α; neutrophils (N), neutrophil elastase (E), CD8+ T-cells (not shown in Fig. 13.1), elastin fragments (F), apoptotic epithelial cells, and reparable damage to alveolar walls and to the extracellular matrix (not shown in Fig. 13.1).

Summary and Conclusions

This chapter has shown how networks of interacting homeostatic equilibrium processes, represented by systems of coupled ODEs, can be simplified to analyze equilibrium responses to exogenous stresses. Applied to a simple model of protease-antiprotease interactions and imbalance widely believed to be important in the pathology of COPD (Fig. 13.1), these methods reduce a complex system of seven ODEs with 18 unknown parameters (Table 13.2) to a simpler system described by a single ODE [Equation (13.2)] whose equilibrium behavior is determined by a quadratic equation with only two unknown parameters. One additional parameter value then determines the corresponding equilibrium values of all six other model variables (formulas in Table 13.2).

Even without estimating parameter values, analysis shows that the model predicts two possible qualitative behaviors for the network in Fig. 13.1, depending on whether the *NetInflow* curve [described by Equation (13.2) and illustrated in Fig. 13.4] cuts the x-axis only once, from above (yielding a unique stable equilibrium, as in Fig. 13.2), or twice, once from below and once from above (the situation shown in Fig. 13.4). In the latter case, a transient upward shift in the *NetInflow* curve during smoking will displace the initial equilibrium to a new, stable, COPD equilibrium with elevated levels of several variables that remain elevated even after smoking ceases. We propose this as a conceptual model for the protease-antiprotease imbalance in human COPD etiology.

Appendix A: Equilibrium in Networks of Homeostatic Processes

Representing Biological Knowledge by Networks of Flow Processes

Consider a network of flow processes, each represented by an ordinary differential equation (ODE) of the form

$$dX/dt = \text{rate of inflow to } X \text{ at time } t - \text{rate of outflow from } X \text{at time } t.$$

X is interpreted as the level of some quantity in some biological compartment (which we will also call *X*). Changes in the level of *X* are conceived of as resulting from inflows to and/or outflows from this compartment. The levels of some quantities may affect the rates of change (via inflows and outflows) in the levels of other quantities. Directing an information arrow (indicating a causal dependency relation, i.e., that the quantity at the arrow's tail affects the quantity at its head) from each compartment to the flows that it affects creates a network of flow processes.

The logical structure of causal dependencies among processes in the network can be depicted in a directed graph where nodes represent the levels of individual flow processes (compartments) and arrows between nodes are information arrows. It is traditional in systems biology diagrams to use two types of information arrows to distinguish visually between upregulating ($\rightarrow$) and downregulating (–|) influences; thus, $Y \rightarrow X$ signifies that an increase in *Y* causes an increase in *X* (all else being held equal), while *Y* –|*X* indicates that an increase in *Y* blocks, inhibits, or reduces the level of *X*. In our framework of dynamic processes, it is useful to accomplish something similar by representing the *level* of each variable as a node (e.g., a box) in the directed graph, attaching an inflow arrow and an outflow arrow to each compartment (represented by a node), and then directing information arrows specifically to the appropriate inflow and outflow arrows that they affect. (See Fig. 13.5.) Such flow processes can easily be represented in the Systems Biology Markup Language (SBML) or in continuous simulation modeling environments such as STELLA®/ITHINK® or ACSL. We use ITHINK® diagrams in this chapter because ITHINK has a well-developed (and fairly well-known) graphical notation for displaying compartments, flows, and information arrows.

To deal more specifically with biological systems consisting of networks of processes that normally maintain themselves in homeostatic equilibrium, but that may be perturbed by exposures, we specialize the preceding very general SBML framework in the following ways:

1. Each compartment is assumed to have a *normal equilibrium level.* Without loss of generality, we choose units so that this level is 1.
2. The inflow into a compartment consists of two components: a baseline inflow that originates from sources exogenous to the system being modeled. and a component that may depend on the levels of other state variables (i.e., other compartment levels) within the system being modeled. We will call the latter component

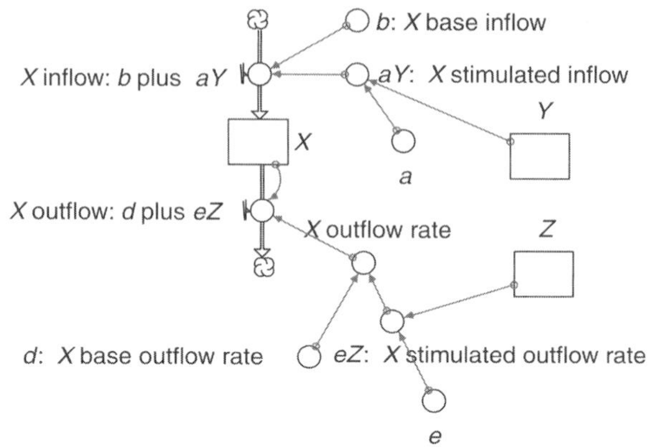

Fig. 13.5 ITHINK® diagram for the process described by Equation (13.3). *Thick arrows* denote flows, *thin arrows* are information arrows, *boxes* represent state variables (compartments), *circles* represent quantities calculated from the quantities that point into them via information *arrows*

the "stimulated inflow" into the model (since it is inflow that may be stimulated by the current levels of other processes/compartments).

3. Similarly, the total outflow from a compartment consists of two components: (1) the product of the current compartment level times a baseline outflow rate and (2) the product of the current compartment level times a stimulated outflow rate that may depend on the current levels of other processes/compartments. Unlike inflows, which typically originate from outside a compartment, outflows drain the level of the compartment itself, and hence are limited by the current contents of the compartment (i.e., zero level in the compartment implies a zero net outflow). Therefore, it is convenient to express outflows in terms of products of the compartment level and outflow rates (which are measured in units of the fraction of the compartment contents cleared per unit time by each of the two outflow components, baseline and stimulated).
4. In normal homeostatic equilibrium, when all compartment levels equal 1, the inflow into each compartment equals the outflow from that compartment.

These assumptions are made in order to model networks of processes that may be coupled to each other (via dependencies of stimulated inflow or outflow rates in some compartments on levels in other compartments) but that ordinarily maintain themselves in a stable, homeostatic equilibrium, signified by all compartments being at their nominal levels of 100%. Thus, the outflow from one compartment is not required to be shown as entering into one or more other compartments, nor are flows assumed necessarily to be conserved within the modeled system. (Such traditional compartmental modeling of conserved flows can be captured if desired, by directing information arrows from the outflows of compartments to the inflows of their downstream compartments, and performing the appropriate calculations to allocate conserved flows.) Our networks are intended not primarily for such traditional compartmental modeling, but rather for modeling coupled homeostatic processes in which levels and/or flows may be perturbed by exogenous exposures and may influence each other's rates (typically by upregulating inflows or outflows, thus upregulating or downregulating, respectively, resulting compartment contents).

Example: ODE and ITHINK® Representations of a Single Process

Equation (13.3) is a typical ODE description for a single process, compartment *X*, with inflow and outflow rates dependent on the levels of two other compartments, *Y* and *Z*, respectively.

$$dX/dt = b + aY - dX - eXZ = (b + aY) - (d + eZ)X. \tag{13.3}$$

X denotes the level of the compartment (a state variable of the model), *b* is the baseline inflow into compartment *X* (independent of the levels of other variables in the model), *Y* is the level of an inflow-stimulating variable, *Z* is the level of an outflow-stimulating variable, and *a* and *e* are coupling coefficients (interpreted as potency factors for stimulating the production and removal of *X*, respectively) showing how strongly the inflow and outflow rates for *X* are stimulated per unit of *Y* and per unit of *Z*, respectively. Of course, other functions could be used to relate the levels of other variables to the inflow and outflow rates of *X*, and any of the nonnegative coefficients *a, b, d,* and *e* could be zero for some processes. However, with the scale for each variable selected to satisfy the constraint that normal (unperturbed) equilibrium is defined as $X = Y = Z = 1$, the coefficients must satisfy $b + a = d + e$, so that inflow and outflow are balanced at equilibrium. Figure 13.5 shows an ITHINK® diagram representing the process for *X* in Equation (13.3).

Note that if *Y* is exogenously increased to a new value $Y^* > 1$, then *X* adjusts to a new, increased level at which the inflow and outflow are again balanced. The condition for equilibrium is that *X* no longer changes:

$$dX/dt = b + aY - dX - eXZ = 0, \text{ or } b + aY^* = (d + eZ)X^*,$$

implying that the new equilibrium value of *X* is

$$X^* = (b + aY^*)/(d + eZ). \tag{13.4}$$

If *Z* remains fixed at its equilibrium value of 1, then the new equilibrium value of *X* after the process adjusts to the new value Y^* will be greater than 1. It is easy to see from (13.3) that *X* will increase if it is below X^* (since inflow then exceeds outflow) and will decrease if it is above X^* (since outflow then exceeds inflow), so the value of X^* given by Equation (13.4) is a unique, globally stable equilibrium. In particular, changes in the level of *X* caused by temporary exogenous changes in the levels of *Y* or *Z* will be reversed after *Y* and *Z* are returned to their original values of 1. Likewise, a one-time exogenous change in the level of *X* itself (e.g., an injection or withdrawal that increases or decreases the value of *X* to a new level different from

1) will not sustain itself: Without continued intervention, the system will return to its unique stable equilibrium at $X = Y = Z = 1$ from any starting values of these variables. This captures the idea that X is a homeostatic process. We will investigate how networks of such processes, interacting through coupling coefficients (such as a and e in this example), behave in response to exogenous stresses.

Reducing Chains of Coupled Processes to Simpler Equivalents

Suppose that several processes are linked together to form a mechanistic pathway, with changes in the level of each compartment driving changes in the level of the next. Figure 13.6 presents a simple example, which can be depicted more tersely as $P \to F \to M \to N$. Here, each process is coupled to its successor by an information arrow that shows that the level of one affects the inflow to the next. For brevity, the baseline and stimulated components of the inflow and outflow are not shown separately for each process, but should be understood as being present implicitly, in the formulas determining the sizes of the inflows and outflows. Explicitly, the following set of equations corresponds to Fig. 13.6 (treating P as exogenously specified):

$$\begin{aligned} dF/dt &= b_F + a_{FP}P - d_F F, \\ dM/dt &= b_M + a_{MF}F - d_M M, \\ dN/dt &= b_N + a_{NM}M - d_N N. \end{aligned}$$

Here, b_j is the baseline "birth" flow of new material into compartment j in the absence of stimulation from other compartments, d_j is the baseline fractional "death" rate out of compartment j (per unit in compartment j per unit time; thus, the outflow from j is d_jX_j if there is no stimulated outflow), and a_{jk} is the stimulated production rate of material into compartment j per unit of material in compartment k (if k stimulates production of j, shown by an information arrow from j to the inflow of k).

Superficially, it appears that there are nine parameters (three for each equation) that might have to be estimated from the data. But the constraint that all rates of

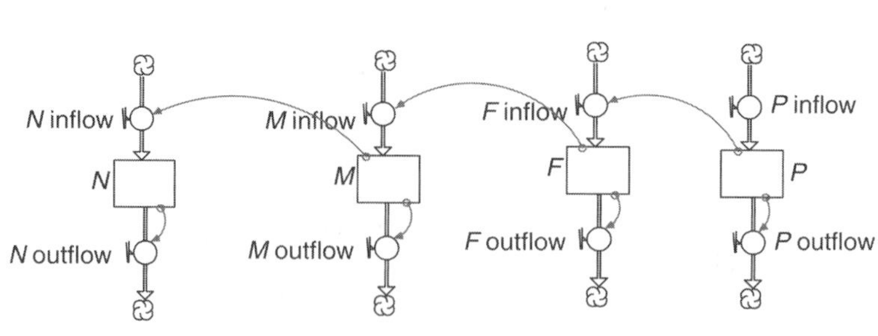

Fig. 13.6 Example of a chain of coupled processes

change are zero when all compartment levels are 1 reduces the degrees of freedom considerably, yielding the following normalized system with six parameters:

$$dF/dt = b_F + a_{FP}P - (b_F + a_{FP})F,$$
$$dM/dt = b_M + a_{MF}F - (b_M + a_{MF})M,$$
$$dN/dt = b_N + a_{NM}M - (b_N + a_{MN})N.$$

Suppose that our goal is to predict the new equilibrium levels in all compartments following an exogenous change in P from its initial equilibrium value of 1 to a new value, P^*. After adjustment, the new equilibrium values (indicated by asterisks) must again cause all time derivatives for state variables to vanish:

$$P = P^*,$$
$$dF/dt = 0 \rightarrow b_F + a_{FP}P = (b_F + a_{FP})F \rightarrow F^* = w_F + (1 - w_F)P^*,$$
$$\text{where } w_F = b_F/(b_F + a_{FP}),$$
$$dM/dt = 0 \rightarrow b_M + a_{MF}F = (b_M + a_{MF})M \rightarrow M^* = w_M + (1 - w_M)P^*,$$
$$\text{where } w_M = b_M/(b_M + a_{MF}),$$
$$dN/dt = 0 \rightarrow b_N + a_{NM}M = (b_N + a_{MN})N \rightarrow N^* = w_N + (1 - w_N)P^*,$$
$$\text{where } w_N = b_N/(b_N + a_{NM}).$$

(We assume that each compartment in the chain has a nonzero inflow and a nonzero outflow rate, so that all w_j are finite.) Thus, a single reduced parameter, w_j, suffices to predict the new equilibrium level in compartment j from the new equilibrium level in its predecessor. The number of parameters to be estimated is now one less than the number of compartments (three parameters, for the example in Fig. 13.6), assuming that the new value of the first compartment is exogenously specified. Clearly, this holds for any number of compartments in a chain constructed as in Fig. 13.6.

More generally, to predict the new equilibrium level in a specific compartment (e.g., the last one in a chain of any length) when the level in the first compartment is exogenously set to a new value (e.g., to predict N^* for any P^*), only one reduced parameter is needed. If the compartments are numbered from 0 for the first to n for the last, with the flow in compartment j being governed by the ODE in Equation (13.5),

$$dX_j/dt = b_j + a_{j,j-1}X_{j-1} - d_jX_j, \tag{13.5}$$

then the general relation between equilibrium levels in consecutive compartments is

$$X_j^* = w_jX_{j-1}{}^* + (1 - w_j), \text{ where } w_j = b_j/(b_j + a_{j,j-1}). \tag{13.6a}$$

This can be rearranged as

$$X_j^* - 1 = w_j(X_{j-1}{}^* - 1). \tag{13.6b}$$

This difference equation, with initial condition X_0 specified, has the obvious solution (as can be verified by repeated substitution)

$$X_j^* = w_j w_{j-1} \ldots w_1 (X_0^* - 1) + 1. \tag{13.7}$$

For the example in Fig. 13.6, $N^* = w_N w_M w_F (P^* - 1) + 1$. Thus, a *single* reduced parameter, $w_j w_{j-1} \ldots w_1$, which may be interpreted as the ratio of $(X_j^* - 1)$ to $(X_0^* - 1)$ for any value of $X_0^* \neq 1$, suffices to predict the new equilibrium values of compartment j for any other value of X_0^*. This ratio can be estimated by simple linear regression of $(X_j^* - 1)$ vs. $(X_0^* - 1)$ (with zero intercept) if several measured pairs of values of X_0^* and corresponding values of X_j^* are available; the values of the individual w_j (or of the structural parameters b_j, $a_{j,j-1}$, and d_j) need not be known.

If each compartment in a chain regulates its successor's outflow, rather than its inflow, via the ODE

$$dX_j/dt = b_j - d_j X_j - e_{j,j-1} X_j X_{j-1} = b_j - (d_j + e_{j,j-1} X_{j-1}) X_j, \tag{13.8}$$

then the normalization that $dX_j/dt = 0$ when $X_j = X_{j-1} = 1$ implies that $b_j = d_j + e_{j,j-1}$. If X_{j-1} is changed to a new value $X_{j-1}^* \neq 1$, then X_j will adjust until equilibrium is restored, with

$$dX_j/dt = b_j - (d_j + e_{j,j-1} X_{j-1}^*) X_j^* = 0.$$

Solving for the new equilibrium level in compartment j yields

$$X_j^* = b_j/(d_j + e_{j,j-1} X_{j-1}^*) = 1/[w_j X_{j-1}^* + (1 - w_j)] = 1/[w_j (X_{j-1}^* - 1) + 1], \tag{13.9}$$

where $w_j = e_{j,j-1}/b_j = e_{j,j-1}/(d_j + e_{j,j-1})$. Again, a single reduced parameter, w_j, determines the new equilibrium level in each compartment from that in its predecessor. It can be estimated by nonlinear regression from (X_{j-1}^*, X_j^*) pairs, if they are available. It is not necessary to estimate structural parameters, b_j, d_j , and $e_{j,j-1}$, for each compartment to predict how its contents will change in response to changes in the equilibrium level of its predecessor.

Figure 13.5 and Equation (13.4) describe a compartment X with inflow stimulated by one predecessor, Y, and outflow stimulated by another predecessor, Z. Several special cases are of interest in real networks. If the baseline (unstimulated) inflow and outflow are both zero (so that $dX/dt = aY - eXZ$), then Equation (13.4) and the requirement that $X = Y = Z = 1$ in normal equilibrium imply the simple equilibrium relation

$$X^* = Y^*/Z^*. \tag{13.10}$$

If there is only stimulated input but not stimulated outflow (so that $dX/dt = aY - dX$), then Equation (13.4) and the requirement that $dX/dt = 0$ when $Y = X = 1$ imply that

$$X^* = Y^*. \tag{13.11}$$

If there is only stimulated output but not stimulated input (so that $dX/dt = b - eXZ$), then the equilibrium ($b = eXZ$) and normalization ($b = e$) conditions imply that

$$X^* = 1/Z^*. \tag{13.12}$$

Equations (13.7) and (13.9)-(13.12) provide formulas for propagating new equilibrium levels along causal chains and through simple junctions where levels of predecessors regulate the inflow and/or outflow of a process. Since these equations are local [each determines the equilibrium level for a compartment from the equilibrium level(s) of its immediate predecessor(s)], they can easily be extended to more complex network topologies in which process nodes can have multiple (>2) predecessors and/or successors. For example, a process with stimulated inflows depending on the levels of $m \geq 2$ parents via the ODE

$$dX_j/dt = b_j + a_1 Y_1 + a_2 Y_2 + \ldots + a_m Y_1 - d_j X_j$$

can be analyzed in exactly the same way as Equation (13.5), with $a_{j,j-1}X_{j-1}$ replaced by $a_1 Y_1 + a_2 Y_2 + \ldots + a_m Y_1$.

Chapter 14
Value of Information (VOI) in Risk Management Policies for Tracking and Testing Imported Cattle for BSE

Previous chapters have discussed and illustrated several methods for building quantitative risk assessment (QRA) models for complex and uncertain systems, including systems with dramatically nonlinear responses (such as the COPD risk model in Chapter 13, where sufficient exposure switches the entire system of interacting feedback-control processes from normal to diseased behavior). Techniques that have proved useful for QRA modeling of complex, uncertain, and potentially nonlinear systems include the following:

- *Information-theory and data-mining algorithms* (Chapters 6 and 7) for identifying potential causal relations (including nonlinear and multivariate ones with high-order interactions) in large multivariate data sets.
- *Upper-bounding methods* (Chapters 8, 9, 12) for using available information about causal pathways, even if very incomplete (e.g., biomarker data for complex diseases), to estimate upper bounds for the preventable fractions of disease. Analogous methods can be used to quantify upper bounds on the preventable fractions of failures in complex reliability and engineering systems, fraudulent transactions in financial systems, defective parts in manufacturing systems, accidents in traffic networks, crimes in social systems, missed opportunities for mutual gains in economic systems, and so forth.
- *Identification of a discrete set of possible risks* (Chapters 10 and 11) that are implied by different assumption sets and/or that are consistent with available knowledge and data about uncertain causal mechanisms.
- *Systems dynamics analysis and simulation* (Chapters 11, 12, 13), together with probability bounds (such as Markov's inequality) that relate the deterministic analysis of mean values to bounds on the probable values of stochastic processes.
- *Comparative statics* analysis and *reduction of complex models* (Chapter 13) to simpler ones that make the same predictions of equilibrium behaviors in response to changes in inputs.

This chapter and Chapters 15 and 16 illustrate the following additional techniques for QRA modeling in systems where uncertainty arises from the actions of one or more intelligent decision makers.

L.A. Cox, Jr., *Risk Analysis of Complex and Uncertain Systems*, International Series in Operations Research & Management Science 129,
DOI 10.1007/978-0-387-89014-2_14,

- *Decision tree, sequential decision optimization,* and *value of information analysis* (this chapter).
- *Game theory and hierarchical optimization modeling* (Chapters 15 and 16) of the behaviors of intelligent attackers and intelligent defenders of a facility or infrastructure target. These models also support optimization of the allocation of defensive resources, taking into account the attacker's best response.
- *Mathematical optimization and phase-transition modeling* (Chapter 16) for predicting the resilience of complex systems (e.g., telecommunications networks) to deliberate attacks and for designing systems to make them resilient to attack.

The main practical purpose of QRA modeling is to *support better risk management decisions* by predicting the probable consequences of different actions or interventions. This chapter and the following two discuss how to use QRA models to improve risk management for systems in which uncertainty about consequences results largely from uncertainty about how other decision makers will respond to possible future events. This chapter studies the economic risk to the United States from mad cow disease (bovine spongiform encephalitis, BSE) caused by cattle imported from Canada. It considers two main risk management scenarios: with and without a system in place to track the origin of cattle discovered to have BSE. Chapters 15 and 16 discuss protecting specific targets and telecommunications networks, respectively, against deliberate attacks by terrorists.

An important lesson from these chapters is that *risk models can guide and improve current risk management decisions even if information about their inputs is highly uncertain.* The relevant question for risk managers is usually not, "Do we have adequate models and input information on which to base current decisions?" but rather, "Can the information and models that we have now be used to improve the expected utility of current decisions?" If the answer to this latter question is yes, then current information, however imperfect, has positive *value of information* (VOI), even if there is considerable room for improvement. For example, even if future information is expected to be much more complete and accurate than current information, it is still possible and important to decide what (if anything) to do now, given whatever is known *now*, assuming that future actions will be optimized with respect to what will be known *then*. Some critics mistakenly believe that QRA requires unrealistically complete information, or high confidence and certainty about cause and effect, before it can be used to guide and improve decisions. This misperception has provided one motivation for the development and use of potentially inferior decision processes (defined as ones that make less desired consequences more likely), such as the concern-driven alternatives to QRA discussed in Chapter 1, or the scoring and ranking methods in Chapters 4 and 5. This chapter illustrates how to optimize current decisions in the presence of large uncertainties about possible futures.

Testing Canadian Cattle for Bovine Spongiform Encephalitis (BSE)

From October 1996 to March 31, 2004, Canada tested the brains of 2,769 targeted cattle for BSE, finding it in one cow, in May 2003. (The cow had been condemned at slaughter and did not enter the human food chain.) Canadian cattle targeted for testing in 1996–2004 included animals with neurological signs or emaciation that were submitted through provincial slaughter facilities and by field veterinarians, as well as samples from cattle submitted to provincial diagnostic laboratories for post-mortem examination. If, based on European experience, targeted animals had been estimated to be not more than about 60 times more likely to have BSE than nontargeted animals as a base case (e.g., Doherr et al., 2001), then a base case prevalence rate of BSE among nontargeted Canadian cattle of about (1/2,768)*(1/60) = 6.0E-6 could have been estimated in March 2004. [The United States Department of Agriculture in its September 18, 2007, rule allowing older (over 30 months old), higher-risk cattle to be imported from Canada into the United States used a Canada-specific estimate of 2.4–6.8 BSE cases per million adult cattle, with an expected value of 3.9E-6 (http://edocket.access.gpo.gov/2007/07-4595.htm).] Chebyshev's inequality (see Chapter 2, Table 2.1) for a binomial model (or a mixture-of-binomials model, which has a smaller variance for the same mean) then implies that it is statistically almost certain that some BSE-positive cattle will be imported into the United States among the first few million Canadian cattle imported, if BSE is prevalent at rates on the order of several per million.

Since 2004, close to 200,000 additional Canadian cattle have been tested. Because BSE cases continue to be found regularly as of this writing (2008), it is a statistical near-certainty that further testing will discover further Canadian BSE cases, as the prevalence rate does not yet appear to be declining. Therefore, rather than attempting to pursue a moving target, we will analyze BSE risk management policies based primarily on data available as of 2004. Subsequent data confirming the continued prevalence of BSE in Canada strengthen the conclusions based on the earlier data.

In December 2003, a second dairy cow from Alberta, imported into the United States to the state of Washington, was also diagnosed with BSE. (As of 2008, this is the only case ever discovered in the United States of infectious BSE of the type found in the United Kingdom and subsequently found in Canada following imports from the United Kingdom.) Following a prompt, thorough investigation by the U.S. Department of Agriculture (USDA) and the Canadian Food Inspection Agency (CFIA), the USDA's APHIS Veterinary Services (VS) issued an "Explanatory Note" in February 2004, concluding that its previous risk analysis of the risks from Canadian cattle and beef products imported into the United States remained unchanged by the new case, and that the risks remained low. As stated in the note:

> Both of the BSE cases of Canadian origin occurred in cattle born before the feed ban was implemented. They were both older than 30 months of age when they were diagnosed as infected. Infection presumably occurred prior to or around the time the Canadian feed ban was enacted. The finding of an imported case in a cow greater than 30 months of age has

> little relevance to an analysis of risk under the proposed mitigation measures, beyond the implications for BSE prevalence in Canada. The proposed rule was not in effect in 2001 when the imported case, which was more than 4 years old at the time, entered the United States. Under the proposed conditions, the animal would not have been allowed entry into the United States. [*Note:* The USDA's (2007) rule does allow over-30-month-old cattle to be imported from Canada into the United States.] Therefore, we continue to consider the import controls in the proposed rule to be effective and the results of the analysis unchanged. (USDA, 2004)

From a statistical perspective, the detection of two BSE cases from Alberta in less than eight months in 2003, and two more in less than a month at the end of 2004 and the beginning of 2005 (among fewer than 1,300 animals tested in Alberta over that interval), raises the question of what the true prevalence of BSE in Canadian cattle may be at present. The statistical inference problem is complicated by the fact that the cow in Washington state was not detected as part of Canada's routine sampling program, and the probability that such cattle will be detected once they have been imported into the United States is not known. From a risk management perspective, the key question is what actions, if any, the United States should take now in light of the uncertainty about the true prevalence rate of BSE among Canadian cattle now and in the future. This decision problem is made more challenging by high economic stakes and by scientific uncertainties regarding BSE sources, reservoirs, and dynamics. As noted by the USDA's Animal and Plant Health Inspection Service in a February 2004 position statement entitled "Official diagnosis of Chronic Wasting Disease (CWD) should be performed exclusively by Federal and State regulatory agency laboratories," even false positives might be economically damaging: "In the case of a disease like BSE, a false positive could be devastating, costing the U.S. economy billions of dollars in unnecessary domestic and international market disruption from which it could take years to recover." Subsequent reporting by the USDA of unconfirmed BSE cases that turned out to be false positives, starting in July 2004, suggests that such market impacts can occur quickly.

The QRA of BSE risks is fraught with many scientific uncertainties and complexities, including the following:

- Uncertain roles of horizontal and vertical transmission (if any) within herds,
- Unknown existing and potential BSE reservoirs in Canada and the United States,
- Transmission dynamics within and between different reservoirs,
- Differences in susceptibility among individual cattle of the same age,
- The shape of the age-vs.-infectivity curve for cattle,
- The distribution of infectivity and differences in virulence among new BSE cases,
- The latency period until clinical expression; the possibility of subclinical cases (Thackray et al., 2003; Hill and Collinge, 2003); definition of clinical BSE expression,
- The potential for clustering of rare events within geographic areas, processing plants, affected populations, etc.
- Error rates and compliance failure rates (such as mislabeling, etc.) in Canada and the United States,

- Possible heterogeneity of the basic reproductive rate for BSE in different geographic areas or for different strains of BSE, different types of cattle, etc.,
- Detection probabilities per case, given the target and sampling schemes used,
- Uncertainty of inferred cattle age measurements (e.g., from dentition, etc.),
- Variability and accuracy in testing methods for BSE detection.

With so many unknowns, predictive modeling can be highly uncertain. Real-world data on observed cases of BSE can therefore potentially be especially valuable for improving estimates of the true BSE prevalence. However, the two BSE cases from Alberta detected in 2003 support alternative interpretations, ranging from (a) the first beginnings of a wave of BSE cases to (b) the last remnants of a problem from the 1980s and 1990s that has already been fixed and that, by chance, escaped detection until 2003 and (c) possibly scenarios in between. (The data available as of 2004 did not reveal a unique correct interpretation, although confirmation of over a dozen BSE cases by 2008 would subsequently make clear that Canada's BSE problem was starting, not ending, in 2003.)

This creates a dilemma for both health and economic risk management. On the one hand, experience since 2003 has shown that discovery of imported BSE cases in the United States can dramatically reduce U.S. beef exports, even if the infected animals originated in Canada. If the true prevalence of BSE in Canadian cattle shipped to the United States were known to be as high as several per million, then prevention of cattle imports from Canada would be necessary to prevent BSE from being imported into the United States. On the other hand, if the prevalence of BSE in Canadian cattle were driven down by one or more orders of magnitude, to zero or to U.S. levels, then the advantages of resumed trade could be gained by allowing unrestricted imports, without incurring a substantial risk of additional BSE cases.

Given the high economic stakes and the uncertainties about the prevalence of BSE in Canadian cattle (and, for that matter, U.S. cattle), it has been difficult to determine what policies would best promote U.S. and international interests – what policies would be optimal, based on a solid analytic foundation. Options range from tightening or loosening current import policies to gathering more information first – for example, by tracking all imported cattle and testing all Canadian cattle in the United States – and then using this information and the results of future sampling to decide when, whether, and how to change import restrictions. To discover which of these (or other) options is most desirable, it is necessary to compare their conditional probability distributions of gains and losses.

This chapter illustrates the application of constructive decision-analytic techniques, including value of information (VOI) calculations (Yokota and Thompson, 2004), to quantify and compare the potential economic values of different risk management and information-seeking options available to the United States for managing the uncertain risks of BSE originating in Canada. The analysis focuses mainly on a near-term decision – whether to require Canadian cattle in the United States to be identified, permanently marked, and tracked to provide information about their origins in case future BSE cases are found – and on the economic consequences of different potential futures whose probabilities can be affected by these near-term decisions. This focus reflects the facts that economic consequences will probably dominate near-term policy decisions, are easier to estimate from available information than possible human health risks, and provide an analytic framework that can be extended to include health risk considerations if desired. By explicitly representing key uncertainties and assessing the probable consequences of alternative current decisions under several scenarios, the decision-analytic framework presented here may prove useful to policy analysts and decision makers in considering how best to assess and manage the highly uncertain risks of BSE in the United States from imported cattle.

Methods and Data

Decision-analytic risk management proceeds through the following steps:

1. Identify a set of *alternative decision rules* or options to be compared. A decision rule specifies the actions to be taken at each time, given the information available at that time. It may be thought of as a plan that specifies what to do under different contingencies.
2. Identify the *consequences* of concern, which the actions may affect.
3. Identify the *probabilities* of different consequences, for each decision rule. This typically requires considering different *scenarios* or assumption sets describing alternative ways in which current uncertainties might be resolved. These are also called *states of nature*. Often, there is no objective, uniquely correct way to determine the prior probabilities of alternative scenarios. Then conservative assumptions (tending to favor the status quo) and sensitivity analyses (in which various prior probabilities of scenarios are assumed) may be used to determine how robust the conclusions and decision recommendations from the analysis are to variations in scenario probabilities.
4. Identify the *optimal decision rule,* defined as the one with the most desirable probability distribution of consequences, given the current information and assuming that future actions will be made optimally given future information.
5. Identify and recommend an *optimal current action*, determined by the optimal decision rule.

This framework is explained in detail in Raiffa (1968) and Clemen (1996).

Formulation of the Risk Management Decision Problem as a Decision Tree

The decision rules compared in this chapter are structured as follows (see Fig. 14.1). First, an initial ("Stage 1") decision must be made either to track Canadian cattle in the United States ("*Track CA imports*") or not to track them ("*Do not track CA imports*"). The main purpose of the decision analysis is to compare the probable

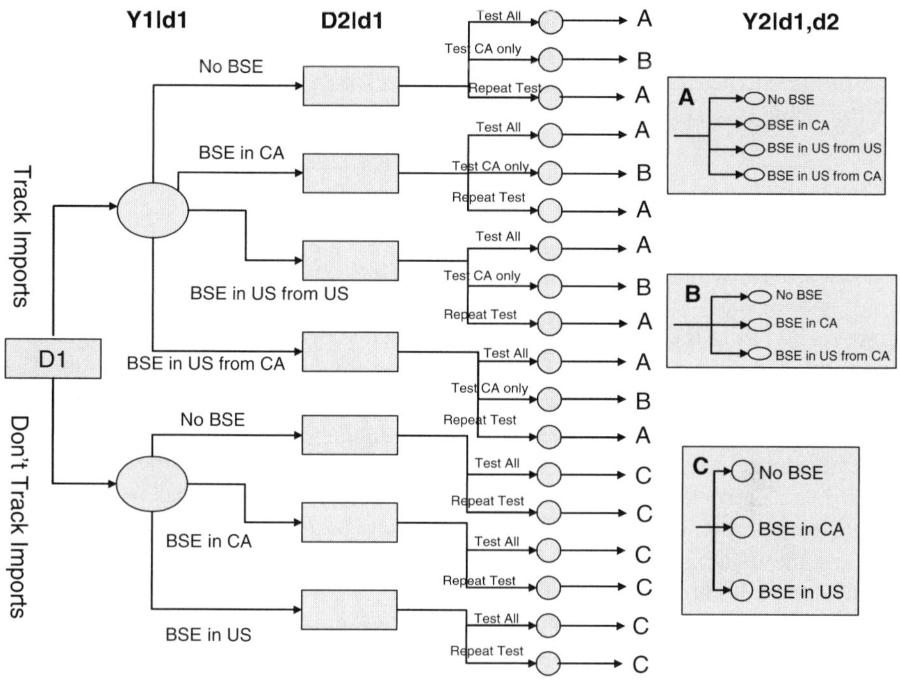

Fig. 14.1 Decision tree for BSE tracking and testing policy
Notation for decision problem components

- D1 = Stage 1 choice set = {Track Imports, Do Not Track Imports}
- {Y1 | d1} = information sets of possible outcomes based upon the Stage 1 decision d1 ε D1
- {D2 | d1} = Stage 2 choice set, given the Stage 1 decision d1 ε D1
- {Y2 | d1,d2} = information sets of possible outcomes after decisions d1 ε D1 and d2 ε {D2 | d1}

consequences to the United States of these two alternative initial actions. Following this Stage 1 decision, additional information will be obtained from ongoing sampling programs in the United States and Canada that perform tests for BSE on both symptomatic (e.g., "downer" cattle) and randomly selected healthy-appearing cattle at slaughter. If the Stage 1 decision was "Track CA imports," then any of the following informative events may be observed over a specified following time period (e.g., one year):

- No new BSE cases detected,
- BSE case of Canadian origin detected in the United States,
- BSE case of U.S. origin detected in the United States,
- BSE case of Canadian origin detected in Canada.

(If several of the last three events occur in a year, we focus on the first to occur as the event of interest.) The probabilities of these events depend on both the unknown true prevalence rates of BSE in the United States and Canadian herds (i.e., on which scenario or state of nature is correct) and also on the sampling plans and tests used to examine the herds. If the Stage 1 decision is "Don't track CA imports," then the four possible observations for the next period are aggregated to only the following three:

- No new BSE cases detected,
- New BSE case detected in Canada,
- New BSE case detected in the United States.

In reality, as in the case of the Washington state cow, forensic efforts might successfully identify the origin of a BSE case even without new tracking measures. The effect of a Stage 1 decision to track imports is then to increase the probability that the origin of a new case can be determined. The formal analysis treats the Track vs. Do not track decisions as providing vs. failing to provide, respectively, the information needed to identify the origin (Canadian or not) of any new BSE case, while recognizing that partial tracking via ear tags, brands, and tattoos may already be available. (Indeed, the tracking issue is confined to Canadian cattle because Mexican cattle in the United States are already well identified.)

After the Stage 1 decision, and given updated information about any new BSE cases, a subsequent ("Stage 2") decision must be made about whether to sell and process healthy-appearing cattle without first requiring them to be tested for BSE ("No required test") vs. requiring all U.S. cattle to be tested for BSE before being sold or processed ("Test all") vs. requiring only all Canadian cattle in the United States to be tested for BSE before being sold or processed ("Require testing for CA cattle only"). The last option is available only if the Stage 1 decision was "Track CA imports." In addition to any required testing, some cattle will continue to be sampled and tested according to a USDA test program, and this is not affected by the Stage 1 and Stage 2 decisions. The Stage 2 decision presumably will be made to obtain the most desirable outcome possible, given the information available then. For example, if a new BSE case is detected in the United States and its origin cannot be ascertained, then the Stage 2 decision might be "Test all" U.S. cattle at slaughter, to reduce export and domestic consumption losses (if the economic benefits outweigh the costs of testing), whereas if the origin of the case is known to be Canadian and the Stage 1 decision was to "Track CA imports," then the best Stage 2 decision might be "Require testing for CA cattle only."

After Stage 1 and Stage 2 decisions have been made and the future information has been obtained, it becomes possible to evaluate how much beef consumption, if any, has been lost in export and domestic markets due to BSE cases and risk management responses, and how much the Stage 1 and Stage 2 decisions cost to implement. A goal for rational risk management decision making *today* is to anticipate how current decisions change probable *future* total costs (i.e., the sum of implementation costs and costs from lost domestic and export sales), as they will eventually be assessed in hindsight. Each Stage 1 decision, in conjunction with optimized Stage 2 decisions given future information, determines a probability distribution for total cost. Rational risk management requires making the choice today that induces the most desirable probability distribution for total costs, as they eventually will be evaluated in the future.

Figure 14.1 presents a decision tree model summarizing the logical structure of the decision problem. In this tree, a *decision rule* specifies which outgoing branch to follow at each decision node (represented by a rectangular node in Fig. 14.1). "Repeat test" refers to the action of doing nothing other than to continue the routine BSE sampling and testing programs. The notation for Stage 1 and Stage 2 decisions ("choice sets") and observed outcome events ("information sets") listed at the bottom of Fig. 14.1 allows the same framework to be expanded to include additional decisions, scenarios, and information events if so desired to increase the resolution of the problem description. However, the relatively simple, aggregate descriptions of possible decisions and futures in Fig. 14.1 suffice to carry out the decision analysis calculations; analogous calculations can be performed for more detailed descriptions.

Estimated Economic Consequences of Detecting Additional BSE Cases

To finish describing the decision problem, it is necessary to estimate the economic costs associated with each terminal node (i.e., "leaf" node) at the tips of Fig. 14.1. Only the direct costs of implementing the different Stage 1 decisions and of reduced beef sales in case of detection of new BSE cases will be considered, as a first approximation to the full societal costs. (A refined analysis could estimate economic multiplier effects and reductions in consumer surplus from reduced domestic sales, which would increase their impacts further. However, sensitivity analysis suggests that the main conclusions, which are dominated by loss-of-export-related impacts, would not be changed by these refinements.) The decision model incorporates the following three types of cost: tracking costs, testing costs, and market costs. Tracking costs represent the cost of permanently marking each live cow coming into the United States, including labor and materials. Testing costs represent the costs per BSE test, including kits, labor, shipping, holding, laboratory facilities, and expenses. Market costs represent market losses (or gains) associated with each second-stage outcome as a function of all that occurred up to that point. Baseline values for each of these costs are estimated next. These are then varied to obtain sensitivity analyses.

Market Impacts

The main economic impacts on the United States of discovering a new BSE case are assumed to be as follows for the baseline scenario.

- If a new BSE case of *unknown origin* is discovered in the United States, then both domestic demand and remaining exports of U.S. beef will immediately decline. Following the discovery of the BSE-positive cow in Washington state in 2003, U.S. exports declined by approximately 50%. For the baseline scenario, we assume that the discovery of a new BSE case of unknown origin in the United States will result in a further loss of \$12.27 billion per year in cattle sales, corresponding to a 25% assumed reduction in consumer demand. The situation where full testing identifies a BSE case of known U.S. origin in the United States provides a similar loss.

- If a new BSE case is found in the United States that is not specifically known to be of Canadian origin, but subsequent full testing does not find a similar case, a smaller loss of \$6.14 billion per year will occur.
- If a new BSE case is discovered *in Canada*, then U.S. exports may increase to replace decreased Canadian exports. The magnitude of this effect is estimated as a gain of \$1.382 billion per year in the base case.
- If a new BSE case *known to be of Canadian origin* is discovered in the United States, and if Canadian cattle are then removed from U.S. exports and from the food supply, the net impact on the United States is a loss of \$2.683 billion per year in the base case, primarily from additional lost exports. (The U.S. domestic markets responded only relatively slightly to the Canadian BSE cases discovered in 2003, suggesting that the main economic impacts come from the closing of export markets to U.S. beef.)

Table 14.1 summarizes the baseline economic impacts for each of the possible futures (i.e., branches through the decision tree to a leaf node) in Fig. 14.1. The appendix provides the supporting rationale and data for the estimated market impacts.

Tracking Costs

Table 14.1 also shows the estimated costs of tracking and testing cattle that are included in the model. Annual cattle-tracking costs are calculated by multiplying an estimated unit cost per animal by the number of live cattle imported annually into the United States from Canada. In 2002, prior to any BSE detections, this was 1,538,715 cattle (http://cattle.guelph.on.ca/statistics/livetrade-withus.html). The annual cost of tracking any newly imported cattle is estimated as \$10 to cover tags, labor, and compliance checks. The baseline total annual tracking costs for such cattle, assuming a return to 2002 levels of imports, are thus $1{,}538{,}715 \times \$10 = \$15{,}538{,}715$. (Part of this cost may initially be borne by Canadian producers, but it is included in the model as the cost results directly from a *Track Imports* policy and may ultimately be passed on to U.S. consumers.) The costs of locating and then tracking Canadian cattle already in the United States are more difficult to estimate; they are addressed in the Sensitivity Analysis and Discussion sections.

Testing Costs

The Stage 1 testing costs in the United States are obtained by multiplying a unit test cost per animal by the size of the assumed Stage 1 sample size. In FY2004, the USDA tested 20,543 cattle (http://www.usda.gov/Newsroom/0105.04.html). In the wake of

Table 14. 1 Economic impact estimates

Stage 1		Stage 2		Economic impacts			
Decision	Outcome	Decision	Outcome	Market impacts	Tracking costs	Testing costs	Total economic impact
Track imports	No BSE	Test all	No BSE	0	–30,774,300	–1,099,200,000	–1,129,974,300
			BSE in CA	1,382,000,000	–30,774,300	–1,099,200,000	252,025,700
			BSE in U.S. from U.S.	–12,270,000,000	–30,774,300	–1,099,200,000	–13,399,974,300
			BSE in U.S. from CA	–2,863,000,000	–30,774,300	–1,099,200,000	–3,992,974,300
		Test CA only	No BSE	0	–30,774,300	–47,361,450	–78,135,750
			BSE in CA	1,382,000,000	–30,774,300	–47,361,450	1,303,864,250
			BSE in U.S. from CA	–2,863,000,000	–30,774,300	–47,361,450	–2,941,135,750
		Repeat test	No BSE	0	–30,774,300	–2,400,000	–33,174,300
			BSE in CA	1,382,000,000	–30,774,300	–2,400,000	1,348,825,700
			BSE in U.S. from U.S.	–12,270,000,000	–30,774,300	–2,400,000	–12,303,174,300
			BSE in U.S. from CA	–2,863,000,000	–30,774,300	–2,400,000	–2,896,174,300
	BSE in CA	Test all	No BSE	1,382,000,000	–30,774,300	–1,099,200,000	252,025,700
			BSE in CA	1,382,000,000	–30,774,300	–1,099,200,000	252,025,700
			BSE in U.S. from U.S.	–12,270,000,000	–30,774,300	–1,099,200,000	–13,399,974,300
			BSE in U.S. from CA	–2,863,000,000	–30,774,300	–1,099,200,000	–3,992,974,300
		Test CA only	No BSE	1,382,000,000	–30,774,300	–47,361,450	1,303,864,250
			BSE in CA	1,382,000,000	–30,774,300	–47,361,450	1,303,864,250
			BSE in U.S. from CA	–2,863,000,000	–30,774,300	–47,361,450	–2,941,135,750

Table 14. 1 (continued)

Stage 1		Stage 2			Economic impacts		
Decision	Outcome	Decision	Outcome	Market impacts	Tracking costs	Testing costs	Total economic impact
		Repeat test	No BSE	1,382,000,000	–30,774,300	–2,400,000	1,348,825,700
			BSE in CA	1,382,000,000	–30,774,300	–2,400,000	1,348,825,700
			BSE in U.S. from U.S.	–12,270,000,000	–30,774,300	–2,400,000	–12,303,174,300
			BSE in U.S. from CA	–2,863,000,000	–30,774,300	–2,400,000	–2,896,174,300
	BSE in U.S. from U.S.	Test all	No BSE	–6,140,000,000	–30,774,300	–1,099,200,000	–7,269,974,300
			BSE in CA	–6,140,000,000	–30,774,300	–1,099,200,000	–7,269,974,300
			BSE in U.S. from U.S.	–12,270,000,000	–30,774,300	–1,099,200,000	–13,399,974,300
			BSE in U.S. from CA	–6,140,000,000	–30,774,300	–1,099,200,000	–7,269,974,300
		Test CA only	No BSE	–12,270,000,000	–30,774,300	–47,361,450	–12,348,135,750
			BSE in CA	–12,270,000,000	–30,774,300	–47,361,450	–12,348,135,750
			BSE in U.S. from CA	–12,270,000,000	–30,774,300	–47,361,450	–12,348,135,750
		Repeat test	No BSE	–6,140,000,000	–30,774,300	–2,400,000	–6,173,174,300
			BSE in CA	–6,140,000,000	–30,774,300	–2,400,000	–6,173,174,300
			BSE in U.S. from U.S.	–12,270,000,000	–30,774,300	–2,400,000	–12,303,174,300
			BSE in U.S. from CA	–6,140,000,000	–30,774,300	–2,400,000	–6,173,174,300
	BSE in U.S. from CA	Test all	No BSE	0	–30,774,300	–1,099,200,000	–1,129,974,300
			BSE in CA	0	–30,774,300	–1,099,200,000	–1,129,974,300
			BSE in U.S. from U.S.	–12,270,000,000	–30,774,300	–1,099,200,000	–13,399,974,300

Table 14. 1 (continued)

Stage 1		Stage 2		Economic impacts			
Decision	Outcome	Decision	Outcome	Market impacts	Tracking costs	Testing costs	Total economic impact
			BSE in U.S. from CA	–2,863,000,000	–30,774,300	–1,099,200,000	–3,992,974,300
		Test CA only	No BSE	0	–30,774,300	–47,361,450	–78,135,750
			BSE in CA	0	–30,774,300	–47,361,450	–78,135,750
			BSE in U.S. from CA	–2,863,000,000	–30,774,300	–47,361,450	–2,941,135,750
		Repeat test	No BSE	0	–30,774,300	–2,400,000	–33,174,300
			BSE in CA	0	–30,774,300	–2,400,000	–33,174,300
			BSE in U.S. from U.S.	–12,270,000,000	–30,774,300	–2,400,000	–12,303,174,300
			BSE in U.S. from CA	–2,863,000,000	–30,774,300	–2,400,000	–2,896,174,300
Don't track	No BSE	Test all	No BSE	0	0	–1,099,200,000	–1,099,200,000
			BSE in CA	0	0	–1,099,200,000	–1,099,200,000
			BSE in U.S.	–12,270,000,000	0	–1,099,200,000	–13,369,200,000
		Repeat test	No BSE	0	0	–2,400,000	–2,400,000
			BSE in CA	0	0	–2,400,000	–2,400,000
			BSE in U.S.	–12,270,000,000	0	–2,400,000	–12,272,400,000
	BSE in CA	Test all	No BSE	0	0	–1,099,200,000	–1,099,200,000
			BSE in CA	0	0	–1,099,200,000	–1,099,200,000
			BSE in U.S.	–12,270,000,000	0	–1,099,200,000	–13,369,200,000
		Repeat test	No BSE	0	0	–2,400,000	–2,400,000

Table 14. 1 (continued)

Stage 1		Stage 2		Economic impacts			
Decision	Outcome	Decision	Outcome	Market impacts	Tracking costs	Testing costs	Total economic impact
			BSE in CA	0	0	–2,400,000	–2,400,000
			BSE in U.S.	–12,270,000,000	0	–2,400,000	–12,272,400,000
	BSE in U.S.	Test all	No BSE	–6,140,000,000	0	–1,099,200,000	–7,239,200,000
			BSE in CA	–6,140,000,000	0	–1,099,200,000	–7,239,200,000
			BSE in U.S.	–12,270,000,000	0	–1,099,200,000	–13,369,200,000
		Repeat test	No BSE	–6,140,000,000	0	–2,400,000	–6,142,400,000
			BSE in CA	–6,140,000,000	0	–2,400,000	–6,142,400,000
			BSE in U.S.	–12,270,000,000	0	–2,400,000	–12,272,400,000

the December 2003 finding of a BSE-positive cow, the annual number of cattle sampled will probably be at least doubled, to around 40,000, in addition to one-time, much larger sampling efforts (http:// usda.mannlib.cornell.edu/reports/nassr/livestock/pct-bb/catl0104.pdf). The per-animal test unit cost is estimated as \$30 (http://www.meatnews.com/index.cfm?fuseaction=article&artNum=7345). The Stage 1 annual testing costs in the United States are therefore estimated as \$30 × 40,000 = \$1,200,000.

Approximately 36.6 million cattle were slaughtered in the United States in 2003 (USDA, 2004). If each animal is tested at slaughter for a unit cost of \$30, then the baseline total annual U.S. testing cost in Stage 2 for "Test All" is approximately \$30 per animal × 36.6 M animals per year = \$1.098 billion per year. The corresponding cost for the "Test Canadian-origin cattle only" is estimated by assuming that the Canadian-origin portion of the U.S. herd is approximately in steady state, that is, the number of Canadian origin cattle slaughtered annually is equal to the number imported. Thus, testing costs are 1,538,715 animals per year × \$30 per animal tested = \$45,161,450 per year. Canadian testing costs are not included in our U.S. policy model, but the numbers of animals tested are included since they affect the probability of detecting new BSE cases.

Scenario Probabilities

The probable consequences of current decisions, specifically, whether to Track Canadian cattle imports, depend on whether and where BSE is detected next. The probabilities of the different economic consequences in Table 14.1, i.e., of different rows, given the choices of Stage 1 and Stage 2 decisions, are modeled via the variables and formulas in Table 14.2.

These formulas are based on a simple, approximate binomial model, in which only the average probability of detecting BSE per animal tested is used (for each of Canada and the United States separately) and details of interanimal variability are ignored. (In practice, the outcome probabilities in the table are renormalized to sum to 1, since ignoring the possibility of multiple BSE discoveries in the same year may lead to slight departures from 1.)

The probabilities p_{USUS}, p_{USCA}, and p_{CA} are estimates of the probabilities of finding one or more BSE positive cattle among each batch of 1,000 tested. (Probability per 1,000 is more convenient than probability per animal, given the small probabilities involved, but either could be used.) Uncertainty about the correct values of these probabilities is modeled by using five possible scenarios or "states of nature," shown in Table 14.3. Columns 2–4 show the values of p_{USUS}, p_{USCA}, and p_{CA} for each of the five scenarios.

The values in Table 14.3 for each scenario are averages for the entire U.S. and Canadian herds. The rationale for these values is as follows. Past testing suggests that the BSE rate in cattle of U.S. origin is likely very low or zero, since no confirmed cases have been discovered to date. The BSE rate in cattle of Canadian

Table 14. 2 Probability model and notation for scenario outcomes

Variables	Meaning	Formulas and baseline values
P_I	Proportion of U.S. cattle tested that were imported from CA	1,538,715/36.6 M = 4.0%
p_{CA}	Probability a Canadian animal tested is positive for BSE	See scenarios in Table 14.3
n_{CA}	Number of Canadian tests performed	See scenarios in Table 14.3
p_{US}	Probability a U.S. animal test is positive for BSE	$p_{USUS}{}^{*}(1–P_I) + p_{USCA}{}^{*}P_I$
n_{US}	Number of U.S. tests performed	$n_{USUS} + n_{\text{ USCA}}$
p_{USUS}	Probability U.S. testing of an animal of U.S. origin is positive	See scenarios in Table 14.3
p_{USCA}	Probability U.S. testing of an animal of CA origin is positive	See scenarios in Table 14.3
n_{USUS}	Number of U.S. tests of animals of U.S. origin	38,400 per year
n_{USCA}	Number of U.S. tests of animals of CA origin	1,600 per year
	Pr{no new BSE cases}	$(1 - p_{CA})^{n_{CA}}(1 - p_{US})^{n_{US}}$
	Pr{case in CA}	$\approx (1 - (1 - p_{CA})^{n_{CA}})$
	Pr{case in U.S.}	$\approx (1 - (1 - p_{US})^{n_{US}})$
	Pr{case is of U.S. origin | U.S. case}	$\approx 1 - (1 - p_{USUS})^{n_{USUS}}$
	Pr{case in U.S. of Canadian origin | a U.S. case}	$\approx 1 - (1 - p_{USCA})^{n_{USCA}}$
	Pr{case in U.S. of U.S. origin}	$\approx (1 - (1 - p_{US})^{n_{US}})^{*} \frac{1-(1-p_{USUS})^{n_{USUS}}}{1-(1-p_{USUS})^{n_{USUS}}+1-(1-p_{USCA})^{n_{USCA}}}$
	Pr{case in U.S. of CA origin}	$\approx (1 - (1 - p_{US})^{n_{US}})^{*} \frac{1-(1-p_{USCA})^{n_{USCA}}}{1-(1-p_{USUS})^{n_{USUS}}+1-(1-p_{USCA})^{n_{USCA}}}$

Table 14.3 Scenario definitions and BSE detection rates (per 1,000 animals tested)

States of nature	P_{USUS}	P_{USCA}	P_{CA}	Prior estimate
1	0	0	0	0.2
2	0	1.00E-04	1.00E-04	0.2
3	0	1.00E-06	1.00E-06	0.2
4	1.00E-06	1.00E-04	1.00E-04	0.2
5	1.00E-06	1.00E-06	1.00E-06	0.2

origin may be zero (if there are no new cases to be discovered), very low, or relatively high, with zero being perhaps somewhat less likely than the others, given the two BSE cases detected in 2003. The value corresponding to "relatively high" (1E-4 per 1,000) is consistent with the rate provided by the World Organization for Animal Health (http://www.oie.int/eng/info/en_esb.htm), which shows a 2003 incidence rate for Canada of 0.33 per million. The "very low" rate (1E-6 per 1,000 animals) is a plausible high-end estimate for the United States that considers the large number of cattle slaughtered annually (~36.6 M) without any cases detected thus far. We combined these considerations into the five scenarios shown. (Each scenario may also be viewed as the centroid of a cluster representing all possible scenarios that are closer to it than to any of the other four, in which case the discretization of all possible scenarios into only these five represents the relatively low degree of resolution permitted by the current data.)

The selection of scenario prior probabilities is potentially controversial. We adopt the following bounding approach to avoid needless controversy. If the main conclusion from the analysis is that the status quo is justified (i.e., tracking of Canadian cattle imports is not recommended because the incremental costs exceed the value of the information provided), then little justification may be needed. By contrast, if the analysis shows that a change from the status quo to "Track Canadian imports" is recommended (because the value of the tracking information exceeds the costs of acquiring it), then more justification may be needed to persuade stakeholders to adopt the conclusion. Therefore, we will pick values of highly uncertain inputs (such as the scenario probabilities) to favor the status quo, so that if the analysis still recommends a change, the result will be relatively strongly supported despite uncertainties in the model inputs. (This intentional bias toward the status quo is not strictly rational but recognizes the reality that any recommended changes from the status quo may require an additional burden of robustness.) Given that the limited available evidence favors the hypothesis that the Canadian BSE prevalence is higher than the U.S. BSE prevalence (as in scenarios 2 and 4), and that these scenarios imply relatively high information values for tracking Canadian cattle, we will use a *uniform distribution* of scenario probabilities as a conservative (i.e., status quo favoring) prior distribution, thereby giving more relative weight to scenarios 1 and 5 (no difference between U.S. and Canadian cattle) than the available data might suggest. The uniform prior also represents a maximum-entropy prior, and in this sense imposes as few assumptions as possible.

In the current situation of limited BSE testing, animals that are considered most likely to have BSE are targeted. Testing data from Europe suggest that the BSE rate among this subpopulation is 60 times greater than that of the general cattle population. This factor is applied to the probabilities in Table 14.3 to obtain the probabilities of positive test results among sampled cattle in Stage 1 with limited testing. The sampling factor will be subject to sensitivity analysis.

Let $s_i \varepsilon S$ represent the state of nature, i, with initial probability ps_i. Then

- $P(y1|\, d1,\, s_i)$ = probability of event $y1$ occurring, given that the first-stage decision was $d1$ and the state of nature is s_i, and
- $P(y1|\, d1) = \sum_{i=1}^{5} P(y1|d1, s_i)ps_i$ = unconditional probability for event $y1$ given decision $d1$.

Second-Stage Probabilities via Bayes' Rule

The states of nature provide a basis for computing second-stage probabilities via Bayesian updating. The first-stage outcomes $\{y1|\, d1\}$ provide information regarding the likelihood of the states of nature, allowing us to revise the estimates, ps_i. Specifically,

$$ps_i' = \frac{P(y1|d1, s_i)ps_i}{\sum_{j=1}^{5} P(y1|d1, s_j)ps_j}.$$

Then, similarly to the first stage,

$$P(y2|d1, d2) = \sum_{i=1}^{5} P(y2|d1, d2, s_i)ps_i'.$$

The conditional probabilities $P(y2 \mid d1, d2, s_i)$ are computed using the binomial formulas from Stage 1, but with the test quantities n_{USUS}, n_{USCA}, and n_{CA} revised. In particular, if $d2$ indicates full testing (of all cattle or all cattle from Canada), then n_{USUS}, n_{USCA} will be greatly increased to reflect full vs. partial testing. Second, the probabilities p_{USUS}, p_{USCA}, and p_{CA} may be quite different from those in Stage 1. In Stage 1, the testing regime targets "downer" cattle and others considered most likely to have BSE. In Stage 2, under full testing, the probabilities of a positive batch are diluted by less likely animals and therefore may be much (e.g., 60-fold) lower.

Solution Algorithms

The decision tree in Fig. 14.1, together with the quantitative data in Tables 14.1, 14.2, and 14.3 that populate it, specify the base case risk management decision problem to be solved. A standard dynamic programming algorithm (Raiffa, 1968) provides the solution and allows variations of the problem with different input values to be solved to yield sensitivity analyses and to characterize the robustness of model recommendations to uncertainties in the input values. We used the TreePlan™ decision tree software package for Excel™ to solve the decision optimization problem for the base case and for sensitivity runs.

Results

Optimal Decision Rule for the Base Case

Under the baseline assumptions in Tables 14.1, 14.2, and 14.3, the expected net cost of "Track Imports" is \$10,385,294 per year while the expected cost of "Do Not Track Imports" is \$90,045,020 per year. Thus, the expected net economic value of the information provided by tracking is \$79,658,726 per year, reflecting the much higher probability of large market losses when imports are not tracked, as BSE cases of Canadian origin in the United States are not distinguished from, and so have the same economic impact as, BSE cases of U.S. origin. The optimal decision rule for the base case is as follows: *Track Canadian cattle imports, then continue limited sampling in Stage 2 no matter what occurs.* In other words, the benefit from tracking in this case does *not* come from avoiding the cost of 100% testing of U.S. cattle, since this is too expensive to undertake. Rather, it comes from the reduced loss of U.S. beef sales if the country of origin of a BSE case detected in the United States is Canada and this can be ascertained and announced.

Sensitivity Analysis Results

The base case is of limited interest by itself, since it is not clear how robust the optimal current decision ("Track Canadian imports") is to plausible variations in the inputs. However, the following sensitivity analysis results indicate that this recommended initial decision is very robust to key input uncertainties:

- *Robustness to market benefits estimates.* Suppose that the positive market impacts (of \$1.382 billion) for the United States of another BSE discovery in Canada in some rows of Table 14.2 may have been estimated incorrectly. What degree of error would change the optimal decision from "Track Canadian imports" to "Don't track Canadian imports"? The answer is that the optimal base case decision ("Track CA imports") remains optimal when all positive market outcomes (those with a value of \$1.382 billion in Table 14.2) are multiplied by *any* positive number, whether less than 1 (scaled-down benefit estimate) or greater than 1 (scaled-up benefit estimate). Indeed, the VOI for tracking remains positive for any benefit multiplier >–0.58.

- *Robustness to market loss estimates.* Similarly, if all outcomes with a negative market impact in Table 14.2 are multiplied by *any* positive factor (and, indeed, any factor >–1.57), the optimal Stage 1 decision remains "Track Canadian imports." (The optimal second-stage decision changes as a function of the scaling factor, with full testing of Canadian-only or all-U.S. cattle becoming optimal for some values, but the VOI increases linearly for all positive values of the loss multiplier.)
- *Robustness to targeting efficiency.* Define the *targeting efficiency factor* as the ratio of the probability of a positive BSE test in a targeted animal vs. a purely randomly sampled animal. Its baseline value is 60. The VOI for tracking Canadian imports increases linearly as this factor is increased; it is negative (expected cost > expected benefit) only for values less than 17. European experience suggests that the true value of this factor could be as high as 186 in some areas (based on Swiss data, http://europa.eu.int/comm/food/fs/bse/bse21_en.html), although it varies among countries.
- *Robustness to consumer loss of confidence.* Suppose that baseline negative consequences are multiplied by a "fear factor" when the "Repeat Test" decision is chosen at the second stage and the market impact is negative, to reflect greater-than-estimated consumer fear and adverse reaction (loss of confidence in beef safety) that could occur if BSE is found in the second stage, but only limited sampling (the "Repeat Test" decision in Fig. 14.1) is used. The VOI for tracking Canadian imports increases as this factor is increased, by over 50% when the "fear factor" is 2 (i.e., if the loss of beef sales due to consumer fear is twice as great as estimated in the base case). The optimal Stage 1 decision remains "Track Canadian imports" for *all* positive values of the "fear factor," indicating considerable robustness to uncertainty about how customers would react to further BSE cases. (Interestingly, the optimal Stage 2 decision shifts from "Repeat Test" to "Test All" if the first stage detects BSE in the United States, and the fear factor is greater than about 1.20, as seems quite plausible.)
- *Robustness to tracking costs.* The base case assumes a tracking cost per animal per year of $10. This cost could be as high as $35 while leaving the VOI from tracking greater than zero. Therefore, locating and tracking Canadian cattle already in the United States appears to be worthwhile when the cost is less than $35/head.
- *Robustness to scenario probabilities.* Figure 14.2 shows the results of varying the probability of each of the individual scenarios in Table 14.3 from 0 to 1 while leaving the remaining probability spread evenly among the other four scenarios. The VOI (= desirability index for tracking imports) increases with the probabilities of scenarios 2 and 4 and decreases with the probabilities of scenarios 1, 3, and 5. Scenarios 2 and 4 are those with a high probability for BSE in Canada and low (or zero) probability for BSE in the United States. Scenarios 1, 3, and 5 each have a zero or very low probability of BSE in either country. They have the potential for a negative VOI, but only at high values (exceeding approximately 0.78). All available data are most consistent with scenarios 2 and 4, which imply a positive VOI for tracking Canadian cattle.

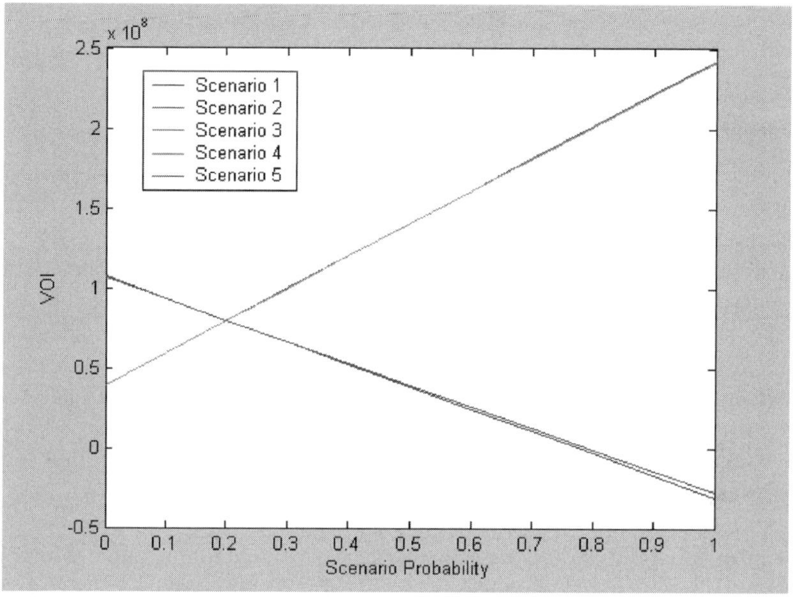

Fig. 14.2 Sensitivity analysis plot for scenario probabilities

In summary, the optimal decision for the base case, "Track Canadian imports," appears to be very robust to a wide range of plausible variations in the input data, as well as to combinations of variations (not shown). Thus, the model's recommendation to begin tracking appears to be well justified, even if some conservative assumptions are made that tend to discount the value of tracking information. Of course, if tracking information is not needed to reliably determine the country of origin, as happened in the Washington state case, then the VOI from tracking would be reduced – potentially to zero, if other data already provided reliable country-of-origin information. However, in practice, the ability to reliably identify the country of origin is often haphazard at best. Similarly, if U.S. consumer confidence and exports do not respond to information that a new BSE case is of non-U.S. origin as modeled in any of these sensitivity analyses, then their conclusions would not necessarily hold. Thus, the current analysis only shows what might be the case, based on the baseline scenarios and the wide range of variations shown. It does not prove that the VOI must be high. But it shows that it may be, under input assumptions that appear (at least to us) to be plausible. This provides an incentive to better understand the key drivers of potential economic value from better tracking information.

The economic value of tracking information in some sensitivity analyses comes primarily from limited export losses if the next case of BSE detected in the United States can be shown to be of Canadian origin, whereas in others, it comes primarily from avoiding the need to test all U.S. cattle, as opposed to just those of Canadian origin, to win back customers. Although the second-stage decisions that benefit from a first-stage decision to track Canadian cattle imports vary across sensitivity analysis cases, most of the sensitivity analyses agree that this is the optimal current decision, even while differing in their precise (Stage 2 planning) reasons.

Impacts of Possible Win-Back of Export Markets

The base case analysis and the assumptions in Table 14.1 are perhaps pessimistic, in that they assume that the losses of U.S. cattle and beef export markets following the discovery of a Canadian-origin BSE case in December 2003 are persistent and irreversible. Depending on the evolution of international risk perceptions and the harmonization of risk management standards and plans, it is possible that aggressive tracking and testing policies in the United States might result in the recovery of some lost export markets. If so, the economic impacts from tracking and testing could dwarf those calculated for the base case. For example, under an assumption that aggressive testing would allow the United States to regain its lost exports (as long as no confirmed BSE case of U.S. origin is discovered), the optimal strategy becomes to immediately start tracking all Canadian cattle and, if a confirmed BSE case of Canadian origin is found, to test all Canadian-origin cattle in the United States prior to export. In this case, the expected net economic value of the information provided by tracking increases to $771,570,514 per year, i.e., by close to an order of magnitude.

Discussion

This chapter has developed and applied a decision-analytic value of information (VOI) framework to quantify the potential economic value of tracking Canadian cattle imported into the United States. The major conclusion is that, based on data available through early 2004, the potential economic value of such information to the United States greatly exceeds its costs for cattle that may be imported in the future (by a ratio of 79,658,726 to 15,538,715 $\approx$ 5 in the base case, and more in many sensitivity analyses). For "legacy" Canadian cattle that have already entered the United States, moving quickly to locate and start tracking them before any additional BSE cases are detected appears to be well justified for many sets of input assumptions, provided that the cost per head is kept within bounds (e.g., up to $35 per head, based on the sensitivity analyses for the base case). If the costs per head are too great to justify locating all legacy animals, then location and tracking efforts should focus on the oldest animals – those with the greatest risk of becoming new BSE cases.

The potential economic value of information from a tracking program is estimated to exceed its costs by more than fivefold if such information can reduce future losses in export and domestic markets and reduce the future testing costs required to reassure or win back customers. Sensitivity analyses indicate that this conclusion is somewhat robust to many technical, scientific, and market uncertainties, including the current prevalence of BSE in the United States and/or Canada and the likely reactions of consumers to possible future discoveries of BSE in the United States and/or Canada. Indeed, the potential value of tracking information is great enough to justify tracking Canadian cattle already in the United States when this can be done for a reasonable cost. If aggressive tracking and testing can win back lost exports, then the VOI of a tracking program may increase to over half a billion dollars per year.

Epilogue and Conclusions

In 2004, a new rule was indeed passed to make imported Canadian cattle easier to identify, based on a combination of identification procedures, permanent markings, and restricted transportation of Canadian-origin cattle within the United States. Rolling the historical viewpoint forward to 2009, uncertainty about the responses of potential trade partners to discoveries of mad cow disease (BSE) continues to be a major component of uncertainty about the economic consequences of future BSE discoveries (and even of past ones, if subsequent risk management is mishandled by commingling high-risk and low-risk cattle). This was demonstrated dramatically in 2008, when the government of South Korea was thrown into crisis by an agreement to import beef from the United States (possibly including BSE-positive beef from Canada). In November 2007, the USDA had removed restrictions on imports of Canadian cattle into the United States, publishing a rule that would allow even the over-30-month-old cattle at a relatively high risk of BSE to be imported into the United States. The USDA's rule making acknowledged concerns expressed in public comments "that the rulemaking would exacerbate the limited access of U.S. beef to world markets and harm the ability of the United States to restore lost export markets," as follows:

> A number of commenters expressed concern that the rulemaking would exacerbate the limited access of U.S. beef to world markets and harm the ability of the United States to restore lost export markets. Commenters stated that imports of Canadian cattle and beef are currently banned by 35 countries, including the important U.S. export markets of the Republic of Korea, Singapore, and Taiwan, and that APHIS [the USDA's Animal and Plant Health Inspection Service] should not consider relaxing its BSE import restrictions in light of ongoing international concerns regarding the safety of Canadian beef and cattle. Other commenters stated that the United States should allow imports only of classes of cattle and beef that U.S. export markets are willing to accept from the United States.

However, the USDA then dismissed these concerns, as follows: "[USDA] Response: . . .However, we [USDA/APHIS] expect any restrictions placed on beef from the United States and Canada by an importing country to become more uniform, as discussed below, and, therefore, for the rule to have little effect on U.S. beef export markets." The near-revolt in South Korea in 2008 showed that the "little effect on U.S. beef export markets" predicted by the USDA might turn out to be larger than expected.

The Canadian BSE story continues to unfold. A partial chronology of significant events in 2008, while this book was being written, included the following.

- May 13, 2008: President Lee Myung-bak of South Korea announced that the United States agreed to halt imports if a BSE outbreak occurs in the United States.
- June 23, 2008: The Canadian Food Inspection Agency confirmed the 13th case of BSE since the national surveillance program started in 2003.
- July 3, 2008: A federal judge in South Dakota, presiding in a lawsuit brought against the USDA by a cattle producers' advocacy group, R-CALF USA, ordered the USDA to reconsider its rule allowing over-30-month-old cattle to be imported from Canada.
- July 28, 2008: A story in *The Chicago Tribune* reported that "Despite persistent fears of mad cow disease in Canadian beef, the Department of Agriculture has failed to properly track hundreds of Canadian cattle coming into the United States, the department's inspector general has concluded. The inspector general's audit, completed in March but only recently made public, said that some of the imported cattle did not have proper identification or health records despite federal regulations requiring them. ... About 1.1 million cattle were imported into the U.S. from Canada in the fiscal year ending in September 2006, the period covered by the audit. The audit mainly faulted Agriculture's Animal Plant Health Inspection Service, or APHIS, for failing to properly check records as the cattle crossed the Canadian border" (http://www.chron.com/disp/story.mpl/nation/5908457.html).

In light of such developments, VOI analysis appears to still be relevant for making decisions now about how far to enforce compliance with tracking and testing requirements. At least some undetected BSE cases are likely to be imported from Canada, based on imports of several million cattle and a Canadian prevalence rate estimated at several BSE cases per million. The risk of finding new cases of BSE in the United States is steadily increasing as imports from Canada continue. VOI analysis can help to identify present actions to help manage the economic risks created by these imports.

The analysis in this chapter has focused on potential economic consequences, and risk management options for possibly mitigating them, if another BSE case is discovered in the United States. That some BSE cases might possibly increase the risk of vCJD among people eating beef or exposed to beef products reinforces the conclusions from the economic analysis, insofar as they make it even more important to be able to identify the origin of any new BSE cases quickly, to reduce the possible human health risks as soon and as fully as possible.

That tracking and testing are imperfect has sometimes been advanced as a qualitative reason for restricting or rejecting them. The quantitative comparisons carried out in our sensitivity analyses suggest that this reasoning is not necessarily justified: measures that help to identify the origins and prevalence of BSE cases can have a high information value for improving future risk management decisions and creating additional risk management options, even if they are less than perfect.

Appendix: Market Impact Assumptions and Calculations

This appendix lists the main assumptions used in the market impact calculations referred to in the text. (This appendix is based primarily on work by economist J. van Sickle.) Table 14.1 summarizes the expected economic impacts for each set of potential future events, i.e., each branch through the decision tree to a leaf node in Fig. 14.1. The impacts are calculated starting from a baseline situation in which BSE has already been discovered in Canadian cattle (May 2003) and the impacts of the discovery have been absorbed in the market. The impacts also assume that BSE has been discovered in a Canadian animal in the United States (December 2003), resulting in a roughly 50% reduction of the previous U.S. export market.

Major assumptions for estimating the expected market outcomes following these events are listed below.

1. If the United States tracks all imports and tests all animals and no subsequent BSE case is discovered, then the exports lost following the discovery of the BSE-positive cow in Washington state in 2003 will gradually be regained. Analysis of a proposed rule for designating minimal risk regions presented by the USDA in 2004 indicates that producer surplus would decline \$1.91 billion as a result of U.S. producers losing 50% of their export market (excluding Canada and Mexico). The impact on producer revenues from this decline is estimated to be \$2.864 billion (\$1.02 billion from lower values on beef continuing to be sold and \$1.84 billion on the value of beef no longer produced because of lower prices). It is assumed that tracking and testing will restore confidence in the international community and allow that market to be restored.
2. If the United States chooses to track all imports and test only Canadian animals, then it will only gain back half of the export market lost in the baseline model. Therefore, the impact of this scenario is assumed to be half of that estimated in assumption 1, i.e., \$1.432 billion.
3. If a case of BSE is discovered in Canada, it is assumed that loss of confidence in the Canadian beef supply will increase demand for U.S. beef in domestic and international markets by another \$1.382 billion (roughly half of the estimated gains the United States experienced following the first case of BSE in Canada). The gains the United States realized from the first discovery of BSE in Canada were estimated from the USDA study on the proposed rule for minimal risk regions. The USDA estimated that the reintroduction of all beef from Canada into the U.S. market would result in a decline in producer surplus of \$1.545 billion. The impact on producer revenues from this decline is \$2.765 billion (\$1.375 billion price impact and a \$1.39 billion quantity impact). Because most of the U.S. revenue gains to be expected from a subsequent discovery of BSE in Canada were realized in the first discovery, revenue gains to U.S. producers are assumed to be half of that value (\$1.382 billion).
4. A discovery of BSE in a U.S. animal inside the U.S. border (Stage 2 outcome) is expected to cause a decline in domestic demand for U.S. beef and also result in the loss of most remaining exports. A study completed by Jin et al. (2004)

indicated that a discovery of BSE in a U.S. animal was expected to cause a 25% decline in domestic demand for U.S. beef, equal to $12.27 billion in producer revenues from our base model analysis. This result, combined with the loss of export markets equal to $2.864 billion (see discussion in assumption 1), results in a combined impact on producer revenues of $15.14 billion.

5. A discovery of BSE in a U.S. animal inside the U.S. border as a Stage 1 outcome followed by rigorous testing (test all) with no additional discoveries of BSE in the United States is assumed to have half the impact of a Stage 2 outcome of BSE in the United States It is assumed in this scenario that rigorous testing with no additional discoveries of BSE will mitigate the impacts resulting from a decline in domestic demand. It is assumed that consumer demand will decline by half that of a Stage 2 outcome of BSE in the United States ($6.14 billion) and that all remaining export markets will be lost ($2.863 billion), resulting in a total impact of $9.003 billion.
6. A discovery of BSE in a Canadian animal in the United States is assumed to cause a decline in U.S. producer revenues of $1.432 billion. This result is assumed to occur from a loss equal to one half the remaining export earnings from beef (see discussion in assumption 1).

The $4.246 billion impact for Stage 1 No BSE and Stage 2 BSE in Canada is a combination of assumptions 1 and 2.

Chapter 15
Improving Antiterrorism Risk Analysis

Several important risk analysis methods now used in setting priorities for protecting U.S. infrastructures against terrorist attacks are based on the formula *risk* = *threat* × *vulnerability* × *consequence*. This chapter identifies potential limitations in such methods that limit their ability to guide resource allocations to optimize risk reductions. After considering specific examples for the Risk Analysis and Management for Critical Asset Protection (RAMCAP™) framework used by the Department of Homeland Security, we address fundamental limitations of the product formula. These include its failure to adjust for correlations among its components, the nonadditivity of risks estimated using the formula, its inability to use risk-scoring results to allocate defensive resources optimally, and the intrinsic subjectivity and ambiguity of the *threat*, *vulnerability*, and *consequence* numbers.

Trying to assess probabilities for the actions of intelligent antagonists directly, instead of modeling how they pursue their goals adaptively in light of available information and experience, can produce ambiguous or mistaken risk estimates. Fortunately, recent work demonstrates that two-level (or few-level) hierarchical optimization models can provide a useful alternative to *risk* = *threat* × *vulnerability* × *consequence* scoring rules, and also to probabilistic risk assessment (PRA) techniques that ignore rational planning and adaptation by attackers. In such two-level optimization models, Defender predicts Attacker's best response to Defender's actions and then chooses actions anticipating the best responses. Such models appear valuable as practical approaches to antiterrorism risk analysis (Bier and Azaiez, 2009).

The *Risk* = *Threat* × *Vulnerability* × *Consequence* Framework

In April 2007, the Department of Homeland Security (DHS) released a risk-based performance standard for the security of chemical facilities in the United States (www.dhs.gov/xlibrary/assets/IP˙ChemicalFacilitySecurity.pdf). The new standard estimates risks by means of the following formula:

$$risk = threat \times vulnerability \times consequence. \tag{15.1}$$

L.A. Cox, Jr., *Risk Analysis of Complex and Uncertain Systems*,
International Series in Operations Research & Management Science 129,
DOI 10.1007/978-0-387-89014-2_15,

This standard is based on the Risk Analysis and Management for Critical Asset Protection (RAMCAP^{TM}) framework (RAMCAP^{TM} Framework, 2006). Table 15.1 shows the explanations provided for the terms in this formula, as well as for "conditional risk," i.e., risk when intent to attack is assumed (i.e., conditioned on *threat* = 1).

RAMCAP^{TM} models the actions of rational adversaries using a "reasonable worst case" as follows:

> The worst reasonable case consequence should consider that the adversary is intelligent and adaptive and will attempt to optimize or maximize the consequences of a particular attack scenario. ... Rational judgment is necessary in defining the reasonable worst case. ... [T]here is a gaming aspect to the decisions of the adversary... [s]uch models have not yet been developed.... (RAMCAP^{TM} Framework, pp. 28 and 45)

One purpose of this chapter is to show how the concept of "reasonable worst case" can be made more precise in some applications by assuming that *intelligent attackers optimize* (and, where necessary, adapt in light of new information)

Table 15.1 RAMCAP^{TM} terminology

Term	RAMCAP^{TM} definition
Risk	The potential for loss or harm due to the likelihood of an unwanted event and its adverse consequences. It is measured as the combination of the probability and consequences of an adverse event, i.e., threat. When the probability and consequences are expressed numerically, the expected risk is computed as the product of those values with uncertainty considerations.... In security, risk is based on the analysis and aggregation of three widely recognized factors: threat, vulnerability and consequence.
Conditional risk	A measure of risk that focuses on consequences, vulnerability, and adversary capabilities, but excludes intent. It is used as a basis for making long-term risk management decisions. The adversary capabilities, countermeasures, and residual vulnerability are often combined into a measure of likelihood of adversary success.
Consequence	The outcome of an event occurrence, including immediate, short- and long-term, direct, and indirect losses and effects. Loss may include human casualties, monetary and economic damages, and environmental impact, and may also include less tangible and therefore less quantifiable effects, including political ramifications, decreased morale, reductions in operational effectiveness, or other impacts.
Threat	Any indication, circumstance, or event with the potential to cause the loss of, or damage to, an asset or population. In the analysis of risk, threat is based on the analysis of the intention and capability of an adversary to undertake actions that would be detrimental to an asset or population.
Vulnerability	Any weakness in an asset's or infrastructure's design, implementation, or operation that can be exploited by an adversary. Such weaknesses can occur in building characteristics, equipment properties, personnel behavior, locations of people, equipment and buildings, or operational and personnel practices.

their attack plans to maximize the expected damage achieved. However, modeling the optimizing behaviors of attackers requires risk assessment models different from Equation (15.1). The following sections survey some important limitations on attempts to directly estimate $risk = threat \times vulnerability \times consequence$ for purposes of allocating defensive resources, without modeling intelligent planning and optimization by attackers.

RAMCAP™ Qualitative Risk Assessment

Before considering the more fundamental limitations of Equation (15.1), we first consider some flaws in the specific implementation of the equation in RAMCAP™. RAMCAP™ proposes two options for risk assessment, which it calls "qualitative" and "quantitative," although both are based on semiquantitative (ordered categorical) ratings of *threat*, *vulnerability*, and *consequence*. The "qualitative" option (which might also be called semiquantitative) categorizes economic consequences using the following rating scale: 0 = \$0–25 million loss, 1 = \$25–50 million, 2 = \$50-100 million, . . ., 13 = \$102,401 million and above. Fatalities and injuries are scored similarly: 0 = 0–25 fatalities, 1 = 25–50 fatalities, . . . , 13 = 102,401 fatalities or more, with a similar rating scale for the number of injuries. (The severity of injuries is not included in the injury score.) (The RAMCAP™ tables actually leave small gaps between intervals, e.g., \$0–25 million, \$26-50 million, \$51-100 million, etc., so that consequences such as \$25.4 million or \$50.7 million do not fall in any category. We assume that "26" includes values greater than 25 and less than 26, and similarly for other gaps.)

Vulnerability is assessed similarly, using a "likelihood of attack success scale" that assigns a score of 0 to success probabilities below 0.0312, 1 to probabilities from 0.0312 to 0.0625, 2 to probabilities from 0.0625 to 0.125, 3 to probabilities from 0.125 to 0.25, 4 to probabilities from 0.25 to 0.5, and 5 to probabilities above 0.5. (Probabilities in "bin" 5 are further subdivided into 0.5–0.75, 0.75–0.9, and greater than 0.9.)

The RAMCAP™ documentation suggests using event tree analysis to estimate the likelihood of attack success. In event tree analysis, different possible sequences of events are represented by corresponding sequences of nodes in a tree (a directed acyclic graph in which each node has a unique parent, except for the first or "root" node, which represents an initiating event such as "Attack attempted"). Nodes represent events, multiple arcs branching from a node represent different possible outcomes of a random event, and the probability of each terminal node (each "leaf" of the tree) is the product of the conditional probabilities of the arcs along the unique path leading from the root node to it. To use event trees to estimate attack success probabilities, the conditional probabilities for the arcs in the tree must be estimated. This begs the question of how such probabilities are to be determined, especially for "events" that represent attacker actions. Several researchers have commented that modeling actions as random variables is inadequate for representing the purposive (and opportunistic) behaviors of intelligent attackers (Golany et al., 2009; Bier and Azaiez, 2009; National Research Council, 2008).

Finally, a "conditional risk matrix" (i.e., a risk matrix assuming that an attack takes place) assigns overall conditional risk scores to pairs of consequence and vulnerability scores via the formula

$$\textit{conditional risk score} = \textit{consequence score} + \textit{vulnerability score}. \tag{15.2}$$

This reflects the identity

$$\log(vulnerability \times consequence) = \log(vulnerability) + \log(consequence) \tag{15.3}$$

because the scales used to rate consequence and vulnerability are logarithmic. (There also appears to be an implicit independence assumption that allows vulnerability and consequence scores to be assessed separately, which might not be realistic.)

In general, the qualitative risk rating does not provide adequate information to guide resource allocation. For example, it assigns the same qualitative risk score ("5") to (a) a 100% probability of zero fatalities (quantitative risk = 0, qualitative risk = 5 + 0 = 5) and (b) a 20% probability of 100 fatalities (qualitative risk = 3 + 2 = 5). Similarly, a zero probability of a \$100 billion loss is given the same risk score ("11") as a certainty of a \$1 billion loss. Such anomalies arise because consequence scores and vulnerability scores are summed to get risk scores; thus, even if one score is zero, the risk score (unlike the quantitative risk) can be nonzero. The scoring also can assign relatively small scores to relatively large risks. For example, a 0.10 probability of 100 deaths (expected value = 10 deaths) would have a smaller risk score (4) than a 0.26 probability of 26 deaths (expected value = 6.76 expected deaths, risk score = 5).

Limitations of RAMCAP™ for Quantitative Risk Assessment

RAMCAP™'s "quantitative" scoring (which might also be called semiqualitative) is also based on Equation (15.1). Vulnerability and consequence numbers are calculated as the arithmetic average of the upper and lower values of the "bins" (the value ranges in the preceding "qualitative" approach) for the attack's success probability and the consequence of a successful attack, respectively. All quantities are interpreted as expected values.

The RAMCAP™ Framework states that an advantage of using the above formula with a defined set of scales for vulnerability and consequence is that "The risk associated with one asset can be added to others to obtain the aggregate risk for an entire facility... [and] can be aggregated and/or compared across whole industries and economic sectors. This is precisely the goal of DHS." However, such summation is, in general, mathematically incorrect, as shown in the following examples.

Moreover, it lets facility owners manipulate risk estimates up or down, depending on preferences. It is unable to distinguish among some risks (limited resolution) and can give incorrect estimated risk rankings. The following examples illustrate these limitations.

Example: Distortions Due to Use of Arithmetic Averages on Logarithmic Scales

For the following two risks:

- A: (*vulnerability* = 0.25, *consequence* = \$400 M),
- B: (*vulnerability* = 1, *consequence* = \$60 M),

the formula *conditional risk* = *vulnerability* × *consequence* implies that A has a larger conditional risk than B (\$100 M vs. \$60 M). However, RAMCAP™ would assign a vulnerability of (0.125 + 0.25)/2 = 0.1875 and a consequence of (200 M + 400 M)/2 = 300 M to A, implying an estimated conditional risk of 0.1875*300 M = \$56.25 M for A. It would assign a vulnerability of (0.9 + 1)/2 = 0.95 and a consequence of (50 M + 100 M)/2 = \$75 M to B, implying an estimated conditional risk of 0.95*\$75 M = \$71.25 M for B. Thus, it reverses the correct ranking of these two risks.

Example: Limited Resolution

RAMCAP™ quantitative risk assessment assigns the same quantitative conditional risk estimate of [(0.125 + 0.25)/2]*[(\$0 + \$25 M)/2] = \$2.34 M to a facility with (*vulnerability* = 0.15, *consequence* = \$1 M) and a facility with (*vulnerability* = 0.25, *consequence*= \$25 M). Yet many stakeholders might view these as significantly different risks.

Example: Manipulating Vulnerability Estimates by Aggregating Attack Scenarios

Suppose that a facility can be attacked via any of six separate approaches (generically called front, back, left, right, top, and bottom). The success probability for each approach (if attempted) is 0.07, and these success probabilities are independent of each other. (This is for illustration only; such independence assumptions may dangerously oversimplify reality.) If one approach is attempted and fails, others may be attempted. The consequence of a successful attack is the same in each case. Thus, the sum of the conditional risks is proportional to the sum of the vulnerabilities contributed by the different approaches: 6*[(0.0625 + 0.125)/2] = 0.5625.

To achieve a lower vulnerability number, a facility owner can simply represent each possible attack path (e.g., "attack via front," "attack via back," etc.) as one possible aspect of the same overall attack ("attack via some approach"). The overall attack has success probability: Pr(attack succeeds) = 1 – Pr(all approaches fail) = 1 – $(1 - 0.07)^6 = 0.353$, which is coded as $(0.25 + 0.5)/2 = 0.375$. Thus, the facility owner can claim a vulnerability estimate of either 0.5625 or 0.375, whichever he prefers, by choosing to calculate the vulnerability of the facility as the sum of the vulnerabilities from different attack paths or as the vulnerability from the overall probability of a successful attack. The claim that "The risk associated with one asset can be added to others to obtain the aggregate risk for an entire facility" is not true in general, as the correct probabilistic formulas, such as Pr(attack succeeds) = 1 – Pr(all approaches fail), are not additive.

Example: Nonadditive Vulnerabilities

Suppose that an attack on a facility succeeds if and only if both of two activities, *A* and *B*, are successfully completed. *A* can be accomplished in either of two ways, *A*1 or *A*2, and *B* can be achieved in either of two ways, *B*1 or *B*2. *A* must be completed before *B* can be attempted. Thus, the four possible successful attack sequences are (*A*1, *B*1), (*A*1, *B*2), (*A*2, *B*1), (*A*2, *B*2). If each of *A*1, *A*2, *B*1, and *B*2 independently has a 0.4 success probability, then the overall probability of a successful attack is Pr(*A* is completed)*Pr(*B* is completed | *A* is completed) = [1 – Pr(*A*1 and *A*2 both fail)]*[1 – Pr(*B*1 and *B*2 both fail)] = $(1 - 0.6{*}0.6)^2 = 0.41$. But summing the vulnerabilities contributed by each of the four attack scenarios (*A*1, *B*1), (*A*1, *B*2), (*A*2, *B*1), (*A*2, *B*2), each having success probability 0.4*0.4 = 0.16, yields a total vulnerability estimate of 4*0.16 = 0.64, larger than the correct probability, 0.41.

Example: Product of Expected Values Not Equal to Expected Value of Product

Suppose that *threat* and *vulnerability* are each uniformly distributed between 0 and 1 and that *consequence* is uniformly distributed between 0 and \$25 million. RAMCAP™ applies Equation (15.1) to the expected values of the quantities on its right side, yielding *risk* = (0.5)*(0.5)*(\$12.5 M) = \$3.125 M. However, if *threat* is perfectly positively correlated with each of *vulnerability* and *consequence* (with *vulnerability* = *threat* and *consequence* = \$25 M**threat*), then the correct value of *risk* is \$6.3 million. If *threat* is perfectly negatively correlated with each of *vulnerability* and *consequence* [with *vulnerability* = 1 – *threat* and *consequence* = \$25 M*(1 – *threat*)], then the correct value of *risk* is only about \$2 million. Thus, correlations among the components substantially affect the correct value of *risk*.

More simply, suppose that each of *T*, *V*, and *C* is equally likely to be 0 or 1. (For simplicity, here *C* is scaled to vary from 0 to 1.) Then the product of their expected values is 0.5*0.5*0.5 = 0.125. But if the variables are interdependent, with the value of C determining the values of T and V via the equations $T = C$ and $V = C$, then the expected value of their product, TVC, is 0.5, not 0.125. On the other hand, if $T = C$ and $V = 1 - C$, then the expected value of TVC is 0. Thus, multiplying expected values gives an answer (0.125) that may be very different from the correct expected value of the product, taking into account the dependencies among the components, which may be as low as 0 or as high as 0.5. Because the formula *risk* = *threat* × *vulnerability* × *consequence* ignores dependencies among components, the numerical value of *risk* that it produces (with all terms being expected values, as in RAMCAP) may be either higher than the correct value (e.g., 0.125 instead of 0) or lower than the correct value (e.g., 0.125 instead of 0.5).

In practice, positive correlations may arise if intelligent attackers are more likely to attack targets with high *vulnerability* and *consequence* values, or if larger targets are both more valuable and more vulnerable. Then Equation (15.1), with terms interpreted as expected values, may underestimate risks.

Risk Rankings Are Not Adequate for Resource Allocation

After a risk assessment has been completed, how can its results be used to improve risk management decision making? One common answer is that risk managers should rank-order estimated risks from highest to lowest based on the risk assessment results, then allocate risk management resources to risk-reducing countermeasures (e.g., for vulnerability reduction or consequence mitigation) from the top of the list down, until available resources have been spent. For example, a Department of Energy (DOE) report states that

> The risk values are then determined and ranked from the highest to the lowest producing a relative risk ranking. Obviously, resources should be used to reduce the vulnerabilities or mitigate the consequences from the highest ranked threat scenarios first. In the National Strategy for Homeland Security, it is stated "Protecting America's critical infrastructures thus require that we determine the highest risks. ..." In planning security upgrades at Brookhaven National Laboratory, a select committee was established and this relative risk ranking concept was used for ordering the upgrade schedule. (Indusi, 2003)

However, allocating resources "to reduce the vulnerabilities or mitigate the consequences from the highest-ranked threat scenarios first" may *not* be an effective way to allocate defensive resources to reduce risks. Effective risk management requires considering risk *reductions* achieved by different allocations of limited resources. These are not necessarily implied by the sizes of different risks, because countermeasures may only partly reduce risks and because costs of different countermeasures constrain which ones can be implemented with the available budget. In addition, if the optimal portfolio of risk-reducing activities requires diversifying defensive investment across multiple types of threat scenarios, then any priority rule that ranks all instances of one type above all instances of another is obviously inconsistent with an optimal (diversified) investment (see Chapter 5).

Example: Priority Ranking May Not Support Effective Resource Allocation

Setting: The following counterexample is repeated (with minor modifications) from Chapter 4. Suppose that an agency must allocate a limited budget to reduce the risks of terrorist attacks. Three opportunities have been identified to reduce risks:

- Act A reduces risk from 100 to 80. It costs $30.
- Act B reduces risk from 50 to 10. It costs $40.
- Act C reduces risk from 25 to 0. It costs $20.

Here, "risk" is measured on a scale such as expected casualties in the event of an attack, i.e., it is a conditional risk. (This example can also be constructed so that all three acts start from the same base level of risk, say 50, and A, B, and C reduce risk by 20, 40, and 25, respectively. Using different base levels allows for the possibility that options A, B, and C protect different subpopulations.) The goal for resource allocation is to achieve the largest possible total risk reduction for the available budget.

Problem: What priority ranking of A, B, and C achieves the largest risk reduction from the allocation of limited funds, if resources are allocated from the top of this priority list down until they are exhausted?

Solution: No priority ranking exists that answers this question. Instead, *the correct answer depends on the budget.* For a budget of $49, the largest feasible risk reduction is achieved by funding B, so the best priority order puts B first. If the budget is $50, then funding A and C achieves the greatest risk reduction, so B should be ranked last. At $60, the best investment is to fund B and C, so now A should be ranked last. Thus, *no rank ordering of A, B, and C optimizes resource allocations independent of the budget.* For example, no possible rank order is optimal for budgets of both $49 and $50.

The difficulty illustrated here is that *resource-constrained investments in reducing risks cannot, in general, be optimized using priority rankings.* Resource allocation problems that can be solved this way have been characterized (Bertsimas and Nino-Mora, 1996), but selecting a portfolio of expensive risk-reducing activities to maximize the risk reduction achieved is not a problem of this type.

Some Fundamental Limitations of *Risk* = *Threat* × *Vulnerability* × *Consequence*

Some of the limitations discussed in the preceding sections are specific to RAMCAP™ and can be overcome fairly easily. Instead of using arithmetic averages on a log scale, one could use geometric averages. Instead of discretizing or "binning" estimated values of the *vulnerability* and *consequence* attributes, one could use point estimates without binning. Instead of simply multiplying expected values, one could adjust for covariance [as in the formulas $E(XY) = E(X)E(Y) + \text{Cov}(X, Y)$ or $E(XYZ) = E(X)E(Y)E(Z) + E(X)\text{Cov}(Y, Z) + E(Y)\text{Cov}(X, Z) + E(Z)\text{Cov}(X, Y) + E\{[(X - E(X)][Y - E(Y)][Z - E(Z)]\}$] or use Monte Carlo simulation to estimate the mean of a product of (possibly correlated) factors. However, other difficulties appear to be intrinsic to Equation (15.1). This section discusses limitations that appear to be impossible to avoid.

"Threat" Is Not Necessarily Well Defined

Equation (15.1) assumes that a *threat* probability number exists, at least in principle, reflecting the probability of an attack in a stated interval of time. However, if the attack probability in that interval depends on the assessed *threat* number, then any estimate of *threat* may be self-defeating. This occurs if an attacker's response to the *threat* estimate (or to the defender's actions based on it) invalidates the estimate. In general, any threat estimate that does not model how attackers respond to the threat estimates (and resulting defender actions) may be unreliable. This holds no matter how the threat estimates are derived, e.g., whether by Bayesian, frequentist, or other (e.g., game-theoretic) threat assessment. Moreover, even apart from the need to model attacker responses, threat numbers may be inherently ambiguous when the assumptions and knowledge on which they are based are not explicitly stated.

Example: Self-Defeating Threat Predictions

Suppose that two players, Attacker and Defender, engage in the following game.

Stage 1: Defender estimates the *threat* (= attack probability), *vulnerability*, and *consequence* values for each of M facilities. Defender identifies the N top-ranked (highest *threat* × *vulnerability* × *consequence* values) facilities, where $N < M$ reflects Defender's resource constraints.

Stage 2: Attacker randomly selects $K > 0$ of the other $(M - N)$ facilities to attack, with probabilities proportional to their *vulnerability* × *consequence* values (and independent of their estimated *threat* probability numbers). (K reflects Attacker's resource constraints.)

In this setting, assigning a high enough *threat* value to a facility to place it in the top N facilities guarantees that it will not be attacked (true *threat* and *risk*= 0, conditioned on estimated *threat* and *risk* being sufficiently high). Thus, estimating a threat as high makes the true threat low. The concept of a threat as a static probability number that is "out there" to be estimated is fundamentally inadequate for protecting against informed, intelligent attackers if the threat estimate itself affects the threat being estimated. Although this example has used a deliberately simple pair of decision rules for Defender and Attacker, it illustrates that Attacker's strategy may exploit Defender's own threat estimates and ranking, thus undermining their predictive validity.

Example: Ambiguity of Threat Estimates

Suppose that an adversary will attack a facility within the next year if and only if he knows that the attack will succeed with probability at least 0.8. Suppose that a perfectly reliable and well-calibrated vulnerability assessment expert announces that the probability that an attack on the facility will succeed (assuming, for purposes of the vulnerability assessment, that it is attempted) is 0.33, based on knowledge that an attack can succeed if and only if the adversary has a secret weapon (or inside help, etc.) and the probability of obtaining this required advantage in the next year is only 0.33. A threat assessment expert knowing these probability numbers might well conclude that the facility will not be attacked in the next year (since $0.33 < 0.8$). But, in reality, the probability of an attack on this facility in the next year is 0.33, the probability that the adversary will gain the secret weapon and then attack. (A 100% chance of a 33% chance of success at the time the adversary makes a go-no go decision would indeed guarantee no attack, but a 33% chance of a 100% chance of success does not.)

"Vulnerability" Can Be Ambiguous and Difficult to Calculate via Event Trees

The concept of "vulnerability" as the conditional probability that an attack succeeds, given that it is attempted, is vague about *how* and *why* the attack is conducted and what contingency plans an attacker uses if some setbacks are encountered. These details can greatly affect the calculated vulnerability values.

Example: Ambiguity of Attack Success Probabilities Elicited from Experts, Due to Unspecified Implications of an Attack

Suppose that we ask an expert in terrorism risk analysis for (a) the probability that a certain facility will be attacked within the next year (the "threat" to that facility) and (b) the probability that an attack will succeed in destroying or compromising the facility, assuming that an attack is made (the "vulnerability" of the facility). The expert knows (although we do not) that an attack will be made in the next year if and only if the potential attackers first obtain a device that is guaranteed to make the attack successful. Any attack made without that device will fail. The probability that the device is obtained in the next year is 1/3. Based on this knowledge, the answer to question (a) is that the threat is 1/3. But the correct answer to question (b) depends on exactly how the expert interprets the question. If she interprets "assuming that an attack is made" to mean "assume that an attack is made, whether or not the attackers have the device," then the probability that the attack will succeed is only 1/3 (the probability that the device is obtained prior to the attack). But if she interprets "assuming that an attack is made" to imply that the attackers will necessarily have the device (since otherwise they would not attack), then the probability that the attack will succeed is 1. Which answer, 1/3 or 1, the expert gives in response to question (b) depends entirely on her assumptions about the implications of the assumption that an attack is made. Since no uniquely correct interpretation or set of implications is specified as part of the question, there is no unique correct answer. Using such elicited probabilities to allocate resources makes the allocation depend on interpretations of ambiguous hypothetical conditions, rather than on how terrorists will actually behave.

Suggesting that event tree analysis can be used to estimate vulnerability numbers begs the question of exactly *how* the plans, contingency plans, and adaptive responses (and, if necessary, replanning) of intelligent agents should be modeled, so that they can be represented as event trees. The following examples show that treating activities of intelligent attackers as random variables in a standard event tree is generally not adequate for modeling how determined, intelligent attackers learn from failure and adaptively modify their plans and behaviors. Risk assessments that do not model such features of intelligent attacks can underestimate risks and misallocate resources.

Example: Ambiguous Elicited Probabilities of Successful Attack

For simplicity, suppose that an attacker moves among only four states, *Ready*, *Succeed*, *Fail*, and *Interdicted*. The attacker starts in the *Ready* state. From that state, all he can do is attempt an attack. The attack consists of a single activity, which either succeeds (with probability s) or fails (with probability $1 - s$) when attempted. A failed attempt leads to a probability c of being interdicted ("caught"). If this is avoided, then the attacker eventually returns to *Ready*, perhaps after a delay. This example considers how this simple process might be represented and analyzed via probabilistic risk assessment (PRA), which seeks to characterize risk via the frequency (or probability) and severity of possible adverse consequences (Bier and Azaiez, 2009).

Problem: If the attack success and interdiction probabilities are $s = 0.2$ and $c = 0.1$, respectively, then what is the probability that this situation will end with a successful attack ("vulnerability")?

Solution: The answer depends completely on exactly how the two probabilitiess $s = 0.2$ and $c = 0.1$ are interpreted, and on the apparently subtle philosophical distinction between aleatory and epistemic probabilities. One possible answer is that the probability of a successful attack is $s = 0.2$. This is correct if the success probability of $s = 0.2$ is interpreted as an *epistemic probability* reflecting uncertainty about whether the attacker can successfully complete the attack activity once it is attempted. (For example, an attack team that rams a car loaded with gasoline into an airport building may have a 20% probability of discovering that this successfully compromises the integrity of the building and kills a large number of people and an 80% probability of discovering that it does not work very well, and just blows up the car and kills the driver. In this case, attempting the attack will resolve the uncertainty. A success probability of $s = 0.2$ would then be the answer to the question.)

A different answer is correct if the success probability is interpreted as an *aleatory* (or "stochastic") *probability*. In this interpretation, failure on one attempt does not reveal that success is impossible: it only means that the attacker was not lucky on that particular attempt. After returning to the *Ready* state, another attempt could be made, and the probability of success would then be $s = 0.2$ again. (For example, this is the interpretation that might be most relevant if "success" of an attack depends on random factors such as the direction of the wind, traffic conditions, or the chance presence or absence of police or detectors at the time of the attack.)

If the aleatory probability interpretation is used, then the situation moves among different states according to the Markov transition diagram in Fig. 15.1. There are thus two possible eventual outcomes (absorbing states) of this process: *Succeed* or *Interdicted*. The probability of eventual success starting from the *Ready* state, denoted by p, satisfies the recursive equation

$$p = s + (1 - s)(1 - c)p$$

That is, it is the sum of the probability of immediate success starting from *Ready*, which is s, plus [the probability of returning to *Ready,* which is $(1 - s)(1 - c)$] $\times$ [probability of eventual success starting from *Ready*, which is p.] The solution is

$$p[1 - (1 - s)(1 - c)] = s, \text{ or } p = s/[s + c(1 - s)].$$

For the particular numerical values in this example, $s = 0.2$ and $c = 0.1$, the probability of a successful attack being completed before interdiction is

$$p = s/[s + c(1 - s)] = 0.2/(0.2 + 0.1^{*}0.8) = 0.714.$$

Thus, an aleatory interpretation of the probabilities $s = 0.2$ and $c = 0.1$ leads to a predicted risk of 0.714, significantly higher than the predicted risk of 0.2 if an epistemic interpretation is used.

$$\begin{array}{ccc} & s & \\ & \textit{Ready} \rightarrow \textit{Succeed} & \\ (1-s) & \downarrow\uparrow & (1-c) \\ & \textit{Fail} \rightarrow \textit{Interdicted} & \\ & c & \end{array}$$

State-transition diagram

Fig. 15.1 Transition diagram for a simple stochastic attack process. Arrows indicate possible transitions among the four states (*Ready, Succeed, Fail, Interdicted*). Each transition arrow is labeled with its probability of occurrence as follows: s = probability that *Succeed* follows *Ready*; $1 - s$ = probability that *Fail* follows *Ready*; c = probability that *Interdicted* follows *Fail*; $1 - c$ = probability that *Ready* follows *Fail*

In most practical applications of Equation (15.1), interpretations of probability numbers (aleatory or epistemic) are left to subject matter experts. Hence, it can be unclear how users of such models are supposed to interpret the probabilities produced and whether the risks quantified by the model are correct. For example, defining vulnerability as "the conditional probability that an attack succeeds, given that it is attempted" would be inadequate, as it is silent about whether "probability" is to be interpreted as aleatory or epistemic – *and this difference matters.*

Example: Event Trees vs. Decision Modeling for Actions of Intelligent Attackers

Even if all probabilities are known and are aleatory, standard event tree modeling that ignores rational decision making by an attacker may lead to conclusions and recommendations importantly different from those generated by models that account for the optimizing behavior of intelligent attackers. Treating attackers as optimizers and calculating their best responses to different conditions may allow a defender with limited resources to achieve larger risk reductions than can be produced by any model [including Equation (15.1)] that ignores details of how intelligent attackers adapt their plans as information becomes available before and during the course of an attack. "Best-response" models represented as two-level or few-level hierarchical optimization problems (e.g., in which Defender calculates Attacker's best responses to various conditions, and then chooses defensive investments to minimize the damage from Attacker's best response) are typically far easier to formulate, solve, and understand than full game-theoretic analyses, yet are adequate for much practical work in counterterrorism and infrastructure protection (Brown et al., 2006).

Problem setting: Suppose that an attack on a facility succeeds if and only if Attacker successfully completes both of two activities, A and B. It costs Attacker 1 unit (on some scale) to attempt each activity. Activities A and B have respective success probabilities of 0.8 and 0.5, when and if each is attempted. The benefit to Attacker of a successful attack is 10 (on a scale comparable to the one used for costs). Attacker acts intelligently to maximize the expected net value. He can afford to initiate one attack per year, and the facility is one of 100 similar facilities that are equally likely to be targeted; thus, the probability that it will be attacked in a year is assumed by Defender to be *threat* = 0.01.

Defender, with a limited budget, must choose which of two expensive countermeasures to implement: option 1 reduces the success probability for attack activity A from 0.8 to 0.24, while option 2 reduces the success probability for activity B from 0.5 to 0.145.

Problem: (a) Which countermeasure should Defender implement, option 1 or option 2 (assuming that he can afford only one)? (b) How much risk reduction will this achieve?

Solution based on traditional event tree analysis: The probability that an attack succeeds is the probability that both A and B succeed. This is 0.8*0.5 = 0.4 in the absence of intervention; 0.24*0.5 = 0.12 if option 1 is implemented; and 0.8*0.145 = 0.116 if option 2 is implemented. Therefore, the decision *implement* *option 2* maximizes predicted risk reduction. It reduces the predicted conditional risk of a successful attack (conditioned on an attack being attempted) from 0.4 to 0.116.

Solution accounting for intelligent decision making by adversary: An intelligent adversary maximizes the expected net benefit of an attack by trying the activities in order of decreasing ratio of failure-probability-per-unit-cost until no remaining activity has positive expected value. [This is a simple example of an *index policy* for dynamic optimization (Cox, 1990; Denardo et al., 2004; Sethuraman and Tsitsikilis, 2007)]. Thus, Attacker's optimal strategy if Defender does nothing to reduce vulnerabilities is to first attempt activity B (having failure-probability-per-unit-cost ratio of 0.5/1 = 0.5) and then attempt activity A (having failure-probability-per-unit-cost ratio of 0.2/1 = 0.2) if B succeeds. This attack sequence has an expected net benefit to Attacker of (–1 to attempt B) + (0.5 probability that B succeeds)*[(–1 to attempt A) + (0.8 probability that A succeeds)*(10 benefit if B & A both succeed)] = –1 + .5*(–1 + 0.8*10) = 2.5. Since this is positive, it is worthwhile for Attacker to undertake an attack.

If option 2 is implemented, then the optimal attack still attempts activity B first (with failure-probability-per-unit-cost ratio of 0.855) and then, if B succeeds, attempts A (with failure-probability-per-unit-cost ratio of 0.2/1 = 0.2), for an expected cost of (–1 to attempt B) + (0.145 probability that B succeeds)*[(–1 to attempt A) + (0.8 probability that A succeeds)*(10 benefit if B & A both succeed)] = –1 + 0.145*(–1 + 0.8*10) = 0.015. Again, an attack is worthwhile.

However, if option 1 is implemented, then it is best for Attacker to attempt activity A first (with failure-probability-per-unit-cost ratio of 0.76/1 = 0.76) and then, if A succeeds, attempt B (with failure-probability-per-unit-cost ratio of 0.5/1 = 0.5), for an expected net benefit to Attacker of (–1 to attempt A) + (0.24 probability that A succeeds)*[(–1 to attempt B) + (0.5 probability that B succeeds)*(10 benefit if B & A both succeed)] = –1 + 0.24*(–1 + 0.5*10) = –0.04. Since the expected net benefit of this attack is negative, *the attacker has no attractive attack opportunity left at this target location.* [Trying B first and then A if B succeeds would have an expected value of (–1 to attempt B) + (0.5 probability that B succeeds)*[(–1 to attempt A) + (0.24 probability that A succeeds)*(10 benefit if B & A both succeed)] = –1 + 0.5*(–1 + 0.24*10) = –0.30.]

Therefore, the optimal defensive strategy, taking into account the attacker's intelligent responses to different countermeasures, is to *implement option 1*. Doing so reduces the risk from 0.4 to 0. This differs significantly from the recommendations and predicted risk reduction made above using event tree analysis, which models Attacker's behavior using random variables instead of optimized decisions.

The principle illustrated in this example, that *modeling an intelligent attacker's intelligent (optimizing, adaptive) behavior leads to different recommendations and risk estimates from traditional event tree analysis*, also applies to expected utility-maximizing attackers with exponential utility functions and to attack opportunities with general precedence constraints (Denardo et al., op cit.) The lesson here is that event trees can be used, but they do not represent or solve Attacker's and Defender's key planning and decision problems. These must therefore be solved using other principles, such as the index policy illustrated above.

Optimization of decisions based on their probable consequences, taking into account the actions of opponents, is essential for realistic QRA. This example has illustrated a very simple case of *two-level hierarchical optimization* in which Defender can evaluate each feasible alternative course of action (i.e., implement option 1, option 2, both, or neither) by calculating Attacker's best response to it. This enables Defender, in turn, to choose a best course of action for reducing risk, taking Attacker's best response into account. Military operations research experts have developed two-level optimization for leader-follower (Stackelberg) games into a powerful planning and resource allocation technique that appears to be practical for many counterterrorism and infrastructure protection risk management applications (Brown et al., 2006). Hierarchical optimization dispenses with the *threat* component, conceived of as a single number to be estimated, and goes beyond simplistic estimates of vulnerability in Equation (15.1). Instead, it focuses on predicting and controlling attacker behaviors via the incentives created by defensive investments. Hierarchical optimization can produce strategies (perhaps mixed) for the opponents as outputs, rather than requiring them as inputs. Such modeling avoids both the (potentially unrealistic) idealizations of game theory (Bier and Azaiez, 2009) and the limitations of Equation (15.1).

Example: Probabilistic vs. Decision Modeling for Actions of Intelligent Attackers

This example reinforces the previous one by showing that, even in the absence of any complicated optimization calculations, an attacker who simply "follows the path of least resistance" by picking activities that are most likely to succeed in completing the attack may behave very differently from what a simple probabilistic risk assessment (PRA) model of an attack would predict (Golany et al., 2009). As a result, the best choice of countermeasures may differ from what would be recommended using PRA calculations that treat attacker activities as random variables rather than as outcomes of intelligent planning and adaptation to intelligence about countermeasures.

Suppose that public health is threatened if some logical combination of events takes place. For clarity, consider this small example, with only four basic events:

- A = a specific infectious agent is introduced into a building via air intake,
- B = the infectious agent is introduced into the building via drinking water,
- C = the agent escapes detection until after building occupants are exposed and some become infected,
- D = one or more people become ill or die as a result of the undetected exposure.

The probability of D over the next year can be calculated as follows:

$$\Pr(D) = \Pr(D|C)^*\Pr(C) = \Pr(D|C)^*[\Pr(C|B)^*\Pr(B) + \Pr(C|A)^*\Pr(A) - \Pr(C|A\&B)^*\Pr(B|A)^*\Pr(A)].$$

In this expression, each of the initiating events A and B would typically be assigned an estimated annual frequency of occurrence. $\Pr(C \mid B)$ and $\Pr(C \mid A)$ reflect any detection countermeasures that are in place. $\Pr(D \mid C)$ reflects the virulence, time to diagnosis, and availability of effective treatments.

If the numbers are $\Pr(A) = 0.01$, $\Pr(B) = 0.02$, $\Pr(C \mid A) = 0.5$, $\Pr(C \mid B) = 0.4$, $\Pr(D \mid C) = 0.2$, and $\Pr(A \,\&\, B) = \Pr(B \mid A)^*\Pr(A)$ is small enough to ignore, then the annual probability predicted for D would be estimated as

$$\begin{aligned}\Pr(D) &= \Pr(D|C)^*[\Pr(C|B)^*\Pr(B) + \Pr(C|A)^*\Pr(A)]\\ &= 0.2^*[0.4^*0.02 + 0.5^*0.01] = 0.0026.\end{aligned}$$

A countermeasure that reduces the nondetect probability for airborne agents from 0.5 to 0.25 would reduce predicted risk to 0.0021 [i.e., by (0.0026 – 0.0021)/0.0026 = 19.2%]. A countermeasure that reduces the nondetect probability for waterborne agents from 0.4 to 0.2 would reduce the predicted risk to 0.0018 (i.e., by 30.8%). Installing both countermeasures reduces the predicted risk to 0.0013, i.e., by half.

If the initiating events are caused by intelligent attackers with good intelligence about what countermeasures have been implemented, then the risk assessment changes dramatically. Absent countermeasures, such an attacker who has sufficient resources to afford an average of 0.01 + 0.02 = 0.03 attacks per facility-year at this location would focus on the path of least resistance or greatest success probability, i.e., the airborne route. Doing so creates an annual risk of 0.2*[0.4 * 0 + 0.5*0.03] = 0.0030. A countermeasure that reduces the nondetect probability for airborne agents from 0.5 to 0.25 would cause the attacker to shift the attack toward waterborne attacks. After the countermeasure has been implemented, the new risk becomes 0.2*[0.4 * 0.03 + 0.25*0] = 0.0024, corresponding to only a 7.7% reduction in risk. Implementing a countermeasure that reduces the nondetect probability for waterborne agents from 0.4 to 0.2 would leave the optimal attack strategy (airborne route) and resulting risk unchanged at 0.0030. Implementing both countermeasures would reduce the risk to 0.0015 (or possibly to zero, if a success probability of only 0.0015 is too low to justify the attacker's investment of resources).

In summary, for the same total frequency of initiating events (i.e., 0.03 per year), the intelligent attack gives a slightly higher predicted risk than a fault tree model (0.0030 instead of 0.0026). However, the two approaches give very different predictions for the effects of alternative risk management countermeasures. For example, a countermeasure that reduces the nondetect probability for waterborne agents from 0.4 to 0.2 would reduce the predicted risk to 0.0018 in the fault tree model, but would leave it unchanged at 0.0030 in the intelligent-attacker model. The two models have opposite implications for which countermeasure to implement if only one can be afforded. The purely probabilistic model predicts that the larger risk reduction is achieved by the countermeasure that reduces the nondetect probability for the waterborne route, while the intelligent-attacker model predicts that implementing this countermeasure will leave the risk unchanged. Thus, it is important to use decision-optimizing models rather than purely probabilistic models if initiating events do, in fact, result from the efforts of intelligent agents who adapt their attacks to exploit relative weaknesses in defenses. The reason is that intelligent agents can use information as it becomes available to continually update their strategies (for example, to follow the changing path of least resistance until no attractive path forward remains). This type of adaptation to exploit changing situations does not arise in standard PRA models for applications without intelligent agents.

"Consequence" Can Be Ambiguous and/or Subjective

In many applications, the *consequence* term in Equation (15.1) is interpreted as a single number. (Of course, the number can be drawn from a distribution, or arrived at via multiattribute modeling.) This number may represent the midpoint of a range or the mean of a probability distribution for multiple possible consequences. However, in general, there is no unique "correct" way to represent an uncertain consequence by a single number (see Chapter 4). Such numbers are inherently either subjective (if they are calculated for a particular, stated risk attitude) or ambiguous (otherwise). For example, recall from Chapter 4 that three prospects (e.g., attack scenarios), A, B, and C, having normally distributed consequences (on some value scale) with respective means of 1, 2, and 3 and respective variances of 0, 1, and 2, can have oppositely ordered certainty equivalent values (with A being ranked either highest or lowest), depending on the assessor's subjective risk attitude.

Discussion and Conclusions

The concepts of *threat*, *vulnerability*, and *consequence* as numbers that experts can estimate and use in calculating risk are problematic for assessing risks from intelligent adversaries. *Threat* estimates may be self-defeating if attackers use intelligence about the defender's own threat estimates to help decide where and when to attack. *Vulnerability*, meaning the probability that an attack succeeds if it is attempted, may depend on the attacker's ability to dynamically replan and continue the attack when obstacles are encountered. The information needed to predict what an intelligent attacker will do and how likely he is to succeed must include such contingent actions and therefore is not well represented by a single vulnerability number. Rather, *vulnerability* is perhaps better represented as a model for calculating the conditional probability that an attack will succeed if the attacker plans optimally (and then reoptimizes throughout the course of the attack as new information becomes available). The correct answer to "How likely is it that an attack will succeed?" typically depends on the attacker's response to the defender's preparations. Attempting to assess vulnerability holistically, or by standard PRA techniques, without an explicit analysis of the attacker's best responses, can produce misleading risk estimates and poor risk management recommendations.

Finally, when the consequences of an attack are very uncertain (perhaps depending on factors such as the wind direction and speed at the time of attack and the ability to warn and/or evacuate people quickly and effectively), the use of *consequence* numbers may be inherently subjective and/or ambiguous. Even in principle, there may be no set of *consequence* numbers that different rational risk managers (having different risk attitudes) agree with.

These limitations suggest that Equation (15.1) ignores some key problems. It forces practitioners to try to use and interpret numbers that have no clear conceptual definitions and that do not model the planning, learning, and adaptive replanning of intelligent attackers. Although expert probability elicitation techniques can certainly be applied even to poorly defined or meaningless quantities, simply eliciting numerical values does not resolve the conceptual limitations of the *threat*, *vulnerability*, and *consequence* numbers, nor model what real attackers will do to achieve their goals, nor make the elicited values useful for predicting risks or for priority setting and resource allocation.

What would be better? Perhaps the most important improvement is to model attacker actions not as random events, but as results of intelligent, goal-directed choices that are responsive to information about defenses and about the success or failure of attempted actions. Several technical options for such modeling have been developed, including the following.

- *Decision tree analysis* generalizes event tree analysis by allowing choice nodes as well as chance nodes in the tree representing attack sequences. At each choice node, the attacker can decide what to do next based on all the information available at that node (i.e., the sequence of event outcomes and actions that led to it). Intelligent attacks are modeled by assuming that, at each choice node, the attacker chooses the outgoing arc that maximizes the conditional expected utility starting from that node. A decision tree goes beyond *TVC* by allowing $T =$ "probability of attack" and $V =$ "probability of success" to evolve as future information items (events or actions) become known. Thus, *threat* and *vulnerability* are not simply numbers, but stochastic processes with numerical values that changs contingent on different information. Risk is not determined by multiplying T, V, and C, but by optimizing via backward dynamic programming in a decision or game tree. Off-the-shelf decision tree software products provide extensive support for commercial-quality quantitative risk analysis using decision trees (von Winterfeldt and O'Sullivan, 2006; NRC, 2008).
- *Probabilistic activity AND-OR networks.* A practical limitation in formulating decision trees is that they can be very large when there are many choices to be made. (For example, if N activities must be completed to accomplish a certain goal, and if they can be attempted in any order, then there is a choice of $N!$ possible orders in which to try them.) One solution to this problem (Cox, 1990) is to represent attack opportunities as stochastic activity networks in which nodes represent activities, arcs represent precedence constraints, and each activity has a specified probability of being successfully completed if it is attempted and requires a specified amount of time (or other resources) to attempt. The attacker uses an optimal index policy (Denardo et al., 2004) to continually choose the most promising remaining path forward, given the outcomes of activities

attempted so far, and abandons the attack only when no attractive path forward (with positive expected value to the attacker) remains.
- *Project planning models of terrorist attacks* apply off-the-shelf project planning and risk analysis software to represent the interdependent tasks that terrorists must successfully undertake to complete planned attacks, to plan in what order to attempt these activities, and to calculate the overall probability of successful completion from the success probabilities for individual tasks (Rosoff and von Winterfeldt, 2007).
- *Hierarchical optimization* focuses on optimizing allocations of defensive resources, assuming that attackers will then adopt "best responses" to the allocations. This suggests two- (or more-) level hierarchical optimization, as in the simple examples in the previous section and the more complex real-world applications in Brown et al. (2006). Synergies or other interactions among defensive options magnify the value of optimizing allocations, rather than using holistic scores or priority orders to allocate defensive resources.

These and other approaches, such as game-theory models (Bier and Azaiez, 2009), typically do not even attempt to assess the *threat*, *vulnerability*, and *consequence* numbers or scores as inputs. Instead, they focus on modeling how intelligent attackers can best exploit opportunities to do damage, and how defenders can allocate defensive resources to minimize the damage that attackers can do, assuming that the attackers will take full advantage of remaining weaknesses.

Regardless of which technical options are used, treating attackers as intelligent opportunists, rather than as random variables, appears to be essential for overcoming key limitations of purely probabilistic modeling (Bier and Azaiez, 2009; NRC, 2008; Golany et al., 2009). Rather than trying to *assess* risks holistically based on probability judgments about what damage attackers may do (e.g., through expert judgments of *threat*, *vulnerability*, and *consequence* scores), it is more useful – and technically practical – to focus on *optimizing defenses,* assuming that attackers will respond by optimizing their attacks accordingly.

Chapter 16
Designing Resilient Telecommunications Networks

How can telecommunications networks be designed to withstand deliberate attacks by intelligent agents, possibly working in teams? This chapter continues discussing quantitative risk assessment (QRA) for systems with intelligent adversaries by reviewing progress in methods for designing communications networks that are resilient to attacks – that is, that are able to quickly and automatically reroute traffic around affected areas to maintain communications with little or no interruption. Current network architectures, routing and restoration protocols, and design methods already suffice to protect networks against the loss of any single link or node, so the main focus for defending against deliberate attacks is on the design of networks that can reroute traffic even when multiple simultaneous failures occur.

We first survey deterministic optimization for designing networks that can reroute all traffic after loss of any k links or nodes, where k is an integer reflecting the attacker's ability to do simultaneous damage. Optimization heuristics for $k = 1$ and $k = 2$ are relatively well developed, but the design of networks that are resilient to k link cuts or node deletions, for arbitrary k, is an open problem. Next, we consider simple probability models for failures in packet-switched data networks caused by attacks on the most loaded nodes. At least in highly simplified (not necessarily realistic) models, such networks are inherently resilient to attacks if and only if they have enough spare capacity at each node (typically about $10k\%$ more than would be required in the absence of attacks, if the attacks are focused on the k most heavily loaded node). Networks with less than this critical amount of extra node capacity are predicted to be vulnerable to cascading failures following a successful attack.

Even though progress in the deterministic optimization and probability modeling of telecommunications networks suggests that it may be possible to meet the technical challenges of designing highly resilient networks, network operators and owners may lack incentives to invest in network protection. We present examples showing that traditional game-theoretic results on the Pareto inefficiency (and nonuniqueness) of Nash equilibria and other solution concepts apply forcefully to jointly owned telecommunications networks. In such settings, network owners and users might all benefit from institutions that enable them to mutually commit to larger investments in network resiliency than they would otherwise make.

L.A. Cox, Jr., *Risk Analysis of Complex and Uncertain Systems*,
International Series in Operations Research & Management Science 129,
DOI 10.1007/978-0-387-89014-2_16,

Introduction: Designing Telecommunications Infrastructure Networks to Survive Intelligent Attacks

Three possibilities for designing telecommunications networks that will provide reliable communications, even if terrorists or other intelligent adversaries attack the network infrastructure, are as follows:

- *Physical defenses* seek to "harden" selected network nodes and links enough to make attacks undesirable, perhaps by burying cables more deeply or reinforcing buildings against attacks.
- *Information defenses* use secrecy, deception, and randomization (Brown et al., 2006) about facility locations and interconnection plans to increase the costs of planning successful attacks.
- *Resilient networks,* the main focus of this chapter, provide enough flexibility, redundancy, and rapid recovery (or "self-healing") capability so that any affordable attempts to disrupt traffic instead result in the automatic rerouting of traffic and uninterrupted service. This chapter surveys recent ideas, methods, and models for designing such resilient networks to protect telecommunications infrastructure against deliberate attacks.

Throughout this chapter, networks are understood to consist of nodes and links. Nodes represent locations where network elements carry out the switching, grooming (i.e., aggregation from lower- to higher-bandwidth signals), regeneration and amplification of optical signals, and routing of traffic. Network elements consist of equipment such as add-drop multiplexers (ADMs) and remotely configurable optical cross-connect switches (OXCs). Links, such as fiber optic cables and microwave links, carry signals between nodes. More specifically, *terminating equipment* for creating and receiving traffic coded as modulated laser light is situated at the nodes. Any other equipment that is located on the links themselves (e.g., regenerator huts, where optical signals are strengthened for transmission over long distances) will be treated as part of the links. The key question addressed in this chapter is: How should network topologies be designed and how should traffic be routed and switched to make communication among nodes resilient to link and node failures, whether accidental or resulting from deliberate attacks?

Background: Diverse Routing, Protection Paths, and Protection Switching

Modern fiber optic telecommunications networks are vulnerable to failures of *links* (i.e., individual fibers or entire fiber cables carrying bundles of fibers) and failures of *nodes* (e.g., due to failures of terminating equipment or to fires in central offices). Historically, major fiber breaks have occurred several times per year in the United States, usually because of backhoes accidentally digging through buried fiber conduits. Node failures are uncommon, but do occur. For example, a 1988 fire at the Hinsdale central office in Illinois, which was a hub for multiple long-distance companies, led to multiple simultaneous outages. This event taught the telecommunications industry the importance of avoiding the possibility of such single-point failures. However, deliberate attacks might increase the frequency of failures, especially, coordinated simultaneous failures intended to disrupt the flow of telecommunications network traffic. The resilience of networks to such coordinated failures requires additional analysis.

Automated Protection Switching (APS) for Packets and Light Paths

Telecommunications networks rely on *automated protection switching* (APS) to automatically reroute traffic when failures of nodes or links occur. Two quite different types of traffic can be rerouted automatically: *data packets* and *light paths.* Packets are used to carry information in data networks and in many mobile and wireless networks. Light paths are dedicated sequences of links and equipment, together with optical channel assignments for each link, that carry very high-bandwidth optical signals (typically consisting of many lower-level channels multiplexed together and encoded as modulated laser light) through optical networks. For purposes of resilient network design, the main difference is that packets can follow different routes through a network, although constraints on latency times for video and other applications may limit the diversity of routes used by consecutive packets from the same session. By contrast, a light path must provide dedicated end-to-end connectivity; thus, if a link fails, each light path that has been using it must promptly switch to another path with the same origin and destination nodes for all traffic in order to avoid any interruption of service.

Demands Consist of Origins, Destinations, and Bandwidth Requirements

It is often convenient to describe optical network traffic in terms of *demands*, with each demand being specified by its origin node, destination node, and size (meaning the amount of bandwidth that it requires, typically measured in units of optical channels). The digital hierarchy of optical channels used in these networks provides discrete standard sizes of bandwidth, ranging from a single optical channel (an "OC-1") up to 192 channels (OC-192) or more, in standard multiples (OC-1, OC-3, OC-12, OC-48, and OC-192). Smaller channel sizes (measured in bandwidth units such as STS-1, DS-3, and DS-1s) are also available and are used in the access networks that feed into optical backbone networks, but will not be considered further here. The design issues considered in this chapter apply to all levels of granularity.

At any moment, an optical network is provisioned to carry a given set of demands. This is done by assigning each demand to a specific light path that includes the desired origin and destination nodes and that has sufficient capacity (i.e., bandwidth) to carry this demand in addition to other demands assigned to it.

Multiple Levels of Protection for Demands

In addition to its origin, destination, and size, each demand has a required level of *protection* that determines what should happen in the event of a network failure event. Unprotected demands are routed on a best-effort basis, and they may be dropped if a node or link fails. Each protected demand has an associated contingency plan that allows the possibility of rerouting if such a failure condition occurs.

Different degrees of protection are available. For example, a very strong form of protection reserves *dedicated protection capacity* for each demand on a "protection" (or "backup") light path with all links disjoint from those on the primary light path that normally carries the demand. (For maximum protection, all nodes other than the origin and destination may also be required to be disjoint from those on the primary path.) If the primary path is interrupted at any point, the entire demand is immediately switched onto the reserved protection path. Such protection typically provides a physically diverse route (unless primary and backup links happen to pass through the same conduit, which can be avoided with additional design effort), with adequate capacity already reserved, for each protected demand. Recovery following a failure is therefore extremely fast, since protection paths are precomputed and no further computation is needed to figure out what to do when a failure occurs.

Example: Protection and Self-Healing Services in SONET Rings

Many existing legacy synchronous optical network (SONET) ring networks in metropolitan areas embody this principle of fully redundant protection capacity. Traffic is routed around an optical fiber ring in one primary direction (for unidirectional SONET rings), with a duplicate signal being sent around in the other direction for protection. Add-drop multiplexers (ADMs) positioned at nodes on the fiber ring (e.g., local wire centers) insert ("add") or receive ("drop") traffic in time slot channels within each optical channel. (An optical channel typically uses a specific wavelength, or color, of laser light. This wavelength, called a "lambda," is modulated at a high frequency to transmit information. Wavelengths in the vicinity of 1,530–1,560 nanometers can travel great distances through modern fiber optic glass before having to be amplified or regenerated. These wavelengths are partitioned into optical channels separated by guard bands of unused wavelengths. Different time slot channels within wavelengths are assigned to the laser pulses coding the information for different demands.) If a link between two adjacent nodes on the ring fails, traffic still flows between all origin-destination pairs of nodes using the reverse direction of routing. Thus, restoration is almost instantaneous (less than 50 milliseconds), and SONET ring owners can offer "self-healing" services.

However, this arrangement is costly, in that half the capacity of the ring is used only by an unneeded duplicate signal unless a failure occurs. (In a bidirectional

SONET ring, protected signals are normally sent in only one direction and preemptable lower-priority traffic may use the other direction. If a failure occurs, the preemptable traffic is dropped and the direction of flow for all protected traffic is reversed, occupying the bandwidth that had been used for the preemptable traffic.) In addition, of course, a ring is vulnerable to a coordinated attack that produces two simultaneous link failures, creating two ring segments that are no longer connected. No node in one segment can then send a signal to any node in the other.

Weaker forms of protection do not require the entire protection path for a demand to be disjoint from the primary path. Instead, when a link fails, other nonfailed links in the primary path may continue to be used, with traffic being switched locally around the failed link. Such local rerouting to restore traffic flows, called *restoration*, uses capacity more efficiently (and is less disruptive) than protection path switching, since a relatively small set of flows can be rerouted based on the specific location of a failure. Moreover, if it is assumed that only one or at most a few links will ever fail simultaneously, then the same protection paths and capacity can be reserved to handle the traffic from multiple primary paths. This leads to *shared-capacity protection plans*, where multiple primary paths are assigned the same protection capacity under the assumption that not all of them will require it simultaneously. Intelligent attackers may try to exploit such less-than-full redundancy to crash part of the network by causing more traffic to be switched onto protection paths than their capacity can hold. However, several contemporary network designs are resilient to any small set of failures.

Example: Resilient Packet Ring (RPR) Protection Scheme

Resilient packet rings (RPRs) use Ethernet packets, rather than dedicated light paths, to carry demands. They achieve fast recovery from fiber cuts, comparable to SONET ring self-healing services, while providing more efficient and flexible utilization of available capacity (bandwidth on links) since packets from the same session can follow different routes. RPRs are organized as sets of intersecting small fiber rings (called "ringlets"), which provide for highly flexible routing of traffic. As in a SONET ring, protected traffic can be "steered" or "wrapped" around a ringlet to avoid a failed link. ("Steering" redirects traffic in the event of a node or link failure by informing all nodes of the change in connectivity and letting the routing algorithms update packet-routing decision tables based on the new topology. "Wrapping" simply treats the last node prior to the failure point as a new origin and forwards packets on to their destinations from there.) In addition, since each packet contains its destination node address (coded into the packet header data), this information can be used to route packets through the network while avoiding congested (or failed) links, using concurrent information about the currently least-congested routes as well as information about the priority classes for demands (e.g., to avoid slow delivery of consecutive packets in applications such as video that are sensitive to transmission delays). The intersecting-rings design allows many possible routes between origin and destination nodes, making the network "resilient" to failures in one or a few nodes or links.

A Simple Two-Stage Attacker-Defender Model

Most traditional network designs have considered a network to be "survivable" if all protected traffic can be rerouted and carried without service interruption following the loss of any single link (and/or node). Multiple simultaneous failures are usually treated as negligibly rare events. Of course, this can change dramatically when the failures are intelligently planned and coordinated instead of occurring at random. Multiple-failure events may then become realistic possibilities, and network designs must be revised accordingly to defend against such events.

The vulnerability of a telecommunications network to intelligent attacks can be assessed with the help of the following two-stage attacker-defender model, in which an intelligent defender designs a network to withstand attack by an intelligent attacker with limited resources, anticipating that the attacker will optimally exploit any weaknesses.

Stage 1 (Defender's move): The network operator ("Defender") decides on all of the following:

- *Network topology:* What nodes are joined by what fiber links?
- *Fiber capacity:* How much bandwidth is installed on each link? (This typically involves deciding what terminating equipment to put at the nodes at the two ends of the link.)
- *Equipment locations and quantities:* What equipment (referred to generically as "network elements") should be placed where in the network? This step specifies the locations of add-drop multiplexers (ADMs), regenerators and amplifiers to boost optical signals attenuated by passage through fiber, and wavelength conversion equipment (typically based on remotely configurable tunable lasers) to change the wavelengths used to carry specific optical signals, thus allowing more efficient use of wavelengths that would otherwise be underutilized. (Regenerators may be placed on links, but other network elements are placed at nodes; thus, this decision is mainly about what equipment to place at each site.)
- *Network interconnection:* Which fibers connect which network elements?
- *Admission control policy:* When new demands arrive dynamically, as requests for given amounts of connection bandwidth between two points (or among multiple points, for broadcast demands) for stated time intervals, the admission control policy determines which requests to accept and reserves primary paths and restoration capacity (for protected demands) to accommodate them. This chapter does not address dynamic admission control, but takes demands as given and known.
- *Traffic routing:* How is each demand routed through the network? Which channels on which wavelengths on which fibers carry each demand (for dedicated routes)? If the signal is converted from one channel or wavelength to another along the way, then which cards (electronics boards that plug into slots in the shelves of equipment racks) in which slots of which shelves in which equipment bays receive, convert, and send the signal? (Such detailed route planning is required to assure that sufficient ports and capacity are available to handle the assigned traffic.) In other words, what is the *primary path* for each demand?

Assignments of traffic to primary paths must be feasible, as determined by capacity constraints (including the availability of cards and ports), interconnect plans, and signal power budgets and distance constraints [required to keep the attenuation of optical signals and the optical signal-to-noise ratio (OSNR) within the tolerance of the signal processing equipment]. For packetized traffic, what decision rules, algorithms, or routing tables are used to route packets through the network?

- *Restoration plans:* For protected demands, how will each demand be rerouted if nodes, network elements, or links on its primary path fail? For packet networks, how will packets be rerouted if congestion or failure is encountered, while still meeting constraints on delivery times, i.e., "latency"?

Stage 2 (Attacker's move): Once Defender's move has been completed, Attacker may cut any k links simultaneously. Here, k is an integer reflecting Attacker's ability to launch simultaneous attacks on multiple links simultaneously. A variation is to allow attacks on nodes instead of, or in addition to, links. ("Simultaneous" cuts are cuts completed within a short enough time so that the last is completed before the first is repaired.)

To model intelligent attacks in this framework, one might assume that the attacker has some particular goal in mind (e.g., to disconnect one or more specific target nodes or subsets of nodes, perhaps representing strategic command and control centers, from the rest of the network). The defender seeks to reserve sufficient protection capacity and diversity of paths in Stage 1 so that these targets can withstand any k-cut attack in Stage 2. This means that, after any k links are cut, all protected demands can be rerouted on the reserved capacity. In particular, the network will withstand an attack in which Attacker optimizes some objective function (which may be unknown to the defender) to select which k links to cut. Formulating the problem as protecting against *any* k cuts makes it unnecessary to speculate more deeply about exactly how the attacker decides which links to target.

Results for Networks with Dedicated Routes ("Circuit-Switched" Networks)

This section considers the problem of designing attack-resistant circuit-switched optical networks. In a circuit-switched network, each demand is normally carried by a dedicated light path. The following special cases of this framework have been investigated.

Designing Networks to Withstand a Single ($k = 1$) Link Cut

The case $k = 1$ is the classic "survivable network" design problem of greatest interest when failures are random and are rare enough so that the probability of multiple simultaneous failures can be ignored. Designs that protect against any single link failure (such as "self-healing" SONET ring designs) also trivially protect against any single deliberate link cut.

Solving the multiple interdependent decision problems in Stage 1 even for $k = 1$ can lead to very large-scale, computationally intractable optimization problems (Tornatore et al., 2002) that include as subproblems well-known difficult (NP-complete) combinatorial optimization problems such as the bandwidth-packing problem (Laguna and Glover, 1993; Parker and Ryan, 1994; Villa and Hoffman, 2006) and multicommodity network flow problems (one for each failure scenario) (Rajan and Atamturk, 2004). However, several solution heuristics and meta-heuristics have proved to be highly effective in solving real-world survivable network design problems with $k = 1$ (Soriano et al., 1998). For example, Rajan and Atamturk (2004) compare (a) a relatively naïve two-stage *hierarchical design*, in which demands are first routed ignoring survivability requirements, and then additional capacity is added to links to allow traffic to be rerouted in the event of any single link failure; and (b) a more sophisticated *joint design* that simultaneously routes demands along primary paths and reserves sufficient additional capacity to allow all traffic to be rerouted following any link failure. Heuristics for solving the joint design problem (discussed next) achieve on the order of 20% cost savings compared to the hierarchical designs when capacity is expensive.

Although the literature on capacitated survivable network design problems is now vast (e.g., Soriano et al., 1998), many practical design strategies use a combination of a few key heuristic ideas, including the following.

(a) *Consider only a subset of all possible primary and protection paths.* These are typically generated using a greedy or fast (low-order polynomial-time) heuristic (e.g., Dacomo et al., 2002), such as paths or rings constructed using modified shortest-path or minimal spanning tree algorithms. For networks with general topologies (i.e., mesh networks), it is common practice to search for rings or directed cycles that "cover" the nodes and links of the network, meaning that each node and each link belong to at least one ring. These rings are then used to form protection paths. Finding minimal-cost ring covers in which each protection link is used exactly once can be accomplished easily for planar networks (since then each ring corresponds to a face of the graph), but is NP-hard in general. However, effective (polynomial-time) heuristics are available for various ring-covering problems (including "double-cycle covers" in which each ring appears in exactly two rings, being covered by a cycle in each direction exactly once) (Dacomo et al., 2002).
(b) *Iteratively improve feasible solutions* by using local (myopic) search and optimization procedures to add or delete capacities to maintain feasibility while seeking to reduce costs.
(c) *Ignore or relax difficult (e.g., integer-capacity) constraints* at first, so that useful initial approximate solutions can be obtained using linear programming. The solutions to the relaxed problem serve as starting points for myopic improvement heuristics.

(d) *Use multiple random starting solutions and evolve the population of candidate solutions* [using meta-heuristics such as simulated annealing, greedy randomized search procedure (GRASP), Tabu Search (Laguna and Glover, 1993) and genetic algorithms] to discover lower-cost feasible ones.

The survivable network design heuristics of Rajan and Atamturk (2004) illustrate several of these ideas. To obtain a tractably small number of possibilities to evaluate, their solution heuristic only evaluates a subset of directed cycles as candidate paths for restoring traffic flows following a link failure. The cycles are generated and evaluated within an iterative "column-and-cut generation" heuristic optimization algorithm that uses a variant of a shortest-path calculation to quickly generate candidate directed cycles to consider as restoration paths. This restriction allows formulation of mixed-integer programs (MIPs) that describe the capacitated survivable network design problem. The decision variables are the amounts of capacity added to each link. The constraints require that enough capacity be added to carry all traffic (i.e., demands) in the absence of failure and also following any single link failure. The objective function to be minimized is the total cost of all the capacity installed in the network. The resulting MIPs can be solved approximately in a matter of hours using a linear programming relaxation, strengthened with valid inequalities representing the survivability requirements.

Many other researchers have also proposed integer linear programming (ILP) formulations of the capacitated survivable network design problem with $k = 1$, differing according to the details of the technologies modeled, such as whether it is assumed that light paths must use the same wavelengths on all links or instead allow for wavelength conversion at nodes. Current solution heuristics typically give solutions with costs that are close (often within about 3%) of the theoretical optimum on benchmark cases where the exact solution is known or for which useful bounds are available (usually based on branch-and-bound solvers with relatively long run times).

A theoretical result of Brightwell et al. (2001) nicely captures both the computational intractability of exact solutions and the availability of good approximate solution heuristics for the design of survivable networks with $k = 1$. The result is expressed for a particularly simple model in which the attacker seeks to cut a link to cause at least some traffic between a specified source node s and destination node t to become unroutable using the remaining capacitated links in the network. To prevent this, the defender is allowed to add additional discrete increments of capacity to the links (e.g., corresponding to additional terminating equipment). The defender knows the costs of adding different amounts of capacity to the links and seeks a minimum-cost set of capacity additions that will allow all protected traffic to be rerouted between s and t following any single link cut. In this framework, Brightwell et al. establish that

(a) The problem can be solved easily (in polynomial time) using an algorithm based on shortest paths if the discreteness of capacity additions is ignored.

(b) The problem is computationally intractable (NP-hard) when constraints are enforced, reflecting the discrete nature of capacity expansion decisions (due to the fact that only a whole number of discrete capacity additions can be made on any link).
(c) Despite this theoretical intractability, a simple polynomial-time heuristic based on modified least-cost path assignments provides solutions that are not more than 15/14 times more expensive than the exact (but perhaps too hard to compute) cost-minimizing solution.

Designing Networks to Withstand k = 2 Link Cuts

In light of the maturity and widespread practical deployment of survivable network designs for $k = 1$ cuts, an intelligent attacker attempting to disconnect one or more target nodes from a modern network by severing links will typically have to make at least two simultaneous cuts (e.g., one for each direction in a SONET ring). Choi et al. (2002) address the design of networks that remain connected when any two edges fail. Such networks, which must be 3-connected, must therefore also have three link-disjoint paths between any two nodes (by Menger's theorem in graph theory). These link-disjoint paths can be identified quickly (using a Ford-Fulkerson max-flow algorithm). They are used to create a primary backup path and a secondary backup path. Any single link failure automatically triggers switching of affected traffic to the primary backup path, while a subsequent link failure then triggers switching to the secondary backup path.

Choi et al. consider several variations of this basic idea, involving different amounts of signaling after a failure event (to alert nodes to the change in network topology) and using different information about path failure vs. specific link failures. Based on an evaluation of restoration possibilities in three real-world networks (ARPANET, NJ LATA network, and a national network), they conclude that "It is possible to achieve almost 100% recovery from double-link failures with a modest increase in backup capacity." Thus, although much more can be done (e.g., to allow for dynamic restoration based on specific information about failed links rather than precomputed protection paths, or to minimize restoration capacity costs), it appears that designing networks to protect against $k = 2$ simultaneous cuts is practical.

Results for the General Case of k Cuts

Brightwell et al. have extended some of their results for $k = 1$ to larger values of k as follows. The problem of finding a minimum-cost set of protective capacity reservations (using link-disjoint paths) so that a specific demand can be restored following any simultaneous deletion of at most k links can be solved in polynomial time using a linear programming or ellipsoid algorithm, provided that the requirement that the capacities be integer amounts is ignored. (Recall that the demand is identified by a specific origin node s, a specific destination node t, and a bandwidth requirement T.) However, the problem is strongly NP-hard (even for $k = 1$) when the integer constraint is enforced. Despite this theoretical complexity, good approximate solutions can be found with relatively little computational effort using a successive shortest-path method (involving solving several shortest-path problems) for minimum-cost flow problems. For some specific formulations of the minimum-cost capacity-reservation problem, it can be shown that the solution found by ignoring integer constraints is no more than 6/5 times more expensive than the true optimal solution with integer constraints enforced, but is much easier to find.

In summary, it is practical to design networks that are resilient enough to protect any single specified demand against the failure of any k arcs or nodes. The next step is to investigate extensions of this encouraging result from a single demand to all protected demands. This will require quantifying the costs of protecting different amounts of demand against simultaneous attacks of different sizes, $k \geq 1$. The work reviewed in this section provides the beginnings of a theory of resilient network design for arbitrary k, but much remains to be done to develop practical resilient designs for arbitrary $k > 2$.

Statistical Risk Models and Results for Scale-Free Packet Networks

This section considers statistical properties of failures in packet-switched networks when intelligent attacks are modeled by assuming that the attacker always targets the most heavily loaded node(s). In packet networks (e.g., the Internet), failures can be caused by overloading nodes, that is, by forcing more traffic to be routed through a node (e.g., a router) than it has the capacity to handle. In this case, packets dropped at the overloaded node must be rerouted, shifting additional load onto nearby nodes. This may trigger a spreading cascade of congestion and node failures.

Although such a cascading failure process might be difficult to model mathematically in detail, methods adopted from statistical physics – especially, the statistical mechanics of phase transitions – provide suggestive insights (at least in highly simplified and idealized models) into both the probability that a deliberate attack on the most-linked node(s) will succeed in inducing a cascade of node failures, and also the probable size of resulting network service outages. Precise results have been derived for *scale-free networks*, i.e., networks in which the probability that a randomly selected node is connected to x other nodes is proportional to $x^{-\gamma}$ (a Yule-Simon power-law distribution). Such networks arise through *preferential attachment* growth processes in which "the rich probably grow richer," i.e., each new node added to the growing network forms links to previously existing nodes with probabilities proportional to how many links they already have. These (and some other) random growth processes can generate networks in which a small fraction of nodes are major hubs with many links, while most nodes have relatively few links.

Empirically, scale-free networks with γ values between 2 and 3 have been reported to describe well both the statistical pattern of World Wide Web page links and the statistical distribution of connections among routers (i.e., nodes) in the Internet (as well as other networks, such as air traffic networks and metabolic and proteomic networks in systems biology). A study of 284,805 Internet routers revealed a value of $\gamma = 2.3$, with the probability that a randomly selected node has $x = 1$ neighbor being $\Pr(x = 1) = 0.53$ and the probability that it has 100 neighbors being $\Pr(x = 100) = 2.5 \times 10^{-5}$ (Deccio et al., http://dna.cs.byu.edu/papers/pdf/survive.pdf.) However, the relevance of such statistical measures, and of the scale-free model, to real networks has been challenged, in part by showing that networks with completely different resilience to attacks may have identical scale-free network parameters (Doyle et al., 2005).

Qualitatively, it has frequently been claimed that scale-free networks tend to be highly resistant to random failures, yet very vulnerable to targeted attacks. This is a direct consequence of the fact that they typically have a tiny proportion of very heavily connected nodes, so that crashing (in effect, deleting from the network) a few of the most heavily loaded nodes can cause most of the network to collapse through cascading failures (Albert et al., 2000). For example, Deccio et al. (op cit.) found that randomly deleting 1% of the nodes in a scale-free network (based on real Internet router data) would leave over 97% of the original nodes operating normally, but targeting the top 1% of most heavily connected nodes for removal would reduce the size of the functioning network to less than 60% of its original size. Similarly, Crucitti et al. (2004) found that the average time required to send packets through a network in many simulated examples with 2,000 nodes and 10,000 edges was much more sensitive to intelligently targeted attacks than to random failures. Complete network failure (inability to deliver packets) occurred when about 0.4 of the nodes were removed in a targeted way, but not when 90% or more of nodes were removed randomly.

In practice, of course, even carefully planned and targeted attacks are more likely to bring down only a few nodes simultaneously than a high proportion of all nodes in a network. It is therefore desirable to understand the risk of network failure from an attack that succeeds in knocking out only one or a few most heavily loaded nodes. To this end, it is convenient to adopt the following explicit (though admittedly highly simplified and unrealistic) model of how network traffic loads are initially allocated among nodes and how they are redistributed following a successful attack (Zhao et al., 2004).

1. In each time period, each node sends one unit of traffic to each other node via the shortest path between them. (Ties are resolved arbitrarily.)
2. The total load on any node is the total amount of traffic per unit time passing through it. By assumption 1, this is just the number of shortest paths passing through it. For scale-free networks, it can be deduced that node loads follow a power-law distribution of the form $L(x) \sim x^{\eta}$, where $\eta > 0$ is a scaling exponent and $L(x)$ is the load on a node that is connected to x neighbors (Zhao et al., 2004).
3. The total capacity of a node is the maximum load that it can process.

4. A node fails if and only if the traffic load sent to it exceeds its capacity.
5. If a node fails, then all traffic is rerouted using the new set of shortest paths available after the failed node is deleted from the network.
6. The attacker chooses to attack the node with the largest number of links (i.e., neighbors).
7. The defender purchases sufficient capacity at each node to carry its entire original load plus some additional fraction α of that load: *capacity allocated to node i* = *(1 +α)*(load at node i)*, where $\alpha \geq 0$ is called the *tolerance parameter*.

This simple (not necessarily realistic) model leads to the following striking qualitative result. There is a *critical value* of the tolerance parameter, denoted α_c (a function of the scaling exponents in the power laws for the load distribution and the degree distribution, η and γ, respectively, and of the number of nodes in the network) such that

- If sufficient node capacities are allocated so that $\alpha > \alpha_c$, then an attack on a single node in a large network has probability close to zero of crashing the network. More quantitatively, the ratio of the numbers of nodes in the largest connected component of the network before the attack and after the attack (and after any cascade of node failures that it triggers) will be approximately 1.
- On the other hand, if $\alpha < \alpha_c$, then a successful attack that deletes the single most loaded node in a network has a high probability (close to 1) of crashing the network. In other words, the ratio of the numbers of nodes in the largest connected component of the network before and after the attack (and after any cascade of node failures that it triggers) will be approximately 0 with high probability.
- Theoretical calculations and numerical simulations show that the numerical value of the critical value of the tolerance parameter is approximately $\alpha_c \approx 0.10$.

The critical value of the tolerance parameter may be interpreted as a statistical margin of safety at which a statistical *phase transition* occurs between the network being resilient to deletion of any single node (meaning that other nodes remain connected, with high probability) and the network being vulnerable to deletion of a single node (meaning that it triggers a cascade of failures that end up disconnecting most nodes).

Many variations of this basic phase-transition model have been explored, and investigations of more realistic models and of the detailed statistics of how networks break apart into disconnected fragments (including the distributions of their numbers and sizes) in the vicinity of the phase-transition point α_c are ongoing (e.g., Duenas-Osorio et al., 2004). For example, numerical simulations suggest that an attacker who attacks the $k > 1$ most linked nodes simultaneously can induce collapse of the network at higher values of the tolerance parameter. The critical value of the tolerance parameter increases approximately in proportion to the number of simultaneous attacks, k (Zhao et al., op cit., Fig. 4). Attacks on edges as well as on nodes have been studied and scale-free networks have been modified to include different types of clustering (e.g., Holme et al., 2002). Dynamic models, where node deletions by an attacker and node additions or repairs by a defender take place over the same time interval, have also been examined. These lead to the conclusion that the evolving network remains highly connected (it has a giant component) with high probability (approaching 1 in large networks) if the rate at which the attacker can delete nodes is sufficiently small compared to the rate at which the defender can repair or replace damaged nodes (Flaxman et al., 2007). Recently, the effects of attacker information on the vulnerability of scale-free networks have been investigated. One conclusion is that concealing information about a small fraction of nodes from the attacker can greatly improve the ability of networks to withstand intelligent (but incompletely informed) attacks (Gallos et al., 2005; Wu et al., 2007).

These more sophisticated (but still not necessarily realistic) models share with simpler ones the key qualitative property of phase transitions, in which sufficient capacities (and/or repair rates) protect against large-scale failures with high probability (approaching 1 in large networks), while insufficient capacities (or repair rates) allow the possibility that a small set of attacks (possibly on only one node) can trigger a cascade of failures that leave most of the network disconnected. Although a great deal of additional quantitative detail has been uncovered through theoretical analysis and numerical simulation of attack-defense processes near the phase-transition boundary, and although this remains a very active area of research, the crucial qualitative insight that there is often a well-defined transition from low vulnerability to high vulnerability as the load increases relative to capacity is one of the most valuable and robust lessons to emerge from the study of attack and defense in scale-free networks.

The existence of such a phase transition suggests that many network risk management decision problems can be simplified to the task of keeping networks in the low-vulnerability regime. This is done by maintaining sufficient margins of safety (e.g., greater than α_c) so that the network will be able to withstand attacks. Simulation and detailed mathematical analysis of phase transitions provide practical methods for quantitative assessment of the required safety measures. For the specific attack model of Zhao et al. (2004) discussed above, a useful design rule of thumb is that $\alpha > k^* 10\%$ provides an adequate safety margin for withstanding k-link attacks.

Real-World Implementation Challenges: Incentives to Invest in Protection

The preceding sections have described several recent constructive approaches to designing telecommunications networks (circuit-switched or packet-switched) that can withstand targeted attacks at several locations simultaneously. Investing in protection capacity can make it possible to restore protected demands or sessions almost immediately following such an attack. This section asks whether network owners and operators have incentives to actually make such investments. How do interdependencies among the risks and benefits of jointly owned network infrastructure affect incentives to invest in protection?

Telecommunications networks typically have many owners and many users; they exemplify systems in which interactions among players drive the costs and benefits received by each. Many of the best-studied models in game theory may be interpreted as special cases of network games. The following examples are intended to illustrate how some of the most familiar paradigms of cooperative and noncooperative game theory apply to decisions about whether to invest in improved reliability (or resilience, if failures come from deliberate attacks) in telecommunications networks. Throughout this section, the focus is on whether network owners will invest in improved network reliability, without regard for whether potential failures result from deliberate attacks or random failure events; thus, we will use the general term "reliability" to refer to the probability that a system does not fail, without further inquiry into possible reasons for failure.

Example: An N*-Person Prisoner's Dilemma for Network Maintenance*

Setting: Suppose that three different owners own the three different links in the following linear series network:

$$A \rightarrow B \rightarrow C \rightarrow Z.$$

It costs \$1 to maintain each link. Doing so guarantees that the link can reliably transport two units of traffic. Without maintenance, the link deteriorates with probability 0.5 (possibly due to deliberate attacks, but this example can also be applied to random failures); a deteriorated link can reliably transport only one unit of traffic. In this simplified model, each owner decides at the start of the game whether to invest in maintenance, and any deterioration of nonmaintained links takes place immediately after the decision has been made.

The demand for network services is such that customers will use this linear chain network to carry as much traffic as it can carry reliably (one or two units), generating either \$6 or \$12 units of revenue, respectively, which is then shared equally among the link owners, giving each either \$2 or \$4, respectively.

Problem: What is each owner's optimal strategy for investing in maintenance?

Solution: If all players maintain their links, each receives a profit of (\$4 revenue – \$1 maintenance cost) = \$3. A player who does not maintain his own link when both of the other two players maintain theirs also receives $0.5 \times \$4 + 0.5 \times \$2 = \$3$ expected profit. A player who maintains his own link when only one of the other two players does so receives an expected profit of $0.5 \times \$4 + 0.5 \times \$2 - \$1 = \$2$; but if he chooses not to maintain his own link, then his expected profit increases to $0.25 \times \$4 + 0.75 \times \$2 = \$2.5$. A player who maintains his own link when neither other player does likewise receives an expected profit of $0.25 \times \$4 + 0.75 \times \$2 - \$1 = \$1.5$; but if he chooses not to maintain his own link, then his expected profit increases to $0.125 \times \$4 + 0.875 \times \$2 = \$2.25$. Thus, *no matter what the other players do, each player maximizes his own expected profit by not maintaining his own link.* Doing otherwise might decrease, but cannot increase, his expected profit.

Discussion: When all players follow the unique dominant strategy of not investing in maintenance, each receives an expected profit of only \$2.25 (or \$2 per day in the long run, if the play is repeated on consecutive days until all links fail). But if all players maintained their links, each would receive a profit of \$3. Thus, *the dominant strategy is Pareto-inefficient:* all players could do better by choosing a different (dominated) strategy. This is like a three-player version of the Prisoner's Dilemma, in that the unique dominant strategy for all players gives each a lower payoff than would one of the dominated strategies. A similar result holds in analogous games with $N > 3$ players; i.e., the unique Nash equilibrium is to underinvest in maintenance.

This example suggests how the presence of multiple owners in a network can lead to free-riding incentives, underinvestment in network maintenance or quality, and Pareto-inefficient provision of network resources to users who would be willing to pay for better service.

Example: Nash Equilibrium Can Be Inadequate for Predicting Investments

Setting: Suppose that Player 1 owns the link from A to B and that Player 2 owns the link from B to C in the simple network A → B → C. A customer is willing to pay up to \$100 per month to use this network to send messages from A to C, provided that the product of the reliabilities of the two links is sufficiently great, say $p_1 p_2 \geq 0.5$. Assume that Player 1 must select a nonnegative reliability p_1 between 0 and 1 (inclusive) for link AB, and that Player 2 must select a nonnegative reliability p_2 between 0 and 1 for link BC. The customer will pay each player \$50 per month if and only if $p_1 p_2 \geq 0.5$, and otherwise will pay 0 and will forego using the network. It costs a player \$50p per month to maintain a reliability level p on her link.

Problem: What reliability level should each player choose for her link?

Solution: This is an instance of the classical two-person bargaining game (isomorphic to the "divide a dollar" game), to which an entire large literature is devoted. *Every* pair of nonnegative reliabilities (p_1, p_2) such that $p_1 p_2 = 0.5$ is a Nash equilibrium, since if one player can credibly insist on maintaining a level p, then the other player's unique best response is to settle for maintaining level $0.5/p$. Thus,*the Nash equilibrium concept is inadequate to predict a unique outcome for this game.* Axiomatic bargaining theory (including the Nash bargaining solution) proposes various normative justifications for specific solutions, especially the symmetric (in dollars, although not necessarily in utilities) solution $(p_1, p_2) = (0.707, 0.707)$, but the confident prediction of a specific outcome is impossible. In some cases, players may commit themselves to inadequate investments in unsuccessful attempts to induce each other to spend more, thus arriving at a Pareto-inefficient outcome in which neither sells capacity to the customer (Roth, 1985).

Example: A Network Collusion Game with an Empty Core

Problem: A customer is willing to pay \$6 to send crucial data from the origin O to the destination D along any highly reliable path. There are three owners, 1, 2, and 3, of network links. Links AB and BE belong to Player 1, links BD and CE belong to Player 2, and link ED belongs to Player 3. Suppose that all links are currently low-reliability, but that any link can be converted to high-reliability by adding expensive terminating equipment at either end. If it costs \$1 to upgrade the links, how should the three players cooperate to make an offer to the customer? (See Fig. 16.1.)

Solution: This is an example of a cooperative game with an empty core: There is no coalition and agreement among its members that is proof against different coalitions being formed that can improve the profits offered to each of its members. Therefore, *no possible coalition is stable.* The customer may wait forever for the owners to stop realigning and make an offer.

For example, suppose that Player 1 approaches Player 2 and proposes that they each invest \$1 (Player 1 to upgrade link AB and 2 to upgrade link BD) and then offer the customer the use of path OABD and split the \$6 from the customer evenly. Under this agreement, Players 1 and 2 each make a profit of \$3 – \$1 = \$2 and Player 3 makes nothing. Dissatisfied with this outcome, Player 3 might approach Player 2 with the suggestion that Player 3 invest in upgrades for both Player 3's own link (ED) and also one of Player 2's (CE). Then the two can team up to offer path OCED to the customers, split the \$6 evenly, and Player 2 will make a profit of \$3 (beating the profit made in the proposal from Player 1) while Player 3 will make a \$1 profit (beating its zero profit under Player 1's proposal). Player 1 will now be excluded and make no profit. (Player 3 might propose many other deals, but this illustrates that Player 3 has both the incentive and the ability to make a deal with Player 2 that both of them would prefer to the one that Player 1 initially proposed.)

Rather than making no profit, Player 1 can do better by offering to team with Player 3. For example, Player 1 might now offer to invest \$2 (one each in links AB and BE), have Player 3 invest in only one (ED), and split the \$6 evenly, so that Player 1 makes a \$1 profit and Player 3 makes a \$2 profit. Player 2 is excluded. But now, Player 2 can repropose Player 1's original offer back to Player 1 (giving each of them a profit of \$2). And so the cycle continues: Whoever is currently left worst off (e.g., excluded from making a profit) by any proposed agreement can make an offer to one of the other players that will increase both their profits (and make someone else the worst off, and hence ready to propose a different deal). There is no

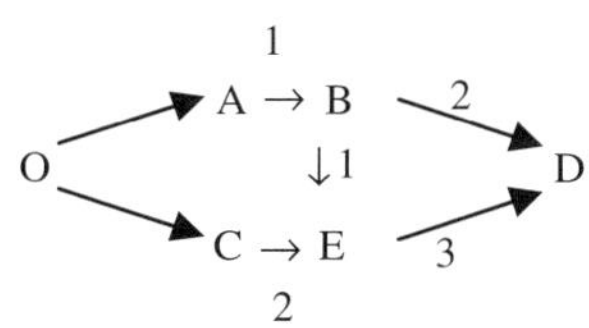

Fig. 16.1 Network showing which player (1, 2, or 3) owns each link

possible agreement that cannot be improved (in the sense of increasing profits for all who participate) by some subset of two players who have both the incentive and the power to do so. Thus, companies trying to find the best deal for themselves by exploring different partnerships will find that there is no solution to this problem.

Example: A Tipping Point

Suppose that Player 1 will invest in security if at least two other players do, that Player 2 will invest in security if at least one other player does, and that Player 3 would like to invest in security, but only if everyone else does. If these preferences are common knowledge, then Player 3 can tip the system from no investment by any player to investment by all players by being the first to invest (thus inducing Player 2, and then Player 1, to follow). If the players do not know each other's preferences, however, then each will wait for someone else to move first, no tipping point will be reached, and the players will fail to make the mutual investment in security that all three of them would prefer to make. Thus, knowing the player's preferences and willingness to pay for investments in security (or network reliability) is not sufficient to predict what will happen: knowledge of what the players know about each other's preferences is also needed.

The examples in this section have led to the same major policy-relevant conclusion by different routes: Even if it is possible to solve the purely technical problems of devising cost-effective networks that are resilient against accidental and/or malicious failure (e.g., using the design heuristics and statistical phase-transition safety margins discussed in earlier sections), *interdependencies among network owners can undermine willingness to invest in network resilience, leading to socially suboptimal (Pareto-inefficient) decisions that invest too little.* In some cases, there may be simple solutions (e.g., to create an artificial "tipping point" by stimulating initial investments in increased security or by providing players with credible information about each other's preferences). However, the incentive, information, and distributed decision-making challenges illustrated in these examples have proved to be enduring and difficult in many applications of game theory. Socially optimal investment in more resilient telecommunications networks may be inhibited by these challenges, even when algorithms for minimizing the costs of protecting against intelligent attacks are well developed.

Summary

Mathematical optimization can be used to design networks that can withstand the simultaneous loss of any k links or nodes, where k is a small integer reflecting the attacker's ability to do simultaneous damage. Solution heuristics for $k = 1$ and $k = 2$ are relatively well developed. However, the optimal (least-cost) design of networks that are resilient to k link cuts or node deletions, for arbitrary k, remains an open problem.

Simple probability and statistical models have been developed that predict the effects of attacks and repairs in large scale-free networks (typical of large data networks, including the Internet), assuming that the attacker always targets the most heavily loaded links or nodes. These models typically exhibit a *phase transition*, in which the networks are inherently resilient to attacks if and only if they have enough spare capacity at each node (about $10k\%$ more than would be required in the absence of attacks, if the attacks are focused on the k most heavily loaded node). Networks with enough extra capacity at each node can absorb the impacts of successful attacks by redistributing the traffic originally routed through failed nodes to surviving neighbors, leaving almost 100% of the network (other than the failed nodes) still connected. On the other hand, networks with less than this critical amount of extra node capacity are vulnerable to cascading failures (similar to spreading blackouts in an electric grid), so that a successful initial attack can lead to the progressive collapse of the network, with the largest connected component remaining after the attack and its aftermath probably consisting of a negligibly small fraction (close to 0%) of all nodes. This threshold-like division between resilient and vulnerable networks makes it relatively simple to design resilient networks: simply include enough extra capacity at each node so that attacks will be absorbed rather than causing cascading failures.

Although optimization and probability models may reveal how to design networks that can withstand intelligent attacks, network operators who own only some portions of a network may lack incentives to actually make the required investments. Strategic incentives can encourage network owners to invest too little in network reliability and resilience. The traditional game-theoretic problems of free riding, Pareto-inefficient Nash equilibria, nonunique Nash equilibria, empty cores, and poor coordination due to private information can be powerful in network investment settings. Each problem can result in underinvestment in network protection, meaning that all players would benefit if all could commit to make greater investments in network reliability and resilience. Regulatory intervention, centralized command and control, or institutional designs facilitating mutual commitments and multiway contracting may be needed to overcome such adverse incentive effects, enabling network owners and users to reap the potential benefits from greater network reliability and resilience.

Epilogue

Earlier chapters of this book illustrated methods for quantifying and comparing the probable health consequences of a few discrete risk management alternatives, such as *ban* or *do not ban* for a food animal antibiotic or *remove* or *do not remove* for various constituents of cigarette smoke. The three final chapters have discussed more complicated risk management problems, including ones with multiple stages of decision making, outcomes that are contingent on both present decisions and future events (such as discovery of BSE), future decisions that are contingent on future information and that are constrained by current choices (such as whether to track cattle origins), and consequences of present decisions that depend on subsequent decisions by determined, intelligent players (e.g., attackers and defenders, or different owners of a network infrastructure). In all these cases, QRA models that include the *optimization* of future decisions suffice to predict (or at least bound, as in the BSE model) the probabilities of different consequences for current decisions. Thus, these QRA models support optimization of current risk management decisions, even for relatively complex, uncertain systems with multiple decision makers and sequential decisions.

More generally, a combination of QRA modeling techniques enables useful prediction and control of probable outputs in complex and uncertain systems. We have illustrated QRA for important real-world applications, including lung cancer, food safety, antimicrobial resistance, economic risks from imported BSE, and terrorist attacks on target locations or telecommunications networks. Highly useful QRA methods include: methods for bounding preventable risks, based on partial knowledge of causal mechanisms; asymptotic probability results that are independent of details (such as phase-transition behaviors in large random networks); simulation and mathematical analysis of system dynamics and equilibria (even when many details of causal pathways are unknown or are described only by high-level empirical relations); and optimization of decisions, assuming that future decisions, perhaps by other players, will also be optimized, contingent on future information available when they are made.

An optimistic conclusion from the QRA models and calculations presented throughout this book is that *QRA modeling is currently practical for a variety of complex, uncertain systems in real-world applications of great practical importance.* Despite large remaining uncertainties about exactly how complex systems work, and about how changing controllable inputs will affect outputs that decision makers care about, QRA enables practitioners to predict key aspects of system responses to changes in controllable inputs. It lets risk analysts identify and eliminate dominated decision options using realistically incomplete and imperfect information. QRA is a practical discipline for improving decisions in a complex and uncertain world.

References

Aardema MJ, MacGregor JT. Toxicology and genetic toxicology in the new era of "toxicogenomics": Impact of "-omics" technologies. *Mutat Res.* 2002 Jan 29;499(1):13–25.

Abboud RT, Vimalanathan S. Pathogenesis of COPD. Part I. The role of protease-antiprotease imbalance in emphysema. *Int J Tuberc Lung Dis.* 2008 Apr;12(4):361–7.

Abshire MK, Devor DE, Diwan BA, Shaughnessy JD Jr, Waalkes MP. *In vitro* exposure to cadmium in rat L6 myoblasts can result in both enhancement and suppression of malignant progression in vivo. *Carcinogenesis.* 1996 Jun;17(6):1349–56.

Ades AE, Kazantzis G. Lung cancer in a non-ferrous smelter: The role of cadmium. *Br J Ind Med.* 1988 Jul;45(7):435–42.

Ahn WK, Bailenson J. Causal attribution as a search for underlying mechanisms: An explanation of the conjunction fallacy and the discounting principle. *Cognit Psychol.* 1996 Aug;31(1): 82–123.

Alberg AJ, Samet JM. Epidemiology of lung cancer. *Chest.* 2003 Jan;123(1 Suppl):21S–49S.

Albert R, Jeong H, Barabási A-L. Error and attack tolerance of complex networks. *Nature.* 2000 Jul 27; 406:378–82. http://www.nature.com/nature/journal/v406/n6794/full/406378a0.html

Alberts B et al. *The Molecular Biology of the Cell,* 4th Ed. Garland Science, New York. 2002.

Alexandrov K, Cascorbi I, Rojas M, Bouvier G, Kriek E, Bartsch H. CYP1A1 and GSTM1 genotypes affect benzo[a]pyrene DNA adducts in smokers' lung: Comparison with aromatic/hydrophobic adduct formation. *Carcinogenesis.* 2002 Dec;23(12):1969–77.

Aliferis C, Tsamardinos I, Statnikov A. HITON: A novel Markov Blanket algorithm for optimal variable selection. *AMIA Annu Symp Proc.* 2003; 21–5.

Allio L, Ballantine B, Meads R. Enhancing the role of science in the decision-making of the European Union. Regulatory *Toxicol Pharmacol.* 2006 Feb;44(1):4–13.

The American Thoracic Society and the European Respiratory Society. Pretreatment evaluation of non-small-cell lung cancer. *Am J Respir Crit Care Med.* 1997 Jul;156(1):320–32.

American Veterinary Medical Association (AVMA). American Veterinary Medical Association Judicious Therapeutic Use of Antimicrobials. Dec 2005. www.avma.org/issues/jtua/default.asp

Andradóttir S, Bier, VM. Applying Bayesian ideas in simulation. *Simulat Pract Theory* 2000;8:253–80.

Andrieu C, de Freitas N, Doucet A, Jordan MI. An introduction to MCMC for machine learning. *Mach Learn.* 2003 Jan–Feb; 50: 5–43. http://citeseer.csail.mit.edu/andrieu03introduction.html

Ang JY, Nachman S. *Campylobacter* Infections. http://www.emedicine.com/ped/topic2697.htm (Last accessed 3-20-2009)

Angoss Software. KnowledgeSeeker 5.0 Program. 2005. http://www.angoss.com

Angulo FJ, Baker NL, Olsen SJ, Anderson A, Barrett TJ. Antimicrobial use in agriculture: Controlling the transfer of antimicrobial resistance to humans. *Semin Pediatr Infect Dis.* 2004 Apr;15(2):78–85.

Animal Health Institute (AHI). Animal Health Companies Meet Increased Market Need for Antibiotics. News Release from the Animal Health Institute, Washington, DC. Oct. 12, 2006. http://www.ahi.org/mediacenter/documents/Antibioticuse2005REVISED.pdf

Apostolakis GE, Kaplan S. Pitfalls in risk calculations. *Reliab Eng Syst Safe.* 1981;2:135–45.

Apostolakis GE, Lemon DM. A screening methodology for the identification and ranking of infrastructure vulnerabilities due to terrorism. *Risk Anal.* 2005 Apr;25(2):361–76.

Apostolakis GE, Pickett SE. Deliberation: Integrating analytical results into environmental decisions involving multiple stakeholders. *Risk Anal.* 1998;18(5):621–34.

Arce M., Daniel G, Sandler, T. Transnational public goods: Strategies and institutions. *Eur J Polit Econ.* 2001;17:493–516.

Ariyoshi N, Miyamoto M, Umetsu Y, Kunitoh H, Dosaka-Akita H, Sawamura Y, Yokota J, Nemoto N, Sato K, Kamataki T. Genetic polymorphism of CYP2A6 gene and tobacco-induced lung cancer risk in male smokers. *Cancer Epidem Biomar Prev.* 2002 Sep;11(9):890–4.

Armitage P, Doll R. The age distribution of cancer and a multi-stage theory of carcinogenesis. *Br J Cancer.* 1954;8:1–12.

Arora D, Bhatia DP, Prasad MA. Exact solution for a two-stage stochastic evolution system. 1993 Mar;47(3):2187–9.

Artzner P, Delbaen F, Eber J-M, Heath D. Coherent measures of risk. *Math Finance.* 1999;9: 203–28.

Arvai JL. Using risk communication to disclose the outcome of a participatory decision-making process: Effects on the perceived acceptability of risk-policy decisions. *Risk Anal.* 2003 Apr;23(2):281–9.

Arvai JL, Gregory R, McDaniels TL. Testing a structured decision approach: Value-focused thinking for deliberative risk communication. *Risk Anal.* 2001 Dec;21(6):1065–76.

Asvadi S, Hayes JA. Acute lung injury induced by cadmium aerosol. II. Free airway cell response during injury and repair. *Am J Pathol.* 1978 Jan;90(1):89–98.

Audoly S, Bellu G, D'Angio L, Saccomani MP, Cobelli C. Global identifiability of nonlinear models of biological systems. *IEEE Trans Biomed Eng.* 2001 Jan;48(1):55–65.

Audoly S, D'Angio L, Saccomani MP, Cobelli C. Global identifiability of linear compartmental models - A computer algebra algorithm. *IEEE Trans Biomed Eng.* 1998 Jan;45(1):36–47.

Australian Government. Australian Greenhouse Office, in the Department of the Environment and Heritage. *Climate Change Impacts & Risk Management: A Guide for Business and Government.* Commonwealth of Australia. Canberra, Australia. 2006. (Last accessed 19 Aug 2007) http://www.greenhouse.gov.au/impacts/publications/pubs/risk-management.pdf

Babusyte A, Stravinskaite K, Jeroch J, Lötvall J, Sakalauskas R, Sitkauskiene B. Patterns of airway inflammation and MMP-12 expression in smokers and ex-smokers with COPD. *Respir Res.* 2007 Nov 14;8(1):81.

Bagley SC, White H, Golomb BA. Logistic regression in the medical literature: Standards for use and reporting, with particular attention to one medical domain. *J Clin Epidemiol.* 2001 Oct;54(10):979–85.

Baird RW. Antibiotic prescribing, controls and antimicrobial resistance: An Australian experience. *APUA Newsl.* 1997;15(4):1–2, 6.

Bajpai R, Waseem M, Khanna AK, Kaw JL. Comparative pulmonary toxicity of cadmium and nickel: Histopathological and bronchoalveolar lavage analysis. *Indian J Exp Biol.* 1999 Jun;37(6):541–5.

Ball DJ. Environmental risk assessment and the intrusion of bias. *Environ Int.* 2002 Dec;28(6): 529–44.

Barbour AD, Chryssaphinou O, Ross M. Compound Poisson approximation in reliability theory. *IEEE Trans Reliab.* 1995;44:398–402.

Barlow RE. *Engineering Reliability.* American Statistical Association and Society for Industrial and Applied Mathematics (ASA-SIAM), ASA-SIAM Series on Statistics and Applied Probability. Philadelphia. 1998.

Baron DP. On the utility-theoretic foundation of mean-variance analysis. *J Finance.* 1997;32(5):1683–97.

Bartholomew MJ, Vose DJ, Tollefson LR, Travis CC. A linear model for managing the risk of antimicrobial resistance originating in food animals. *Risk Anal.* 2005 Feb;25(1): 99–108.

Bartsch H, Rojas M, Alexandrov K, Risch A. Impact of adduct determination on the assessment of cancer susceptibility. *Recent Results Cancer Res.* 1998;154:86–96.

Bastardi A, Shafir E.On the pursuit and misuse of useless information. *J Pers Soc Psychol.* 1998 Jul;75(1):19–32.

Bauer E, Kohavi R. An empirical comparison of voting classification algorithms: Bagging, boosting, and variants. *Mach Learn.* 1999 Jul–Aug; 36 (1–2): 106–39. http://citeseer.ist.psu.edu/bauer99empirical.html

Bedford T, Cooke RM. Probability density decomposition for conditionally dependent random variables modeled by vines. *Ann Math Artif Int.* 2001;32:245–68.

Beierle TC. The quality of stakeholder-based decisions. *Risk Anal.* 2002 Aug;22(4):739–49.

Belongia E. Marshfield Enterococcal Study Public Use Data Set.XLS. Excel spreadsheet dated 11 Sep 2007 provided by Edward Belongia on behalf of the Marshfield Study Group, Marshfield Clinic Research Foundation, Marshfield, WI.

Benartzi S, Thaler RH. Risk aversion of myopia? Choices in repeated gambles and retirement investments. *Manage Sci.* 1999;45(3):365–81.

Berend D, Sapir L. Expert rule versus majority rule under partial information, II. *J Math Decis Sci.* 2002;6(2):79–99.

Berman DW. 2006. *Evaluation of the Approach Recently Proposed for Assessing Asbestos – Related Risk in Eldorado County, California.* Aeolus, Inc. Technical Report, June 30, 2006. Albany, CA. http://www.aeolusinc.com/

Berman DW, Crump KS. Final Draft: Technical Support Document for a Protocol to Assess Asbestos-Related Risk. EPA # 9345. 2003;4–06.

Berry SM, Carroll RJ, Ruppert D. Bayesian smoothing and regression splines for measurement error problems. *J Am Stat Assoc.* 2002 Mar; 97 (457): 160–9. http://citeseer.ist.psu.edu/309111.html

Bertsimas D, Nino-Mora J. Conservation laws, extended polymatroids and multiarmed bandit problems: Polyhedral approach to indexable systems. *Math Oper Res.* 1996;21(2):257–306.

Best J. *Damned Lies and Statistics: Untangling Numbers from the Media, Politicians, and Activists.* University of California Press. Berkeley. 2001 http://www.ucpress.edu/books/pages/9358.html

Beyersmann D. Effects of carcinogenic metals on gene expression. *Toxicol Lett.* 2002 Feb 28;127(1–3):63–8.

Beyersmann D, Hechtenberg S. Cadmium, gene regulation, and cellular signalling in mammalian cells. *Toxicol Appl Pharmacol.* 1997 Jun;144(2):247–61.

Biavasco F, Foglia G, Paoletti C, Zandri G, Magi G, Guaglianone E, Sundsfjord A, Pruzzo C, Donelli F, Facinelli B. VanA-type Enterococci from humans, animals and food: Species distribution, population structure, Tn1546-typing and location, and virulence determinants. *Appl Environ Microbiol.* 2007;73(10):3307–19.

Bier VM. *Illusions of safety.* Paper presented at the Workshop on Organizational Analysis in High Hazard Production Systems: An Academy/Industry Dialogue, Apr 1997, Dedham, MA.

Bier VM. On the state of the art: Risk communication to decision makers. *Reliab Eng Syst Safe.* 2001a;71:151–7.

Bier VM. On the state of the art: Risk communication to the public. *Reliab Eng Syst Safe.* 2001b;71:139–50.

Bier VM, Azaiez MN (Eds.). *Game Theoretic Risk Analysis of Security Threats.* Springer. New York. 2009.

Bier VM, Cox LA Jr. Probabilistic risk analysis for engineered systems. Chapter 15 in Edwards W, Miles R, von Winterfeldt D (Eds.), *Advances in Decision Analysis.* Cambridge University Press. New York. 2007. http://www.cambridge.org/us/catalogue/catalogue.asp?isbn=0521682304

Bier VM, Jang SC. Defaults and incentives in risk-informed regulation. *Hum Ecol Risk Assess.* 1999;5:635–44.

Bier VM, Nagaraj A, Abhichandani V. Protection of simple series and parallel systems with components of different values. *Reliab Eng Syst Safe.* 2005;87:315–23.

Biggs D, de Ville B, Suen E. A method of choosing multiway partitions for classification and decision trees. *J Appl Stat.* 1991;18(1):49–62.

Bitton A, Neuman MD, Barnoya J, Glantz SA. The p53 tumour suppressor gene and the tobacco industry: Research, debate, and conflict of interest. *Lancet.* 2005 Feb 5;365(9458):531–40.

Blaine K, Powell D. Communication of food-related risks. *AgBioForum.* 2001;4:179–85.

Blaser MJ, Sazie E, Williams LP Jr. The influence of immunity on raw milk–associated *Campylobacter* infection. *JAMA.* 1987 Jan 2;257(1):43–6.

Bobbio A, Portinale L, Minichino M, Ciancamerla, E. Improving the analysis of dependable systems by mapping fault trees into Bayesian networks. *Reliab Eng Syst Safe.* 2001;71:249–60.

Borch KH. A note on uncertainty and indifference curves. *Rev Econ Stud.* 1969 Jan;36(1):1–14.

Borchers MT, Wesselkamper SC, Harris NL, Deshmukh H, Beckman E, Vitucci M, Tichelaar JW, Leikauf GD. CD8+ T cells contribute to macrophage accumulation and airspace enlargement following repeated irritant exposure. *Exp Mol Pathol.* 2007 Dec;83(3):301–10.

Borgonovo E, Smith CL, Apostolakis GE, Deriot S, Dewailly J. (2000). Insights from using influence diagrams to analyze precursor events. In Kondo S, Furuta K (Eds.), *Proceedings of PSAM 5, Probabilistic Safety Assessment and Management.* Universal Academy Press. Tokyo. 2000.

Bornstein BH, Emler AC. Rationality in medical decision making: A review of the literature on doctors' decision-making biases. *J Eval Clin Pract.* 2001 May;7(2):97–107.

Borrell-Carrio F, Epstein RM. Preventing errors in clinical practice: A call for self-awareness. *Ann Fam Med.* 2004;2(4):310–6.

Bowles S, Gintis H. The evolution of strong reciprocity: Cooperation in heterogeneous populations. *Theor Popul Biol.* 2004 Feb;65(1):17–28.

Box GE. Robustness in the strategy of scientific model building. In Launer R, Wilkinson G (Eds.), *Robustness in Statistics* (pp. 201–36). Academic Press. New York. 1979.

Boysen G, Hecht SS. Analysis of DNA and protein adducts of benzo[a]pyrene in human tissues using structure-specific methods. *Mutat Res.* 2003 Jan;543(1):17–30.

Brafman RI. A simplifier for propositional formulas with many binary clauses. *IEEE Trans Syst Man Cybern B Cybern.* 2004 Feb;34(1):52–9.

Brand M, Heinze K, Labudda K, Markowitsch HJ. The role of strategies in deciding advantageously in ambiguous and risky situations. *Cogn Process.* 2008 Aug;9(3):159–73.

Breiman L, Friedman J, Olshen R et al. *Classification and Regression Trees.* Wadsworth. Belmont, CA. 1984.

Brenner H, Loomis D. Varied forms of bias due to nondifferential error in measuring exposure. *Epidemiology* 1994;5(5):510–7.

Brightwell G, Orlodo G, Shepherd FB. Reserving resilient capacity in a network. *SIAM J Discrete Math.* 2001;14 (4): 524–39. http://citeseer.ist.psu.edu/730763.html

Brockett PL, Kahane Y. Risk, return, skewness and preference. *Manage Sci.* 1992 Jun;38(6): 851–66.

Brown R. Using Soft Data to Make "Probabilistic Risk Assessments" Realistic. 1999; retrieved July 29, 2005. http://fisher.osu.edu/~butler_267/DAPapers/WP000002.pdf

Brown G, Carlyle M, Salmeron J, Wood K. Defending critical infrastructure. *Interfaces.* 2006 Nov;36(6) 530–44.

Bucklew JA. *Introduction to Rare Event Simulation.* Springer. New York. 2004.

Buechley RW. Epidemiological consequences of an arsenic-lung cancer theory. *Am J Public Health Nations Health.* 1963 Aug;53:1229–32.

Burmaster DE, Anderson PD. Principles of good practice for the use of Monte Carlo techniques in human health and ecological risk assessments. *Risk Anal.* 1994 Aug;14(4):477–81.

Business and Government. Commonwealth of Australia. Canberra, Australia. http://www.greenhouse.gov.au/impacts/publications/pubs/risk-management.pdf. (Last accessed19 Aug – 2007)

Buzby JC, Roberts T, Jordan Lin CT, MacDonald JM. Bacterial Foodborne Disease: Medical Costs and Productivity Losses. USDA, Economic Research Service, Agricultural Economics Report # 741. 1996. http://www.ers.usda.gov/publications/aer741

Byrd DM, Cothern CR. *Introduction to Risk Analysis: A Systematic Approach to Science-Based Decision Making*.ABS Group, Houston. 2000. Chapter 12.

Bywater RJ. Identification and surveillance of antimicrobial resistance dissemination in animal production. *Poult Sci*. 2005 Apr;84(4):644–8.

California Department of Transportation, Federal Highway Administration, California Division. 2007. *Systems Engineering Guidebook for ITS Version 2.0*. www.fhwa.dot.gov/cadiv/segb/views/document/Sections/Section3/3_9_4.htm

Camerer CE, Fehr E. When does "economic man" dominate social behavior? *Science*. 2006 Jan; (311): 47–52.

Campling BG, el-Deiry WS. Clinical implications of p53 mutations in lung cancer. *Methods Mol Med*. 2003;75:53–77.

Canadian Food Inspection Agency. 2004. http://www.inspection.gc.ca/english/anima/heasan/disemala/bseesb/bsefaqe.shtml

Cardinal RN. Neural systems implicated in delayed and probabilistic reinforcement. *Neural Netw*. 2006 Oct;19(8):1277–301.

Carlin BP, Louis TA. *Bayes and Empirical Bayes Methods for Data Analysis,* 2nd ed.Chapman and Hall/CRC. Boca Raton, FL. 2000.

Carroll RJ, Ruppert D, Stefanski LA, Crainiceanu C. *Measurement Error in Nonlinear Models: A Modern Perspective,* 2nd ed. Chapman & Hall. New York. 2006. http://www.stat.tamu.edu/~carroll/eiv.SecondEdition/index.php

Cassin MH, Paoli GM, Lammerding AM. Simulation modeling for microbial risk assessment. *J Food Prot*. 1998 Nov;61(11):1560–6.

Ceci SJ, Bjork RA. Psychological science in the public interest: The case for juried analyses. *Psychol Sci*. 2000 May;11(3):177–8.

CFIA. 2005. www.inspection.gc.ca/english/anima/heasan/disemala/bseesb/200503canadae.shtml#chap2

Chanda S, Dasgupta UB, Guhamazumder D, Gupta M, Chaudhuri U, Lahiri S, Das S, Ghosh N, Chatterjee D. DNA hypermethylation of promoter of gene p53 and p16 in arsenic-exposed people with and without malignancy. *Toxicol Sci*. 2006 Feb;89(2):431–7.

Chang Y-R, Amari SV, Kuo S-Y. Computing system failure frequencies and reliability importance measures using OBDD. *IEEE Trans Comput*. 2004 Jan;53(1):54–68.

Chang KC, Tian Z. Efficient Inference for Mixed Bayesian Networks. 2002. http://citeseer.nj.nec.com/cachedpage/570208/1

Chan-Yeung M, Tan-Un KC, Ip MS, Tsang KW, Ho SP, Ho JC, Chan H, Lam WK. Lung cancer susceptibility and polymorphisms of glutathione-S-transferase genes in Hong Kong. *Lung Cancer*. 2004 Aug;45(2):155–60.

Chartier J, Gabler S. *Risk Communication and Government: Theory and Application for the Canadian Food Inspection Agency*.Canadian Food Inspection Agency. Ottawa. 2001.

Chaudhuri P, Lo W-D, Loh W-Y, Yang C-C. Generalized regression trees. *Statistica Sinica*. 1995; 5: 641–66. http://www.stat.wisc.edu/~loh/treeprogs/guide/grapes.pdf

Chekuri C, Gupta A, Kumar A, Naor J, Raz D. Building Edge-Failure Resilient Networks. 2002. http://citeseer.ist.psu.edu/chekuri02building.html

Chen CL, Hsu LI, Chiou HY, Hsueh YM, Chen SY, Wu MM, Chen CJ; Blackfoot Disease Study Group. Ingested arsenic, cigarette smoking, and lung cancer risk: A follow-up study in arseniasis-endemic areas in Taiwan. *JAMA*. 2004 Dec 22;292(24):2984–90.

Chen Z, McGee M. A Bayesian approach to zero-numerator problems using hierarchical models. *J Data Sci*. 2008;6:261–8.

Chen JJ, Tsai CA, Moon H, Ahn H, Young JJ, Chen CH. Decision threshold adjustment in class prediction. *SAR QSAR Environ Res*. 2006 Jun;17(3):337–52.

Cheng J, Druzdzel MJ. AIS-BN: An Adaptive Importance Sampling Algorithm for Evidential Reasoning in Large Bayesian Networks. 2000. http://citeseer.nj.nec.com/cheng00aisbn.html

Chevillard S, Radicella JP, Levalois C, Lebeau J, Poupon MF, Oudard S, Dutrillaux B, Boiteux S. Mutations in OGG1, a gene involved in the repair of oxidative DNA damage, are found in human lung and kidney tumours. *Oncogene*. 1998 Jun 11;16(23): 3083–6.

Chick SE. Bayesian analysis for simulation input and output. In Andradóttir S, Healy KJ, Withers DH, Nelson BL (Eds.), *Proceedings of the 1997 Winter Simulation Conference* (pp. 253–260). IEEE Press. Washington. 1997.

Chinander KR, Kleindorfer PR, Kunreuther HC. Compliance strategies and regulatory effectiveness of performance-based regulation of chemical accident risks. *Risk Anal.*, 1998;18: 135–43.

Choi BC, Noseworthy AL. Classification, direction, and prevention of bias in epidemiologic research. *J Occup Med.* 1992 Mar;34(3):265–71.

Choi BC, Pang T, Lin V, Puska P, Sherman G, Goddard M, Ackland MJ, Sainsbury P, Stachenko S, Morrison H, Clottey C. Can scientists and policy makers work together? *J Epidemiol Community Health.* 2005 Aug;59(8):632–7.

Choi H, Subramaniam S, Choi, H-A. On Double-Link Failure Recovery in WDM Optical Networks. 2002; http://citeseer.ist.psu.edu/choi02doublelink.html

Churg A, Wright JL. Animal models of cigarette smoke-induced chronic obstructive lung disease. *Contrib Microbiol.* 2007;14:113–25.

Churg A, Zay K, Shay S, Xie C, Shapiro SD, Hendricks R, Wright JL. Acute cigarette smoke-induced connective tissue breakdown requires both neutrophils and macrophage metalloelastase in mice. *Am J Respir Cell Mol Biol.* 2002 Sep;27(3):368–74.

Clemen RT. *Making Hard Decisions: An Introduction to Decision Analysis,* 2nd ed. PWS-Kent. Boston. 1996.

Collignon P. Fluoroquinolone use in food animals [letter]. *Emerg Infect Dis.* Nov 2005. http://www.cdc.gov/ncidod/EID/vol11no11/04-0630_05-0652_05-1022.htm

Colman AM. Cooperation, psychological game theory, and limitations of rationality in social interaction. *Behav Brain Sci.* 2003 Apr;26(2):139–53; discussion 153–98.

Conway K, Edmiston SN, Cui L, Drouin SS, Pang J, He M, Tse CK, Geradts J, Dressler L, Liu ET, Millikan R, Newman B. Prevalence and spectrum of p53 mutations associated with smoking in breast cancer. *Cancer Res.* 2002 Apr 1;62(7):1987–95.

Cover TM, Thomas JA. *Elements of Information Theory.* John Wiley & Sons. Hoboken, NJ. 1991.

Cox LA Jr. A new measure of attributable risk for public health applications. *Manage Sci.* 1985;31(7):800–14.

Cox LA Jr. A probabilistic risk assessment program for analyzing security risks, pp 331–340 in Cox, LA Jr, Ricci PF (Eds.), *New Risks: Issues and Management.* Plenum Press. New York. 1990.

Cox LA Jr. Extending the stochastic two-stage model of carcinogenesis to include self-regulation of the nonmalignant cell population. *Risk Anal.* 1992 Mar;12(1):129–38.

Cox LA Jr. An exact analysis of the multistage model explaining dose-response concavity. *Risk Anal.* 1995;15(3):359–68.

Cox LA Jr. *Risk Analysis: Foundations, Models and Methods.* Springer. New York. 2001.

Cox LA Jr. Mortality associated with foodborne bacterial gastrointestinal infections: Statistical method is worth examining. *BMJ.* 2003 Jun 7;326(7401): 1265. http://bmj.com/cgi/eletters/326/7385/357#29767

Cox LA Jr. Potential human health benefits of antibiotics used in food animals: A case study of virginiamycin. *Environ Int.* 2005a;31(4):549–63.

Cox LA Jr. Some limitations of a proposed linear model for antimicrobial risk management. *Risk Analysis.* 2005b Dec; 25 (6): 1327–32. http://www.blackwell-synergy.com/doi/abs/10. 1111/j.

Cox LA Jr. Detecting causal nonlinear exposure-response relations in epidemiological data. *Dose-Response.* 2006a;4(2):119–32.

Cox LA Jr. *Quantitative Health Risk Analysis Methods: Modeling the Human Health Impacts of Antibiotics Used in Food Animals.* Springer. New York. 2006b.

Cox LA Jr. Enrofloxacin in poultry and human health [letter]. *Emerging Infectious Diseases.* 2006c May;12(5):872–873. http://www.cdc.gov/ncidod/EID/vol12no05/05-1477_06-0305.htm

Cox LA Jr. Does concern-driven risk management provide a viable alternative to QRA? *Risk Anal.* 2007a;27(1): 27–43.

Cox LA Jr. Health risk analysis for risk management decision-making. Chapter 17 in Edwards W, Miles R, von Winterfeldt D (Eds.), *Advances in Decision Analysis*. Cambridge University Press, New York. 2007b. http://www.cambridge.org/us/catalogue/catalogue.asp?isbn=0521682304

Cox LA Jr. What's wrong with risk matrices? *Risk Anal.* 2008 Apr;28(2):497–512.

Cox, LA Jr. Some limitations of "*risk* = *threat* × *vulnerability* × *consequence*" for risk analysis of terrorist attacks. *Risk Anal.* 2008b Dec;28(6):1749–61.

Cox LA Jr, Babayev D, Huber W. Some limitations of qualitative risk rating systems. *Risk Anal.* 2005 Jun;25(3):651–62.

Cox LA Jr., Copeland D, Vaughn M. Ciprofloxacin resistance does not affect duration of domestically acquired campylobacteriosis. *J Infect Dis.* 2005 May 1;191(1):1565–6. http://www.journals.uchicago.edu/JID/journal/contents/v191n9.html

Cox LA Jr, Huber WA. Symmetry, identifiability, and prediction uncertainties in multistage clonal expansion (MSCE) models of carcinogenesis. *Risk Anal.* 2007 Dec;27(6):1441–53.

Cox LA Jr, Popken DA. Bayesian Monte Carlo uncertainty analysis of human health risks from animal antimicrobial use in a dynamic model of emerging resistance. *Risk Anal.* 2004a;24(5):1153–64.

Cox LA Jr, Popken DA. Quantifying human health risks from virginiamycin used in poultry. *Risk Anal.* 2004b;24(1):271–88.

Cox LA Jr, Popken DA. *Asbestos Health Risk Analysis Using EPA Region IX's Data on El Dorado Hills, California.* Cox Associates. Denver 2006a.

Cox LA Jr, Popken DA. Quantifying potential human health impacts of animal antibiotic use: Enrofloxacin and macrolides in poultry. *Risk Anal.* 2006b;26(1):135–46.

Cox LA Jr, Popken DA. Some limitations of aggregate exposure metrics. *Risk Anal.* 2007 Apr;27(2):439–45.

Cox LA Jr, Sanders E. Estimating preventable fractions of disease caused by a specified biological mechanism: PAHs in smoking lung cancers as an example. *Risk Anal.* 2006 Aug;26(4):881–92.

Cox LA Jr., VanSickle JJ, Popken DA, Sahu R. Optimal tracking and testing of U.S. and Canadian herds for BSE: A value-of-information (VOI) approach. *Risk Anal.* 2005;25 (4): 827–40. http://www.blackwell-synergy.com/doi/abs/10. 1111/j. 1539-6924. 2005.00648.x

Csicsaky MJ, Roller M, Pott F. Risk modeling: Which models to choose? *Exp Pathol.* 1989;37(1–4):198–204.

Cui X, Wakai T, Shirai Y, Hatakeyama K, Hirano S. Chronic oral exposure to inorganic arsenate interferes with methylation status of p16INK4a and RASSF1A and induces lung cancer in A/J mice. *Toxicol Sci.* 2006 Jun;91(2):372–81.

Cullen AC, Frey HC. *Probabilistic Techniques in Exposure Assessment: A Handbook for Dealing with Variability and Uncertainty in Models and Inputs.*Plenum. New York. 1999.

Crowley M. Evaluating influence diagrams: where we've been and where we're going. Department of Computer Science. University of British Columbia. August 31, 2004. http://www.cs.ubc.ca/~crowley/academia/papers/aiproj.pdf (Last accessed 3-20-2009)

Crucitti P, Latora V, Marchiori M, Rapisarda A. Error and Attack Tolerance of Complex Networks. 2004. http://citeseer.ist.psu.edu/724744.html.

Dacomo A, De Patre S, Maier G, Pattavina A, Martinelli M. Design of Static Resilient WDM Mesh Networks with Multiple Heuristic Criteria. 2002. http://citeseer.ist.psu.edu/508406.html

DANMAP 2004. http://www.dfvf.dk/Files/Filer/Zoonosecentret/Publikationer/Danmap/Danmap_2004.pdf

Dasgupta P, Kinkade R, Joshi B, Decook C, Haura E, Chellappan S. Nicotine inhibits apoptosis induced by chemotherapeutic drugs by up-regulating XIAP and survivin. *Proc Natl Acad Sci USA.* 2006;103:6332–7.

Deccio CT, Cox S, Smith M, Wan J, Clement M, Snell Q. Survivability of Large Networks in the Presence of Malicious Attacks. 2003. http://dna.cs.byu.edu/papers/pdf/survive.pdf.

Deeks JJ, Dinnes J, D'Amico R, Sowden AJ, Sakarovitch C, Song F, Petticrew M, Altman DG; International Stroke Trial Collaborative Group; European Carotid Surgery Trial Collaborative Group. Evaluating non-randomised intervention studies. *Health Technol Assess.* 2003;7(27):iii–x, 1–173.

Demedts IK, Morel-Montero A, Lebecque S, Pacheco Y, Cataldo D, Joos GF, Pauwels RA, Brusselle GG. Elevated MMP-12 protein levels in induced sputum from patients with COPD. *Thorax*. 2006a Mar;61(3):196–201.

Demedts IK, Demoor T, Bracke KR, Joos GF, Brusselle GG. Role of apoptosis in the pathogenesis of COPD and pulmonary emphysema. *Respir Res*. 2006b Mar 30;7:53.

Denardo EV, Rothblum UG, van der Heyden L. Index policies for stochastic search in a forest with an application to R&D project management. *Math Oper Res*. 2004 Feb;29(1):162–81.

Denissenko MF, Pao A, Tang M, Pfeifer GP. Preferential formation of benzo[a]pyrene adducts at lung cancer mutational hotspots in P53. *Science*. 1996 Oct 18;274(5286):430–2.

Deshpande LM, Fritsche TR, Moet GJ, Biedenbach DJ, Jones RN. Antimicrobial resistance and molecular epidemiology of vancomycin-resistant enterococci from North America and Europe: A report from the SENTRY antimicrobial surveillance program. *Diagn Micr Infect Dis*. 2007;58(2):163–70.

Doherr MG, Heim D, Fatzer R, Cohen CH, Vandevelde M, Zurbriggen A. Targeted screening of high-risk cattle populations for BSE to augment mandatory reporting of clinical suspects. *Prev Vet Med*. 2001 Sep 20;51(1–2):3–16.

Domagała-Kulawik J, Maskey-Warzechowska M, Kraszewska I, Chazan R. The cellular composition and macrophage phenotype in induced sputum in smokers and ex-smokers with COPD. *Chest*. 2003 Apr;123(4):1054–9.

Domagała-Kulawik J, Maskey-Warzechowska M, Hermanowicz-Salamon J, Chazan R. Expression of macrophage surface markers in induced sputum of patients with chronic obstructive pulmonary disease. *J Physiol Pharmacol*. 2006 Sep;57(Suppl 4):75–84.

Donabedian SM, Perri MB, Vager D et al. Quinupristin-dalfopristin resistance in *Enterococcus faecium* isolates from humans, farm animals, and grocery store meat in the United States. *J Clin Microbiol*. 2006;44(9):3361–5.

Dowell AM, Hendershot DC. No good deed goes unpunished: Case studies of incidents and potential incidents caused by protective systems. *Process Safety Progress*. 1997;16:132–9.

Doya K. Modulators of decision making. *Nat Neurosci*. 2008 Apr;11(4):410–6.

Doyle JC, Alderson DL, Li L, Low S, Roughan M, Shalunov S, Tanaka R, Willinger W. The “robust yet fragile” nature of the Internet. *Proc Natl Acad Sci USA*. 2005 Oct 11;102(41):14497–502. Epub 2005 Oct 4.

Dreiseitl S, Ohno-Machado L, Vinterbo S. Evaluating variable selection methods for diagnosis of myocardial infarction. *Proc AMIA Symp*. 1999;246–50.

Druzdzel MJ. Some Useful Properties of Probabilistic Knowledge Representations from the Point of View of Intelligent Systems. 1994. http://citeseer.nj.nec.com/druzdzel94some.html

Duenas-Osorio L, Craig JI, Goodno BJ. Probabilistic Response of Interdependent Infrastructure Networks. Second Annual Meeting of the Asian-Pacific Network of Centers. 2004. http://mceer.buffalo.edu/research/International_Research/ANCER/Activities/2004/osorio_l_maec.pdf

Dyer JS, Jia J. Preference conditions for utility models: A risk-value perspective. *Ann Oper Res*. 1998;80(1):167–82. http://citeseerx.ist.psu.edu/viewdoc/summary?doi=10.1.1.39.5480

Dyer JS, Sarin RK. Measurable multiattribute value functions. *Oper Res*. 1979;27(4):810–22.

Edmond MB, Wallace SE, McClish DK, Pfaller MA, Jones RN, Wenzel RP. Nosocomial bloodstream infections in United States hospitals: A three-year analysis. *Clin Infect Dis*. 1999;29(2):239–44.

Effler P, Ieong MC, Kimura A, Nakata M, Burr R, Cremer E, Slutsker L. Sporadic *Campylobacter jejuni* infections in Hawaii: Associations with prior antibiotic use and commercially prepared chicken. *J Infect Dis*. 2001 Apr 1;183(7):1152–5.

ElAmin A. Foodborne *Campylobacter* infections increase. *Food Production Daily*. 2006 Jan 3. http://www.foodproductiondaily.com/news/ng.asp?id=64828

Elinder CG, Kjellstrom T, Hogstedt C, Andersson K, Spang G. Cancer mortality of cadmium workers. *Br J Ind Med*. 1985 Oct;42(10):651–5.

Ellis B, Wong WH. Learning causal Bayesian network structures from experimental data. *J Am Stat Assoc*. 2008 Jun;103(482):778–89.

Elstein AS. Heuristics and biases: Selected errors in clinical reasoning. *Acad Med.* 1999 Jul;74(7):791–4.

Elting LS, Martin CG, Cantor SB, Rubenstein EB. Influence of data display formats on physician investigators' decisions to stop clinical trials: Prospective trial with repeated measures. *BMJ.* 1999 Jun 5;318(7197):1527–31.

Emmons KM, Wong M, Puleo E, Weinstein N, Fletcher R, Colditz G. Tailored computer-based cancer risk communication: Correcting colorectal cancer risk perception. *J Health Commun.* 2004 Mar–Apr;9(2):127–41.

Enders W, Sandler, T. What do we know about the substitution effect in transnational terrorism? In Silke A, Ilardi G (Eds.), *Researching Terrorism: Trends, Achievements, Failures*. Frank Cass. London. 2004.

Enterline PE, Day R, Marsh GM. Cancers related to exposure to arsenic at a copper smelter. *Occup Environ Med.* 1995 Jan;52(1):28–32.

Eurosurveillance 2005;10 (5): 050526, http://www.eurosurveillance.org/ew/2005/050526.asp

Evans JS, Handley SJ, Hadjchristidis C et al. On the basis of belief in causal and diagnostic conditionals. *Q J Exp Psychol (Colchester).* 2007; 60(5):635–43.

Ezell BC, Haimes YY, Lambert JH. Cyber attack to water utility supervisory control and data acquisition (SCADA) systems. *Mil Oper Res.*2001;6:23–33.

Fan R, Wu MT, Miller D, Wain JC, Kelsey KT, Wiencke JK, Christiani DC. The p53 codon 72 polymorphism and lung cancer risk. *Cancer Epidemiol Biomarkers Prev.* 2000 Oct;9(10): 1037–42.

Fang MZ, Mar W, Cho MH. Cadmium affects genes involved in growth regulation during two-stage transformation of Balb/3T3 cells. *Toxicology.* 2002 Aug 15;177(2–3):253–65.

FAO/WHO. Joint FAO/WHO Expert Consultation on Risk Assessment of Microbiological Hazards in Foods: Risk characterization of *Salmonella* spp. in eggs and broiler chickens and *Listeria* monocytogenes in ready-to-eat foods. FAO Headquarters, Rome, Italy, 30 Apr–4 May 2001. http://www.who.int/foodsafety/publications/micro/en/may2001.pdf

FAO/WHO/OIE. Joint FAO/WHO/OIE Expert Meeting on Critically Important Antimicrobials. Report of a meeting held in FAO, Rome, Italy, 26–30 Nov 2007. FAO, Rome, Italy, and WHO, Geneva, Switzerland. 2008.

FDA, 2003. Guidance for Industry 152 – Evaluating the Safety of Antimicrobial New Animal Drugs with Regard to Their Microbiological Effects on Bacteria of Human Health Concern, 2003;Oct 23. http://www.fda.gov/cvm/guidance/fguide152.pdf

FDA-CVM. Human Health Impact of Fluoroquinolone Resistant *Campylobacter* Attributed to the Consumption of Chicken. U.S. Food and Drug Administration, Center for Veterinary Medicine (revised Jan 2001).

FDA-CVM. Risk assessment of streptogramin resistance in *Enterococcus faecium* attributable to the use of streptogramins in animals: " Virginiamycin Risk Assessment" – Draft for Comment, 23 Nov 2004.

FDA-CFSAN. FDA/Center for Food Safety and Applied Nutrition USDA/Food Safety and Inspection Service Centers for Disease Control and Prevention. Quantitative Assessment of Relative Risk to Public Health from Foodborne *Listeria monocytogenes* Among Selected Categories of Ready-to-Eat Foods. Appendix 3. Sep 2003. http://vm.cfsan.fda.gov/~dms/lmr2-a3.html

Federal Aviation Administration. Introduction to Safety Management Systems (SMA) for Airport Operators (Advisory Circular). U.S. Department of Transportation. Washington, DC. 28 Feb 2007. www.faa.gov/airports_airtraffic/airports/resources/advisory_circulars/media/150-5200-37/150_5200_37.doc (Last accessed 19 Aug 2007)

Federal Highway Administration of the U.S. Department of Transportation. *Risk Assessment and Allocation for Highway Construction Management.* U.S. Department of Transportation, Washington, DC. 2006. http://international.fhwa.dot.gov/riskassess/index.htm (Last accessed 19 Aug 2007)

Fehr E, Rockenbach B. Human altruism: Economic, neural, and evolutionary perspectives. *Curr Opin Neurobiol.* 2004 Dec;14(6):784–90.

Feldman RA. Confounding factors in observational and intervention studies. *Ital J Gastroenterol Hepatol.* 1998 Oct;30(Suppl 3):S248–53.

Ferson S, Donald S. Probability bounds analysis. In Mosleh A, Bari RA (Eds.), *Probabilistic Safety Assessment and Management* (pp. 1203–1208). Springer-Verlag. New York. 1998.

Ferson S, Hajagos JG. Varying correlation coefficients can underestimate uncertainty in probabilistic models. *Reliab Eng Syst Safe.* 2006;91:1461–7.

Flaxman AD, Frieze AM, Vera J. Adversarial deletion in a scale-free random graph process. *Comb Probab Comput.* 2007;16:261–70.

Flynn J, Slovic P, Mertz CK. Gender, race, and perception of environmental health risks. *Risk Anal.* 1994 Dec;14(6):1101–8.

Food and Drug Administration – Center for Veterinary Medicine (FDA-CVM). Review of Agricultural Antibiotics Policies, 2000. http://www.fda.gov/cvm/HRESP106_157.htm#nrdc

Food and Drug Administration – Center for Veterinary Medicine (FDA-CVM). Guidance for Industry # 152: Evaluating the Safety of Antimicrobial New Animal Drugs with Regard to their Microbiological Effects on Bacteria of Human Health Concern. U.S. Dept of Health and Human Services, Food and Drug Administration, Center for Veterinary Medicine, 2003. http://www.fda.gov/cvm/Guidance/fguide152.pdf

Food and Drug Administration – Center for Veterinary Medicine (FDA-CVM). Risk Assessment of Streptogramin Resistance in *Enterococcus faecium* Attributable to the Use of Streptogramins in Animals. Draft for Comment. 23 Nov 2004. http://www.fda.gov/cvm/Documents/SREF_RA_FinalDraft.pdf

Food and Drug Administration, DA Center for Veterinary Medicine. http://www.fda.gov/cvm/Documents/SREF_RA_FinalDraft.pdf (Last accessed 14 Nov 2007)

Fortun J, Coque TM, Martin-Davila P, Moreno L, Canton R., Loza E, Baquero F, Moreno S. Risk factors associated with ampicillin resistance in patients with bacteraemia caused by *Enterococcus faecium. J Antimicrob Chemoth.* 2002;50(6):1003–9.

Franklin WA, Gazdar AF, Haney J, Wistuba II, La Rosa FG, Kennedy T, Ritchey DM, Miller YE. Widely dispersed p53 mutation in respiratory epithelium. A novel mechanism for field carcinogenesis. *J Clin Invest.* 1997 Oct 15;100(8):2133–7.

Freedman DA. Graphical models for causation, and the identification problem. *Eval Rev.* 2004 Aug;28(4):267–93.

Frewer L. The public and effective risk communication. *Toxicol Lett.* 2004 Apr 1;149(1–3):391–7.

Frey BS, Luechinger S. How to fight terrorism: Alternatives to deterrence. *Defence Peace Econ.* 2003;14:237–49.

Frey L, Fisher D, Tsamardinos I, Aliferis C, Statnikov A. Identifying Markov Blankets with Decision Tree Induction. 2003. http://citeseer.ist.psu.edu/frey03identifying.html

Frey R, McNeil AJ, Nyfeler MA. Copulas and credit models. Oct 2001. http://citeseer.ist.psu.edu/frey01copulas.html

Friedgut E. Hunting for sharp thresholds. *Random Struct Algor.* 2005;26(1–2):37–51.

Friedman N, Goldszmidt, M. Discretizing Continuous Attributes While Learning Bayesian Networks. 1996a. http://citeseer.ist.psu.edu/friedman96discretizing.html

Friedman N, Goldszmidt, M. Learning Bayesian Networks With Local Structure. 1996b. http://citeseer.ist.psu.edu/friedman96learning.html

Friedman N, Goldszmidt, M. Sequential Update of Bayesian Network Structure. 1997. http://citeseer.ist.psu.edu/friedman97sequential.html

Friedman CR, Hoekstra RM, Samuel M, Marcus R, Bender J, Shiferaw B, Reddy S, Ahuja SD, Helfrick DL, Hardnett F, Carter M, Anderson B, Tauxe RV; Emerging Infections Program FoodNet Working Group. Risk factors for sporadic *Campylobacter* infection in the United States: A case-control study in FoodNet sites. *Clin Infect Dis.* 2004 Apr 15;38(Suppl 3):S285–96.

Friedman C, Reddy S, Samual M, Marcus R, Bender J, Desai S, Shiferaw B, Helfrick D, Carter M, Anderson B, Hoekstra M, and the EIP Working Group. Risk factors for sporadic *Campylobacter* infections in the United States: A case-control study on FoodNet sites. Second International Conference on Emerging Infectious Diseases. Atlanta, GA. Jul 2000. http://www.cdc.gov/foodnet/pub/iceid/2000/friedman_c.htm

Frimodt-Moller N, Hammerum AM. Food safety revisited. *J Infect Dis.* 2006;194(9):1191–3.

Fugelsang JA, Dunbar KN. Brain-based mechanisms underlying complex causal thinking. *Neuropsychologia.* 2005;43(8):1204–13.

Fugelsang JA, Stein CB, Green AE et al. Theory and data interactions of the scientific mind: Evidence from the molecular and the cognitive laboratory. *Can J Exp Psychol.* 2004;58(2): 86–95.

Fugelsang JA, Thompson VA. A dual-process model of belief and evidence interactions in causal reasoning. *Mem Cognit.* 2003 Jul;31(5):800–15.

Fujieda M, Yamazaki H, Saito T, Kiyotani K, Gyamfi MA, Sakurai M, Dosaka-Akita H, Sawamura Y, Yokota J, Kunitoh H, Kamataki T. Evaluation of CYP2A6 genetic polymorphisms as determinants of smoking behavior and tobacco related lung cancer risk in male Japanese smokers. *Carcinogenesis.* 2004 Dec;25(12):2451–8.

Gairola CG, Wagner GJ. Cadmium accumulation in the lung, liver and kidney of mice and rats chronically exposed to cigarette smoke. *J Appl Toxicol.* 1991 Oct;11(5):355–8.

Gallos LK, Cohen R, Argyrakis P, Bunde A, Havlin S. Stability and topology of scale-free networks under attack and defense strategies. *Phys Rev Lett.* 2005 May 13;94(18):188701.

GAO 1999. *Combating Terrorism: Threat and Risk Assessments Can Help Prioritize and Target Program Investments.* U.S. Government Accounting Office. Washington, DC. http://www.gao.gov/archive/1998/ns98074.pdf (Last accessed 19 Aug 2007)

Garrick BJ, Hall JE, Kilger M., McDonald JC et al. Confronting the risks of terrorism: Making the right decisions. *Reliab Eng Syst Safe* 2004;86:129–76.

Garvin T. Analytical paradigms: The epistemological distances between scientists, policy makers, and the public. *Risk Anal.* 2001 Jun;21(3):443–55.

Gaubin Y, Vaissade F, Croute F, Beau B, Soleilhavoup J, Murat J. Implication of free radicals and glutathione in the mechanism of cadmium-induced expression of stress proteins in the A549 human lung cell-line. *Biochim Biophys Acta.* 2000 Jan 10;1495(1):4–13.

Gelman A, Van Mechelen I, Verbeke G, Heitjan DF, Meulders M. Multiple imputation for model checking: completed-data plots with missing and latent data. *Biometrics.* 2005 Mar;61(1): 74–85.

Geradts J, Fong KM, Zimmerman PV, Maynard R, Minna JD. Correlation of abnormal RB, p16ink4a, and p53 expression with 3p loss of heterozygosity, other genetic abnormalities, and clinical features in 103 primary non-small cell lung cancers. *Clin Cancer Res.* 1999;5: 791–800.

Gerhardsson L, Brune D, Nordberg GF, Wester PO. Distribution of cadmium, lead and zinc in lung, liver and kidney in long-term exposed smelter workers. *Sci Total Environ.* 1986 Apr;50: 65–85.

Gerhardsson L, Englyst V, Lundstrom NG, Sandberg S, Nordberg G. Cadmium, copper and zinc in tissues of deceased copper smelter workers. *J Trace Elem Med Biol.* 2002;16(4):261–6.

Gerrard MB. Risks of hazardous waste sites versus asteroid and comet impacts: Accounting for the discrepancies in U.S. resource allocation. *Risk Anal.* 2000 Dec;20(6):895–904.

Givens GH, Roback PJ. Logarithmic pooling of priors linked by a deterministic simulation model. *J Comput Graph Stat.* 1999;8:452–78.

Glaser U, Kloppel H, Hochrainer D. Bioavailability indicators of inhaled cadmium compounds. *Ecotoxicol Environ Saf.* 1986 Jun;11(3):261–71.

Glazko GV, Koonin EV, Rogozin IB. Mutation hotspots in the p53 gene in tumors of different origin: correlation with evolutionary conservation and signs of positive selection. *Biochim Biophys Acta.* 2004 Aug 12;1679(2):95–106.

Glymour C, Cooper GF. *Computation, Causation & Discovery,* MIT Press. Cambridge. 1999.

Golany B, Kaplan EH, Marmur A, Rothblum UG. Nature plays with dice – terrorists do not: Allocating resources to counter strategic vs. probabilistic risk. *Eur J Oper Res.* 2009;192:198–208.

Goldman R, Enewold L, Pellizzari E, Beach JB, Bowman ED, Krishnan SS, Shields PG. Smoking increases carcinogenic polycyclic aromatic hydrocarbons in human lung tissue. *Cancer Res.* 2001 Sep 1;61(17):6367–71.

Gopalakrishna R, Chen ZH, Gundimeda U. Tobacco smoke tumor promoters, catechol and hydroquinone, induce oxidative regulation of protein kinase C and influence invasion and metastasis of lung carcinoma cells. *Proc Natl Acad Sci USA.* 1994 Dec 6;91(25):12233–7.

Government of Alberta. 2004. http://www1.agric.gov.ab.ca/$department/deptdocs.nsf/all/afs3781?opendocument

Grasseschi RM, Ramaswamy RB, Levine DJ, Klaassen CD, Wesselius LJ. Cadmium accumulation and detoxification by alveolar macrophages of cigarette smokers. *Chest.* 2003 Nov;124(5):1924–8.

Green, PJ. Reversible Jump Markov Chain Monte Carlo Computation and Bayesian Model Determination. 1995. http://citeseer.ist.psu.edu/green95reversible.html

Greenberg MR, Schneider DF. Gender differences in risk perception: Effects differ in stressed vs. non-stressed environments. *Risk Anal.* 1995 Aug;15(4):503–11.

Greenberg MR, Williams B. Geographical dimensions and correlates of trust. *Risk Anal.* 1999 Apr;19(2):159–69.

Greenland S. Modeling and variable selection in epidemiologic analysis. *Am J Public Health.* 1989 Mar;79(3):340–9.

Greenland S. Quantifying biases in causal models: Classical confounding vs collider-stratification bias. *Epidemiology.* 2003 May;14(3):300–6.

Greenland S, Brumback B. An overview of relations among causal modeling methods. *Int J Epidemiol.* 2002 Oct;31(5):1030–7.

Greenland S, Morgenstern H. Confounding in health research. *Annu Rev Public Health.* 2001;22:189–212.

Grimes DA, Schulz KF. Bias and causal associations in observational research. *Lancet.* 2002 Jan 19;359(9302):248–52.

Grose EC, Richards JH, Jaskot RH, Menache MG, Graham JA, Dauterman WC. A comparative study of the effects of inhaled cadmium chloride and cadmium oxide: Pulmonary response. *J Toxicol Environ Health.* 1987;21(1–2):219–32.

Grove WM, Zald DH, Lebow BS, Snitz BE, Nelson C. Clinical versus mechanical prediction: A meta-analysis. *Psychol Assess.* 2000 Mar;12(1):19–30.

Guardabassi L, Dalsgaard A. Occurrence, structure, and mobility of Tn1546-like elements in environmental isolates of vancomycin-resistant enterococci. *Appl Environ Microbiol.* 2004;70(2):984–90.

Guatama T, Van Hulle MM. Surrogate-Based Test for Granger-Causality. 2003. http://citeseer.ist.psu.edu/588339.html

Gurm HS, Litaker DG. Framing procedural risks to patients: Is 99% safe the same as a risk of 1 in 100? *Acad Med.* 2000 Aug;75(8):840–2.

Gustafson P, Le ND. Comparing the effects of continuous and discrete covariate mismeasurement, with emphasis on the dichotomization of mismeasured predictors. *Biometrics.* 2002;58(4): 878–87.

Gyorffy E, Anna L, Gyori Z, Segesdi J, Minarovits J, Soltesz I, Kostic S, Csekeo A, Poirier MC, Schoket B. DNA adducts in tumour, normal peripheral lung and bronchus, and peripheral blood lymphocytes from smoking and non-smoking lung cancer patients: Correlations between tissues and detection by 32P-postlabelling and immunoassay. *Carcinogenesis.* 2004 Jul;25(7):1201–9.

Haas CN. On modeling correlated random variables in risk assessment. *Risk Anal.* 1999 Dec;19(6):1205–14.

Haas CN, Rose JB, Gerba CP. *Quantitative Microbial Risk Assessment.* John Wiley & Sons. New York. 1999.

Hagmayer Y, Waldmann MR. How temporal assumptions influence causal judgments. *Mem Cognit.* 2002 Oct;30(7):1128–37. See also Waldmann MR, Hagmayer Y, Estimating causal strength: The role of structural knowledge and processing effort. *Cognition.* 2001 Nov; 82 (1): 27–58.

Haimes YY, Matalas NC, Lambert JH, Jackson BA, Fellows JFR. Reducing vulnerability of water supply systems to attack.*J Infrastruct Syst.* 1998;4:164–77.

Hammond JS, Keeney RL, Raiffa H. The hidden traps in decision making. *Clin Lab Manage Rev.* 1999 Jan–Feb;13(1):39–47.

Hanrahan J, Hoyen C, Rice LB. Geographic distribution of a large mobile element that transfers ampicillin and vancomycin resistance between *Enterococcus faecium* strains. *Antimicrob Agents Chemot.* 2000;44:1349–51.

Harrell FE. Hmisc Package for R. http://biostat.mc.vanderbilt.edu/twiki/bin/view/Main/Hmisc (Last accessed 3 Oct 2007)

Hart BA, Gong Q, Eneman JD, Durieux-Lu CC. *In vivo* expression of metallothionein in rat alveolar macrophages and type II epithelial cells following repeated cadmium aerosol exposures. *Toxicol Appl Pharmacol.* 1995 Jul;133(1):82–90.

Hart BA, Potts RJ, Watkin RD. Cadmium adaptation in the lung – A double-edged sword? *Toxicology.* 2001 Mar 7;160(1–3):65–70.

Hart BA, Voss GW, Willean CL. Pulmonary tolerance to cadmium following cadmium aerosol pretreatment. *Toxicol Appl Pharmacol.* 1989 Dec;101(3):447–60.

Hartemink AJ, Gifford DK, Jaakkola TS, Young RA. Using graphical models and genomic expression data to statistically validate models of genetic regulatory networks. *Pac Symp Biocomput.* 2001;422–33. http://psb.stanford.edu/psb-online/proceedings/psb01/abstracts/p422.html

Hausken K. Probabilistic risk analysis and game theory. *Risk Anal* 2002;22(1):17–27.

Hautamaki RD, Kobayashi DK, Senior RM, Shapiro SD. Requirement for macrophage elastase for cigarette smoke-induced emphysema in mice. *Science.* 1997 Sep 26;277(5334): 2002–4.

Hayes AW. The precautionary principle. *Arh Hig Rada Toksikol.* 2005 Jun;56(2):161–6.

Hayes JR, English LL, Carr LE, Wagner DD, Joseph SW. Multiple-antibiotic resistance of *Enterococcus* spp. isolated from commercial poultry production environments. *Appl Environ Microbiol.* 2004;70(10): 6005–11.

Hayes JR, English LL, Carter PJ, Proescholdt T, Lee KY, Wagner DD, White DG. Prevalence and antimicrobial resistance of *Enterococcus* species isolated from retail meats. *Appl Environ Microbiol.* 2003;69(12):7153–60.

Hayes JR, Jensen H. Lessons from the Danish Ban on Feed-Grade Antibiotics. 2003. http://www.choicesmagazine.org/current/2003-3-01_print.htm

Hays AM, Srinivasan D, Witten ML, Carter DE, Lantz RC. Arsenic and cigarette smoke synergistically increase DNA oxidation in the lung. *Toxicol Pathol.* 2006;34(4):396–404.

Hazelton WD, Clements MS, Moolgavkar SH. Multistage carcinogenesis and lung cancer mortality in three cohorts. *Cancer Epidemiol Biomar Prev.* 2005 May;14(5):1171–81.

Hazelton WD, Luebeck EG, Heidenreich WF, Moolgavkar SH. Analysis of a historical cohort of Chinese tin miners with arsenic, radon, cigarette smoke, and pipe smoke exposures using the biologically based two-stage clonal expansion model. *Radiat Res.* 2001 Jul;156(1):78–94.

Hazen, G. Multiattribute Structure for QALYS. 2003. http://fisher.osu.edu/~butler_267/DAPapers/WP030018.pdf

Hazen G, Sounderpandian J. Lottery acquisition versus information acquisition: Price and preference reversals. *J Risk Uncertainty.* 1999;18(2):125–36.

Heal G, Kriström B. Uncertainty and climate change. *Environ Resour Econ.* 2002;22:3–39.

Health and Safety Policy Advisory Group. Quantified risk assessment techniques – Part 3: Fault tree analysis – FTA. *Health and Safety Briefing 26c*, Institution of Engineering and Technology. 2004. http://www.theiet.org/publicaffairs/health/hsb26c.pdf

Healy S. Risk as social process: The end of "the age of appealing to the facts"? *J Hazard Mater.* 2001 Sep 14;86(1–3):39–53.

Hecht SS. Tobacco carcinogens, their biomarkers and tobacco-induced cancer. *Nat Rev Cancer.* 2003 Oct;3(10):733–44. Review. Erratum in *Nat Rev Cancer*. 2004 Jan; 4 (1): 84.

Heinrich U. Pulmonary carcinogenicity of cadmium by inhalation in animals. *IARC Sci Publ.* 1992;(118): 405–13.

Heinz G, Ploner M. A SAS macro, S-PLUS library and R package to perform logistic regression without convergence problems. Technical Report 2/2004, Medical University of Vienna, Department of Medical Computer Sciences. http://www.meduniwien.ac.at/user/georg.heinze/programs/logistf (Last accessed 3 Oct 2007).

Helms M, Vastrup P, Gerner-Smidt P, Molbak K.. Short and long term mortality associated with foodborne bacterial gastrointestinal infections: Registry based study. *BMJ.* 2003 Feb 15;326(7385):357.

Henrion M. Propagation of uncertainty by probabilistic logic sampling in Bayes' networks. In Lemmer JF, Kanal LN (Eds.), *Uncertaintyin Artificial Intelligence 2.* North Holland. Amsterdam. 1988.

Henrion M, Pradhan M, Del Favero B, Huang K, Provan G, O'Rorke P. Why Is Diagnosis Using Belief Networks Insensitive to Imprecision in Probabilities? 1996. http://citeseer.ist.psu.edu/henrion96why.html

Hernandez-Boussard TM, Hainaut P. A specific spectrum of p53 mutations in lung cancer from smokers: Review of mutations compiled in the IARC p53 database. *Environ Health Perspect.* 1998 Jul;106(7):385–91. Erratum in *Environ Health Perspect.* 1998 Sep; 106 (9):A421.

Hertwig R, Barron G, Weber EU, Erev I. Decisions from experience and the effect of rare events in risky choice. *Psychol Sci.* 2004 Aug;15(8):534–9.

Hertz-Picciotto I, Hu SW. Contribution of cadmium in cigarettes to lung cancer: An evaluation of risk assessment methodologies. *Arch Environ Health.* 1994 Jul–Aug;49(4):297–302.

Hertz-Picciotto I, Smith AH. Observations on the dose-response curve for arsenic exposure and lung cancer. *Scand J Work Environ Health.* 1993 Aug;19(4):217–26.

Heyes A. Implementing environmental regulation: Enforcement and compliance. *J Regul Econ.* 2000;17:107–29.

Hill AF, Collinge J. Subclinical prion infection in humans and animals. *Br Med Bull.* 2003;66: 161–70.

Hiraki A, Matsuo K, Hamajima N, Ito H, Hatooka S, Suyama M, Mitsudomi T, Tajima K. Different risk relations with smoking for non-small-cell lung cancer: comparison of TP53 and TP73 genotypes. *Asian Pac J Cancer Prev.* 2003 Apr–Jun;4(2):107–12.

Hirao T, Nelson HH, Ashok TD, Wain JC, Mark EJ, Christiani DC et al. Tobacco smoke-induced DNA damage and an early age of smoking initiation induce chromosome loss at 3p21 in lung cancer. *Cancer Res.* 2001;61:612–5.

Hiratsuka H, Satoh S, Satoh M, Nishijima M, Katsuki Y, Suzuki J, Nakagawa J, Sumiyoshi M, Shibutani M, Mitsumori K, Tanaka-Kagawa T, Ando M. Tissue distribution of cadmium in rats given minimum amounts of cadmium-polluted rice or cadmium chloride for 8 months. *Toxicol Appl Pharmacol.* 1999 Oct 15;160(2):183–91.

Hiroshima K, Iyoda A, Shibuya K, Hoshino H, Haga Y, Toyozaki T et al. Evidence of neoangiogenesis and an increase in the number of proliferating cells within the bronchial epithelium of smokers. *Cancer.* 2002;95:1539–45.

Hoeting J, Madigan D, Raftery A, Volinsky C. Bayesian model averaging. *Stat Sci.* 1999;14: 382–401.

Hoffman FO, Hammonds JS. Propagation of uncertainty in risk assessments: The need to distinguish between uncertainty due to lack of knowledge and uncertainty due to variability. *Risk Anal.* 1994 Oct;14(5):707–12.

Hoffmann K, Heidemann C, Weikert C, Schulze MB, Boeing H. Estimating the proportion of disease due to classes of sufficient causes. *Am J Epidemiol.* 2006 Jan 1;163(1):76–83.

Hoffmann D, Hoffmann I. The changing cigarette, 1950– 1995. *J Toxicol Environ Health.* 1997 Mar;50(4):307–64.

Hogg JC. Why does airway inflammation persist after the smoking stops? *Thorax.* 2006 Feb;61(2):96–7.

Holcomb DL, Smith MA, Ware GO, Hung YC, Brackett RE, Doyle MP. Comparison of six dose-response models for use with food-borne pathogens. *Risk Anal.* 1999 Dec;19(6):1091–100.

Holland RH, Wilson RH, Acevedo AR, Mccall MS, Clark DA, Lanz HC. The cigarette smoke-arsenic-cancer of the lung problem. *Acta Unio Int Contra Cancrum.* 1959;15:608–11.

Holme P, Kim BJ, Yoon CN, Han SK. Attack vulnerability of complex networks. *Phys Rev E Stat Nonlin Soft Matter Phys.* 2002 May;65(5 Pt 2):056109. http://arxiv.org/ftp/cond-mat/papers/0202/0202410.pdf

Hoogervorst EM, de Vries A, Beems RB, van Oostrom CT, Wester PW, Vos JG, Bruins W, Roodbergen M, Cassee FR, Vijg J, van Schooten FJ, van Steeg H. Combined oral benzo[a]pyrene and

inhalatory ozone exposure have no effect on lung tumor development in DNA repair-deficient Xpa mice. *Carcinogenesis.* 2003 Mar;24(3):613–9.
Hope BK. Assessment of risk to terrestrial receptors using uncertain analysis – A case study. *Hum Ecol Risk Assess.* 1999;5:145–70.
Horton HJ, Kleinman KP. Much ado about nothing: A comparison of missing data methods and software to fit incomplete data regression models. *Am Stat.* 2007;61(1):79–90.
Hou SM, Falt S, Nyberg F. Glutathione S-transferase T1-null genotype interacts synergistically with heavy smoking on lung cancer risk. *Environ Mol Mutagen.* 2001;38(1):83–6.
Houghton AM, Quintero PA, Perkins DL, Kobayashi DK, Kelley DG, Marconcini LA, Mecham RP, Senior RM, Shapiro SD. Elastin fragments drive disease progression in a murine model of emphysema. *J Clin Invest.* 2006 Mar;116(3):753–9.
Huettel SA, Stowe CJ, Gordon EM, Warner BT, Platt ML. Neural signatures of economic preferences for risk and ambiguity. *Neuron.* 2006 Mar 2;49(5):765–75.
Hung RJ, Boffetta P, Brockmoller J, Butkiewicz D, Cascorbi I, Clapper ML, Garte S, Haugen A, Hirvonen A, Anttila S, Kalina I, Le Marchand L, London SJ, Rannug A, Romkes M, Salagovic J, Schoket B, Gaspari L, Taioli E. CYP1A1 and GSTM1 genetic polymorphisms and lung cancer risk in Caucasian non-smokers: A pooled analysis. *Carcinogenesis.* 2003 May;24(5): 875–82.
IARC. Monographs on the evaluation of carcinogenic risks to humans. Beryllium, cadmium, mercury and exposures in the glass manufacturing industry. International Agency for Research on Cancer. *Lyon.* 1993;58:119–238.
Ibrahim JG, Chen M-H, Lipsitz SR, Herring AH. Missing-data methods for generalized linear models: A comparative review. *J Am Stat Assoc.* 2005 Mar;100(469):332–46.
Iizuka T, Ishii Y, Itoh K, Kiwamoto T, Kimura T, Matsuno Y, Morishima Y, Hegab AE, Homma S, Nomura A, Sakamoto T, Shimura M, Yoshida A, Yamamoto M, Sekizawa K. Nrf2-deficient mice are highly susceptible to cigarette smoke-induced emphysema. *Genes Cells.* 2005 Dec;10(12):1113–25.
Indusi, JP. 2003. Terrorist Protection Planning Using a Relative Risk Reduction Approach. Brookhaven National Laboratory. Upton, NY. BNL-71383–2003-CP. http://www.pubs.bnl.gov/documents/25368.pdf
Infanger G. Dynamic asset allocation strategies using a stochastic dynamic programming approach. Chapter 5 in Zenios SA, Ziemba WT (Eds.), *Handbook of Assets and Liability Management, Volume 1.* North Holland. New York. 2006.
Institute of Medicine (IOM). *Human Health Risks with the Subtherapeutic Use of Penicillin or Tetracyclines in Animal Feed.* Report by the Committee of Human Health Risk Assessment of Using Subtherapeutic Antibiotics in Animal Feeds, Institute of Medicine, IOM-88–89. National Academy Press. Washington, DC. 1989.
Institute of Medicine (IOM) of the National Academies. *Clearing the Smoke.* 2001. http://www.iom.edu/report.asp?id=5424
International Agency for Research on Cancer. IARC Monographs on the Evaluation of Carcinogenic Risks to Humans, Volume 83. Tobacco Smoke and Involuntary Smoking. Lyon, France: IARC 2004.
Irarrazabal CE, Rojas C, Aracena R, Marquez C, Gil L. Chilean pilot study on the risk of lung cancer associated with codon 72 polymorphism in the gene of protein p53. *Toxicol Lett.* 2003 Sep 15;144(1):69–76.
ISO. 1995. International Standard: ISO 10312 International Standard Ambient Air - Determination of Asbestos Fibers - Direct Transfer. Transmission Electron Microscopy Method, 1995-0501.
Issel LM, Narasimha KM. Creating complex health improvement programs as mindful organizations: From theory to action. *J Health Organ Manag.* 2007;21(2–3):166–83.
Izzotti A, Cartiglia C, Longobardi M, Bagnasco M, Merello A, You M, Lubet RA, De Flora S. Gene expression in the lung of p53 mutant mice exposed to cigarette smoke. *Cancer Res.* 2004 Dec 1;64(23):8566–72.
Janson S. Coupling and Poisson Approximation. 1994. http://citeseer.ist.psu.edu/janson94coupling.html

Jardine C, Hrudey S, Shortreed J, Craig L, Krewski D, Furgal C, McColl S. Risk management frameworks for human health and environmental risks. *J Toxicol Environ Health B Crit Rev.* 2003 Nov–Dec;6(6):569–720.

Jarmalaite S, Kannio A, Anttila S, Lazutka JR, Husgafvel-Pursiainen K. Aberrant p16 promoter methylation in smokers and former smokers with nonsmall cell lung cancer. *Int J Cancer* 2003 Oct 10;106(6):913–918.

Jarup L, Bellander T, Hogstedt C, Spang G. Mortality and cancer incidence in Swedish battery workers exposed to cadmium and nickel. *Occup Environ Med.* 1998 Nov;55(11): 755–9.

Järup L, Pershagen G, Wall S. Arsenic exposure, smoking, and lung cancer in smelter workers – a case-control study. *Am J Epidemiol.* 1991 Sep 15;134(6):545–551. (Erratum in: *Am J Epidemiol* 1992 Nov 1;136(9):1174).

Jaynes ET. *Probability Theory: The Logic of Science.* Cambridge University Press. New York. 2003.

Jenkins-Smith H, Kunreuther H. Mitigation and benefits measures as policy tools for siting potentially hazardous facilities: Determinants of effectiveness and appropriateness. *Risk Anal.* 2001 Apr;21(2):371–82.

Jia J, Dyer JS, Butler JC. Measures of perceived risk. *Manage Sci.* 1999 Apr; 45(4):519–32. http://fisher.osu.edu/~butler_267/DAPapers/WP970005.pdf

Jin HJ, Skriptchenko A, Koo WW. The Effects of the BSE Outbreak in the United States on the Beef and Cattle Industry. Center for Agricutural Policy and Trade Studies Special Report 03–4 (13 pp.). Jan 2004.

Joe H. *Multivariate Models and Dependence Concepts.* Chapman and Hall. London. 1997.

Johnson BB. Gender and race in beliefs about outdoor air pollution. *Risk Anal.* 2002 Aug;22(4):725–38.

Jonas E, Schulz-Hardt S, Frey D et al. Confirmation bias in sequential information search after preliminary decisions: An expansion of dissonance theoretical research on selective exposure to information. *J Pers Soc Psychol.* 2001;80(4): 557–71.

Jonas E, Schulz-Hardt S, Frey D. Giving advice or making decisions in someone else's place: The influence of impression, defense, and accuracy motivation on the search for new information. *Pers Soc Psychol Bull.* 2005 Jul;31(7):977–90.

Jonas E, Schulz-Hardt S, Frey D, Thelen N. Confirmation bias in sequential information search after preliminary decisions: An expansion of dissonance theoretical research on selective exposure to information. *J Pers Soc Psychol.* 2001 Apr;80(4):557–71.

Jones SR, Atkin P, Holroyd C, Lutman E, Batlle JV, Wakeford R, Walker P. Lung cancer mortality at a UK tin smelter. *Occup Med* (Lond). 2007 Jun;57(4):238–45.

Jones RN, Deshpande LM. Are *Enterococcus faecalis* strains with *vat(E)* in poultry a reservoir for human streptogramin resistance? *vat(E)* occurrence in human enterococcal bloodstream infections in North America (SENTRY Antimicrobial Surveillance Program, 2002). *Antimicrob Agents Chemother.* 2004;48(1):360–1.

Jones ME, Draghi DC, Thornsberry C, Karlowsky JA, Sahm DF, Wenzel RP. Emerging resistance among bacterial pathogens in the intensive care unit – A European and North American surveillance study (2000–2002). *Ann Clin Microbiol Antimicrob.* 2004;3(14). http://www.pubmedcentral.nih.gov/articlerender.fcgi?artid=509280

Jones P, Edmonds Y. Risk-based strategies for allocating resources in a constrained environment. *J Homeland Security.* 2008 Mar. www.homelandsecurity.org/newjournal/Articles/displayArticle2.asp?article=171

Jones PE, Roelofsma PH. The potential for social contextual and group biases in team decision-making: Biases, conditions and psychological mechanisms. *Ergonomics.* 2000 Aug;43(8):1129–52.

Joseph P, Muchnok TK, Klishis ML, Roberts JR, Antonini JM, Whong WZ, Ong T. Cadmium-induced cell transformation and tumorigenesis are associated with transcriptional activation of c-fos, c-jun, and c-myc proto-oncogenes: Role of cellular calcium and reactive oxygen species. *Toxicol Sci.* 2001 Jun;61(2):295–303.

Jouini M, Clemen RT. Copula models for aggregating expert opinions. *Oper Res.* 1996;44: 444–57.

Jourenkova-Mironova N, Wikman H, Bouchardy C, Voho A, Dayer P, Benhamou S, Hirvonen A. Role of glutathione S-transferase GSTM1, GSTM3, GSTP1 and GSTT1 genotypes in modulating susceptibility to smoking-related lung cancer. *Pharmacogenetics.* 1998 Dec;8(6): 495–502.

Kable JW, Glimcher PW. The neural correlates of subjective value during intertemporal choice. *Nat Neurosci.* 2007 Dec;10(12):1625–33.

Kang SH, Kodell RL, Chen JJ. Incorporating model uncertainties along with data uncertainties in microbial risk assessment. *Regul Toxicol Pharmacol.* 2000 Aug;32(1):68–72.

Kaplan S, Garrick BJ. On the quantitative definition of risk. *Risk Anal.* 1981;1(1):11–27.

Kassenborg HD, Smith KE, Vugia DJ, Rabatsky-Ehr T, Bates MR, Carter MA, Dumas NB, Cassidy MP, Marano N, Tauxe RV, Angulo FJ; Emerging Infections Program FoodNet Working Group. Fluoroquinolone-resistant *Campylobacter* infections: Eating poultry outside of the home and foreign travel are risk factors. *Clin Infect Dis.* 2004 Apr 15;38(Suppl 3):S279–84.

Keeney RL, Raiffa H. *Decisions with Multiple Objectives: Preferences and Value Trade-Offs.* Wiley. New York. 1976.

Keiding N, Budtz-Jorgensen E. The Precautionary Principle and statistical approaches to uncertainty. *Int J Occup Med Environ Health.* 2004;17(1):147–51.

Keohane NO, Zeckhauser RJ. The ecology of terror defense. *J Risk Uncertainty.* 2003;26:201–29.

Kieke AL, Borchardt MA, Kieke BA et al. Use of streptogramin growth promoters in poultry and isolation of streptogramin-resistant *Enterococcus faecium* from humans. *J Infect Dis.* 2006 Nov 1;194(9): 1200–8.

Kim H, Kwon YM, Kim JS, Lee H, Park JH, Shim YM et al., Tumor-specific methylation in bronchial lavage for the early detection of non-small-cell lung cancer. *J Clin Oncol* 2004;22:2363–2370.

Kim DH, Nelson HH, Wiencke JK, Zheng S, Christiani DC, Wain JC, Mark EJ, Kelsey KT. p16(INK4a) and histology-specific methylation of CpG islands by exposure to tobacco smoke in non-small cell lung cancer. *Cancer Res.* 2001 Apr 15;61(8):3419–3424.

Klinke A, Renn O. A new approach to risk evaluation and management: Risk-based, precaution-based, and discourse-based strategies. *Risk Anal.* 2002 Dec;22(6):1071–94.

Knoch D, Pascual-Leone A, Meyer K, Treyer V, Fehr E. Diminishing reciprocal fairness by disrupting the right prefrontal cortex. *Science.* 2006 Nov 3;314(5800):829–32.

Knutson B, Wimmer GE, Kuhnen CM, Winkielman P. Nucleus accumbens activation mediates the influence of reward cues on financial risk taking. *Neuroreport.* 2008 Mar 26;19(5):509–13.

Kollmeier H, Seemann J, Wittig P, Rothe G, Muller KM. Cadmium in human lung tissue. *Int Arch Occup Environ Health.* 1990;62(5):373–377.

Kraan B, Cooke R. The effect of correlations in uncertainty analysis: Two cases. In Cooke R (Ed.), *Technical Committee Uncertainty Modeling: Report on the Benchmark Workshop Uncertainty/Sensitivity Analysis Codes.* European Safety and Reliability Association. Delft. 1997.

Kuhn I, Iversen A, Finn M, Greko C, Burman LG, Blanch AR, Vilanova X, Manero A, Taylor H, Caplin J, Dominguez L, Herrero IA, Moreno MA, Mollby R. Occurrence and relatedness of vancomycin-resistant enterococci in animals, humans, and the environment in different European regions. *Appl Environ Microbiol.* 2005;71(9):5383–90.

Kuilman T, Michaloglou C, Vredeveld LC, Douma S, van Doorn R, Desmet CJ, Aarden LA, Mooi WJ, Peeper DS. Oncogene-induced senescence relayed by an interleukin-dependent inflammatory network. *Cell.* 2008 Jun 13;133(6):1019–1031.

Kunreuther H, Heal G. Interdependent security. *J Risk Uncertainty.* 2003;26:231–49.

Kurowicka D, Cooke R. *Distribution-free continuous Bayesian belief nets.* Paper presented at the Fourth International Conference on Mathematical Methods in Reliability: Methodology and Practice, Santa Fe, NM. June 2004.

Lacke C. *Decision analytic modeling of colorectal cancer screening policies.* Unpublished doctoral dissertation, North Carolina State University. 1998.

Ladd K. El Dorado Hills Naturally Occurring Asbestos Multimedia Exposure Assessment El Dorado Hills, California. Preliminary Assessment and Site Inspection Report Interim Final. 2005.

Lag M, Westly S, Lerstad T, Bjornsrud C, Refsnes M, Schwarze PE. Cadmium-induced apoptosis of primary epithelial lung cells: Involvement of Bax and p53, but not of oxidative stress. *Cell Biol Toxicol.* 2002;18(1):29–42.

Laguna M, Glover F. Bandwidth packing: A TABU search approach. *Manage Sci.* 1993;39: 492–500.

Lakdawalla D, Zanjani G. Insurance, self-protection, and the economics of terrorism. *J Public Econ* 2005;89:1891–905.

Lam SS, Schaubroeck J. Improving group decisions by better pooling information: A comparative advantage of group decision support systems. *J Appl Psychol.* 2000 Aug;85(4):565–73.

Lamm SH, Engel A, Penn CA, Chen R, Feinleib M. Arsenic cancer risk confounder in southwest Taiwan data set. *Environ Health Perspect.* 2006 Jul;114(7):1077–82.

Lan Q, He X, Costa DJ, Tian L, Rothman N, Hu G, Mumford JL. Indoor coal combustion emissions, GSTM1 and GSTT1 genotypes, and lung cancer risk: A case-control study in Xuan Wei, China. *Cancer Epidemiol Biomarkers Prev.* 2000 Jun;9(6):605–8.

Lan Q, Mumford JL, Shen M, DeMarini DM, Bonner MR, He X, Yeager M, Welch R, Chanock S, Tian L, Chapman RS, Zheng T, Keohavong P, Caporaso N, Rothman N. Oxidative damage-related genes AKR1C3 and OGG1 modulate risks for lung cancer due to exposure to PAH-rich coal combustion emissions. *Carcinogenesis.* 2004 Nov;25(11):2177–81.

Lange K. *Applied Probability.* Springer. New York. 2003.

Lauritzen SL, Spiegelhalter DJ. Local computations with probabilities on graphical structures and their application to expert systems. *J Royal Stat Soc B.* 1998;50:157–224.

Le Calvez F, Mukeria A, Hunt JD, Kelm O, Hung RJ, Taniere P et al. TP53 and KRAS mutation load and types in lung cancers in relation to tobacco smoke: Distinct patterns in never, former, and current smokers. *Cancer Res.* 2005;65:5076–83.

Le Marchand L, Donlon T, Lum-Jones A, Seifried A, Wilkens LR. Association of the hOGG1 Ser326Cys polymorphism with lung cancer risk. *Cancer Epidemiol Biomarkers Prev.* 2002 Apr;11(4):409–12.

Leavis HL, Bonten MJ, Willems, RJ. Identification of high-risk enterococcal clonal complexes; global dispersion and antibiotic resistance. *Curr Opin Microbiol.* 2006;9(5):454–60.

Leavis HL, Willems RJ, Top J, Spalburg E, Mascini EM, Fluit AC, Hoepelman A, de Neeling AJ, Bonten MJ. Epidemic and nonepidemic multidrug-resistant *Enterococcus faecium. Emerg Infect Dis* 2003;9(9):1108–15.

Leavis HL, Willems RJ, van Wamel WJ, Schuren FH, Caspers MP, Bonten MJ. Insertion sequence-driven diversification creates a globally dispersed emerging multiresistant subspecies of *E. faecium. PLoS Pathogens* 2007; 3(1) e7. doi:10.1371/journal.ppat.0030007

Lee PM. *Bayesian Statistics: An Introduction,* 3rd ed. Arnold. London. 2004.

Lee D, Seo H. Mechanisms of reinforcement learning and decision making in the primate dorsolateral prefrontal cortex. *Ann NY Acad Sci.* 2007 May;1104:108–22.

Lee CE, Zembower TR, Fotis MA, Postelnick MJ, Greenberger PA, Peterson LR, Noskin GA. The incidence of antimicrobial allergies in hospitalized patients: Implications regarding prescribing patterns and emerging bacterial resistance. *Arch Intern Med.* 2000;160(18):2819–22.

Lee-Feldstein A. Cumulative exposure to arsenic and its relationship to respiratory cancer among copper smelter employees. *J Occup Med.* 1986 Apr;28(4):296–302.

Leighton JK. Application of emerging technologies in toxicology and safety assessment: Regulatory perspectives. *Int J Toxicol.* 2005 May–Jun;24(3):153–5.

Lemon SC, Roy J, Clark MA, Friedmann PD, Rakowski W. Classification and regression tree analysis in public health: Methodological review and comparison with logistic regression. *Ann Behav Med.* 2003 Dec;26(3):172–81.

Leveson NG. A systems-theoretic approach to safety in software-intensive systems. *IEEE Trans. Dependable Secur Comput.* 2004 Jan;1(1):66–86.

Levitin G. Optimal multilevel protection in series-parallel systems. *Reliab Eng Syst Safe.* 2003;81:93–102.

Levitin G, Dai Y, Xie M., Poh KL. Optimizing survivability of multi-state systems with multi-level protection by multi-processor genetic algorithm. *Reliab Eng Syst Safe.* 2003;82:93–104.

Levitin G, Lisnianski A. Optimal separation of elements in vulnerable multi-state systems. *Reliab Eng Syst Safe.* 2001;73:55–66.

Lewis SJ, Cherry NM, Niven RM, Barber PV, Povey AC. GSTM1, GSTT1 and GSTP1 polymorphisms and lung cancer risk. *Cancer Lett.* 2002 Jun 28;180(2):165–71.

Lewis PD, Parry JM. In silico p53 mutation hotspots in lung cancer. *Carcinogenesis.* 2004 Jul;25(7):1099–107.

Li D, Firozi PF, Chang P, Wang LE, Xiong P, Sturgis EM, Eicher SA, Spitz MR, Hong WK, Wei Q. *In vitro* BPDE-induced DNA adducts in peripheral lymphocytes as a risk factor for squamous cell carcinoma of the head and neck. *Int J Cancer.* 2001a Aug 1;93(3):436–40.

Li D, Firozi PF, Wang LE, Bosken CH, Spitz MR, Hong WK, Wei Q. Sensitivity to DNA damage induced by benzo(a)pyrene diol epoxide and risk of lung cancer: A case-control analysis. *Cancer Res.* 2001 Feb 15; 61(4):1445–50. Erratum in *Cancer Res* 2001b Apr 15;61(8):3531.

Li Y, Zhang Z, Xu Y, Xiong S, Ni W, Chen S. TNF-alpha up-regulates matrix metalloproteinase-9 expression and activity in alveolar macrophages from patients with chronic obstructive pulmonary disease. *J Huazhong Univ Sci Technolog Med Sci.* 2006;26(6):647–50.

Liao WT, Lin P, Cheng TS, Yu HS, Chang LW. Arsenic promotes centrosome abnormalities and cell colony formation in p53 compromised human lung cells. *Toxicol Appl Pharmacol.* 2007 Dec 1; 225(2):162–70.

Lin S-W. *Designing incentive systems for risk-informed regulation.* Unpublished doctoral dissertation, University of Wisconsin-Madison. 2004.

Lin P, Hsueh YM, Ko JL, Liang YF, Tsai KJ, Chen CY. Analysis of NQO1, GSTP1, and MnSOD genetic polymorphisms on lung cancer risk in Taiwan. *Lung Cancer.* 2003 May;40(2):123–9.

Little MP. Are two mutations sufficient to cause cancer? Some generalizations of the two mutation model of carcinogenesis of Moolgavkar, Venzon, and Knudson, and of the multistage model of Armitage and Doll. *Biometrics.* 1995 Dec;51(4):1278–91.

Little MP, Haylock RG, Muirhead CR. Modelling lung tumour risk in radon-exposed uranium miners using generalizations of the two-mutation model of Moolgavkar, Venzon and Knudson. *Int J Radiat Biol.* 2002 Jan;78(1):49–68.

Little RJA, Rubin DB. *Statistical Analysis with Missing Data.* John Wiley & Sons. New York. 1987.

Liu Y, Gao W, Siegfried JM, Weissfeld JL, Luketich JD, Keohavong P. Promoter methylation of RASSF1A and DAPK and mutations of K-ras, p53, and EGFR in lung tumors from smokers and never-smokers. *BMC Cancer.* 2007 May 3;7:74.

Liu Y, Lan Q, Siegfried JM, Luketich JD, Keohavong P. Aberrant promoter methylation of p16 and MGMT genes in lung tumors from smoking and never-smoking lung cancer patients. *Neoplasia* 2006;8:46–51.

Livernois J, McKenna CJ. Truth or consequences: Enforcing pollution standards with self-reporting. *J Public Econ.* 1999;71:415–40.

Lloyd GR, Brereton RG, Faria R, Duncan JC. Learning vector quantization for multiclass classification: Application to characterization of plastics. *J Chem Inf Model.* 2007 Jul–Aug;47(4):1553–63.

Lodovici M, Akpan V, Giovannini L, Migliani F, Dolara P. Benzo[a]pyrene diol-epoxide DNA adducts and levels of polycyclic aromatic hydrocarbons in autoptic samples from human lungs. *Chem Biol Interact.* 1998 Nov 27;116(3):199–212.

Loewenstein G, Rick S, Cohen JD. Neuroeconomics. *Ann Rev Psychol* 2008;59:647–72.

Löfdahl JM, Wahlström J, Sköld CM. Different inflammatory cell pattern and macrophage phenotype in chronic obstructive pulmonary disease patients, smokers and non-smokers. *Clin Exp Immunol.* 2006 Sep;145(3):428–37.

Loh W-Y, 2002. Regression trees with unbiased variable selection and interaction. *Statistica Sinica* 2002;12:361–86. http://www3.stat.sinica.edu.tw/statistica/j12n2/j12n21/j12n21.htm

Loriot MA, Rebuissou S, Oscarson M, Cenee S, Miyamoto M, Ariyoshi N, Kamataki T, Hemon D, Beaune P, Stucker I. Genetic polymorphisms of cytochrome P450 2A6 in a case-control study on lung cancer in a French population. *Pharmacogenetics.* 2001 Feb;11(1):39–44.

Luebeck EG, Heidenreich WF, Hazelton WD, Paretzke HG, Moolgavkar SH. Biologically based analysis of the data for the Colorado uranium miners cohort: Age, dose and dose-rate effects. *Radiat Res.* 1999 Oct;152(4):339–51.

Lundström NG, Englyst V, Gerhardsson L, Jin T, Nordberg G. Lung cancer development in primary smelter workers: a nested case-referent study. *J Occup Environ Med.* 2006 Apr;48(4):376–380.

MacIntyre CR, Seccull A, Lane JM, Plant A. Development of a risk-priority score for category A bioterrorism agents as an aid for public health policy. *Mil Med.* 2006 Jul;171(7):589–94.

Maclure M. Multivariate refutation of aetiological hypotheses in non-experimental epidemiology. *Int J Epidemiol.* 1990 Dec;19(4):782–7.

Maclure M.Taxonomic axes of epidemiologic study designs: A refutationist perspective. *J Clin Epidemiol.* 1991;44(10):1045–53.

Macovei L, Zurek L. Ecology of antibiotic resistance genes: Characterization of enterococci from houseflies collected in food settings. *Appl Environ Microbiol.* 2006;72(6):4028–35.

Maeno T, Houghton AM, Quintero PA, Grumelli S, Owen CA, Shapiro SD. CD8+ T cells are required for inflammation and destruction in cigarette smoke-induced emphysema in mice. *J Immunol.* 2007 Jun 15; 178(12):8090–6.

Maier G, Pattavina A, De Patre S, Martinelli M. Optical network survivability: Protection techniques in the WDM layer. *Photonic Netw Commun.* 2002 Jul;4(3):251–69. http://citeseer.ist.psu.edu/maier02optical.html

Mancini NM, Bene MC, Gerard H, Chabot F, Faure G, Polu JM, Lesur O. Early effects of short-time cigarette smoking on the human lung: A study of bronchoalveolar lavage fluids. *Lung.* 1993;171(5):277–91.

Mark DH. Interpreting the term selection bias in medical research. *Fam Med.* 1997 Feb;29(2): 132–6.

Markowitz HM. *Portfolio Selection: Efficient Diversification of Investments.* John Wiley & Sons. New York. 1959.

Martins J, Mendes RV. Neural networks and logical reasoning systems: A translation table. *Int J Neural Syst.* 2001 Apr;11(2):179–86.

Martuzzi M. Science, policy, and the protection of human health: A European perspective. *Bioelectromagnetics.* 2005;Suppl 7:S151–6.

Mass MJ, Wang L. Arsenic alters cytosine methylation patterns of the promoter of the tumor suppressor gene p53 in human lung cells: A model for a mechanism of carcinogenesis. *Mutat Res.* 1997 Jun;386(3):263–77.

Mather FJ, White LE, Langlois EC, Shorter CF, Swalm CM, Shaffer JG, Hartley WR. Statistical methods for linking health, exposure, and hazards. *Environ Health Perspect.* 2004 Oct;112(14):1440–5. http://ehp.niehs.nih.gov/members/2004/7145/7145.html

McGowan, LL, Jackson CR, Barrett JB, Hiott LM, Fedorka-Cray. 2006. Prevalence and antimicrobial resistance of enterococci isolated from retail fruits, vegetables, and meats. *J Food Prot* 69(12):2976–2982.

McKenna IM, Gordon T, Chen LC, Anver MR, Waalkes MP. Expression of metallothionein protein in the lungs of Wistar rats and C57 and DBA mice exposed to cadmium oxide fumes. *Toxicol Appl Pharmacol.* 1998 Dec;153(2):169–78.

Mechanic LE, Marrogi AJ, Welsh JA, Bowman ED, Khan MA, Enewold L et al. Polymorphisms in XPD and TP53 and mutation in human lung cancer. *Carcinogenesis.* 2005;26:597–604.

Meehl PE. *Clinical vs. Statistical Prediction: A Theoretical Analysis and a Review of the Evidence.* University of Minnesota Press. Minneapolis. 1954.

Meehl PE. Causes and effects of my disturbing little book. *J Pers Assess.* 1986;50:370–5.

Meehl PE. When shall we use our heads instead of the formula? *J Couns Psychol.* 1993;4:81–9.

Meeuwissen AMH., Bedford T, Minimally informative distributions with given rank correlation for use in uncertainty analysis. *J Stat Comput Simulat.* 1997;57:143–74.

Merrick JRW, van Dorp JR, Singh A. Analysis of correlated expert judgments from pairwise comparisons. *Decision Anal.* 2005;2:17–29.

Meuwissen R, Berns A. Mouse models for human lung cancer. *Genes Develop.* 2005;19:643–64.

Miao XS, Bishay F, Chen M, Metcalfe CD. Occurrence of antimicrobials in the final effluents of wastewater treatment plants in Canada. *Environ Sci Technol.* 2004 Jul 1;38(13):3533–41.

Michor F, Nowak MA, Frank SA, Iwasa Y. Stochastic elimination of cancer cells. *Proc Biol Sci.* 2003 Oct 7;270(1528):2017–24.

Minna JD, Fong K, Zochbauer-Muller S, Gazdar AF. Molecular pathogenesis of lung cancer and potential translational applications. *Cancer J.* 2002 May–Jun;8(Suppl 1):S41–6.

Mitchell C, Decker C. Applying risk-based decision-making methods and tools to U.S. Navy antiterrorism capabilities. *J Homeland Security*. 2004 Feb. www.homelandsecurity.org/journal/Articles/Mitchell_Decker.html

Miyamoto JM. Quality-adjusted life years (QALY) utility models under expected utility and rank dependent utility assumptions. *J Math Psychol.* 1999 Jun;43(2):201–37.

Molet S, Belleguic C, Lena H, Germain N, Bertrand CP, Shapiro SD, Planquois JM, Delaval P, Lagente V. Increase in macrophage elastase (MMP-12) in lungs from patients with chronic obstructive pulmonary disease. *Inflamm Res.* 2005 Jan;54(1):31–6. http://www.ncbi.nlm.nih.gov/pubmed/15723202?ordinalpos=1&itool=EntrezSystem2.PEntrez.Pubmed.Pubmed_ResultsPanel.Pubmed_RVDocSum

Montaño M, Beccerril C, Ruiz V, Ramos C, Sansores RH, González-Avila G. Matrix metalloproteinases activity in COPD associated with wood smoke. *Chest.* 2004 Feb;125(2):466–72.

Moolgavkar SH. The multistage theory of carcinogenesis and the age distribution of cancer in man. *J Natl Cancer Inst.* 1978 Jul;61(1):49–52.

Moolgavkar SH, Luebeck EG, Krewski D, Zielinski JM. Radon, cigarette smoke, and lung cancer: A re-analysis of the Colorado Plateau uranium miners' data. *Epidemiology.* 1993 May;4(3):204–17.

Morgan JM, Burch HB, Watkins JB. Tissue cadmium and zinc content in emphysema and bronchogenic carcinoma. *J Chronic Dis.* 1971 Jul;24(2):107–10.

Mori H, Nose T, Ishitani K, Kasagi S, Souma S, Akiyoshi T, Kodama Y, Mori T, Kondo M, Sasaki S, Iwase A, Takahashi K, Fukuchi Y, Seyama K. Phosphodiesterase 4 inhibitor GPD-1116 markedly attenuates the development of cigarette smoke-induced emphysema in senescence-accelerated mice P1 strain. *Am J Physiol Lung Cell Mol Physiol.* 2008 Feb;294(2): L196–204.

Morla M, Busquets X, Pons J, Sauleda J, MacNee W, Agusti AG. Telomere shortening in smokers with and without COPD. *Eur Respir J.* 2006;27:525–8.

Morris MW, Moore PC, Sim DL. Choosing remedies after accidents: Counterfactual thoughts and the focus on fixing "human error." *Psychon Bull Rev.* 1999 Dec;6(4):579–85.

Motwani R, Raghavan P. *Randomized Algorithms.* Cambridge University Press. New York. 1995.

Motykiewicz G, Faraglia B, Wang LW, Terry MB, Senie RT, Santella RM. Removal of benzo(a)pyrene diol epoxide (BPDE)-DNA adducts as a measure of DNA repair capacity in lymphoblastoid cell lines from sisters discordant for breast cancer. *Environ Mol Mutagen.* 2002;40(2):93–100.

Mukherjee JJ, Gupta SK, Kumar S, Sikka HC. Effects of cadmium(II) on (+/–)-anti-benzo[a]pyrene-7,8-diol-9,10-epoxide-induced DNA damage response in human fibroblasts and DNA repair: A possible mechanism of cadmium's cogenotoxicity. *Chem Res Toxicol.* 2004 Mar;17(3):287–93.

Murphy KP. Learning Bayes Net Structure from Sparse Data Sets. 2001. http://citeseer.ist.psu.edu/murphy01learning.html

Murray BE. Vancomycin-resistant enterococcal infections. *New Eng J Med.* 2000;342(10):710–21.

Nardini L, Aparicio L, Bandoni JA., Tonelli SM. Regional risk associated with the transport of hazardous materials. *Latin Am Appl Res.* 2003;33:213–8.

National Academies of Science (NAS). *The Use of Drugs in Food Animals: Benefits and Risks.* National Academies Press. Washington, DC. 1999. http://www.nap.edu/books/0309054346/html/18.html

National Antimicrobial Resistance Monitory System (NARMS). NARMS Retail Meat Annual Reports. 2002–2005. http://www.fda.gov/cvm/narms_pg.html#Publications

National Research Council (NRC). Committee on Methodological Improvements to the Department of Homeland Security's Biological Agent Risk Analysis. *Department of Homeland Security Bioterrorism Risk Assessment: A Call for Change.* National Academies Press. Washington, DC. 2008. http://www.nap.edu/catalog.php?record_id=12206#toc

Navarro Silvera SA, Rohan TE. Trace elements and cancer risk: A review of the epidemiologic evidence. *Cancer Causes Control.* 2007 Feb;18(1):7–27.

Nawrot T, Plusquin M, Hogervorst J, Roels HA, Celis H, Thijs L, Vangronsveld J, Van Hecke E, Staessen JA. Environmental exposure to cadmium and risk of cancer: A prospective population-based study. *Lancet Oncol.* 2006 Feb;7(2):119–26.

Nelson R. *Probability, Stochastic Processes, and Queuing Theory: The Mathematics of Computer Performance Modeling.* Springer-Verlag. New York, NY. 1995.

Nelsen RB. *An Introduction to Copulas.* Springer-Verlag. New York. 1999.

Nelson JM, Smith KE, Vugia DJ, Rabatsky-Ehr T, Segler SD, Kassenborg HD, Zansky SM, Joyce K, Marano N, Hoekstra RM, Angulo FJ. Prolonged diarrhea due to ciprofloxacin-resistant *Campylobacter* infection. *J Infect Dis.* 2004 Sep 15;190(6):1150–7.

Ng KL, Hamby DM.. Fundamentals for establishing a risk communication program. *Health Phys.* 1997 Sep;73(3):473–82.

NIST. DRAFT CVSS v2.10 Equations (last revised 20 Mar 2007). http://nvd.nist.gov/cvsseq2.htm (Last accessed 27 Aug 2008)

Nowak MA, Michor F, Iwasa Y. Genetic instability and clonal expansion. *J Theor Biol.* 2006;241:26–32.

Oberauer K. Reasoning with conditionals: A test of formal models of four theories. *Cognit Psychol.* 2006;53(3):238–83.

Oberdorster G. Pulmonary deposition, clearance and effects of inhaled soluble and insoluble cadmium compounds. *IARC Sci Publ.* 1992;(118):189–204.

Oberdorster G, Cherian MG, Baggs RB. Correlation between cadmium-induced pulmonary carcinogenicity, metallothionein expression, and inflammatory processes: A species comparison. *Environ Health Perspect.* 1994 Sep;102(Suppl 3):257–63.

O'Brien BJ, Sculpher MJ. Building uncertainty into cost-effectiveness rankings: Portfolio risk-return tradeoffs and implications for decision rules. *Med Care.* 2000 May;38(5): 460–8.

O'Hanlon M, Orszag P, Daalder I, Destler M, Gunter D, Litan R, Steinberg J. *Protecting the American Homeland.* Brookings Institution. Washington, DC. 2002.

Ohmura Y, Yamagishi T. Why do people reject unintended inequity? Responders' rejection in a truncated ultimatum game. *Psychol Rep.* 2005 Apr;96(2):533–41.

Ohtani N, Yamakoshi K, Takahashi A, Hara E. The p16INK4a-RB pathway: molecular link between cellular senescence and tumor suppression. *J Med Invest.* 2004 Aug;51(3–4):146–153.

Ólafsson S, Kim J. Simulation optimization. In Yücesan E, Chen CH, Snowdon JL, Charnes JM (Eds.), *Proceedings of the 2002 Winter Simulation Conference* (pp. 79–84). 2002.

Olivier M, Hussain SP, Caron de Fromentel C, Hainaut P, Harris CC. TP53 mutation spectra and load: A tool for generating hypotheses on the etiology of cancer. *IARC Sci Publ.* 2004;(157):247–70.

Over DE, Hadjichristidis C, Evans JS et al. The probability of causal conditionals. *Cognit Psychol.* 2007;54(1):62–97.

Owens DK, Shachter RD, Nease RF Jr. Representation and analysis of medical decision problems with influence diagrams. *Med Decision Making.* 1997 Jul–Sep;17(3):241–262.

Paakko P, Kokkonen P, Anttila S, Kalliomaki PL. Cadmium and chromium as markers of smoking in human lung tissue. *Environ Res.* 1989 Aug;49(2):197–207.

Palmer KC, Snider GL, Hayes JA. Cellular proliferation induced in the lung by cadmium aerosol. *Am Rev Respir Dis.* 1975 Aug;112(2):173–9.

Pan H, Califano J, Ponte JF, Russo AL, Cheng KH, Thiagalingam A, Nemani P, Sidransky D, Thiagalingam S. Loss of heterozygosity patterns provide fingerprints for genetic heterogeneity in multistep cancer progression of tobacco smoke-induced non-small cell lung cancer. *Cancer Res.* 2005 Mar 1;65(5):1664–9.

Panov SZ. Molecular biology of the lung cancer. *Radiol Oncol.* 2005;39:197–210.

Park J, Chen L, Tockman MS, Elahi A, Lazarus P. The human 8-oxoguanine DNA N-glycosylase 1 (hOGG1) DNA repair enzyme and its association with lung cancer risk. *Pharmacogenetics.* 2004 Feb;14(2):103–9.

Park IW, Wistuba II, Maitra A, Milchgrub S, Virmani AK, Minna JD, Gazdar AF. Multiple clonal abnormalities in the bronchial epithelium of patients with lung cancer. *J Natl Cancer Inst.* 1999 Nov 3;91(21):1863–8.

Parker M, Ryan J. A column generation algorithm for bandwidth packing. *Telecommun Syst.* 1994;2:185–96.

Parsons DJ, Orton TG, D'Souza J, Moore A, Jones R, Dodd CE. A comparison of three modeling approaches for quantitative risk assessment using the case study of *Salmonella* spp. in poultry meat. *Int J Food Microbiol.* 2005 Jan 15;98(1):35–51.

Paté-Cornell E, Guikema S. Probabilistic modeling of terrorist threats: A systems analysis approach to setting priorities among countermeasures. *Mil Oper Res.* 2002;7:5–20.

Patrick ME, Christiansen LE, Waino M, Ethelberg S, Madsen H, Wegener HC. Effects of climate on incidence of *Campylobacter* spp. in humans and prevalence in broiler flocks in Denmark. *Appl Environ Microbiol.* 2004 Dec;70(12):7474–80.

Patton DE. The ABCs of risk assessment. *EPA J.* 1993;19(1):10–5. http://www.bethel.edu/~kisrob/hon301k/readings/risk/RiskEPA/riskepa1.html

Patton BS, Cast WR, Kocher ME, Matthews JO, Griffith RW, Hurd S, Dickson JS. The Effect of Swine Production System on Bacterial Prevalence and Antibiotic Resistance. Poster Presentation at IAFP (International Association for Food Protection) 2006, Calgary, Alberta, Canada, 13–16 Aug 2006. Abstract available at http://www.foodprotection.org/meetingsEducation/IAFP%202006/IAFP%2006%20Poster%20Abstracts.pdf

Paz-Elizur T, Krupsky M, Blumenstein S, Elinger D, Schechtman E, Livneh Z. DNA repair activity for oxidative damage and risk of lung cancer. *J Natl Cancer Inst.* 2003 Sep 3;95(17): 1312–9.

Pearl J. Causal Inference in the Health Sciences: A Conceptual Introduction. Health Services *Outcomes Res Methodol* 2:189–220, 2002. http://citeseer.ist.psu.edu/599949.html

Pearson CA, Prozialeck WC. E-Cadherin, beta-Catenin and cadmium carcinogenesis. *Med Hypotheses.* 2001 May;56(5):573–81.

Pedersen CS, Satchell SE. 1999. Choosing the Right Risk Measure: A Survey. http://citeseer.ist.psu.edu/520923.html

Pekoz E. A compound Poisson approximation inequality. *J Appl Probab* 2006;43:282–288.

Perrow C. *Normal Accidents: Living with High-Risk Technologies*. Basic Books. NY. 1984.

Pfanzagl J. A general theory of measurement – Applications to utility *Naval Research Logistics Quarterly*. 1959. 6(4):283–294.

Pfeifer GP, Denissenko MF, Olivier M, Tretyakova N, Hecht SS, Hainaut P. Tobacco smoke carcinogens, DNA damage and p53 mutations in smoking-associated cancers. *Oncogene.* 2002 Oct 21;21(48):7435–51.

Pfeifer GP, Hainaut P. On the origin of G $\rightarrow$ T transversions in lung cancer. *Mutat Res.* 2003 May 15; 526(1–2):39–43.

Phillips I, Casewell M, Cox T, De Groot B, Friis C, Jones R, Nightingale C, Preston R, Waddell J. Does the use of antibiotics in food animals pose a risk to human health? A critical review of published data. *J Antimicrob Chemother.* 2004 Jan;53(1):28–52. See Table 2, http://jac.oupjournals.org/cgi/content/full/53/1/28/DKG483TB2

Phillips CV, Goodman KJ. The missed lessons of Sir Austin Bradford Hill. *Epidemiol Perspect Innov.* 2004 Oct 04;1(1):3.

Phimister JR, Bier VM, Kunreuther HC. (Eds.) *Accident Precursor Analysis and Management: Reducing Technological Risk Through Diligence*. National Academies Press. Washington, DC. 2004.

Plous S. *The Psychology of Judgment and Decision Making.* McGraw Hill. New York. 1993.

Postmes T, Spears R, Cihangir S. Quality of decision making and group norms. *J Pers Soc Psychol.* 2001 Jun;80(6):918–30.

Pott F, Heinrich U. Relative significance of different hydrocarbons for the carcinogenic potency of emissions from various incomplete combustion processes. *IARC Sci Publ.* 1990;(104): 288–97.

Potts RJ, Bespalov IA, Wallace SS, Melamede RJ, Hart BA. Inhibition of oxidative DNA repair in cadmium-adapted alveolar epithelial cells and the potential involvement of metallothionein. *Toxicology.* 2001 Mar 21;161(1–2):25–38.

Potts RJ, Watkin RD, Hart BA. Cadmium exposure down-regulates 8-oxoguanine DNA glycosylase expression in rat lung and alveolar epithelial cells. *Toxicology.* 2003 Mar 3; 184(2–3): 189–202.

Pourahmad J, O'Brien PJ, Jokar F, Daraei B. Carcinogenic metal induced sites of reactive oxygen species formation in hepatocytes. *Toxicol in Vitro.* 2003 Oct–Dec;17(5–6):803–10.

Pugh DM. The EU precautionary bans of animal feed additive antibiotics. *Toxicol Lett.* 2002 Mar 10;128 (1–3):35–44.

R Development Core Team. R: A language and environment for statistical computing. R Foundation for Statistical Computing, Vienna, Austria. 2007. ISBN 3-900051-07-0. http://www.R-project.org (Last accessed 3 Oct 2007)

Raftery AE, Givens GH, Zeh JE. Inference from a deterministic population dynamics model for bowhead whales. *J Am Stat Assoc.* 1995;90:402–15.

Raftery AE, Hoeting J, Volinsky C et al. Bayesian Model Averaging Functions – Version 3.03. http://cran.r-project.org/src/contrib/Descriptions/BMA.html (August 10, 2008 date last accessed).

Raftery AE, D Madigan, JA Hoeting. Bayesian model averaging for linear regression models. *J Am Stat Assoc.* 1997;92:179–91.

Raftery AE, Zheng Y. Discussion: Performance of Bayesian model averaging. *J Am Stat Assoc.* 2003;98:931–8.

Ragland DR. Dichotomizing continuous outcome variables: Dependence of the magnitude of association and statistical power on the cutpoint. *Epidemiology* 1992;3(5):434–40.

Rahman I, MacNee W. Lung glutathione and oxidative stress: Implications in cigarette smoke-induced airway disease. *Am J Physiol.* 1999 Dec;277(6 Pt 1):L1067–88.

Rai SN, Krewski D. Uncertainty >*Uncertainty* and variability analysis in multiplicative risk models. *Risk Anal.* 1998 Feb;18(1):37–45.

Raiffa H. *Decision Analysis: Introductory Lectures on Choices Under Uncertainty.* Addison-Wesley. Reading, MA. 1968.

Rajan D, Atamturk A. A directed cycle-based column-and-cut generation method for capacitated survivable network design. *Networks.* 2004 Apr; 43 (4): 201–11. http://citeseer.ist.psu.edu/760600.html

RAMCAP™ Framework. 2006. www.asme-iti.org/RAMCAP/RAMCAP_Framework_2.cfm

Ramirez DC, Gimenez MS. Induction of redox changes, inducible nitric oxide synthase and cyclooxygenase-2 by chronic cadmium exposure in mouse peritoneal macrophages. *Toxicol Lett.* 2003 Nov 30;145(2):121–32.

Regan HM, Hope BK, Ferson S. Analysis and portrayal of uncertainty in a food-web exposure model. *Human Ecol Risk Assess.* 2002;8:1757–77.

Renew JE, Huang CH. Simultaneous determination of fluoroquinolone, sulfonamide, and trimethoprim antibiotics in wastewater using tandem solid phase extraction and liquid chromatography-electrospray mass spectrometry. *J Chromatogr A.* 2004 Jul 9;1042(1–2):113–21.

Renfroe NA, Smith JL. Whole Building Design Guide: Threat/Vulnerability Assessments and Risk Analysis. National Institute of Building Sciences. Washington, DC. 2007. http://www.wbdg.org/design/riskanalysis.php (Last accessed 19 Aug 2007)

Rice LB. Antimicrobial resistance in gram-positive bacteria. *Am J Infect Control.* 2006;34(5 Suppl 1):S11–9, discussion S64–73.

Rice LB. Emergence of vancomycin resistant enterococci. *Emerg Infect Dis.* 2001;7(2):183–7.

Richardson S, Green PJ. On Bayesian analysis of mixtures with an unknown number of components. 1997. http://citeseer.ist.psu.edu/richardson97bayesian.html

Riedel F. Dynamic coherent risk measures. *Stoch Process Appl.* 2004 Aug;112(2):185–200.

Rios Insua D, Ruggeri F (Eds.). *Robust Bayesian Analysis.* Springer-Verlag. New York. 2000.

Ritter G, Wilson R, Pompei F, Burmistrov D. The multistage model of cancer development: some implications. *Toxicol Ind Health.* 2003 Oct;19(7–10):125–145.

Roback PJ, Givens GH. Supra-Bayesian pooling of priors linked by a deterministic simulation model. *Commun Stat Simulat Comput.* 2001;30:447–76.

Robinson, DA. Infective Dose of *Campylobacter Campylobacter jejuni* in milk. *BMJ.* 1981;282:1584.

Rodin SN, Rodin AS. On the origin of p53 G:C → T:A transversions in lung cancers. *Mutat Res.* 2002 Oct 31;508(1–2):1–19.

Rodin SN, Rodin AS. On the excess of G → T transversions in the p53 gene in lung cancer cell lines. Reply to Pfeifer and Hainaut. *Mutat Res.* 2004 Jan 12;545(1–2):141–4; author reply 145–6; discussion 147.

Rodrigues J, Poeta P, Martins A, Costa D. The importance of pets as reservoirs of resistant *Enterococcus* strains, with special reference to vancomycin. *J Vet Med B.* 2002;49(6):278–80.

Roelen, ALC, Wever R, Hale AR, Goossens LHJ, Cooke RM, Lopuhaa R, Simons M, Valk PJL. Causal modeling for integrated safety at airports. In Bedford T, van Gelder P (Eds.), *Proceedings of ESREL 2003, The European Conference on Safety and Reliability.* 2003;2:1321–7.

Rojas M, Marie B, Vignaud JM, Martinet N, Siat J, Grosdidier G, Cascorbi I, Alexandrov K. High DNA damage by benzo[a]pyrene 7,8-diol-9,10-epoxide in bronchial epithelial cells from patients with lung cancer: Comparison with lung parenchyma. *Cancer Lett.* 2004 Apr 30;207(2):157–63.

Romano JP, Wolf M. Exact and approximate stepdown methods for multiple hypothesis testing. *J Am Stat Associat.* 2005 Mar;100(469):94–108.

Rosoff H, von Winterfeldt D. A risk and economic analysis of dirty bomb attacks on the ports of Los Angeles and Long Beach. *Risk Analysis*, 2007 June;27(3):533–546(14).

Ross SM. *Stochastic Processes,* 2nd Ed., Wiley, New York. 1996.

Roth AE (Ed.). *Game-Theoretic Models of Bargaining.* Cambridge University Press. New York. 1985.

Rothman KJ, Greenland S. Causation and causal inference in epidemiology. *Am J Public Health.* 2005;95(Suppl 1):S144–50.

Royston P, Altman DG, Sauerbrei W. Dichotomizing continuous predictors in multiple regression: A bad idea. *Stat Med.* 2005;25(1):127–41.

Rubinstein RY, Kroese DP. *The Cross-Entropy Method: A Unified Approach to Combinatorial Optimization, Monte-Carlo Simulation and Machine Learning.* Springer-Verlag. New York. 2004.

Ruden C. Principles and practices of health risk assessment under current EU regulations. *Regul Toxicol Pharmacol.* 2006 Feb;44(1):14–23.

Rufino R, da Costa CH, de Souza HS, Madi K, Lape e Silva JR. Induced sputum and peripheral blood cell profile in chronic obstructive pulmonary disease. *J Bras Pneumol.* 2007 Oct;33(5):510–8.

Ruscio J. Information integration in child welfare cases: An introduction to statistical decision making. *Child Maltreatment.* 1998a;3:143–56.

Ruscio J. The perils of post-hockery. *Skeptical Inquirer.* 1998b;22:44–8.

Ruscio J. Risky business: Vividness, availability, and the media paradox. *Skeptical Inquirer.* 2000a;24:22–6.

Ruscio J. The role of complex thought in clinical prediction: Social accountability and the need for cognition. *J Consult Clin Psychol.* 2000b;68:145–54.

Ruscio J. *Clear Thinking with Psychology: Separating Sense from Nonsense.* Wadsworth. Pacific Grove, CA. 2002.

Ruscio J. Holistic judgment in clinical practice. *SRMHP Home* / Spring-Summer 2003; 2(1).

Rushworth MF, Behrens TE. Choice, uncertainty and value in prefrontal and cingulate cortex. *Nat Neurosci.* 2008 Apr;11(4):389–97.

Russell SM. The effect of airsacculitis on bird weights, uniformity, fecal contamination, processing errors, and populations of *Campylobacter* spp. and *Escherichia coli*. *Poult Sci.* 2003 Aug;82(8):1326–31.

Russell S, Binder J, Koller D, Kanazawa K. Local Learning in Probabilistic Networks with Hidden Variables. 1995. http://citeseer.ist.psu.edu/russell95local.html

Russo JE, Schoemaker PJH. *Decision Traps: The Ten Barriers toBrilliant Decision-Making and How to Overcome Them.* Simon and Schuster. New York. 1989.

Sakumi K, Tominaga Y, Furuichi M, Xu P, Tsuzuki T, Sekiguchi M, Nakabeppu Y. Ogg1 knockout-associated lung tumorigenesis and its suppression by Mth1 gene disruption. *Cancer Res.* 2003 Mar 1;63(5):902–5.

Salagovic J, Kalina I, Stubna J, Habalova V, Hrivnak M, Valansky L, Kohut A, Biros E. Genetic polymorphism of glutathione S-transferases M1 and T1 as a risk factor in lung and bladder cancers. *Neoplasma.* 1998;45(5):312–7.

Samuelson PA. *Foundations of Economic Analysis.* Harvard University Press. Cambridge. 1947 (enlarged edition, 1983).

Sanchez A, Cuesta JA. Altruism may arise from individual selection. *J Theor Biol.* 2005 Jul 21;235(2):233–40.

Sanchez-Cespedes M, Decker PA, Doffek KM, Esteller M, Westra WH, Alawi EA et al. Increased loss of chromosome 9p21 but not p16 inactivation in primary non-small cell lung cancer from smokers. *Cancer Res.* 2001;61:2092–6.

Sanders JW et al. An observational clinic-based study of diarrheal illness in deployed United States military personnel in Thailand: Presentation and outcome of *Campylobacter* infection. *Am J Trop Med Hyg.* 2002 Nov;67(5):533–8.

Sanfey AG, Rilling JK, Aronson JA, Nystrom LE, Cohen JD. The neural basis of economic decision-making in the Ultimatum Game. *Science.* 2003 Jun 13;300(5626):1755–8.

Savitz DA, Greenland S, Stolley PD, Kelsey JL. Scientific standards of criticism: A reaction to "Scientific standards in epidemiologic studies of the menace of daily life," by A.R. Feinstein. *Epidemiology.* 1990 Jan;1(1):78–83.

Schafer JL. *Analysis of Incomplete Multivariate Data.* Chapman and Hall. New York. 1997.

Schneider J, Bernges U, Philipp M, Woitowitz HJ. GSTM1, GSTT1, and GSTP1 polymorphism and lung cancer risk in relation to tobacco smoking. *Cancer Lett.* 2004 May 10;208(1):65–74.

Schoen A, Beck B, Sharma R, Dubé E. Arsenic toxicity at low doses: Epidemiological and mode of action considerations. *Toxicol Appl Pharmacol.* 2004 Aug 1;198(3):253–67.

Schollnberger H, Manuguerra M, Bijwaard H, Boshuizen H, Altenburg HP, Rispens SM, Brugmans MJ, Vineis P. Analysis of epidemiological cohort data on smoking effects and lung cancer with a multi-stage cancer model. *Carcinogenesis.* 2006 Jul;27(7):1432–44.

Schönberg T, Daw ND, Joel D, O'Doherty JP. Reinforcement learning signals in the human striatum distinguish learners from nonlearners during reward-based decision making. *J Neurosci.* 2007 Nov 21;27(47):12860–7.

Schulz-Hardt S, Frey D, Luthgens C, Moscovici S. Biased information search in group decision making. *J Pers Soc Psychol.* 2000 Apr;78(4):655–69.

Schwartz JA, Chapman GB. Are more options always better? The attraction effect in physicians' decisions about medications. *Med Decis Making.* 1999 Jul–Sep;19(3):315–23.

Schweder T, Hjort NL. Bayesian synthesis or likelihood synthesis – What does Borel's paradox say? *Forty-Sixth Report of the International Whaling Commission* (pp. 475–579). International Whaling Commission. Cambridge, UK . 1996.

Sechen S. The review of animal production drugs by FDA. *FDA Veterinarian* 2006;21 (1): 8–11. http://www.fda.gov/cvm/Documents/FDAVetVolXXINo1.pdf

Sethuraman J., Tsitsikilis JN. Stochastic search in a forest revisited. *Math Oper Res.* 2007 Aug;32(3):589–93.

Shafir E, LeBoeuf RA. Rationality. *Annu Rev Psychol.* 2002;53:491–517.

Shapiro SD, Goldstein NM, Houghton AM, Kobayashi DK, Kelley D, Belaaouaj A. Neutrophil elastase contributes to cigarette smoke-induced emphysema in mice. *Am J Pathol.* 2003 Dec;163(6):2329–35.

Sharafkhaneh A, Hanania NA, Kim V. Pathogenesis of emphysema: From the bench to the bedside. *Proc Am Thorac Soc.* 2008 May 1;5(4):475–7.

Sheldon A, Sproule R. Variance as a proxy for risk: The case of the binomial distribution. *Statistician.* 1997;46(3):317–22.

Shih CM, Wu JS, Ko WC, Wang LF, Wei YH, Liang HF, Chen YC, Chen CT. Mitochondria-mediated caspase-independent apoptosis induced by cadmium in normal human lung cells. *J Cell Biochem.* 2003 May 15;89(2):335–47.

Shimoda R, Nagamine T, Takagi H, Mori M, Waalkes MP. Induction of apoptosis in cells by cadmium: Quantitative negative correlation between basal or induced metallothionein concentration and apoptotic rate. *Toxicol Sci.* 2001 Dec;64(2):208–15.

Shin HJ, Lee BH, Yeo MG, Oh SH, Park JD, Park KK, Chung JH, Moon CK, Lee MO. Induction of orphan nuclear receptor Nur77 gene expression and its role in cadmium-induced apoptosis in lung. *Carcinogenesis.* 2004 Aug;25(8):1467–75.

Shinmura K, Yokota J. The OGG1 gene encodes a repair enzyme for oxidatively damaged DNA and is involved in human carcinogenesis. *Antioxid Redox Signal.* 2001 Aug;3(4): 597–609.

Shipley B. *Cause and Correlation in Biology. A User's Guide to Path Analysis, Structural Equations and Causal Inference.* Cambridge University Press. New York. 2000. http://callisto.si.usherb.ca:8080/bshipley/my%20book.htm

Shukla GS, Chiu J, Hart BA. Cadmium-induced elevations in the gene expression of the regulatory subunit of gamma-glutamylcysteine synthetase in rat lung and alveolar epithelial cells. *Toxicology.* 2000 Oct 26;151(1–3):45–54.

Siegrist M, Cvetkovich GT, Gutscher H. Shared values, social trust, and the perception of geographic cancer clusters. *Risk Anal.* 2001 Dec;21(6):1047–53.

Sikkink SK, Liloglou T, Maloney P, Gosney JR, Field JK. In-depth analysis of molecular alterations within normal and tumour tissue from an entire bronchial tree. *Int J Oncol.* 2003 Mar;22(3):589–95.

Silander T, Myllymäki P. A simple approach for finding the globally optimal Bayesian network structure. UAI 2006, Proceedings of the 22nd Conference in Uncertainty in Artificial Intelligence, July 13–16 2006. AUAI Press. Cambridge, MA, USA. http://www.informatik.uni-trier.de/~ley/db/conf/uai/uai2006.html#SilanderM06 (Last accessed, 3-20-09)

Singpurwalla ND, Wilson SP. *Statistical Methods in Software Reliability: Reliability and Risk.* Springer-Verlag. New York. 1999.

Slebos RJ, Habets GG, Evers SG, Mooi WJ, Rodenhuis S. Allele-specific detection of K-ras oncogene expression in human non-small-cell lung carcinomas. *Int J Cancer.* 1991;48:51–6.

Slovic P, Finucane ML, Peters E, MacGregor DG. Risk as analysis and risk as feelings: Some thoughts about affect, reason, risk, and rationality. *Risk Anal.* 2004 Apr;24(2):311–22.

Smith DJ. *Reliability, Maintainability and Risk: Practical Methods for Engineers Including Reliability Centered Maintenance and Safety-Related Systems,* 7th Ed. Elsevier. New York. 2005.

Snow ET, Sykora P, Durham TR, Klein CB. Arsenic, mode of action at biologically plausible low doses: What are the implications for low dose cancer risk? *Toxicol Appl Pharmacol.* 2005 Sep 1;207(2 Suppl):557–64.

Sonis J. A closer look at confounding. *Fam Med.* 1998 Sep;30(8):584–8.

Sorahan T, Esmen NA. Lung cancer mortality in UK nickel-cadmium battery workers, 1947–2000. *Occup Environ Med.* 2004 Feb;61(2):108–16.

Sorahan T, Lancashire RJ. Lung cancer mortality in a cohort of workers employed at a cadmium recovery plant in the United States: An analysis with detailed job histories. *Occup Environ Med.* 1997 Mar;54(3):194–201.

Sorensen M, Autrup H, Tjonneland A, Overvad K, Raaschou-Nielsen O. Glutathione S-transferase T1 null-genotype is associated with an increased risk of lung cancer. *Int J Cancer.* 2004 Jun 10;110 (2): 219–24. www.ulb.ac.be/polytech/smg/publications/Preprints/FullText/Fortz98_12.ps.gz

Soriano P, Wynants C, Seguin R, Labbe M, Gendreau M, Fortz, B. Design and dimensioning of survivable SDH/SONET networks. In Sanso B,, Soriano P (Eds.). *Telecommunications Network Planning* (pp. 147–168). Kluwer Academic Publishers. Eindhoven, the Netherlands. 1998.

Sozzi G, Pastorino U, Moiraghi L, Tagliabue E, Pezzella F, Ghirelli C et al. Loss of FHIT function in lung cancer and preinvasive bronchial lesions. *Cancer Res.* 1998;58:5032–37.

Spanakis NE, Gorgoulis V, Mariatos G, Zacharatos P, Kotsinas A, Garinis G et al. Aberrant p16 expression is correlated with hemizygous deletions at the 9p21–22 chromosome region in non-small cell lung carcinomas. *Anticancer Res.* 1999;19(3A):1893–99.

Spruill MD, Song B, Whong WZ, Ong T. Proto-oncogene amplification and overexpression in cadmium-induced cell transformation. *J Toxicol Environ Health A.* 2002 Dec 27;65(24): 2131–44.

St Clair DK, Jordan JA, Wan XS, Gairola CG. Protective role of manganese superoxide dismutase against cigarette smoke-induced cytotoxicity. *J Toxicol Environ Health.* 1994 Oct;43(2): 239–49.

Staller A, Sloman SA, Ben-Zeev T. Perspective effects in nondeontic versions of the Wason selection task. *Mem Cognit.* 2000;28(3):396–405.

Stayner L, Smith R, Thun M, Schnorr T, Lemen R. A dose-response analysis and quantitative assessment of lung cancer risk and occupational cadmium exposure. *Ann Epidemiol.* 1992a May;2(3):177–94.

Stayner L, Smith R, Thun M, Schnorr T, Lemen R. A quantitative assessment of lung cancer risk and occupational cadmium exposure. *IARC Sci Publ.* 1992b; (118): 447–55.

Steinel W, De Dreu CK. Social motives and strategic misrepresentation in social decision making. *J Pers Soc Psychol.* 2004 Mar;86(3):419–34.

Stellman SD, Takezaki T, Wang L, Chen Y, Citron ML, Djordjevic MV, Harlap S, Muscat JE, Neugut AI, Wynder EL, Ogawa H, Tajima K, Aoki K. Smoking and lung cancer risk in American and Japanese men: an international case-control study. *Cancer Epidemiol Biomarkers Prev.* 2001 Nov;10(11):1193–9.

Stern PC, Fineberg HV (Eds.). *Understanding Risk: Informing Decisions in a Democratic Society.* National Academies Press, Washington, DC. 1996.

Stern NJ, Robach MC. Enumeration of *Campylobacter* spp. in broiler feces and in corresponding processed carcasses. *J Food Prot.* 2003 Sep;66(9): 1557–63.

Stewart N, Chater N, Stott HP, Reimers S. Prospect relativity: How choice options influence decision under risk. *J Exp Psychol Gen.* 2003 Mar;132(1):23–46.

Stohs SJ, Bagchi D, Hassoun E, Bagchi M. Oxidative mechanisms in the toxicity of chromium and cadmium ions. *J Environ Pathol Toxicol Oncol.* 2001;20(2):77–88.

Streiner DL. Breaking up is hard to do: The heartbreak of dichotomizing continuous data. *Can J Psychiatry* 2002;47(3):262–6.

Stucker I, Hirvonen A, de Waziers I, Cabelguenne A, Mitrunen K, Cenee S, Koum-Besson E, Hemon D, Beaune P, Loriot MA. Genetic polymorphisms of glutathione S-transferases as modulators of lung cancer susceptibility. *Carcinogenesis.* 2002 Sep;23(9):1475–81.

Sunaga N, Kohno T, Yanagitani N, Sugimura H, Kunitoh H, Tamura T, Takei Y, Tsuchiya S, Saito R, Yokota J. Contribution of the NQO1 and GSTT1 polymorphisms to lung adenocarcinoma susceptibility. *Cancer Epidemiol Biomarkers Prev.* 2002 Aug;11(8):730–8.

Suppola J, Kolho E, Salmenlinna S, Tarkka E, Vuopio-Varkila J, Vaara M. VanA and VanB incorporate into an endemic ampicillin-resistant vancomycin-sensitive *Enterococcus faecium* strain: Effect on interpretation of clonality. *J Clin Microbiol.* 1999;37(12):3934–9.

Swanson NR, Ozyildirim A, Pisu M. A comparison of alternative causality and predictive accuracy tests in the presence of integrated and co-integrated economic variables. 2001. http://citeseer.nj.nec.com/swanson01comparison.html

Swets JA, Dawes RM, Monahan J. Psychological science can improve diagnostic decisions.*Psychol Sci Public Interest.* 2000;1:1–26.

Swiergosz-Kowalewska R. Cadmium distribution and toxicity in tissues of small rodents. *Microsc Res Tech.* 2001 Nov 1;55(3):208–22.

Takenaka S, Oldiges H, Konig H, Hochrainer D, Oberdorster G. Carcinogenicity of cadmium chloride aerosols in W rats. *J Natl Cancer Inst.* 1983 Feb;70(2):367–73.

Takiguchi M, Achanzar WE, Qu W, Li G, Waalkes MP. Effects of cadmium on DNA-(Cytosine-5) methyltransferase activity and DNA methylation status during cadmium-induced cellular transformation. *Exp Cell Res.* 2003 Jun 10;286(2):355–65.

Tam IY, Chung LP, Suen WS, Wang E, Wong MC, Ho KK et al. Distinct epidermal growth factor receptor and KRAS mutation patterns in non-small cell lung cancer patients with different tobacco exposure and clinicopathologic features. *Clin Cancer Res.* 2006;12:1647–53.

Tangen JM, Allan LG. Cue interaction and judgments of causality: Contributions of causal and associative processes. *Mem Cognit.* 2004 Jan;32(1):107–24.

Thackray AM, Klein MA, Bujdoso R. Subclinical prion disease induced by oral inoculation. *J Virol.* 2003 Jul;77(14):7991–8.

Thompson WA, Jr. *Point Process Models with Applications to Safety and Reliability*. Chapman & Hall. London. 1988.

Thompson KM. Variability and uncertainty meet risk management and risk communication. *Risk Anal.* 2002;22:647–54.

Thompson KM, Burmaster DE, Crouch EA. Monte Carlo techniques for quantitative uncertainty analysis in public health risk assessments. *Risk Anal.* 1992 Mar;12(1):53–63.

Thompson KM, Deisler PF Jr., Schwing RC. Interdisciplinary vision: The first 25 years of the Society for Risk Analysis (SRA), 1980–2005. *Risk Anal.* 2005 Dec;25(6):1333–86.

Tobacco Act, Tobacco Reporting Regulations, P.C. 2000– 1040, 21 June, 2000, Schedules 1 and 2, Department of Justice, Canada.

Tomizawa Y, Iijima H, Nomoto T, Iwasaki Y, Otani Y, Tsuchiya S et al. Clinicopathological significance of aberrant methylation of RARbeta2 at 3p24, RASSF1A at 3p21.3, and FHIT at 3p14.2 in patients with non-small cell lung cancer. *Lung Cancer.* 2004;46:305–12.

Top J, Willems R, Blok H, de Regt M, Jalink K, Troelstra A, Goorhuis B, Bonten M. Ecological replacement of *Enterococcus faecalis* by multiresistant clonal Complex-17 *Enterococcus faecium*. *Clin Microbiol Infect.* 2007;13(3):316–9.

Tornatore M, Maier G, Pattavina A. WDM Network Optimization by ILP Based on Source Formulation. INFOCOM 2002. Twenty-First Annual Joint Conference of the IEEE Computer and Communications Societies. Proceedings. IEEE Volume 3, 2002. Pp 1813–1821. http://citeseer.ist.psu.edu/tornatore02wdm.html

Toyooka S, Tokumo M, Shigematsu H, Matsuo K, Asano H, Tomii K, Ichihara S, Suzuki M, Aoe M, Date H, Gazdar AF, Shimizu N. Mutational and epigenetic evidence for independent pathways for lung adenocarcinomas arising in smokers and never smokers. *Cancer Res.* 2006 Feb 1;66(3):1371–5.

Tsamardinos I, Aliferis C, Statnikov A. 2003. Time and Sample Efficient Discovery of Markov Blankets and Direct Causal Relations. http://citeseer.nj.nec.com/tsamardinos03time.html

Tudhope SJ, Finney-Hayward TK, Nicholson AG, Mayer RJ, Barnette MS, Barnes PJ, Donnelly LE. Different mitogen-activated protein kinase-dependent cytokine responses in cells of the monocyte lineage. *J Pharmacol Exp Ther.* 2008 Jan;324(1):306–12.

Tversky A, Kahneman D. Judgment under uncertainty: heuristics and biases. *Science.* 1974 Sep 27;185(4157):1124–31.

Unicomb LE, Ferguson J, Stafford RJ et al. Low-level fluoroquinolone resistance among *Campylobacter jejuni* isolates in Australia. *Clin Infect Dis.* 2006;42(10):1368–74.

U.S. Surgeon General. The Health Consequences of Smoking: A Report of the Surgeon General. 2004. http://www.cdc.gov/tobacco/sgr/index.htm

USDA. Risk Analysis: BSE Risk from Importation of Designated Ruminants and Ruminant Products from Canada into the United States. USDA, APHIS, Veterinary Services, Explanatory Note. Feb 2004.

van Houwelingen AH, Weathington NM, Verweij V, Blalock JE, Nijkamp FP, Folkerts G. Induction of lung emphysema is prevented by L-arginine-threonine-arginine. *FASEB J.* 2008 Jun;22:3403–8.

Varman M, Chatterjee A, Abuhammour W, Johnson WC. Enterococcal infection. *Emedicine.com*. Online article available at http://www.emedicine.com/ped/topic2703.htm (last edited 26 Jul 2006).

Verougstraete V, Lison D, Hotz P. Cadmium, lung and prostate cancer: A systematic review of recent epidemiological data. *J Toxicol Environ Health B Crit Rev.* 2003 May–Jun;6(3): 227–55.

Veterinary Laboratories Agency (VLA). Surveillance Report Avian 2004. *Quart Report* 2004; 7 (4): 9–10. http://www.defra.gov.uk/corporate/vla/science/documents/science-end-survrep-qtlya404.pdf

Vernooy JH, Lindeman JH, Jacobs JA, Hanemaaijer R, Wouters EF. Increased activity of matrix metalloproteinase-8 and matrix metalloproteinase-9 in induced sputum from patients with COPD. *Chest.* 2004 Dec;126(6):1802–10.

Viallefont V, Raftery AE, Richardson S. Variable selection and Bayesian model averaging in case-control studies. *Stat Med.* 2001 Nov 15;20(21):3215–30.

Villa C, Hoffman K. A column-generation and branch-and-cut approach to the bandwidth-packing problem. *J Res Natl Inst Stand Technol.* 2006;111:161–85.

von Winterfeldt D, O'Sullivan TM. Should we protect commercial airplanes against surface-to-air missile attacks by terrorists? *Decis Anal.* 2006;3(2):63–75.

Vose DJ. *Risk Analysis: A Quantitative Guide,* 2nd Ed. John Wiley & Sons. New York. 2000.

Waalkes MP. Cadmium carcinogenesis. *Mutat Res.* 2003 Dec 10;533(1–2):107–20.

Waalkes MP, Diwan BA. Cadmium-induced inhibition of the growth and metastasis of human lung carcinoma xenografts: Role of apoptosis. *Carcinogenesis.* 1999 Jan;20(1):65–70.

Waalkes MP, Diwan BA, Weghorst CM, Ward JM, Rice JM, Cherian MG, Goyer RA. Further evidence of the tumor-suppressive effects of cadmium in the B6C3F1 mouse liver and lung: Late stage vulnerability of tumors to cadmium and the role of metallothionein. *J Pharmacol Exp Ther.* 1993 Sep;266(3):1656–63.

Wagner WE. Choosing ignorance in the manufacture of toxic products. *Cornell Law Rev.* 1997;82:773–855.

Waisberg M, Joseph P, Hale B, Beyersmann D. Molecular and cellular mechanisms of cadmium carcinogenesis. *Toxicology.* 2003 Nov 5;192(2–3):95–117.

Waldmann MR, Hagmayer Y. Seeing versus doing: two modes of accessing causal knowledge. *J Exp Psychol Learn Mem Cogn.* 2005 Mar;31(2):216–27.

Wang J, Deng Y, Cheng J, Ding J, Tokudome S. GST genetic polymorphisms and lung adenocarcinoma susceptibility in a Chinese population. *Cancer Lett.* 2003a Nov 25;201(2):185–93.

Wang Y, Fang J, Leonard SS, Rao KM. Cadmium inhibits the electron transfer chain and induces reactive oxygen species. *Free Radic Biol Med.* 2004 Jun 1;36(11):1434–43.

Wang H, Tan W, Hao B, Miao X, Zhou G, He F, Lin D. Substantial reduction in risk of lung adenocarcinoma associated with genetic polymorphism in CYP2A13, the most active cytochrome P450 for the metabolic activation of tobacco-specific carcinogen NNK. *Cancer Res.* 2003b Nov 15;63(22):8057–61.

Wang D, Zhang W, Bakhai A. Comparison of Bayesian model averaging and stepwise methods for model selection in logistic regression. *Stat Med.* 2004 Nov 30;23(22):3451–67.

Watkin RD, Nawrot T, Potts RJ, Hart BA. Mechanisms regulating the cadmium-mediated suppression of Sp1 transcription factor activity in alveolar epithelial cells. *Toxicology.* 2003 Mar 3;184(2–3):157–78.

Webb M, Riley LW, Roberts RB. Cost of hospitalization for and risk factors associated with vancomycin-resistant *Enterococcus faecium* infection and colonization. *Clin Infect Dis.* 2001; 33(4):445–52.

Weed DL. Interpreting epidemiological evidence: How meta-analysis and causal inference methods are related. *Int J Epidemiol.* 2000 Jun;29(3):387–90.

Weed DL, Gorelic LS. The practice of causal inference in cancer epidemiology. *Cancer Epidemiol Biomarkers Prev.* 1996 Apr;5(4):303–11.

Wegener HC. Antibiotics in animal feed and their role in resistance development. *Curr Opin Microbiol.* 2003 Oct;6(5):439–45.

Wegener HC, Bager F, Aarestrup FM. Surveillance of antimicrobial resistance in humans, food stuffs and livestock in Denmark. *Euro Surveill.* 1997 Mar;2(3):17–9.

Wegener HC, Hald T, Lo Fo Wong D, Madsen M, Korsgaard H, Bager F, Gerner-Smidt P, Mølbak K. Salmonella control programs in Denmark. *Emerg Infect Dis.* 2003 Jul; 9 (7): 774–80. http://www.cdc.gov/ncidod/EID/vol9no7/03-0024.htm

Weick KE, Sutcliffe KM. *Managing the Unexpected – Assuring High Performance in an Age of Complexity*. Jossey-Bass. San Francisco. 2001.

White PA. Use of prior beliefs in the assignment of causal roles: Causal powers versus regularity-based accounts. *Mem Cognit.* 1995 Mar;23(2):243–54.

WHO. Joint FAO/OIE/WHO Expert Workshop on Non-Human Antimicrobial Usage and Antimicrobial Resistance: Scientific Assessment. 2003. http://www.who.int/foodsafety/publications/micro/en/amr.pdf, http://www.who.int/foodsafety/micro/ meetings/nov2003/en/

WHO/FAO. Risk assessments of *Salmonella* in eggs and broiler chickens - Interpretative Summary and Full Report. 2002. http://www.who.int/foodsafety/publications/micro/Salmonella/en/

Willems RJ, Homan W, Top J, van Santen-Verheuvel M, Tribe D, Manzioros X, Gaillard C, Vandenbroucke-Grauls CM, Mascini EM, van Kregten E, van Embden JD, Bonten MJ. Variant *esp* gene as a marker of a distinct genetic lineage of vancomycin-resistant *Enterococcus faecium* spreading in hospitals. *Lancet.* 2001 Mar 17;357(9259):853–5.

Willems RJ, Top J, van Den Braak N, van Belkum A, Endtz H, Mevius D, Stobberingh E, van Den Bogaard A, van Embden JD. Host specificity of vancomycin-resistant *Enterococcus faecium. J Infect Dis.* 2000 Sep;182(3):816–23.

Willems RJ, Top J, van Santen M, Robinson DA, Coque TM, Baquero F, Grundmann H, Bonten MJ. Global spread of vancomycin-resistant *Enterococcus faecium* from distinct nosocomial genetic complex. *Emerg Infect Dis.* 2005;11(6):821–8.

Williamson J. *Bayesian Nets and Causality Philosophical and Computational Foundations.* Oxford University Press. Oxford. 2005.

Wilson S, Samaniego F. *Nonparameteric methods in software reliability.* Paper presented at the Third International Conference on Mathematical Methods in Reliability: Methodology and Practice, June 2002. Trondheim, Norway.

Wisplinghoff H, Bischoff T, Tallent SM, Seifert H, Wenzel RP, Edmond MB. Nosocomial bloodstream infections in U.S. hospitals: Analysis of 24,179 cases from a prospective nationwide surveillance study. *Clin Infect Dis.* 2004 Aug 1;39(3):309–17.

Wistuba II, Behrens C, Virmani AK, Mele G, Milchgrub S, Girard L, Fondon JW 3rd, Garner HR, McKay B, Latif F, Lerman MI, Lam S, Gazdar AF, Minna JD. High resolution chromosome 3p allelotyping of human lung cancer and preneoplastic/preinvasive bronchial epithelium reveals multiple, discontinuous sites of 3p allele loss and three regions of frequent breakpoints. *Cancer Res.* 2000 Apr 1;60(7):1949–60.

Wistuba II, Gazdar AE. Lung Cancer Preneoplasia. 2006. http://arjournals.annualreviews.org/doi/abs/10. 1146/annurev.pathol.1. 110304. 100103

Wistuba II, Gazdar AF, Minna JD. Molecular genetics of small cell lung carcinoma. *Semin Oncol.* 2001 Apr;28(2 Suppl 4):3–13.

Wohlfahrt J, Andersen PK, Melbye M. Multivariate competing risks. *Stat Med.* 1999 May 15;18(9):1023–30.

Wold KS, Byers T, Crane LA, Ahnen D. What do cancer survivors believe causes cancer? (United States). *Cancer Causes Control.* 2005 Mar;16(2):115–23.

Wong W-K. Stochastic dominance theory for location-scale family. *J Appl Math Decis Sci.* 2006; Volume 2006, Article ID 82049, pp 1–19. www.emis.de/journals/HOA/JAMDS/f84d.html?doi=10. 1155/JAMDS/2006/82049&e=ref

Woo G. Quantitative terrorism risk assessment. *J Risk Finance.* 2002;4 (1): 7–14. http://www.rms.com/newspress/quantitative_terrorism_risk_assessment.pdf

Wood F, Griffiths TL, Ghahramani Z. A non-parametric Bayesian method for inferring hidden causes. In *Uncertainty in Artificial Intelligence* (UAI-2006): 536–43. http://www.sigmod.org/dblp/db/conf/uai/uai2006.html; http://citeseer.ist.psu.edu/758370.html

World Health Organisation (WHO). Critically Important Antibacterial Agents for Human Medicine for Risk Management Strategies of Non-Human Use. Report of a WHO Working Group Consultation, 15–18 Feb 2005. Canberra, Australia. http://www.who.int/entity/foodborne_disease/resistance/FBD_CanberraAntibacterial_FEB 2005.pdf

Wreathall J, Nemeth C. Assessing risk: The role of probabilistic risk assessment (PRA) in patient safety improvement. *Qual Saf Health Care.* 2004 Jun;13(3):206–12.

Wright JL, Cosio MG, Churg A. Animal models of chronic obstructive pulmonary disease. *Am J Physiol Lung Cell Mol Physiol.* 2008 Jul;295(1):L1–15.

Wu J, Deng HZ, Tan YJ, Zhu DZ. Vulnerability of complex networks under intentional attack with incomplete information. *J Phys A Math Theor.* 2007;40:2665–71.

Wurbs RA. *Comparative Evaluation of Generalized River/Reservoir System Models.* Texas Water Resources Institute. College Station, TX. 2005.

Yacubian J, Gläscher J, Schroeder K, Sommer T, Braus DF, Büchel C. Dissociable systems for gain- and loss-related value predictions and errors of prediction in the human brain. *J Neurosci.* 2006 Sep 13;26(37):9530–7.

Yamada H, Koizumi S. DNA microarray analysis of human gene expression induced by a non-lethal dose of cadmium. *Ind Health.* 2002 Apr;40(2):159–66.

Yanagawa N, Tamura G, Oizumi H, Takahashi N, Shimazaki Y, Motoyama T. Promoter hypermethylation of tumor suppressor and tumor-related genes in non-small cell lung cancers. *Cancer Sci.* 2003;94:589–92.

Yang P, Bamlet WR, Ebbert JO, Taylor WR, De Andrade M. Glutathione pathway genes and lung cancer risk in young and old age populations. *Carcinogenesis.* 2004 Oct;25(10):1935–44.

Yang CF, Shen HM, Shen Y, Zhuang ZX, Ong CN. Cadmium-induced oxidative cellular damage in human fetal lung fibroblasts (MRC-5 cells). *Environ Health Perspect.* 1997 Jul;105(7):712–6.

Yang P, Sun Z, Krowka MJ, Aubry MC, Bamlet WR, Wampfler JA, Thibodeau SN, Katzmann JA, Allen MS, Midthun DE, Marks RS, de Andrade M. Alpha1-antitrypsin deficiency carriers, tobacco smoke, chronic obstructive pulmonary disease, and lung cancer risk. *Arch Intern Med.* 2008 May 26;168(10):1097–103.

Yi W, Bier VM. An application of copulas to accident precursor analysis. *Manage Sci.* 1998;44: S257–70.

Yim HW, Slebos RJ, Randell SH, Umbach DM, Parsons AM, Rivera MP et al. Smoking is associated with increased telomerase activity in short-term cultures of human bronchial epithelial cells. *Cancer Lett.* 2007;246:24–33.

Yokota F, Thompson KM. Value of information analysis in environmental health risk management decisions: Past, present, and future. *Risk Anal.* 2004 Jun;24(3):635–50.

Zhao L, Park K, Lai Y-C. Attack vulnerability of scale-free networks due to cascading breakdown. *Physical Rev E.* 2004;70:035101.

Zienolddiny S, Ryberg D, Arab MO, Skaug V, Haugen A. Loss of heterozygosity is related to p53 mutations and smoking in lung cancer. *Brit J Cancer.* 2001;84:226–31.

Zochbauer-Muller S, Fong KM, Virmani AK, Geradts J, Gazdar AF, Minna JD. Aberrant promoter methylation of multiple genes in non-small cell lung cancers. *Cancer Res.* 2001;61:249–55.

Index

Q

Early Titles in the
INTERNATIONAL SERIES IN OPERATIONS RESEARCH & MANAGEMENT SCIENCE
Frederick S. Hillier, Series Editor, *Stanford University*

Brill/ *LEVEL CROSSING METHODS IN STOCHASTIC MODELS*
Zsidisin & Ritchie/ *SUPPLY CHAIN RISK: A Handbook of Assessment, Management & Performance*
Matsui/ *MANUFACTURING AND SERVICE ENTERPRISE WITH RISKS: A Stochastic Management Approach*
Zhu/*QUANTITATIVE MODELS FOR PERFORMANCE EVALUATION AND BENCHMARKING: Data Envelopment Analysis with Spreadsheets*
Kubiak/ *PROPORTIONAL OPTIMIZATION AND FAIRNESS**
Bier & Azaiez/ *GAME THEORETIC RISK ANALYSIS OF SECURITY THREATS**
Saigal/ *A MODERN APPROACH TO LINEAR PROGRAMMING*
Nagurney/ *PROJECTED DYNAMICAL SYSTEMS & VARIATIONAL INEQUALITIES WITH APPLICATIONS*
Padberg & Rijal/ *LOCATION, SCHEDULING, DESIGN, AND INTEGER PROGRAMMING*
Vanderbei/ *LINEAR PROGRAMMING*
Jaiswal/ *MILITARY OPERATIONS RESEARCH*
Gal & Greenberg/ *ADVANCES IN SENSITIVITY ANALYSIS & PARAMETRIC PROGRAMMING*
Prabhu/ *FOUNDATIONS OF QUEUEING THEORY*
Fang, Rajasekera & Tsao/ *ENTROPY OPTIMIZATION & MATHEMATICAL PROGRAMMING*
Yu/ *OR IN THE AIRLINE INDUSTRY*
Ho & Tang/ *PRODUCT VARIETY MANAGEMENT*
El-Taha & Stidham/ *SAMPLE-PATH ANALYSIS OF QUEUEING SYSTEMS*
Miettinen/ *NONLINEAR MULTIOBJECTIVE OPTIMIZATION*
Chao & Huntington/ *DESIGNING COMPETITIVE ELECTRICITY MARKETS*
Weglarz/ *PROJECT SCHEDULING: RECENT TRENDS & RESULTS*
Sahin & Polatoglu/ *QUALITY, WARRANTY, AND PREVENTIVE MAINTENANCE*
Tavares/ *ADVANCES MODELS FOR PROJECT MANAGEMENT*
Tayur, Ganeshan & Magazine/ *QUANTITATIVE MODELS FOR SUPPLY CHAIN MANAGEMENT*
Weyant, J/ *ENERGY AND ENVIRONMENTAL POLICY MODELING*
Shanthikumar & Sumita/ *APPLIED PROBABILITY AND STOCHASTIC PROCESSES*
Liu & Esogbue/ *DECISION CRITERIA AND OPTIMAL INVENTORY PROCESSES*
Gal, Stewart & Hanne / *MULTICRITERIA DECISION MAKING: Advances in MCDM Models, Algorithms, Theory, and Applications*
Fox/ *STRATEGIES FOR QUASI-MONTE CARLO*
Hall/ *HANDBOOK OF TRANSPORTATION SCIENCE*
Grassman / *COMPUTATIONAL PROBABILITY*
Pomerol & Barba-Romero/ *MULTICRITERION DECISION IN MANAGEMENT*
Axsäter/ *INVENTORY CONTROL*
Wolkowicz, Saigal & Vandenberghe/ *HANDBOOK OF SEMI-DEFINITE PROGRAMMING: Theory, Algorithms, and Applications*
Hobbs & Meier/ *ENERGY DECISIONS AND THE ENVIRONMENT: A Guide to the Use of Multicriteria Methods*
Dar-El/ *HUMAN LEARNING: From Learning Curves to Learning Organizations*
Armstrong/ *PRINCIPLES OF FORECASTING: A Handbook for Researchers and Practitioners*
Balsamo, Personé & Onvural/ *ANALYSIS OF QUEUEING NETWORKS WITH BLOCKING*
Bouyssou et al./ *EVALUATION AND DECISION MODELS: A Critical Perspective*
Hanne / *INTELLIGENT STRATEGIES FOR META MULTIPLE CRITERIA DECISION MAKING*
Saaty & Vargas/ *MODELS, METHODS, CONCEPTS, AND APPLICATIONS OF THE ANALYTIC HIERARCHY PROCESS*
Chatterjee & Samuelson/ *GAME THEORY AND BUSINESS APPLICATIONS*

Early Titles in the
INTERNATIONAL SERIES IN OPERATIONS RESEARCH & MANAGEMENT SCIENCE
(Continued)

Hobbs et al./ *THE NEXT GENERATION OF ELECTRIC POWER UNIT COMMITMENT MODELS*
Vanderbei/ *LINEAR PROGRAMMING: Foundations and Extensions, 2nd Ed.*
Kimms/ *MATHEMATICAL PROGRAMMING AND FINANCIAL OBJECTIVES FOR SCHEDULING PROJECTS*
Baptiste, Le Pape & Nuijten/ *CONSTRAINT-BASED SCHEDULING*
Feinberg & Shwartz/ *HANDBOOK OF MARKOV DECISION PROCESSES: Methods and Applications*
Ramík & Vlach/ *GENERALIZED CONCAVITY IN FUZZY OPTIMIZATION AND DECISION ANALYSIS*
Song & Yao/ *SUPPLY CHAIN STRUCTURES: Coordination, Information, and Optimization*
Kozan & Ohuchi/ *OPERATIONS RESEARCH/ MANAGEMENT SCIENCE AT WORK*
Bouyssou et al./ *AIDING DECISIONS WITH MULTIPLE CRITERIA: Essays in Honor of Bernard Roy*
Cox/ *RISK ANALYSIS: Foundations, Models and Methods*
Dror, L'Ecuyer & Szidarovszky/ *MODELING UNCERTAINTY: An Examination of Stochastic Theory, Methods, and Applications*
Dokuchaev/ *DYNAMIC PORTFOLIO STRATEGIES: Quantitative Methods and Empirical Rules for Incomplete Information*
Sarker, Mohammadian & Yao/ *EVOLUTIONARY OPTIMIZATION*
Demeulemeester & Herroelen/*PROJECT SCHEDULING: A Research Handbook*
Gazis/ *TRAFFIC THEORY*
Zhu/*QUANTITATIVE MODELS FOR PERFORMANCE EVALUATION AND BENCHMARKING*
Ehrgott & Gandibleux/ *MULTIPLE CRITERIA OPTIMIZATION: State-of-the-Art Annotated Bibliographical Surveys*
Bienstock/ *POTENTIAL FUNCTION METHODS FOR APPROX. SOLVING LINEAR PROGRAMMING PROBLEMS*
Matsatsinis & Siskos/ *INTELLIGENT SUPPORT SYSTEMS FOR MARKETING DECISIONS*
Alpern & Gal/ *THE THEORY OF SEARCH GAMES AND RENDEZVOUS*
Hall/ *HANDBOOK OF TRANSPORTATION SCIENCE – 2nd Ed.*
Glover & Kochenberger/ *HANDBOOK OF METAHEURISTICS*
Graves & Ringuest/ *MODELS AND METHODS FOR PROJECT SELECTION: Concepts from Management Science, Finance, and Information Technology*
Hassin & Haviv/ *TO QUEUE OR NOT TO QUEUE: Equilibrium Behavior in Queueing Systems*
Gershwin et al./ *ANALYSIS & MODELING OF MANUFACTURING SYSTEMS*
Maros/ *COMPUTATIONAL TECHNIQUES OF THE SIMPLEX METHOD*
Harrison, Lee & Neale/ *THE PRACTICE OF SUPPLY CHAIN MANAGEMENT: Where Theory and Application Converge*
Shanthikumar, Yao & Zijm/ *STOCHASTIC MODELING AND OPTIMIZATION OF MANUFACTURING SYSTEMS AND SUPPLY CHAINS*
Nabrzyski, Schopf & Węglarz/*GRID RESOURCE MANAGEMENT: State of the Art and Future Trends*
Thissen & Herder/ *CRITICAL INFRASTRUCTURES: State of the Art in Research and Application*
Carlsson, Fedrizzi & Fullér/ *FUZZY LOGIC IN MANAGEMENT*
Soyer, Mazzuchi & Singpurwalla/ *MATHEMATICAL RELIABILITY: An Expository Perspective*
Chakravarty & Eliashberg/ *MANAGING BUSINESS INTERFACES: Marketing, Engineering, and Manufacturing Perspectives*
Talluri & van Ryzin/ *THE THEORY AND PRACTICE OF REVENUE MANAGEMENT*

Early Titles in the
INTERNATIONAL SERIES IN
OPERATIONS RESEARCH & MANAGEMENT SCIENCE
(Continued)

Kavadias & Loch/*PROJECT SELECTION UNDER UNCERTAINTY: Dynamically Allocating Resources to Maximize Value*
Brandeau, Sainfort & Pierskalla/ *OPERATIONS RESEARCH AND HEALTH CARE: A Handbook of Methods and Applications*
Cooper, Seiford & Zhu/ *HANDBOOK OF DATA ENVELOPMENT ANALYSIS: Models and Methods*
Luenberger/ *LINEAR AND NONLINEAR PROGRAMMING, 2nd Ed.*
Sherbrooke/*OPTIMAL INVENTORY MODELING OF SYSTEMS: Multi-Echelon Techniques, 2nd Ed.*
Chu, Leung, Hui & Cheung/ *4th PARTY CYBER LOGISTICS FOR AIR CARGO*
Simchi-Levi, Wu & Shen/*HANDBOOK OF QUANTITATIVE SUPPLY CHAIN ANALYSIS: Modeling in the E-Business Era*
Gass & Assad/ *AN ANNOTATED TIMELINE OF OPERATIONS RESEARCH: An Informal History*
Greenberg/ *TUTORIALS ON EMERGING METHODOLOGIES AND APPLICATIONS IN OPERATIONS RESEARCH*
Weber/ *UNCERTAINTY IN THE ELECTRIC POWER INDUSTRY: Methods and Models for Decision Support*
Figueira, Greco & Ehrgott/ *MULTIPLE CRITERIA DECISION ANALYSIS: State-of-the-Art Surveys*
Reveliotis/ *REAL-TIME MANAGEMENT OF RESOURCE ALLOCATIONS SYSTEMS: A Discrete Event Systems Approach*
Kall & Mayer/ *STOCHASTIC LINEAR PROGRAMMING: Models, Theory, and Computation*
Sethi, Yan & Zhang/ *INVENTORY AND SUPPLY CHAIN MANAGEMENT WITH FORECAST UPDATES*
Cox/ *QUANTITATIVE HEALTH RISK ANALYSIS METHODS: Modeling the Human Health Impacts of Antibiotics Used in Food Animals*
Ching & Ng/*MARKOV CHAINS: Models, Algorithms, and Applications*
Li & Sun/ *NONLINEAR INTEGER PROGRAMMING*
Kaliszewski/*SOFT COMPUTING FOR COMPLEX MULTIPLE CRITERIA DECISION MAKING*
Bouyssou et al./ *EVALUATION AND DECISION MODELS WITH MULTIPLE CRITERIA: Stepping Stones for the Analyst*
Blecker & Friedrich/ *MASS CUSTOMIZATION: Challenges and Solutions*
Appa, Pitsoulis & Williams/ *HANDBOOK ON MODELLING FOR DISCRETE OPTIMIZATION*
Herrmann/ *HANDBOOK OF PRODUCTION SCHEDULING*

* ***A list of the more recent publications in the series is at the front of the book.*** *